D1083003

APPLIED COMBINATORIAL MATHEMATICS

APPLIED COMBINATORIAL

MATHEMATICS

The Authors

GEORGE PÓLYA

DERRICK H. LEHMER

MONTGOMERY PHISTER, Jr.

JOHN RIORDAN

ELLIOTT W. MONTROLL

N. G. DE BRUIJN

FRANK HARARY

RICHARD BELLMAN

ROBERT KALABA

EDWIN L. PETERSON

LEO BREIMAN

ALBERT W. TUCKER

EDWIN F. BECKENBACH

MARSHALL HALL, Jr.

JACOB WOLFOWITZ

CHARLES B. TOMPKINS

KENNETH N. TRUEBLOOD

GEORGE GAMOW

HERMANN WEYL

Editor

EDWIN F. BECKENBACH
Professor of Mathematics
University of California
Los Angeles

ROBERT E. KRIEGER PUBLISHING COMPANY
MALABAR, FLORIDA
1981

Original Edition 1964
Reprint Edition 1981

Printed and Published by
ROBERT E. KRIEGER PUBLISHING COMPANY, INC.
KRIEGER DRIVE
MALABAR, FLORIDA 32950

Copyright © 1964 by
JOHN WILEY & SONS, INC.
Reprinted by Arrangement

Printed in the United States of America.

Library of Congress Cataloging in Publication Data

Beckenbach, Edwin F.   ed.
    Applied Combinatorial Mathematics.

    Based on the Statewide lecture series on combinatorial mathematics offered by the University of California, University Extension, Engineering and Physical Sciences Division, in 1962.
    Reprint of the edition published by Wiley, New York, in series: University of California engineering and physical sciences extension series.
    Includes bibliographies.
    1. Combinatorial analysis.   2. Electronic digital computers. I. Polya, George, 1887-   II. Title.   III. Series: California. University. University of California engineering and physical sciences extension series.
[QA164.B4 1981]   511'.6   80-12457
ISBN 0-89874-172-6

To Professor Clifford Bell, Head
Physical Sciences Extension
University of California

in recognition of his imaginative and tireless devotion
to the Engineering and Physical Sciences Lecture Series

this book is affectionately dedicated
at the time of his retirement

THE AUTHORS
IN ORDER OF PRESENTATION

GEORGE PÓLYA, Ph.D., Emeritus Professor of Mathematics, Stanford University

DERRICK H. LEHMER, Ph.D., Professor of Mathematics, University of California, Berkeley

MONTGOMERY PHISTER, JR., Ph.D., Chief Engineer, Scantlin Electronics, Inc., and Lecturer in Engineering, University of California, Los Angeles

JOHN RIORDAN, Ph.D., Member of Technical Staff, Bell Telephone Laboratories, Inc., Murray Hill, New Jersey

ELLIOTT W. MONTROLL, Ph.D., Vice President for Research, Institute for Defense Analyses, Washington, D.C.

N. G. DE BRUIJN, Ph.D., Professor of Mathematics, Technical University, Eindhoven, Netherlands

FRANK HARARY, Ph.D., Professor of Mathematics, University of Michigan

RICHARD BELLMAN, Ph.D., Mathematician, The RAND Corporation, Santa Monica, California

ROBERT KALABA, Ph.D., Mathematician, The RAND Corporation, Santa Monica, California

EDWIN L. PETERSON, M.S., Member of Professional Staff, Defense Research Corporation, Santa Barbara, California

LEO BREIMAN, Ph.D., Associate Professor of Mathematics, University of California, Los Angeles

ALBERT W. TUCKER, Ph.D., Professor of Mathematics, Princeton University

EDWIN F. BECKENBACH, Ph.D., Professor of Mathematics, University of California, Los Angeles

MARSHALL HALL, JR., Ph.D., Professor of Mathematics, California Institute of Technology

JACOB WOLFOWITZ, Ph.D., Professor of Mathematics, Cornell University

CHARLES B. TOMPKINS, Ph.D., Professor of Mathematics, University of California, Los Angeles

KENNETH N. TRUEBLOOD, Ph.D., Professor of Chemistry, University of California, Los Angeles

GEORGE GAMOW, Ph.D., Professor of Physics, University of Colorado

HERMANN WEYL, Ph.D., Late Member, Institute for Advanced Study

FOREWORD

Engineering achievement depends on the extent to which knowledge generated through research, in universities, in industry, and in government, knowledge expanded through the use of knowledge in industry, and knowledge handed to us through the ages is utilized effectively and at the proper time.

Modern studies in biological, social, physical, and mathematical sciences are uncovering exciting problems in combinatorial mathematics, a subject that is concerned with arrangements, operations, and selections within a finite or discrete system. It includes problems of systems analysis, information transmission, behavior of neural networks, and many others. These problems are yielding to new attacks, based in part on the availability of high-speed automatic computers. To keep pace with this progress, University Extension, Engineering and Physical Sciences Divisions, offered a Statewide Lecture Series on Applied Combinatorial Mathematics in the spring of 1962. This book is an outgrowth of the lecture series; it presents valuable aspects of the underlying theory and also some significant applications of this increasingly important and vital subject.

FRANCIS E. BLACET
Professor of Chemistry
Dean, Division of Physical Sciences
University of California
Los Angeles

L. M. K. BOELTER
Professor of Engineering
Dean, College of Engineering
University of California
Los Angeles

GEORGE J. MASLACH
Professor of Aeronautical Engineering
Acting Dean, College of Engineering
University of California
Berkeley

PAUL H. SHEATS
Professor of Education
Dean, University Extension
University of California
Statewide

PREFACE

. . . We will therefore refer to this group
of problems as those of organized
complexity.

Warren Weaver, "Science and Complexity,"
American Scientist **36** (1948), 538

In the article quoted above, Warren Weaver first points out that
the physical sciences of the sixteenth, seventeenth, and eighteenth
centuries were largely concerned with the analysis of two-variable,
or few-variable, problems: The relation between distance and gravita-
tional force, between voltage and electric current, between pressure
and the volume of a gas, and so on. These great problems were those
of *simplicity*. The life and social sciences were still largely in the
preliminary, observational stages of the scientific method.

At about the beginning of this century, however, the pendulum
swung far in the other direction, and much scientific progress was
made through statistical techniques in the analysis of problems of
disorganized complexity. The exact solution of a ten-body problem,
say the problem of the motion of ten pool balls on a pool table, can
be quite complicated; but statistical mechanics can give good answers
for average behavior when we are dealing with huge numbers of
molecules or of subatomic particles. Statistical methods also are quite
effective in some aspects of the life sciences, as exemplified by the
general reliability of mortality tables.

This leaves the middle ground of *organized complexity*, which is
largely in the scientific foreground today. The operation of a
petroleum-processing plant or of a military organization might in-

volve hundreds or even thousands of variables, but such a problem is tractable by modern mathematical techniques through the use of high-speed computing machines. In the same way, complex mathematical models of subsystems of human physiology, essential to space-age technology, are showing promise of far-reaching results in the diagnosis and prevention of disease.

For a simple example involving some typical combinatorial problems, let us consider a round-robin tennis tournament with a given number of players and a given number of courts. Is it possible to arrange the schedule so that no player participates twice consecutively? This is an *existence* problem. If there is such a schedule, how do we go about determining it? This is a *construction* or *evaluation* problem. For variety in subsequent tournaments, it might be desirable to list all the different possible schedules. This is an *enumeration* problem. Of all schedules, it might be desirable to hit on the most enjoyable one, as measured, for example, by sustaining interest through having the best players meet each other last. This is an *extremization* problem. Most combinatorial problems are of one or another of these types, although, of course, the distinction is not always precise.

Analytic problems, involving continuous variables, are often solved approximately through the use of digital computing machines. Thus these problems are of concern to the combinatorial mathematician. The first two chapters of this book are definitely machine-oriented; the rest are not. It is only this point that separates these two chapters from the other four chapters of Part 1, for all six chapters are concerned with computation and evaluation, just as all six are concerned with counting and enumeration.

The six chapters of Part 2 are also concerned with computational problems, but now the emphasis has turned toward the determination of the solution that, in some sense, is *best*. Similarly, the six chapters of Part 3 are concerned with these same problems, but with greater emphasis on the construction of examples of which the existence initially was more in doubt. The last portion of Part 3 deals also with problems of physical existence and finally with more philosophical considerations.

Unfortunately, two of the lecturers in the Statewide Lecture Series were unable to spare the time required for the preparation of manuscripts. These would have been concerned with error-correcting codes and network-flow problems, respectively. The former subject, however, is treated in part in the chapter on block designs; for the latter subject, the chapter contributed by the editor is a partial substitute. On the other hand, the material in the Lecture Series has been augmented by the addition of Chapters 5, 6, and 18, for the subjects treated in these chapters were not included in the Series.

The editor wishes to express his gratitude to the authors for their diligent work in preparing the material for publication; to the other Advisory Committee members, John L. Barnes, Clifford Bell, Richard E. Bellman, John C. Dillon, Delbert R. Fulkerson, Harold M. Heming, Magnus R. Hestenes, and Charles B. Tompkins for their efforts and excellent ideas; to the Course Coordinators, Clifford Bell, Julius J. Brandstatter, Robert Goss, and Stanley B. Schock for their efficient handling of lecture arrangements; and to Mrs. Caryl Ruenker for her painstaking secretarial work on the manuscript.

EDWIN F. BECKENBACH

August, 1964

CONTENTS ────────────────────────────

PART 2 CONTROL AND EXTREMIZATION

chapter 11 COMBINATORIAL ALGEBRA OF MATRIX GAMES
AND LINEAR PROGRAMS 320
ALBERT W. TUCKER

chapter 12 NETWORK FLOW PROBLEMS 348
EDWIN F. BECKENBACH

PART 3   CONSTRUCTION AND EXISTENCE

Introduction

GEORGE PÓLYA

The celebrated Leibnitz possessed many real insights by
which he has enriched the sciences, but he had still more and greater
projects for the execution of which the world has waited in vain.

IMMANUEL KANT
Werke 2 (1867), p. 385

Gottfried Wilhelm Leibnitz was the first author, it seems, who used the
term "combinatorial" in the same meaning as we use it today in speaking
of combinatorial analysis, or combinatorial mathematics. Leibnitz was
scarcely twenty years old when he wrote his *Dissertatio de Arte Combinatoria*, which was printed in 1666. The title page promises "applications to
the whole sphere of sciences" and "new germs of the logic of invention."
The summary announces applications to locks, organs, and syllogisms,
to the mixing of colors and to protean verses, to logic, geometry, military
art, grammar, law, medicine, and theology.

In fact, the *Dissertatio* contains, besides a bewildering show of scholastic
erudition, some mathematical results. It explains and solves the basic
combinatorial problems that lead to the binomial coefficients and to the
factorial, but there is not much more. These problems were not so
trivial in 1666 as they are today, but many of Leibnitz's results were
known before him. The mathematical propositions are followed by
applications, almost all of which appear as futile or fantastic to the
modern reader and a few of which appeared so to Leibnitz himself.

The *Dissertation on the Combinatorial Art*, however, was just the
beginning of a great project that Leibnitz entertained through his life.

1

He often mentions this project in his letters and in his printed work, and many notes found among his manuscripts that he left unpublished are concerned with it. Some of these notes have been printed posthumously. We see from them that Leibnitz planned more and more applications of his combinatorial art, or "combinatorics": to coding and decoding, to games, to mortality tables, to the combination of observations. Also, he widened more and more the scope of the subject. He sometimes regards combinatorics as one-half of a general Art of Invention; this half should deal with Synthesis, whereas the other half would deal with Analysis. Combinatorics should deal, he says in another passage, with the same and the different, the similar and the dissimilar, the absolute and the relative, whereas ordinary mathematics deals with the one and the many, the great and the small, the whole and the part. Eventually he assigns to combinatorics the widest scope in regarding it as coinciding, or almost coinciding, with the Characteristica Universalis. He planned "the universal characteristic" as a sort of generalized mathematics, which would deal with everything thinkable and would reduce thinking, by the use of appropriate signs and characters, to a sort of calculation.

Were these projects of Leibnitz mere dreams? There was some substance in his projects and there was, perhaps, some prophetic vision in his dreams. With his Characteristica Universalis he intended to reduce concepts to symbols, symbols to numbers—and, eventually, through symbols and numbers, he intended to subject all concepts to mechanical computation. This project appeared fantastic and absurd to many people with usually sound judgment, but today's computing machines are realizing some portion of the visionary scheme. Leibnitz possessed some elements of mathematical logic, of which he recognized the importance long before anybody else; now, mathematical logic lies somehow on the way to the Characteristica Universalis. It is true that the applications he made of his Ars Combinatoria are trivial, futile, or fantastic; but he certainly foresaw the immense variety of applications and the expanding scope of combinatorics. And so the name of Gottfried Wilhelm Leibnitz, the great mathematician, philosopher, and project monger, may fittingly introduce the present book.

PART IA

COMPUTATION
AND EVALUATION

Maybe it has become too hard for us unless we
are given some outside help, be it even by such
devilish devices as high-speed computing machines.

HERMANN WEYL

*(from an address delivered at the Princeton University Bicentennial
Conference on the Problems of Mathematics, December 17–19, 1946)*

CHAPTER 1

The Machine Tools of Combinatorics

DERRICK H. LEHMER

1.1 INTRODUCTION

It is the purpose of this chapter, as its title implies, to set forth a few of the known devices that enable us to pursue the subject of combinatorial analysis in its rudimentary as well as its more advanced stages. The title, however, is a *double-entendre*, because we shall relate our discussions very closely to the modern digital computer, without which many of our devices would be mere curiosities. Until recently, in fact, most of combinatorial analysis was concerned with the number of different ways that a certain thing could be done. With high-speed flexible computers, we are now prepared to ask: What are these ways and which is optimal? To bring questions like these to a computer primarily designed to do so-called scientific computing involves problems of procedure and even programming that are not entirely straightforward. In trying to make combinatorial mathematics workable, there is also the fundamental difficulty of complexity, which increases multiplicatively or exponentially with any slight increase in the domain of a parameter. Thus there is often the urgent need to keep to a minimum the execution time of basic subroutines. Still more urgent is the use of optimum methods. Factors of 10^{20} in performance are sometimes purchased by a good idea.

In spite of the numerous discouraging problems that confront the combinatorial analyst, there are some inherent advantages for him to appreciate. No longer does he have to simulate the real-number system. Gone are the headaches caused by truncation error, round-off noise, floating-point arithmetic, information fadeout, and divide checks. There is a certain cleanliness to the subject that agrees well with the very nature

5

of a digital computer. Although most of his mathematical friends find
refuge in infinity, the combinatorial analyst finds refuge *from* infinity.

This presentation assumes that the reader is acquainted with computers
and their programming. In fact, we do not hesitate to give procedures or
actual brief coding, in SHARE Assembly Program (SAP) language, of
subroutines to illustrate a procedure. On the other hand, many facts
that are well known to such a reader are mentioned. Of course, some of
the tools we need are basic and therefore very simple; they are easily
discovered with a little native ingenuity. It is hoped that a treatment
along somewhat different lines may add a dimension to the reader's
understanding of even the more obvious aspects of our subject. For
further readings, see [2] to [5].

1.2 REPRESENTATION AND PROCESSING OF DIGITAL INFORMATION

The basic format of information in combinatorial problems is the
vector with integer components. We use the term *vector* in the simple
sense of a one-rowed matrix, or a lattice point in a finite-dimensional
space. We shall need no vector analysis or even geometric interpretation
of vectors.

The representation of integers

The integer components may have various useful representations which
themselves may be vectors. Conversely, the basic vector itself may stand
for an integer. We begin by considering that most familiar representation
of the nonnegative integers, the machine-based *polynomial representation*

$$n = d_0 b^0 + d_1 b^1 + d_2 b^2 + \cdots, \qquad 0 \le d_i < b,$$

where the inequalities on the coefficients d_i are imposed to ensure uniqueness. Because this method came to Europe through Arabia, we write the
d's backward, so that, for $b = 10$, we have

$$30147 = 7 \cdot 10^0 + 4 \cdot 10^1 + 1 \cdot 10^2 + 0 \cdot 10^3 + 3 \cdot 10^4,$$

and for $b = 2$,

$$110101 = 1 \cdot 2^0 + 0 \cdot 2^1 + 1 \cdot 2^2 + 0 \cdot 2^3 + 1 \cdot 2^4 + 1 \cdot 2^5.$$

The arithmetic unit of our machine is based on this representation. By
addition the successor of n is $n + 1$; thus (according to Peano) the machine
can methodically generate the integers by the famous operation $n + 1 \to n$.
Eventually this will cause overflow, but usually our numbers n will be of

moderate size. One important feature of the polynomial representation is that it is easy to recognize which of two numbers is the larger by inspecting their digits. This makes the card sorter sometimes competitive to an expensive computer. The polynomial representation is so familiar in the case $b = 10$ that one is in danger of identifying a number with its set of digits d_i.

Machines ordinarily have either $b = 2$ or $b = 10$, and because their arithmetic units perform both arithmetical and logical operations it is possible to exploit this fact for combinatorial purposes. For these purposes, base-2 machines are superior to base-10 machines. In the latter the digits are coded in binary form, but the user does not have access to the four bits of each digit. Of course, one may deal with decimal numbers having only 0's and 1's as digits, but this is wasteful and sometimes cumbersome. One may go further and devote a whole machine word to a single digit. In fact, this is what one is driven to if, for example, $b = 7$ or $b = 12$. Many of the devices we shall describe imply a binary arithmetic unit.

Another representation of integers, useful in combinatorics, is called the *factorial representation*. Here we have

$$n = a_1 \cdot 1! + a_2 \cdot 2! + a_3 \cdot 3! + \cdots, \qquad 0 \le a_i \le i.$$

The inequalities $0 \le a_i \le i$ are imposed to ensure uniqueness. The a_i's are called the *factorial digits* of n. We can write

$$n = (a_1, a_2, \ldots);$$

for example, we have

$$1,000,000 = (0, 2, 2, 1, 5, 2, 6, 6, 2).$$

Conversion from decimal or binary representation to factorial representation is simple. There are two possible methods. One divides n by 2 and the remainder is a_1. The quotient is next divided by 3 and the remainder is a_2, etc. Alternatively, if $p! \le n < (p + 1)!$, then one may divide n by $p!$ and the quotient is a_p. The remainder is now divided by $(p - 1)!$ and the quotient is a_{p-1}, etc. Conversion from factorial to polynomial representation is equally simple.

Two numbers in factorial representation can be compared for size by simply inspecting their leading (right-hand) digits. Thus the largest number having p digits is $(1, 2, 3, \ldots, p)$. By uniqueness, this number must be $(p + 1)! - 1$. Thus we have the possibly unfamiliar identity

$$1 \cdot 1! + 2 \cdot 2! + 3 \cdot 3! + \cdots + p \cdot p! = (p + 1)! - 1.$$

In Sec. 1.5, we shall see how very useful this factorial representation is in dealing with permutation problems.

Yet another representation, unfamiliar to almost everyone, we call the *combinatorial representation of given nome*. Let $k \geq 1$ be a fixed integer, called the *nome*. Then there is the representation in terms of binomial coefficients,

$$n = \binom{a_1}{1} + \binom{a_2}{2} + \binom{a_3}{3} + \cdots + \binom{a_k}{k},$$

with the side condition

$$0 \leq a_1 < a_2 < \cdots < a_k$$

for uniqueness. If we write

$$n = \{a_1, a_2, \ldots, a_k\},$$

then, for example, with $k = 6$ and $k = 8$, we have

$$1{,}000{,}000 = \{9, 11, 14, 25, 27, 32\}$$

and

$$1{,}000{,}000 = \{0, 1, 3, 6, 12, 18, 23, 25\},$$

respectively.

We shall explain the use of this representation method in connection with combinations of things taken k at a time.

A still wider variant of polynomial representation is the so-called *Chinese* or *modular representation* with its interesting associated arithmetic. Since

$$375 \equiv 1 \ (\text{mod } 2)$$
$$\equiv 0 \ (\text{mod } 3)$$
$$\equiv 0 \ (\text{mod } 5)$$
$$\equiv 4 \ (\text{mod } 7)$$
$$\equiv 1 \ (\text{mod } 11)$$
$$\equiv 11 \ (\text{mod } 13)$$
$$\equiv 1 \ (\text{mod } 17),$$

we could write

$$375 \sim \{1, 0, 0, 4, 1, 11, 1\};$$

similarly, we have

$$243 \sim \{1, 0, 3, 5, 1, 9, 5\}.$$

Adding, subtracting, and multiplying corresponding components of these two vectors, and reducing with respect to the corresponding moduli, give at once

$$375 + 243 \sim \{0, 0, 3, 2, 2, 7, 6\} \sim 618,$$
$$375 - 243 \sim \{0, 0, 2, 6, 0, 2, 13\} \sim 132,$$
$$375 \cdot 243 \sim \{1, 0, 0, 6, 1, 8, 5\} \sim 91{,}125.$$

We note the absence of the carry-propagation problem. This truly parallel arithmetic has aroused some interest among machine designers because a simple set of matrices of cores enables one to perform addition, subtraction, or multiplication in only one pulse time. We note also that multiplication does not produce a noticeably larger vector. In fact, the vector for 375 looks, if anything, smaller than the vector for 243. This points up one of the drawbacks of the system: The seven moduli up to 17 are more than adequate to represent numbers up to the foregoing 91,125, so there is redundancy for checking purposes. Actually, these seven-dimensional vectors give a unique representation for each nonnegative integer less than the product of the moduli, namely

$$2 \cdot 3 \cdot 5 \cdot 7 \cdot 11 \cdot 13 \cdot 17 = 510{,}510.$$

Converting from decimal to Chinese is simple enough. Reversing the process is an old Chinese trick that the reader may wish to rediscover for himself.

The representation of vectors

We turn now to the more complex question of vector representation.

If the vectors in question are short, that is, have low dimension, and have only small integers as components, then it is often advantageous to use *fractional precision* or *packing* to store a whole vector into one machine word. In this way vectors can be sent from one place to another in a minimum of time. Certain operations on the components, such as limited addition, shifting, and Boolean operations, can be done in parallel by means of the machine's logical operations. Some of these ideas are illustrated in the next section. On the debit side, fractional precision often involves "unpacking," or extraction of some of the components of the vector, a process that in some cases is fussy and time-consuming. Vectors with limited components, but with too many components to fit into one machine word, can be stored in two or more consecutive words. Such vectors may be sent from one place to another by use of multiprecise arithmetic subroutines. Some machines have variable word length, but unfortunately these are decimal machines. In the worst case, a single word may be devoted to each component, and then one can take advantage of the machine's address-arithmetic facilities. In many problems the components are restricted to the values 0, 1. This offers a very compact representation. Here a table of values of a two-valued function $f(n)$ can be stored as a long binary number whose nth bit is essentially $f(n)$.

In some problems vectors are long but sparse or with limited components. For such a case, one can program an inversion of information.

The long vector $(a_1, a_2, a_3, \ldots)$ is then replaced by a list of component values. After each such value, there is given a sublist of those integers n for which a_n has this value. For example, the coefficients a_i in the expansion

$$\prod_{m=1}^{\infty} (1 - x^m) = \sum_{n=0}^{\infty} a_n x^n$$

are mostly zero. For the nonzero a_n, we have

$a_n = 1$ for $n = 0, 5, 7, 22, 26, 51, 57, 92, 100, 145, 155, 210, \ldots,$

$a_n = -1$ for $n = 1, 2, 12, 15, 35, 40, 70, 77, 117, 126, 176, 187, \ldots.$

Thus the problem of dealing with the vector $\{a_n\}$ is transformed into a question of set inclusion, as discussed on pages 12 to 15.

Another form for the designation of sparse vectors is the *rank-and-value representation*, where the vector is replaced by a sequence of ordered pairs in which the first member of the pair indicates the rank, or subscript, of a nonzero component, and the second member of the pair gives the value of the component. For example, the vector

$$(0, 3, 0, 0, 0, 1, 0, 0, 0, 4, 0, 0, 0, 0, 1, 0, 0, 0, 0, 0, 5, 9, 0, 0)$$

can be written in one word as

$$(2, 3; 6, 1; 10, 4; 15, 1; 21, 5; 22, 9; 0).$$

The number 0 at the end is used to indicate "end of message." This scheme is particularly helpful when one knows in advance that every vector to be stored has at most a certain small number of nonzero components.

In some problems the question, Have we seen this number before?, is important. This can be answered almost instantly if one uses a registration technique and a certain amount of memory. Suppose that a sequence $\{a_n\}$ of positive integers is being generated, and we wish to ask whether the current value a_n has previously occurred. To begin with, we set to zero a block B of consecutive words of the memory. If $a_n \leq N$ for all n under discussion, and if we can afford to allocate N words to B, then we fetch the kth word of B, where $k = a_n$. If this word is zero, then a_n has not previously occurred. A nonzero number, say 1, is then sent to the kth word of B. If the word fetched is not zero, then a_n has already occurred. In case N is too large, we can use the kth bit of the block B instead of the kth word.

More generally, we may have a reservoir of N words, no one of which is supposed to be used more than h times. Here, registration may be effected by reserving part of each word as a tag. Every time the word is used, a unit (in the appropriate position) is added to the tag. When this tally becomes h, the machine, which inspects the tally before using the word, knows that the word is no longer usable. If $h = 1$, we may also use the sign bit to distinguish new words from used ones.

In random walks without return, we have to ask: Have we seen this vector before? In such cases, it is convenient, if possible, to pack each vector into one word with a tag by means of fractional precision.

Another use of tagging is the device of attaching a tag to a number and then storing the number in the address equal to the tag. In this way, numbers may be called from storage and processed, and the result returned to storage, without our having to remember the source of each number. By tagging the number n with $n +$ const., we can rearrange a sequence of numbers to produce a monotone sequence in a single pass, as illustrated in the first example of the next section.

1.3 APPLICATION OF REPRESENTATIONS

A number of simple combinatorial problems, in which the representation schemes mentioned may be used, will now be discussed. In order to illustrate the solutions, formal procedures with numbered steps will be employed. It is to be understood that one proceeds from step n to step $n + 1$ unless ordered otherwise in step n. We use the notation $C(A)$ to denote the word stored in address A. Occasionally, in addition to formal procedures, we also give short routines written in SAP language, with which the reader might or might not be familiar. We begin with a brief description of a single-pass ordering routine, to which we referred at the end of the preceding section.

Monotone arrangement

Suppose that M positive integers $\leq L$, not necessarily distinct, have been generated and stored in addresses $10,000 + A$, $A = 0(1)(M - 1)$. We wish to put out a list of these numbers arranged in increasing order, eliminating all duplicates. We assume that $M < 10,000$, and that the addresses 20,000 to $20,000 + L$ are free for the time being. The problem is solved by the following formal procedure:

1. Store 0 in address $20,000 + A$ for $A = 1(1)L$.
2. Store 0 in T.
3. Store $C(10,000 + T)$ in address $20,000 + C(10,000 + T)$.
4. Replace T by $T + 1$.
5. If $T \neq M$, go to step 3.

6. Store 1 in T.
7. If $C(20,000 + T)$ is 0, go to step 9.
8. Put out $C(20,000 + T)$.
9. Replace T by $T + 1$.
10. If $T \leq L$, go to step 7.
11. Exit.

In SAP, the code might be

	LXA	KL, 1
	STZ	20001 + L, 1
	TIX	* − 1, 1, 1
	LXA	KM, 1
	CLA	10000 + M, 1
	ADD	CONST
	STA	* + 2
	SUB	CONST
	STO	**
	TIX	* − 5, 1, 1
	WRS	(Address of desired output device)
	LXA	KL, 1
	CLA	20001 + L, 1
	TZE	* + 2
	CPY	20001 + L, 1
	TIX	* − 3, 1, 1
	TRA	EXIT
KL	DEC	L
CONST	DEC	20000
KM	DEC	M

It is easy to modify this procedure so that we put out, with each distinct element of the original set of M numbers, an indication of the number of elements having this value. This frequency-tally tag can be part of the word stored in 20,000 + C, and it is increased by a suitable unit each time this address is revisited.

Set-inclusion problems

The general problem of set inclusion is the following: Given a set S consisting of M nonnegative integers,

$$S: \{S_1, S_2, S_3, \ldots, S_M\},$$

and given a nonnegative integer x, does X belong to S? This fundamental question arises in a number of ways. If nothing more is known about the set S or the number X, there can be no better way of answering the question than the "house-to-house" search through S for an $S_k = X$. On the average, $M/2$ inquiries will have to be made in case $X \in S$, and M inquiries otherwise.

If, however, we apply the routine for a monotone arrangement, just described, to the set S, then we can answer our question in only $\log_2 M$ inquiries at most, somewhat like locating a word in a dictionary. Of course, rearrangement of S could be more expensive than one house-to-house search; but if more than one number X is involved, then the rearrangement is a good investment. In many natural problems, the vector $\{S_K\}$ is already monotone.

In any case, if we know that the monotone vector is stored in addresses $10{,}000 + A$, $A = 0(1)(M - 1)$, then the following formal procedure will answer our set-inclusion question at a cost proportional to $\log_2 M$. Here we use $[\theta]$ to denote the greatest integer $\leq \theta$.

1. $0 \to a$.
2. $M + 1 \to b$.
3. $[(a + b)/2] \to k$.
4. If $X < C(10{,}000 + k)$, go to step 8.
5. If $X = C(10{,}000 + k)$, exit with $X \in S$.
6. $k \to a$.
7. Go to step 9.
8. $k \to b$.
9. If $b - a = 1$, go to exit with $X \notin S$.
10. Go to step 3.

By a succession of bisections, the number X either is identified with one of the members of S or is placed, in value, between two neighboring members of S. It is perhaps worth noting that if $X \in S$, then we have no mere existence statement; the routine exits with the actual $S_k = X$ in hand.

If the set S is fairly permanent and there will be a large number of X's to try, we can inversely prepare a characteristic binary number, of a possibly large number of words, in which the kth bit b_k is 1 or 0 according as $k \in S$ or $k \notin S$. In this way, we have answered all possible set-inclusion questions in advance. Later, when an X arises, we have only to extract and examine b_X. There is now only one step to take; that is, the cost of answering our question is no longer a function of M.

The foregoing characteristic binary number belonging to S can be so stored that the last few bits in each word are reserved for a tag giving the number of 1's in the rest of the word. Routines for obtaining this *bit sum* are described later; it is a tag that can be very helpful in evaluating the function $S(X)$, which gives the number of members of S that do not exceed X.

If we are asking our set-inclusion question about a sequence of X's that are merely consecutive integers, as is often the case in practice, it would be wasteful to use expensive address arithmetic to fetch the same word of our characteristic binary number over and over again. Obviously we should fetch it only once and leave it handy for immediate use. This suggests, at first glance, the following procedure, which uses intentional overflow. We let B stand for the selected binary word, and we suppose that X takes the values $1, 2, 3, \ldots$.

    1. $1 \to X$.
    2. $B \to A$.
    3. $A + A \to A$.
    4. If overflow does not occur, go to step 6.
    5. Go to use X as a member of S and return to step 6.
    6. $1 + X \to X$.
    7. Go to step 3.

This procedure is limited by the fact that the temporary word A soon consists wholly of 0's and so must be replenished from time to time by a new word B. To express the same thing another way, the operations in steps 2 and 3 must be interpreted as multiprecise operations in case the largest member S_M of S exceeds the number of bits in a word, as it usually does.

An interesting phenomenon occurs if between steps 4 and 5 we interpose the following step.

    4a. $A + 2^{-p} \to A$.

This keeps alive the information in the word A, which now becomes cyclic of period p, and we get a periodic pattern of signals from step 4. This simple device is very useful. A number of these electronic wheels with different p's can be used in parallel for Monte Carlo problems, diophantine equations, and Chinese arithmetic.

Let us now consider a particular case of the set-inclusion problem that has several applications. We are given a set S of integers, of which the least is m and the greatest is M, and we are also given a parameter r. It is desired to find all runs of r consecutive integers

$$N - r + 1, N - r + 2, \ldots, N,$$

each of which belongs to S. We attach a cost to answering the question:

Does N belong to S? The problem is to minimize the total cost of determining all these runs. A procedure for this is the following:

1. $m + r - 1 \to N$.
2. If $N > M$, go to step 22.
3. If $N \in S$, go to step 6.
4. $N + r \to N$.
5. Go to step 2.
6. $1 \to \sigma$.
7. $N - 1 \to N$.
8. If $N \in S$, go to step 15.
9. $N + \sigma + 1 \to N$.
10. If $N \notin S$, go to step 4.
11. $\sigma + 1 \to \sigma$.
12. If $\sigma = r$, go to step 18.
13. $N + 1 \to N$.
14. Go to step 10.
15. $\sigma + 1 \to \sigma$.
16. If $\sigma \neq r$, go to step 7.
17. $N + r - 1 \to N$.
18. Go to use run of r members of S; return to 19.
19. $N + 1 \to N$.
20. If $N \notin S$, go to step 4.
21. Go to step 18.
22. Halt.

Bit-sum procedures

There are occasions when one wants to count the number of 1's in a binary word. This sum of the bits is, unfortunately, often referred to as the *bit count*.

Suppose we denote the word in question by W and its address by WORD. One way to obtain the bit sum $b(W)$ is to use intentional overflow, shifting left until a zero word is produced. A formal procedure for this is the following:

1. $0 \to b$.
2. $W + W \to W$.
3. If no overflow occurs, go to step 6.
4. $b + 1 \to b$.
5. Go to step 2.
6. If $W \neq 0$, go to step 2.
7. Put out $b = b(W)$.

In SAP coding, this might be written

```
LXA        ZERO, 1
CLA        WORD
ALS        1
TOV        * + 3
TNZ        * − 2
TRA        EXIT
TXI        * − 4,,1, 1
```

On exit, index register 1 has $b(W)$. (A similar code can be written for counting the sign bit if logical words are used.) The time required to find $b(W)$ by this routine depends on the position of the right-most nonzero bit. The following method has an execution time depending on $b(W)$. It uses the logical AND operation, which we denote by $\oplus$. Thus, for example, we write

$$(1\ 0\ 1\ 1\ 0\ 1) \oplus (1\ 1\ 0\ 1\ 0\ 0) = (1\ 0\ 0\ 1\ 0\ 0).$$

The procedure is as follows:

1. $0 \to b$.
2. If $W = 0$, go to step 6.
3. $b + 1 \to b$.
4. $W \oplus W − 1 \to W$.
5. Go to step 2.
6. Put out $b = b(W)$.

In SAP language, we might write

```
        LXA        ZERO, 1
        CLA        WORD
        TRA        * + 4
        STO        TEMP
        SUB        UNIT
        ANA        TEMP
        TZE        EXIT
        TXI        * − 4, 1, 1
TEMP DEC 0
UNIT DEC 1
```

(For logical words, replace CLA and STO by CAL and SLW.)

This routine is faster than the first in processing sparse words W. If W tends to have more than half of its bits equal to 1, the complement of W may be processed instead.

If one is very serious about obtaining $b(W)$ in the smallest average time, one can employ a table look-up method. For example, if one has a 35-bit word, a table of 128 entries can be quickly generated, giving in address $10,000 + a$ the value $b(a)$, for $a = 0(1)127$. By breaking up a given word W into 5 subwords of 7 bits each, and referring to the table five times, one quickly accumulates the value of $b(W)$.

1.4 SIGNATURES AND THEIR ORDERLY GENERATION

We use the word *signature* to denote a vector having indices as components. For example, in the matrix $\{a_{ij}\}_n$ the vector (i, j) is a signature. In the expansion of the determinant

$$D = \sum (-1)^{P(i_1, i_2, \ldots, i_n)} a_{1i_1} a_{2i_2}, \ldots, a_{ni_n},$$

the vector $(i_1, i_2, \ldots, i_n)$ is another kind of signature. In the cross-classification formula

$$N_0 = \sum_{k=0}^{L} \sum_{1 \leq i_1 < i_2 < \cdots < i_k \leq n} (-1)^k N_{i_1, i_2, \ldots, i_k},$$

the vector $(i_1, i_2, \ldots, i_k)$ is a third kind of signature. In the first example, i and j range independently over the positive integers $\leq n$. In the second example, the indices have the same range, but no two indices have the same value. In the third example, the components of the signature are not only distinct but also strictly monotone. These verbal specifications easily convey what is wanted to other human beings. To convey this simple understanding to a machine takes a small amount of basic coding. We consider a few familiar types of signatures and their orderly generation.

The unrestricted m-cube

In this very common case, the signature $(i_1, i_2, \ldots, i_m)$ is specified by

$$i_j = 0(1)(n - 1) \qquad \text{for } j = 1(1)m;$$

that is, the signature ranges over the n^m lattice points of an m-dimensional cube of side n. This occurs in tables of functions of m variables, multiple sums, etc. We have only to write

$$i_1 + i_2 n + i_3 n^2 + \cdots + i_m n^{m-1} = N$$

to realize that we are dealing here with the m-digit integers written to base n. These may be methodically generated by performing the operation $N + 1 \to N$ and obeying the ordinary carry rules for numbers to base n. The basic procedure is simply the following:

1. $0 \to i_j, \qquad j = 1(1)m.$
2. Use signature $(i_1, i_2, \ldots, i_m)$ and go to step 3.

3. $1 \to j$.
4. $1 + i_j \to i_j$.
5. If $i_j = n$, go to step 8.
6. $0 \to i_k$, $\quad k = 1(1)(j-1)$ (for $j = 1$, this is no operation).
7. Go to step 2.
8. $1 + j \to j$.
9. If $j \leq m$, go to step 4.
10. Halt.

In case we have a noncubical "m-box" in which

$$i_j = 0(1)(n_j - 1) \qquad \text{for } j = 1(1)m,$$

we have to change step 5 to read thus:

5. If $i_j = n_j$, go to step 8.

The orderly generation of factorial digits is the special case in which $n_j = j + 1$.

If our signature is the vector of m independent variables of a dependent symmetric function $f(i_1, i_2, \ldots, i_m)$, our table of f will need only $1/2^m$ of the usual space. This gives the simplicial or monotone signature

$$i_1 \geq i_2 \geq \cdots \geq i_m \geq 0.$$

We can assume that the subscripts have been chosen so that the upper limits n_j satisfy

$$n_1 \geq n_2 \geq n_3 \geq \cdots \geq n_m.$$

To generate these signatures methodically, we need only to change step 6 of our basic procedure to read

6. $i_j \to i_k$, $\quad k = 1(1)(j-1)$.

For strict monotoncity,

$$i_1 > i_2 > \cdots > i_m \geq 0, \qquad n_1 > n_2 > \cdots > n_m,$$

we merely change steps 1 and 6 to read

1. $m - j \to i_j$, $\quad j = 1(1)m$.
6. $j + i_j - k \to i_k$, $\quad k = 1(1)(j-1)$.

Other special conditions

If the component i_j is not to be allowed to assume some values between 0 and n_j, this is easily accommodated. Suppose that there are $V_j < n_j$ permitted values for i_j. These may be stored in V_j consecutive addresses.

We then replace n_j by V_j and deal with the addresses instead. This amounts to changing step 4 to read

4. Take next permissible value for i_j.

Suppose that for the m-cube we impose the side condition that the sum of the components has a fixed value, so that

(1.1) $$i_1 + i_2 + \cdots + i_m = \sigma.$$

In this case, we replace m by $m - 1$ and n by $\sigma + 1$, and generate monotone signatures

$$\delta_1 \geq \delta_2 \geq \cdots \geq \delta_{m-1} \geq 0, \qquad \delta_i \leq \sigma.$$

Then we set

$$i_1 = \delta_{m-1} - 0 \geq 0,$$
$$i_2 = \delta_{m+2} - \delta_{m-1} \geq 0,$$
$$. \qquad . \qquad . \qquad .$$
$$i_{m-1} = \delta_1 - \delta_2 \geq 0,$$
$$i_m = \sigma - \delta_1 \geq 0.$$

Adding, we find that (1.1) is automatically satisfied. Conversely, every solution of (1.1) is obtained in this way, with

$$\delta_j = \sigma - i_m - i_{m-1} - \cdots - i_{m+1-j} \geq \delta_{j+1} \geq 0.$$

If instead of (1.1) we have

$$i_1 + i_2 + \cdots + i_m \leq \sigma',$$

we may apply this procedure with $\sigma = 1, 2, \ldots, \sigma'$.

1.5 ORDERLY LISTING OF PERMUTATIONS

Permutations are fundamental to many combinatorial problems. Applications to optimal arrangements are widespread. By an *orderly listing* of permutations we mean a generation for which it is possible to obtain the kth permutation directly from the number k, and conversely, given a permutation, it is possible to determine at once its rank, or serial number, in the list without generating any of the other permutations. One simple example of problems in which one needs an orderly list is that of generating isolated random permutations for Monte Carlo procedures.

There are $n!$ permutations of n distinct marks and also there are $n!$ $(n - 1)$-digit numbers of the form

$$a_1 1! + a_2 2! + \cdots + a_{n-1}(n - 1)!, \qquad 0 \leq a_i \leq i, i = 1(1)(n - 1).$$

Hence, to make an orderly list of permutations on n marks, we have only to set up a one-to-one correspondence between these $n!$ permutations and the $n!$ signatures

$$(a_1, a_2, \ldots, a_{n-1})$$

for the $(n-1)$-box. There are many different ways of setting up such a correspondence. Here we shall briefly mention four. The problem of coding each of the four methods is left to the natural ingenuity of the reader.

We can suppose without real loss of generality that the n marks being permuted are $0, 1, 2, \ldots, n-1$.

Tompkins-Paige method

The simple permutation on ten marks

$$\begin{pmatrix} 0 & 1 & 2 & 3 & 4 & 5 & 6 & 7 & 8 & 9 \\ 0 & 1 & 2 & 6 & 7 & 8 & 9 & 3 & 4 & 5 \end{pmatrix}$$

may be said to be of *order* 7 and *degree* 3, since only the last 7 marks are disturbed and these have been subjected to an end-around shift of 3 to the left. Each permutation on n marks is the result of compounding $n-1$ simple permutations of order $k+1$ and degree a_k, where

$$0 \le a_k \le k, \qquad k = 1(1)(n-1).$$

These a_k are then factorial digits, and thus the correspondence is established [6].

Derangement method of M. Hall

In any permutation we may ask, for each k, $1 \le k \le n-1$, how many of the k marks less than k actually follow k. This number we may denote by a_k. Clearly we have $0 \le a_k \le k$, so this establishes another correspondence between factorial digits and permutations.

Lexicographical method of D. N. Lehmer

The purpose of this method is to produce an alphabetized list of all the $n!$ permutations on n marks. To find, for example, the millionth permutation on ten marks, we first consider the factorial digits of 1,000,000, which we have already found (see page 7) to be

$$0, 2, 2, 1, 5, 2, 6, 6, 2.$$

These are made the column headings of the array displayed in the box.

$$\begin{array}{r}
2 \\
67 \\
678 \\
2223 \\
56789 \\
111111 \\
2334445 \\
23445556 \\
000000000 \\
(0)111223334
\end{array}$$

The array is formed by successive columns beginning from the left. Each column is formed from its left-hand neighbor by attaching its new column heading, copying those elements that are less than this heading, and increasing all other elements by 1. The last column is the millionth permutation, namely 2783915604, the zeroth permutation being 0123456789. Conversely, starting from the right-most column of our triangular array, we can rediscover the rest of it and so read off the factorial digits of the serial number of the permutation with which we started.

Transposition method of M. B. Wells

This recent method [8] generates the permutations in succession in such a way that each is obtained from its predecessor by a single interchange of two marks. Moreover, in more than half of the interchanges the two marks are already adjacent. Incidentally, this solves an interesting transportation problem.

For each integer m, $0 \leq m < n$, with factorial digits

$$a_1, a_2, \ldots, a_{n-1},$$

we define the function $h = h(m)$ to be the least subscript i for which $a_i \neq i$. To obtain the $(m + 1)$st permutation from the mth one, we carry out one of the following instructions:

1. Interchange the marks in places h and $h - 1$ if h is odd or if h is even and $a_{h+1} < 2$.
2. Otherwise interchange the marks in places h and $h - a_{h+1}$.

Places are numbered from 0 to $n - 1$, and a negative place is interpreted as 0.

For example, for $n = 4$ we have the sequence of permutations shown in Table 1.1.

TABLE 1.1
Transposition Method of Generating Permutations

m	$h(m)$	$P(m)$	m	$h(m)$	$P(m)$	m	$h(m)$	$P(m)$	m	$h(m)$	$P(m)$
0	1	0123	6	1	0231	12	1	0312	18	1	1320
1	2	1023	7	2	2031	13	2	3012	19	2	3120
2	1	1203	8	1	2301	14	1	1032	20	1	2130
3	2	2103	9	2	3201	15	2	0132	21	2	1230
4	1	2013	10	1	3021	16	1	3102	22	1	3210
5	3	0213	11	3	0321	17	3	1302	23	4	2310

Adjacent-mark method of S. M. Johnson

Still more recently, Johnson [1] has proposed another method of generating permutations in which *each* permutation is obtained from its predecessor by the single interchange of two adjacent marks. This could

TABLE 1.2
Adjacent-Mark Method of Generating Permutations

N	P_N	N	P_N	N	P_N	N	P_N
0	1234	6	1342	12	4321	18	2431
1	1243	7	1324	13	3421	19	4231
2	1423	8	3124	14	3241	20	4213
3	4123	9	3142	15	3214	21	2413
4	4132	10	3412	16	2314	22	2143
5	1432	11	4312	17	2341	23	2134

be quite useful in problems in which there is a cost attached to moving the marks. When applied to the marks 1, 2, 3, 4, the method gives the sequence of permutations shown in Table 1.2.

From this, the general procedure is nearly evident. The method is based on a representation, akin to the factorial representation, defined in terms of n by

$$N = n! \left\{ \frac{d_2}{2!} + \frac{d_3}{3!} + \cdots + \frac{d_n}{n!} \right\} = d_n + n d_{n-1} + n(n-1)d_{n-2} + \cdots,$$

where

$$0 \leq d_i < i, \qquad i = 1(1)n.$$

For each N, $0 < N < n!$, there is a $k = k(N)$ such that $d_k \neq 0$, whereas $d_i = 0$ for $i > k$, so that d_k is the last nonzero digit of N. The permutation P_N is formed from P_{N-1} by simply interchanging the mark k with its right-hand or left-hand neighbor in P_{N-1} according to whether $d_{k-1} + (k - 1)d_{k-2}$ is odd or even. The actual place occupied by the mark k in P_N is $a_N(k) + b_N(k)$, where

$$a_N(k) = d_k \quad \text{if } d_{k-1} + (k - 1)d_{k-2} \text{ is odd,}$$

$$= k - 1 - d_k \quad \text{otherwise,}$$

and

$$b_N(k) = 1 \quad \text{if } k = n - 1 \quad \text{and} \quad d_{n-1} + (n - 1)d_{n-2} \text{ is odd,}$$

$$= 1 \quad \text{if } k < n - 1 \quad \text{and} \quad k(d_k + d_{k-1}) \text{ is odd,}$$

$$= 2 \quad \text{if } k < n - 1 \quad \text{and} \quad (k - 1)d_k \text{ is odd,}$$

$$= 0 \quad \text{otherwise.}$$

The method can be realized by the following procedure. The marks being permuted may be arbitrary.

1. $0 \to \delta_i$, $1 \to \epsilon_i$, $1 + i \to \alpha_i$, $i = 1(1)(n - 1)$.
2. Go to use current permutation; return to step 3.
3. $n - 1 \to j$.
4. $\alpha_j - \epsilon_j \to \alpha_j$.
5. If $\alpha_j = 1 + j$, go to step 10.
6. If $\alpha_j = 0$, go to step 10.
7. $\alpha_j + \sum\limits_{i=1+j}^{n-1} \delta_i \to k$.
8. Interchange the kth and $(k + 1)$st marks.
9. Go to step 2.
10. $-\epsilon_j \to \epsilon_j$.
11. $1 - \delta_j \to \delta_j$.
12. $j - i \to j$.
13. If $j \neq 0$, go to step 4.
14. Halt.

The method has connections with the so-called Gray codes of switching theory.

I.6 ORDERLY LISTING OF COMBINATIONS

Before one can begin to perform permutations, in many problems he has first to select the objects to be permuted from a larger population. The problem is then methodically to select k words from n words stored in addresses $10{,}000 + a$, $a = 1(1)n$, in all $\binom{n}{k}$ ways and to deliver each sample of size k to the addresses $20{,}000 + b$, $b = 1(1)k$, for processing. One way to get the machine to do this is given by the following procedure:

1. $1 \to t$.
2. $10{,}001 \to A_1$.
3. $C(A_t) \to C(20{,}000 + t)$.
4. If $t = k$, go to step 12.
5. $A_t + 1 \to A_{t+1}$.
6. If $A_{t+1} = 10{,}001 + n$, go to step 9.
7. $t + 1 \to t$.
8. Go to step 3.
9. $t - 1 \to t$.
10. If $t = 0$, go to step 16.
11. Go to step 13.
12. Exit to process $C(20{,}000 + b)$, $b = 1(1)k$, and return to step 13.
13. $A_t + 1 \to A_t$.
14. If $A_t = 10{,}001 + n$, go to step 9.
15. Go to step 3.
16. Halt.

This procedure has the following lexicographical feature: If the n objects are the n letters of the alphabet, then each k-letter word selected will have its letters in alphabetical order, ready for permuting, and the complete list of $\binom{n}{k}$ words will be in alphabetical order.

I.7 ORDERLY LISTING OF COMPOSITIONS

A *composition* of the positive integer n is a vector of positive integers, the sum of whose components is n. Compositions occur frequently as conditions on multiple sums. We have already, in connection with signatures, considered the case where the compositions have components restricted as to number and size.

If the compositions are quite unrestricted, then there are precisely 2^{n-1} compositions of n. These can be generated methodically by using binary representation as follows. To each of the nonnegative integers less than 2^{n-1}, we make correspond the vector

$$(a_1, a_2, a_3, \ldots),$$

where a_i is 1 plus the number of 0 bits between the $(i-1)$st and ith 1 bit. Thus if $n = 10$, corresponding to the 9-bit binary number

$$(0\ 1\ 1\ 0\ 0\ 0\ 1\ 0\ 1),$$

we have

$$a_1 = 2, \qquad a_2 = 1, \qquad a_3 = 4, \qquad a_4 = 2, \qquad a_5 = 1.$$

It is not a coincidence, but an obvious theorem, that the sum of these components is 10. In fact, consider a rope n units long with $n-1$ knots tied in it at unit intervals. If we paint those knots red that correspond to the unit digits of some $(n-1)$-digit binary number, and then proceed to cut the rope at each red knot, then the sum of the lengths of the pieces is, of course, still n units. It is also obvious that all possible cuttings arise from all possible binary numbers. In some problems the number of parts in the composition, that is, the dimension of the vector, is specified. Suppose that the number is k. This means that we are to use only those $(n-1)$-bit numbers that have precisely $k-1$ bits equal to 1. We can manage to arrange for this restriction in three different ways. First, we can use our bit-sum routine to reject all numbers of $n-1$ bits having bit-sum different from $k-1$. Second, we can use our orderly listing of combinations to select those $k-1$ positions, out of $n-1$, that we make equal to 1. Third, these $k-1$ positions form a strictly monotone vector of $k-1$ components, the largest being less than n, and we have seen how to generate these.

I.8 ORDERLY LISTING OF PARTITIONS

Those compositions of n whose components are monotone nondecreasing are called the *unrestricted partitions* of n. These occur also as conditions of summation of multiple sums. They are much less numerous than compositions. Although it is impractical to produce all the compositions of $n = 35$, the number of partitions of 35 is only 14,883. In general, the number of unrestricted partitions of n is asymptotic to

$$\frac{1}{4n\sqrt{3}} \exp\left[\pi\left(\frac{2n}{3}\right)^{1/2}\right].$$

Thus it is clear than an independent procedure for partitions is desirable. The following is one of several possible choices:

1. $1 \rightarrow m$.
2. $1 \rightarrow a_i$, $i = 1(1)(m-1)$ (no operation in case $m = 1$).
3. $n - \sum_{i=1}^{m-1} a_i \rightarrow a_m$.
4. Exit to use partition $\{a_i\}$, $i = 1(1)m$; return to step 5.
5. $m - 1 \rightarrow T$.
6. If $T = 0$, go to step 12.
7. If $a_m - a_T > 1$, go to step 10.
8. $T - 1 \rightarrow T$.
9. Go to step 6.
10. $a_T + 1 \rightarrow a_i$, $i = T(1)(m-1)$.
11. Go to step 3.
12. $m + 1 \rightarrow m$.
13. If $m \leq n$, go to step 2.
14. Halt.

This procedure generates all the partitions of n, beginning with $n = n$ and ending with $1 + 1 + \cdots + 1 = n$. The index m is the number of *components* in the partition. The partitions, for m fixed, arise in dictionary order. If one wishes only those partitions for which $p \leq m \leq q$, he has only to replace steps 1 and 13 by

1. $p \rightarrow m$.
13. If $m \leq q$, go to step 2.

1.9 THE BACK-TRACK PROCEDURE

A basic combinatorial procedure, to which Walker [7] has given the descriptive name *back-track*, has wide applicability in problems of exhaustive search and enumeration. The general problem is that of constructing sequentially all vectors $\{a_1, a_2, \ldots, a_n\}$ having components a_i taken from a finite set U in such a way that the set of all components of each vector satisfy a stated condition C. The procedure builds each vector, element by element. If

$$a_1, a_2, \ldots, a_{K-1}$$

is a partially constructed vector, then all elements subject to C that are candidates for a_K belong to a certain subset S_K of U that depends, usually, on the elements $a_1, a_2, \ldots, a_{K-1}$ already selected. In case S_K is not empty, we select for a_K the least member of S_K and proceed to determine a_{K+1} from S_{K+1}. In case S_K is empty, we back-track to consider the set S_{K-1} that goes with the partially constructed vector $a_1, a_2, \ldots, a_{K-2}$. If necessary, we may have to back-track still further, etc. The general back-track procedure in complete detail is as follows:

1. $1 \to K$.
2. $S_1 \to S_K$.
3. If S_K is empty, go to step 9.
4. The least member of $S_K \to a_K$.
5. If $K = n$, go to step 14.
6. $K + 1 \to K$.
7. Determine the set S_K.
8. Go to step 3.
9. If $K = 1$, halt.
10. $K - 1 \to K$.
11. Determine S_K.
12. Remove from S_K all elements $\leq a_K$.
13. Go to step 3.
14. Exit to use vector $\{a_1, a_2, \ldots, a_n\}$; return to step 12.

Often this procedure is very much better than the two-stage process of first forming vectors $\{a_1, a_2, \ldots, a_n\}$ from U and then applying C to eliminate the unwanted vectors.

1.10 RANKING OF COMBINATIONS

We have already seen how to program all the combinations of n things taken k at a time in alphabetical order. In some problems involving combinations, it is useful to have a way of assigning rank or a serial number to each combination. We give two connected instances of this.

Example 1.1 A matrix has 20 rows and 12 columns. It is required to process all 12×12 minors of this matrix. Two machines are to be used, one being five times as fast as the other. At what point must the job be split in two so that both machines working simultaneously will finish the job as soon as possible?

Example 1.2 In timing the fast machine in the processing of minors, it is found that in five minutes it processed all the minors from the minor determined by rows

$$1, 2, 4, 5, 8, 10, 11, 12, 13, 15, 16, 18$$

to the minor determined by rows

$$1, 2, 7, 8, 9, 10, 12, 14, 15, 16, 18, 19.$$

How long will the whole job take?

Such problems are easily solved in terms of the concept of combinatorial representation of integers, to which we have already alluded. We need the following facts; their easy proofs will be left to the reader. Throughout the discussion k is a fixed positive integer.

1. Each nonnegative integer n has a unique representation of k terms,

$$n = \binom{b_1}{1} + \binom{b_2}{2} + \binom{b_3}{3} + \cdots + \binom{b_k}{k},$$

in which the "combinatorial digits" b_i are subject to monotoneity:

$$0 \leq b_1 < b_2 < \cdots < b_k.$$

2. The combinatorial digits b_i of n are found recursively as follows:

b_k is the greatest integer such that $\binom{b_k}{k} \leq n$,

b_{k-1} is the greatest integer such that $\binom{b_{k-1}}{k-1} \leq n - \binom{b_k}{k}$,

$\cdot \quad \cdot \quad \cdot \quad \cdot \quad \cdot \quad \cdot \quad \cdot \quad \cdot \quad \cdot \quad \cdot \quad \cdot$

b_r is the greatest integer such that $\binom{b_r}{r} \leq n - \sum_{j=r+1}^{k} \binom{b_j}{j}$,

$\cdot \quad \cdot \quad \cdot \quad \cdot \quad \cdot \quad \cdot \quad \cdot \quad \cdot \quad \cdot \quad \cdot \quad \cdot$

b_1 is the greatest integer such that $b_1 \leq n - \sum_{j=2}^{k} \binom{b_j}{j}$.

3. Let $(a_1, a_2, a_3, \ldots, a_k)$ be a combination of the n integers

$$0, 1, 2, \ldots, n - 1.$$

We may assume that

$$0 \leq a_1 < a_2 < \cdots < a_k < n.$$

Let $R_k(a_1, a_2, \ldots, a_n)$, satisfying

$$0 \leq R_k(a_1, a_2, \ldots, a_k) < \binom{n}{k},$$

denote the rank or serial number of the combination in the alphabetized list of all $\binom{n}{k}$ possible combinations. Then

$$R_k(a_1, a_2, \ldots, a_k) = \binom{n}{k} - 1 - \sum_{j=1}^{k} \binom{n - 1 - a_{k-j+1}}{j},$$

so that $n - 1 - a_{k-j+1}$, $j = 1(1)k$, are the unique combinatorial digits b_j of

$$\binom{n}{k} - 1 - R_k(a_1, a_2, \ldots, a_k).$$

We are now in a position to answer the questions of Examples 1.1 and 1.2.

For Example 1.1 we may assume that the slow machine starts at the beginning of the problem. Since there are altogether

$$\binom{20}{12} = 125{,}970$$

minors to process, the fast machine must be set to start with the minor having serial number

$$\frac{1}{6}\binom{20}{12} = 20{,}995 = R.$$

Hence we need the combinatorial digits of the number

$$\binom{20}{12} - 1 - 20{,}995 = 104{,}974.$$

These are found to be

$$\{b_j\} = \{1\ 2\ 3\ 4\ 6\ 7\ 8\ 13\ 14\ 17\ 18\ 19\},$$

from which we obtain

$$\{a_j\} = \{19 - b_{13-j}\} = \{0\ 1\ 2\ 5\ 6\ 11\ 12\ 13\ 15\ 16\ 17\ 18\}.$$

Therefore, the fast machine should begin by processing the minor whose rows are numbered

$$1, 2, 3, 6, 7, 12, 13, 14, 16, 17, 18, 19.$$

For Example 1.2, the serial number of the first minor is found as follows. The rows of this minor correspond to the vector

$$\{a_j\} = \{0, 1, 3, 4, 7, 9, 10, 11, 12, 14, 15, 17\},$$

and therefore

$$\{b_j\} = \{19 - a_{13-j}\} = \{2, 4, 5, 7, 8, 9, 10, 12, 15, 16, 18, 19\},$$

$$R_{12}\{a_i\} = \binom{20}{12} - 1 - \sum_{j=1}^{12}\binom{b_j}{j} = 29{,}936.$$

Similarly, the second minor has serial number 42,921. Hence 42,921 − 29,936, or 12,985, minors were processed in 5 minutes. The whole run will therefore require 104,975/12,985 5-minute intervals, or about 40 minutes, 25 seconds. Incidentally, we now see that the timing test took place between 11 minutes, 32 seconds, and 16 minutes, 32 seconds, after starting.

MULTIPLE-CHOICE REVIEW PROBLEMS

1. The factorial representation of 794,283 is

(a) (1, 1, 0, 0, 3, 4, 5, 1, 2),
(b) (1, 1, 0, 0, 4, 4, 5, 1, 2),
(c) (1, 1, 0, 0, 2, 4, 5, 1, 2),
(d) (1, 1, 0, 0, 1, 4, 5, 1, 2).

2. In modular representation, {1, 2, 0, 4, 10, 3, 5} + {0, 2, 2, 4, 2, 3, 10} is

(a) {1, 4, 2, 8, 12, 6, 15},
(b) {1, 1, 2, 1, 1, 6, 15},
(c) {1, 4, 2, 8, 10, 6, 10},
(d) none of the above.

3. In logical addition, (0, 1, 0, 1) $\oplus$ (0, 1, 1, 0) is

(a) (1, 1, 0, 0),   (b) (0, 1, 0, 0),
(c) (0, 1, 1, 1),   (d) none of the above.

4. The number of unrestricted partitions of 7 is

(a) 5,   (b) 10,   (c) 15,   (d) 20.

5. With 0 1 2 3 4 5 6 7 8 9 as zeroth permutation in a lexicographical list, the serial number of the permutation 8 5 9 2 1 7 4 6 0 3 is

(a) 4,297,417,   (b) 3,141,592,
(c) 2,938,406,   (d) 3,251,722.

REFERENCES

1. Johnson, S. M., *Generation of Permutations by Adjacent Transposition*, RM-3177-PR, The RAND Corporation, Santa Monica, Calif., 1962; *Math. Comp.* **17** (1963), 282–285.
2. Lehmer, D. H., "The Sieve Problem for All-Purpose Computers," *Math. Tables and Other Aids to Computation,* **7** (1953), 6–14.
3. Lehmer, D. H., "Combinatorial Problems with Digital Computers," *Proc. Fourth Canadian Mathematical Congress,* 1957, 160–173, University of Toronto Press, Toronto, Canada, 1960.
4. Lehmer, D. H., "Discrete Variable Methods in Numerical Analysis," *Proc. Internat. Congress Math.,* 1958, 545–552, Cambridge University Press, New York, 1960.
5. Lehmer, D. H., "Teaching Combinatorial Tricks to a Computer," Chapter 15 in R. Bellman and M. Hall, Jr. (editors), *Combinatorial Analysis, Proc. Sympos. Appl. Math.,* **10**, 179–193, American Mathematical Society, Providence, Rhode Island, 1960.

6. Paige, L. J., and C. B. Tompson, "The Size of the 10 × 10 Orthogonal Latin Square Problem," Chapter 5 in R. Bellman and M. Hall, Jr. (editors), *Combinatorial Analysis, Proc. Sympos. Appl. Math.*, **10**, 71–83, American Mathematical Society, Providence, Rhode Island, 1960.

7. Walker, R. J., "An Enumerative Technique for a Class of Combinatorial Problems," Chapter 7 in R. Bellman and M. Hall, Jr. (editors), *Combinatorial Analysis, Proc. Sympos. Appl. Math.*, **10**, 91–94, American Mathematical Society, Providence, Rhode Island, 1960.

8. Wells, M. B., "Generation of Permutations by Transposition," *Math. Comp.*, **15** (1961), 192–195.

CHAPTER 2

Techniques for Simplifying Logical Networks

MONTGOMERY PHISTER, Jr.

2.1 INTRODUCTION

Digital systems, from the largest, multimillion-dollar general-purpose computers down to the few-hundred-dollar laboratory frequency counters, are constructed from standard building blocks with a common property: The inputs to and outputs from these blocks have two states. Since the mathematics of logic is also primarily two-valued, these building blocks have been called *logical elements*, the art of organizing them has been called *logical design*, and the systems themselves are called *logical networks*. Since the late 1940's, an ever-increasing effort has gone into attempting to convert the art of logical design into a science, with the objectives of making the designer's work easier, his products more efficient, and his profession more respectable. In this chapter, we shall attempt to describe the current state of this activity.

There are basically only two kinds of logical elements: memory elements and decision elements. A *memory element* stores one unit of information and thus can at any time exist in precisely one of two states. For convenience, and without loss of generality, we shall here discuss only *synchronous systems*, in which the state of a memory element is significant only at discrete instants of time known as *bit times* or *clock-pulse times*. Various kinds of memory elements differ from one another in cost, speed with which they can change from one state to another, power-handling capacity, and logical properties. We shall be interested only in this last characteristic. The simplest memory element is a *delay flip-flop*; its logical properties are indicated in Fig. 2.1. The element has one input, D_Q, and two outputs, Q and $\bar{Q}$. The outputs are always the opposite of one another, so that if $Q = 1$, then $\bar{Q} = 0$, and vice-versa. If $Q = 1$, then the flip-flop is said to be *set*, or in its 1 state. If $Q = 0$, then it is *reset*, or in its 0 state. At any bit time, the input D_Q is either 0 or 1. As the table of Fig. 2.1 indicates, if D_Q is 0 at any bit time n, then Q is 0 (reset) at time $n + 1$; if $D_Q = 1$ at any bit time n, then Q is 1 (set) at time $n + 1$. Later in the chapter, we shall discuss the use of memory elements with other logical properties.

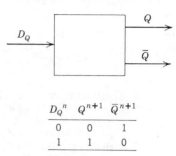

$D_Q{}^n$	Q^{n+1}	$\bar{Q}^{n+1}$
0	0	1
1	1	0

Fig. 2.1   Delay flip-flop.

A *decision element* is a component whose output is some specified function of its inputs, but which introduces no delay. Of the variety of decision elements available, probably the most widely used are the AND gate and the OR gate. The output of an AND gate is 1 if and only if all its inputs are 1; the output of an OR gate is 1 if and only if one or more of its inputs are 1. The properties of the simplest

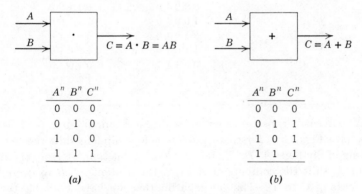

A^n	B^n	C^n
0	0	0
0	1	0
1	0	0
1	1	1

(a)

A^n	B^n	C^n
0	0	0
0	1	1
1	0	1
1	1	1

(b)

Fig. 2.2   AND gate (a) and OR gate (b).

(two-input) AND and OR gates are illustrated in Fig. 2.2. Note that the circuits can be defined by listing what their outputs will be for each of the four possible states of the two input lines. Note also that the output of each circuit is available at the same bit time the inputs are present— there is no storage or delay of information.

A *digital system* is a network of decision and/or memory elements. The most interesting of these contain both kinds of elements and are called *sequential circuits*. To be able to refer to something specific in the following discussion, let us describe a particular sequential circuit. Suppose we have available a binary communication channel X (perhaps a signal from outer space or an encoded message from an enemy agent), and we want to monitor it for three classes of 5-bit patterns. Class 1 will cause output line Z_1 to be 1 whenever either of the two patterns shown in the first column of Table 2.1 has appeared on line X. The bit patterns that cause output lines Z_2 and Z_3 to be 1 are more complicated in that they contain more than two members, as shown in the table.

TABLE 2.1
Monitoring of a Sequential Circuit for Three Classes of Five-Bit Patterns

$Z_1 = 1$ for any of these patterns	$Z_2 = 1$ for any of these patterns	$Z_3 = 1$ for any of these patterns
0 0 0 0 1	0 0 0 0 1	0 0 0 0 0
1 0 0 0 1	1 0 0 0 1	1 0 0 0 0
	0 0 1 0 1	0 1 0 0 0
	1 0 1 0 1	1 1 0 0 0
	0 1 1 0 1	0 0 1 0 0
	1 1 1 0 1	1 0 1 0 0
	0 0 0 1 0	0 1 1 0 0
	1 0 0 1 0	1 1 1 0 0
	0 1 0 1 0	0 0 0 1 0
	1 1 0 1 0	1 0 0 1 0
	0 0 1 1 1	0 1 0 1 0
		1 1 0 1 0
		0 0 1 1 1

We can easily construct a circuit to do the required job in the following way (see Fig. 2.3). First, we provide a four-flip-flop shift register consisting of flip-flops A, B, C, and D. At each bit time we store the bit on X in A, shift the contents of A to B, the contents of B to C, and the contents of C to D. Thus at each bit time we have available the last five bits that have appeared on line X: The most recent is on X itself, the previous bit is in A, the next earlier is in B, the next before that is in C, and the least recent of the five is in D. The circuits necessary to perform this shifting are very simple:

(2.1) $$D_A = X$$
(2.2) $$D_B = A$$
(2.3) $$D_C = B$$
(2.4) $$D_D = C$$

Now let us see what logic is necessary to recognize the patterns for which Z_1 must equal 1. Suppose the pattern ... 110100001 ... appears on X. Then the input line and flip-flops will take on the states shown in Table 2.2, on page 36.

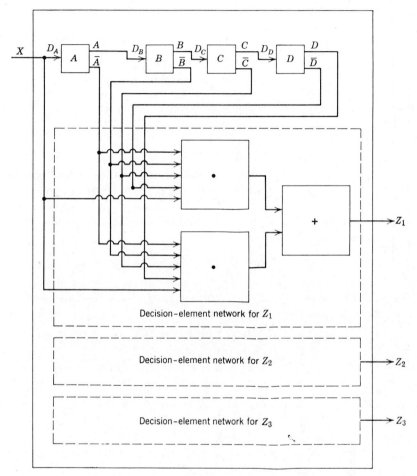

Fig. 2.3 A sequential circuit.

At the ninth bit time, the first of the two patterns that stimulate Z_1—the pattern 00001—appears in flip-flops D, C, B, and A, and on line X. Note that the pattern is shown backward because the bit that appeared first has shifted to the right to flip-flop D. We can recognize this pattern with a single AND gate (see Fig. 2.3) having inputs X, $\bar{A}$, $\bar{B}$, $\bar{C}$, and D, for it is only when all five inputs to this gate are 1, that is,

TABLE 2.2
Input Line X with Flip-Flops
A, B, C, D

Bit time	X	A	B	C	D
1	1	.	.	.	.
2	1	1	.	.	.
3	0	1	1	.	.
4	1	0	1	1	.
5	0	1	0	1	1
6	0	0	1	0	1
7	0	0	0	1	0
8	0	0	0	0	1
9	1	0	0	0	0

when $X = 1$, $A = 0$, $B = 0$, $C = 0$, and $D = 0$, that its output will be 1. Similarly, we can recognize the other Z_1 pattern, 10001, by providing an AND gate with inputs X, $\bar{A}$, $\bar{B}$, $\bar{C}$, and D. Since Z_1 must equal 1 when either one or the other of these circuits has output 1, we combine their outputs in an OR gate, with the result that

$$(2.5) \qquad Z_1 = X\,\bar{A}\,\bar{B}\,\bar{C}\,D + X\,\bar{A}\,\bar{B}\,\bar{C}\,D.$$

In an exactly similar way, although the circuits are too complicated to reproduce conveniently in Fig. 2.3, we can provide decision-element circuits for Z_2 and Z_3:

$$(2.6) \qquad \begin{aligned} Z_2 = {}& X\,\bar{A}\,\bar{B}\,\bar{C}\,D + X\,\bar{A}\,\bar{B}\,\bar{C}\,D + X\,\bar{A}\,B\,\bar{C}\,D \\ &+ X\,\bar{A}\,B\,\bar{C}\,D + X\,\bar{A}\,B\,C\,\bar{D} + X\,\bar{A}\,B\,C\,D \\ &+ \bar{X}\,A\,\bar{B}\,\bar{C}\,D + \bar{X}\,A\,\bar{B}\,\bar{C}\,D + \bar{X}\,A\,B\,C\,D \\ &+ \bar{X}\,A\,\bar{B}\,C\,D + X\,A\,B\,\bar{C}\,D, \end{aligned}$$

$$(2.7) \qquad \begin{aligned} Z_3 = {}& \bar{X}\,\bar{A}\,B\,C\,D + \bar{X}\,\bar{A}\,B\,C\,D + \bar{X}\,\bar{A}\,\bar{B}\,C\,D \\ &+ \bar{X}\,\bar{A}\,\bar{B}\,C\,D + \bar{X}\,\bar{A}\,B\,\bar{C}\,D + \bar{X}\,\bar{A}\,B\,\bar{C}\,D \\ &+ \bar{X}\,\bar{A}\,B\,C\,D + \bar{X}\,\bar{A}\,B\,C\,D + \bar{X}\,A\,\bar{B}\,\bar{C}\,D \\ &+ \bar{X}\,A\,\bar{B}\,\bar{C}\,D + \bar{X}\,A\,\bar{B}\,C\,D + \bar{X}\,A\,B\,\bar{C}\,D \\ &+ X\,A\,B\,\bar{C}\,\bar{D}. \end{aligned}$$

Equations (2.1) to (2.7) thus represent the complete design of our sequential circuit.

In general, a sequential circuit may differ from the circuit in Fig. 2.3 in that it might contain a larger number of inputs, a larger or smaller number of outputs, and a larger or smaller number of memory elements. For any such circuit, however, the designer is faced with certain problems. In increasing order of complexity, he may ask the following.

1. What is the cheapest circuit, composed of delay flip-flops and AND-OR gates, that is equivalent to a given circuit?

2. What is the cheapest circuit that is equivalent to a given circuit, provided the designer may take his choice of types of decision and memory elements to implement the design?

3. What is the cheapest circuit that can be found to perform a given function?

The first two problems are problems of analysis; they will be discussed in the two parts into which this chapter has been divided. The last is a problem of synthesis; it will be briefly discussed in Sec. 2.8.

One difficulty inherent in all three problems is that of finding a suitably realistic definition of the word *cheapest*. Presumably the designer would like to minimize a function that includes the costs of development, tooling, packaging, checkout, and maintenance, in addition to the costs of specific decision and memory elements. In practice, no one has even tried to do more than minimize the number of AND-OR gates and the number of memory elements necessary, and these two costs have been minimized only individually; no one has been able to minimize their sum, except by exhaustively trying all combinations.

SIMPLIFYING A GIVEN SEQUENTIAL CIRCUIT

2.2 SIMPLEST AND-OR GATES

A great deal has been written on the problem of simplifying AND-OR networks, with the definition of *simplest* usually taken to mean the circuit with fewest inputs to gates. The circuit that determines Z_1 in Fig. 2.3, for example, contains two AND gates with five inputs each and an OR gate with two inputs. Two times 5, plus 2, that is, 12, is thus a measure of its complexity. In a similar way, we find from (2.6) and (2.7) that Z_2 has a count of 66 and Z_3 a count of 78, so that the entire sequential circuit, excluding flip-flops, has a count of 156. Standard procedures are available for minimizing this count if we allow no more than two gates in series in a circuit.

One of the simplest of these procedures (see [5]) makes use of a *Veitch diagram* or *Karnaugh map*. In Fig. 2.4a a five-variable diagram is shown that is suitable for use in our five-variable problem. Note that groups of squares are labeled with the five variables in such a way that one square

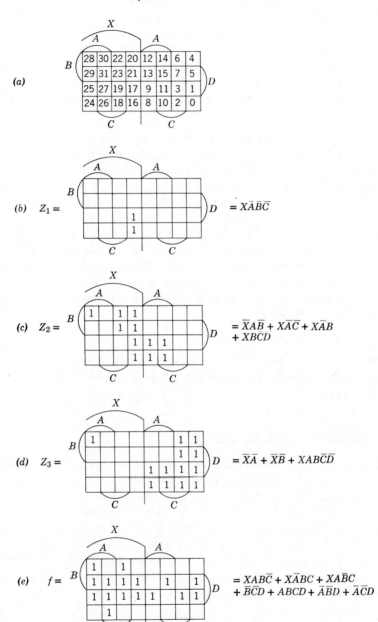

Fig. 2.4 Five-variable Veitch diagrams.

corresponds to each of the $2^5 = 32$ different products of the five variables. Each square is given a number as follows: First, write the variables in a prescribed sequence, for example, $X\,A\,B\,C\,D$. Next, see whether the square in question is in group X. If it is not, put a bar over the X; if it is, leave X unchanged. In either event, repeat the process for the variable A, putting a bar over the A if the square is not in group A. Repeat for B, C, and D. The resulting term is a description of the AND gate corresponding to that square. Now replace each unbarred letter in this term by a 1, and each barred letter by a 0. The resulting binary number, converted to a decimal number, is the label for that square. For example, look at square 26 in Fig. 2.4a. It is in group X, in group A, outside group B, in group C, and outside group D. Therefore it represents the term $X\,A\,\bar{B}\,C\,\bar{D}$. Replacing with 0's and 1's, we get 11010, which is decimal 26. All numbers in Fig. 2.4a are found in this fashion.

Now let us look at equation (2.5) for Z_1. It is composed of two terms, $X\,\bar{A}\,\bar{B}\,\bar{C}\,\bar{D}$ (10000, or 16) and $X\,\bar{A}\,\bar{B}\,\bar{C}\,D$ (10001, or 17). In Fig. 2.4b we have plotted Z_1 by marking squares 16 and 17 with 1's. In a similar way, Z_2 and Z_3 are plotted in Fig. 2.4c, d.

The Veitch-diagram simplification procedure is based on the fact that groups of 2, or 4, or 8, ..., or 2^n, squares, grouped adjacent to one another in certain ways horizontally and vertically, can be combined and represented by terms containing 1, or 2, or 3, ..., or n letters less than the number required to describe a single square. For example, the two squares marked 1 in Fig. 2.4b are adjacent vertically; they are both in X, both in $\bar{A}$, both in $\bar{B}$, and both in $\bar{C}$ (one is in D, the other in $\bar{D}$). They can thus be defined by

$$(2.8) \qquad\qquad Z_1 = X\,\bar{A}\,\bar{B}\,\bar{C}.$$

Next looking at Z_2, we find that square 10 ($\bar{X}\,A\,\bar{B}\,C\,\bar{D}$) combines with the three squares 8, 9, and 11, with which it forms a larger square, to give the term $\bar{X}\,A\,\bar{B}$. Thus if we write

$$Z_2 = \bar{X}\,A\,\bar{B} + \cdots,$$

we have taken care of four of the eleven terms in Z_2. Next look at square 16 in Z_2. It combines with the three vertically above it (17, 21, and 20) to form $X\,\bar{A}\,\bar{C}$, so we write

$$Z_2 = \bar{X}\,A\,\bar{B} + X\,\bar{A}\,\bar{C} + \cdots.$$

Similarly, square 23 combines with 22, 20, and 21 to add $X\,\bar{A}\,B$; thus,

$$Z_2 = \bar{X}\,A\,\bar{B} + X\,\bar{A}\,\bar{C} + X\,\bar{A}\,B + \cdots.$$

The fact that squares 20 and 21 are included both in $X\,\bar{A}\,\bar{C}$ and in $X\,\bar{A}\,B$

makes no difference; it merely means that the outputs of these two AND gates will both be 1 if $X \bar{A} B \bar{C} \bar{D} = 1$ or if $X \bar{A} B \bar{C} D = 1$, and either one or both will make $Z_2 = 1$ in the final OR gate. We have now included every marked square except 28, which has no other 1's immediately adjacent horizontally or vertically with which to combine. It will, however, combine horizontally across the $X = 1$ group with square 20, for both are in X, both in B, both in $\bar{C}$, and both in $\bar{D}$. Adding (or rather, ORing) this final term, we get

$$(2.9) \qquad Z_2 = \bar{X} A \bar{B} + X \bar{A} \bar{C} + X \bar{A} B + X B \bar{C} \bar{D}.$$

Finally, and in quite a similar fashion, we find

$$(2.10) \qquad Z_3 = \bar{X} \bar{A} + \bar{X} \bar{B} + X A B \bar{C} \bar{D}.$$

Now (2.8), (2.9), and (2.10) represent the simplest AND-OR forms for Z_1, Z_2, and Z_3. The reader should verify that they have a count of only 33 compared with 156 for (2.5), (2.6), and (2.7).

In using Veitch diagrams to simplify switching functions, it is not enough simply to look for large numbers of 1's that can be covered by a single term. The dangers inherent in such a procedure are indicated in Fig. 2.4e, in which we have plotted a function that contains eight 1's on the left in $X D$. Nevertheless, the simplest expression for f does *not* contain $X D$, and may be found as follows. Look at each individual marked square, determining all possible ways it can be combined with other marked squares. If a 1 combines only with a single group, that group is an *essential term* and must appear in the final simplest expression for the function. The correct simplification procedure requires that all marked squares be examined for essential terms first, and that then any remaining 1's be covered as efficiently as possible. In Fig. 2.4e, examination of squares 17, 21, 19, 23, 27, 31, 25, and 29 will reveal that each combines into term XD *and also* into one other term. For example, 29 combines with 28 to form $X A B C$, and 21 combines with 1, 5, and 17 to form $\bar{A} \bar{C} D$. Squares 28, 22, 26, 9, 15, 3, and 5 each combine in only one group of terms, and these essential terms cover all 1's and comprise the simplest expression for the function, namely

$$(2.11) \qquad f = X A B \bar{C} + X \bar{A} B C + X A \bar{B} C + \bar{B} \bar{C} D$$
$$+ A B C D + \bar{A} \bar{B} D + \bar{A} \bar{C} D.$$

2.3 MULTIPLE-OUTPUT FUNCTION SIMPLIFICATIONS

In finding the simplest expressions for Z_1, Z_2, and Z_3 in Sec. 2.2, we considered these three functions independently, ignoring the fact that

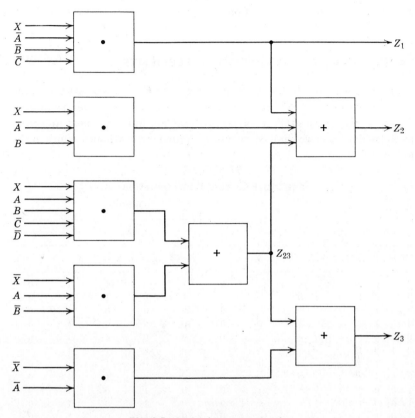

Fig. 2.5   Multiple-output circuit.

they were all part of a single circuit. It is possible (see [1]) to take advantage of their common parts and obtain still further simplification, as follows.

Referring again to Fig. 2.4b, c, d, we make the following observations:

Z_2 and Z_3 have squares 28, 8, 9, 10, and 11 in common.

Z_1 and Z_2 have squares 16 and 17 in common.

We can make use of these facts to produce a further savings in the following way, and the resulting circuit is shown in Fig. 2.5. If we define Z_{23} by

$$(2.12) \qquad Z_{23} = X\,A\,B\,\bar{C}\,D + \bar{X}\,A\,\bar{B},$$

then we have

$$(2.13) \qquad Z_1 = X\,\bar{A}\,\bar{B}\,\bar{C},$$

$$(2.14) \qquad Z_2 = Z_1 + Z_{23} + X\,\bar{A}\,B,$$

$$(2.15) \qquad Z_3 = Z_{23} + \bar{X}\,\bar{A}.$$

The circuit now has a count of only 24, compared with 33 for (2.8), (2.9), and (2.10).

2.4 ELIMINATION OF MEMORY ELEMENTS

Let us now return to the sequential circuit of Fig. 2.3 and ask: Can the function performed by this circuit be achieved with fewer than four flip-flops? Huffman [3] first solved this problem in a very neat and complete fashion, and we shall carry out his quite systematic procedure in the following.

TABLE 2.3
Sequential Circuit Representation

Present state					Next state		Z_1		Z_2		Z_3	
A	B	C	D	Label	$X=0$	$X=1$	$X=0$	$X=1$	$X=0$	$X=1$	$X=0$	$X=1$
0	0	0	0	a	a	i	0	1	0	1	1	0
0	0	0	1	b	a	i	0	1	0	1	1	0
0	0	1	0	c	b	k	0	0	0	0	1	0
0	0	1	1	d	b	k	0	0	0	0	1	0
0	1	0	0	e	c	l	0	0	0	1	1	0
0	1	0	1	f	c	l	0	0	0	1	1	0
0	1	1	0	g	d	m	0	0	0	1	1	0
0	1	1	1	h	d	m	0	0	0	1	1	0
1	0	0	0	i	e	n	0	0	1	0	1	0
1	0	0	1	k	e	n	0	0	1	0	1	0
1	0	1	0	l	f	p	0	0	1	0	1	0
1	0	1	1	m	f	p	0	0	1	0	1	0
1	1	0	0	n	g	r	0	0	0	1	0	1
1	1	0	1	p	g	r	0	0	0	0	0	0
1	1	1	0	r	h	s	0	0	0	0	0	0
1	1	1	1	s	h	s	0	0	0	0	0	0

To begin the process, the circuit must be described in general terms, with all references to flip-flops eliminated so that we are concerned only with a circuit that has certain states. First, we draw up the sequential-circuit pattern shown in Table 2.3. The left-hand columns of this table we label "Present state." Each row in those columns corresponds to a particular state in which the circuit can be at any bit time; and since the circuit contains four flip-flops, there must be sixteen rows. The center columns we label "Next state." For each possible value of the input variable (or variables) X, we now enter, on each row, the name of the state to which the circuit will next go as a result of that value. The last columns are labeled

"Outputs." For each value of the input variable, we must enter the proper value of each output variable.

We start by defining the sixteen states, corresponding to the sixteen discrete combinations of flip-flops A, B, C, and D. Each of these states is labeled with a letter of the alphabet, as shown. We now proceed to fill in the next-state entries by remembering that the circuit is a shift register: At each bit time, X shifts to A, A to B, B to C, and C to D. Suppose that we are in state a (see Fig. 2.6), so that $A = B = C = D = 0$, and suppose that $X = 0$. At the next bit time, each of the 0's will have shifted to the

X	A	B	C	D	State
0	0	0	0	0	a
1	0	0	0	0	a
1	1	0	0	0	i
0	1	1	0	0	n
1	0	1	1	0	g
0	1	0	1	1	m
	0	1	0	1	f

Fig. 2.6 Derivation of sequential-circuit table entries.

right, and therefore at the next bit time we will still be in state a. Therefore in row a of Table 2.3, under $X = 0$, we enter an a in the next-state column. Suppose next that we are in state a, but that $X = 1$. At the following bit time, the 1 will shift from X to A, the 0 in A to B, the 0 in B to C, and the 0 in C to D, as shown in Fig. 2.6. We will thus be in state $A = 1$, $B = C = D = 0$, which is state i. We therefore enter an i in row a under column $X = 1$ in the next-state part of the table. Now suppose $X = 1$ again. At the following bit time, the 1 in X will have shifted to A, the 1 in A to B, the 0 in B to C, and the 0 in C to D. We are thus in state 1100, or state n, and we enter an n in the next state column on row i for $X = 1$. The other entries are determined in a similar fashion.

Table 2.3 can now be completed by determining the output entries. Referring back to our original description of the circuit, we find that we want $Z_1 = 1$ whenever we have just received the pattern 00001 or 10001. We recognize the first of these when four 0's have successively shifted through into A, B, C, and D, and then $X = 1$. This means we are in state a and $X = 1$, and so at that entry in the Z_1 portion of the output part of the table we enter a 1. The second pattern we will recognize by the fact that a 1 has shifted in followed by three 0's (so that $A = B = C = 0$ and $D = 1$), and that X then equals 1. Therefore there must be a 1 in state b under column Z_1 for $X = 1$. These are the only combinations for which $Z_1 = 1$, so all the other entries in the Z_1 columns can be made 0. The entries in the Z_2 and Z_3 columns are determined in a similar way.

We are now ready to try to eliminate one or more flip-flops, that is, to eliminate enough states from Table 2.3 that the remaining states can be distinguished from one another with three or fewer flip-flops, and still constitute a circuit that is the same, from all outward appearances, as our original circuit. Such a circuit can contain at most $2^3 = 8$ states.

What do we mean by "the same, from all outward appearances"? Clearly the "inward appearance" of the two circuits will be different, for one contains four flip-flops and the other three or less. We mean that, if we connect the inputs of both circuits together, put any conceivable sequence of 0's and 1's on X, and monitor the outputs of both circuits, we shall always find outputs Z_1, Z_2, and Z_3 of one circuit exactly the same as the corresponding outputs of the other circuit. If this is the case, we shall say that the two circuits are *equivalent*.

With the foregoing considerations in mind, we can easily specify the rules for finding a simpler circuit equivalent to that described by Table 2.3.

1. Each state in the equivalent circuit must correspond to one or more states in the original circuit. Our job is thus to find *sets* of states in the original circuit, each of which is equivalent to a single state in the equivalent circuit.

2. A set of two or more states in the original circuit can correspond to a single state in the equivalent circuit only if all members of the set have the same outputs for each possible input combination. For example, states s_1 and s_2 in our original circuit can correspond to state S in the equivalent circuit only if, whether the input X is 0 or 1, the outputs Z_1, Z_2, and Z_3 are the same from states s_1 and s_2 as from state S. Applying this rule to Table 2.3, we see that states a and b provide the same outputs whether $X = 0$ or $X = 1$, and therefore may perhaps correspond to a new state A in the equivalent circuit. For the same reason, states c and d may correspond to an equivalent state C, states $e, f, g,$ and h to a new state E, states $i, k, l,$ and m to a new state I, and states $p, r,$ and s to a new state P. State n in the original circuit has a different output combination from each other state in the original table, and so must, all by itself, have a corresponding new state N in the equivalent circuit. Thus tentatively we can note the associations shown in Table 2.4.

3. In step (2) we have shown that, if our new equivalent circuit has states $A, C, E, I, N,$ and P, equivalent to the sets of states in the original circuit as shown in Table 2.4, then any input will result in the correct output on Z_1, Z_2, and Z_3 at *that* bit time. We must now be sure that any *continuing sequence* of inputs will continue to provide the same outputs

from the two circuits. We do this by following through to see what the equivalent next state is for each value of X applied to the original circuit. Referring again to Table 2.3, we see that, if we are in state a or b, and $X = 0$, we stay in state a. Therefore, if our equivalent circuit is in state A,

TABLE 2.4
Proposed and Original Circuit States

Proposed circuit state	Set of original circuit states
A	(a, b)
C	(c, d)
E	(e, f, g, h)
I	(i, k, l, m)
N	(n)
P	(p, r, s)

and $X = 0$, it will stay in state A. If we are in state a or b, and $X = 1$, we go to state i. Therefore, if our equivalent circuit is in state A, and $X = 1$, then this circuit will go to state I. States a and b are thus equivalent to A in their next-state action as well as in their output combinations. We repeat this analysis for each of the states in our original table, as shown in Table 2.5.

TABLE 2.5
Next-State Analysis of Proposed Circuit

Proposed circuit state	Original circuit state	Next state in proposed circuit	
		$X = 0$	$X = 1$
A	(a, b)	(A, A)	(I, I)
C	(c, d)	(A, A)	(I, I)
E	(e, f, g, h)	(C, C, C, C)	(I, I, I, I)
I	(i, k, l, m)	(E, E, E, E)	(N, N, P, P)
N	(n)	(E)	(P)
P	(p, r, s)	(E, E, E)	(P, P, P)

All the sets of original states except (i, k, l, m) have the property that either value of the input will result in the same next state in the proposed circuit. If, however, we are in state i, and $X = 1$, then we go to state n, which is the proposed next state N; but if we are in state m, and $X = 1$, then we go to state p, which is included in the proposed next state P. Since P and N give different outputs, our assumption that i, k, l, and m can be equivalent to a common state I is false. The pairs (i, k) and (l, m) may still be equivalent to a couple of new states, but not to one another. Let us assume that (i, k) correspond to state I', and (l, m) to state L'. We now repeat the next-state analysis, as shown in Table 2.6.

TABLE 2.6
Revised Next-State Analysis of Proposed Circuit

Proposed circuit state	Set of original circuit states	Next state in proposed circuit	
		$X = 0$	$X = 1$
A	(a, b)	(A, A)	(I', I')
C	(c, d)	(A, A)	(I', I')
E	(e, f, g, h)	(C, C, C, C)	(L', L', L', L')
I'	(i, k)	(E, E)	(N, N)
L'	(l, m)	(E, E)	(P, P)
N	(n)	(E)	(P)
P	(p, r, s)	(E, E, E)	(P, P, P)

TABLE 2.7
Reduced Sequential Circuit

Present state	Next state		Z_1		Z_2		Z_3	
	$X = 0$	$X = 1$	$X = 0$	$X = 1$	$X = 0$	$X = 1$	$X = 0$	$X = 1$
A	A	I	0	1	0	1	1	0
C	A	I	0	0	0	0	1	0
E	C	L	0	0	0	1	1	0
I	E	N	0	0	1	0	1	0
L	E	P	0	0	1	0	1	0
N	E	P	0	0	0	1	0	1
P	E	P	0	0	0	0	0	0

We now see that all sets are consistent and that our equivalent circuit must have seven states. The sequential-circuit table for the equivalent circuit is shown in Table 2.7, in which the primes are omitted from I and L for simplicity.

The fact that there are only seven states in our reduced table implies that only three memory elements are required to mechanize it—and that we have achieved our goal of eliminating a memory element. To construct the circuit, we must specify three flip-flops, say, V, W, and Y, and assign a different binary combination of V, W, and Y to each of the seven states in Table 2.7. Suppose we make the assignment shown in Table 2.8, where we simply write the seven binary numbers 000 to 110 and assign them to states A through P. We now have a table, the "Next-state" portion of which tells what values V, W, and Y should assume at each bit time as a function of their states and the state of X at the previous bit time. We shall now see how to write the input logic for the three delay flip-flops, V, W, and Y, from this table.

TABLE 2.8
Reduced Sequential Circuit with Flip-Flops

Label	Flip-flop			$X = 0$			$X = 1$			Z_1		Z_2		Z_3	
	V	W	Y	V	W	Y	V	W	Y	$X = 0$	$X = 1$	$X = 0$	$X = 1$	$X = 0$	$X = 1$
A	0	0	0	0	0	0	0	1	1	0	1	0	1	1	0
C	0	0	1	0	0	0	0	1	1	0	0	0	0	1	0
E	0	1	0	0	0	1	1	0	0	0	0	0	1	1	0
I	0	1	1	0	1	0	1	0	1	0	0	1	0	1	0
L	1	0	0	0	1	0	1	1	0	0	0	1	0	1	0
N	1	0	1	0	1	0	1	1	0	0	0	0	1	0	1
P	1	1	0	0	1	0	1	1	0	0	0	0	0	0	0

(Present state | Next state | Output)

We begin, in Fig. 2.7a, by drawing a Veitch-diagram map for the four variables X, V, W, and Y, and plotting the seven states A, C, E, I, L, N, and P on it. Each state appears twice, once for $X = 0$ and once for $X = 1$. For example, in Table 2.8 we see that state I is specified by $V = 0$, $W = Y = 1$. Therefore we find the two squares on the Veitch diagram of Fig. 2.7a for which $V = 0$, $W = Y = 1$, and label them both I. Continuing in this way, we ultimately find two squares that do not correspond to any of the seven states. This combination, for which $V = W = Y = 1$, is the eighth state of the three flip-flops. Since we need only seven states, we shall arrange that V, W, and Y are never all 1 at any bit time. We enter a $\emptyset$ in each of these last squares to signify that these configurations of X, V, W, and Y can never occur, and that we can use those squares where desired to simplify our logic.

We are now ready to plot, in Fig. 2.7b, the next state of V, that is, V^{n+1}, as a function of X^n, V^n, W^n, and Y^n. We do this by referring back to Table 2.8 and finding the next-state entries for which $V = 1$. There are five such entries: If at time n we have $X = 1$ and we are in state E, I, L, N, or P, then we find that $V^{n+1} = 1$. We therefore enter 1's in Fig. 2.7b on the $X = 1$ side of the diagram in squares E, I, L, N, and P. For all other squares, we have $V^{n+1} = 0$, and for convenience we leave them blank.

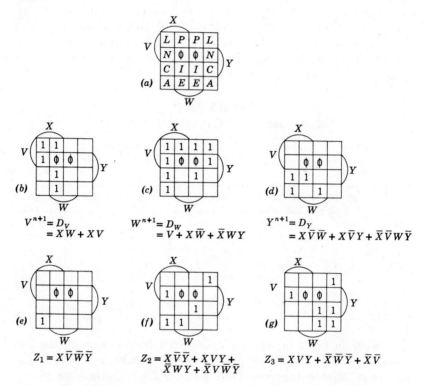

Fig. 2.7 Derivation of input equations for circuit of Table 2.8.

The plot in Fig. 2.7b of V^{n+1} can also be considered as a plot of the input to a delay flip-flop V, for if we want V at time $n + 1$ to be 1, we see, on referring back to Fig. 2.1, that D_V at time n must equal 1. Therefore if we write the simplest Boolean expression for Fig. 2.7b, we are also writing the simplest input equation for flip-flop V. In finding the simplest expression, we must keep in mind that flip-flops V, W, and Y can never all be 1, and therefore that the squares labeled Ø may be regarded as containing either 0's or 1's, whichever results in a simpler equation. The simplest expression is

(2.16) $$D_V = XW + XV.$$

Similarly, the simplest expressions for D_W and D_Y are found from the Veitch-diagram plots of W^{n+1} and Y^{n+1}; they are

$$(2.17) \qquad D_W = V + X \, \bar{W} + \bar{X} \, W \, Y,$$

$$(2.18) \qquad D_Y = X \, \bar{V} \, \bar{W} + X \, \bar{V} \, Y + \bar{X} \, \bar{V} \, W \, \bar{Y}.$$

We can also use Table 2.8 and Fig. 2.7a to plot Z_1, Z_2, and Z_3, as shown in Figs. 2.7e, f, and g. Their simplest expressions are

$$(2.19) \qquad Z_1 = X \, \bar{V} \, \bar{W} \, \bar{Y},$$

$$(2.20) \qquad Z_2 = X \, \bar{V} \, \bar{Y} + X \, V \, Y + \bar{X} \, W \, Y + \bar{X} \, V \, \bar{W} \, \bar{Y},$$

$$(2.21) \qquad Z_3 = X \, V \, Y + \bar{X} \, \bar{W} \, \bar{Y} + \bar{X} \, \bar{V}.$$

Now (2.16) to (2.21) require gates with a total of 59 inputs. We can employ still fewer gates by making use of the multiple-output simplification procedures described previously. These give

$$(2.22) \qquad D_V = X \, W + X \, V,$$

$$(2.23) \qquad D_W = V + X \, \bar{W} + Y_1,$$

$$(2.24) \qquad Y_1 = \bar{X} \, W \, Y,$$

$$(2.25) \qquad D_Y = X \, \bar{V} \, Y + \bar{X} \, \bar{V} \, W \, \bar{Y} + Z_1,$$

$$(2.26) \qquad Z_1 = X \, \bar{V} \, \bar{W} \, \bar{Y},$$

$$(2.27) \qquad Z_2 = X \, \bar{V} \, \bar{Y} + Y_1 + Y_2,$$

$$(2.28) \qquad Y_2 = X \, V \, Y + \bar{X} \, V \, \bar{W} \, \bar{Y},$$

$$(2.29) \qquad Z_3 = Y_2 + \bar{X} \, \bar{V}.$$

This set of gates has only 47 inputs. We have thus reduced a circuit containing 4 flip-flops (2.1) to (2.4) and 24 gating inputs (2.12) to (2.15) to one requiring only 3 flip-flops and 47 gating inputs. Since flip-flops are generally more expensive than gates, this represents a net simplification.

2.5 ALTERNATIVE METHODS OF ASSIGNING FLIP-FLOP STATES

The gate count for the circuit we now have stems entirely from the assignment of flip-flop states we made in Table 2.8. Another assignment—and there are 40,319 others—would result in a different count, perhaps better, perhaps worse. Hartmanis [2] has suggested a procedure that will sometimes lead to a good choice of states, although it is not guaranteed to do so. He begins with the suggestion that we ignore the output logic for the circuit and attempt to find an assignment that will simplify the input logic to the memory elements. He argues that the input logic for a given memory element is likely to be simple if it is *not* a function of *all* the other memory elements. Thus, if a circuit contains three flip-flops Q_1, Q_2, and Q_3, then in general we have an assignment of the form

$$Q_1^{n+1} = Q(Q_1, Q_2, Q_3, X).$$

If we can find an assignment of the form

$$Q_1^{n+1} = Q'(Q_2, X),$$

then the resulting input logic for Q_1 is likely to be simple. Obviously this is not generally true, since, for example, $Q_1^{n+1} = Q_1 Q_2 Q_3 X$ requires only four inputs, whereas $Q_1^{n+1} = \bar{Q}_2 \bar{X} + Q_2 X$ requires six; but it seems a reasonable thing to try. Applied to the circuit of Table 2.7, his procedure may be described as follows.

Ignoring circuit outputs, we want to arrange our seven states in four groups so that no group contains more than two members, and so that the next-state entries for the two states in a group are again members of a common group, whether the input X is 0 or 1. If we can find four such groups, we can assign the four combinations of flip-flops V and W to them, use flip-flop Y to distinguish between the two states in a group, and have the desired result that the next state of V and W will be independent of Y.

To find four such groups, we try every combination of pairs of groups, noting for each pair what other pairs of terms must be grouped together in order that next-state entries shall be members of a common group. The results are indicated in Table 2.9. Referring to Table 2.7, we can see how the entries of Table 2.9 are derived. The entry in the upper left-hand corner specifies what conditions must be satisfied if A and C are to be grouped together. Since both A and C go to state A when $X = 0$ and to state I when $X = 1$, no special grouping conditions are required, and we place a check mark in the square of Table 2.9 corresponding to the grouping of A and C. Next consider the possibility of grouping states A and E. Since A stays in state A, but E goes to state C when $X = 0$, the collection of A and E into a group implies that A and C must also be in the same group. We therefore enter AC in the square corresponding to the grouping of A and E. Furthermore, since A goes to state I and E to state L when $X = 1$, the grouping of A and E implies that I and L must also be grouped together. We therefore add IL to the square corresponding to the grouping of A and E. Similarly, we write the implications of each possible group of two states and thus complete Table 2.9.

We now examine Table 2.9 to identify all possible consistent groupings of states. Choosing the pair AC as one obvious first choice, because it implies nothing about any other groupings, perhaps we try EI next. Referring to Table 2.9, we see that if EI are in the same group, then so must CE and also IL be in the same group. Since we started with E and I together and found that C must be grouped with E, and L with I, then C, E, I, and L must all be together. A quick survey of Table 2.9 with this in mind leads to the conclusion that E and I being in the same group implies that *all* states are in the same group.

TABLE 2.9
Determination of State Groupings

Other member	One member of pair					
	A	C	E	I	L	N
C	√					
E	AC IL	AC IL				
I	AE IN	AE IN	CE LN			
L	AE IP	AE IP	CE LP	NP		
N	AE IP	AE IP	CE LP	NP	√	
P	AE IP	AE IP	CE LP	NP	√	√

If we start with AC again, but try adding IL as a group instead of EI, we fare better. Having I and L in the same group implies that N and P must be in the same group, but a look in the table for the implications of an NP grouping indicates that there are none. We can therefore call our four groups AC, E, IL, and NP, and meet all necessary conditions. Stated in another way, we could build a four-state machine with next-state entries as those shown in Table 2.10. We could make the flip-flop assignments as shown, and write input equations for flip-flops V and W that would be functions of X, V, and W alone.

We can now add another flip-flop Y as shown in Table 2.11 and use it to distinguish between A and C, I and L, and N and P. The addition of

TABLE 2.10
Next-State Pattern for a Four-State Machine with Pair Groupings

	Present state		Next state	
Label	Flip-flop		$X = 0$	$X = 1$
	V	W		
AC	0	0	AC	IL
E	0	1	AC	IL
IL	1	0	E	NP
NP	1	1	E	NP

TABLE 2.11
Hartmanis's Procedure for Choosing State Assignments

	Present state			Next state						Output					
Label	Flip-flop			$X = 0$			$X = 1$			Z_1		Z_2		Z_3	
	V	W	Y	V	W	Y	V	W	Y	$X=0$	$X=1$	$X=0$	$X=1$	$X=0$	$X=1$
A	0	0	0	0	0	0	1	0	1	0	1	0	1	1	0
C	0	0	1	0	0	0	1	0	1	0	0	0	0	1	0
E	0	1	0	0	0	1	1	0	0	0	0	0	1	1	0
I	1	0	1	0	1	0	1	1	1	0	0	1	0	1	0
L	1	0	0	0	1	0	1	1	0	0	0	1	0	1	0
N	1	1	1	0	1	0	1	1	0	0	0	0	1	0	1
P	1	1	0	0	1	0	1	1	0	0	0	0	0	0	0

flip-flop Y will obviously not change the input logic for flip-flops V and W, and so the logic for V and W is independent of Y—the result that Hartmanis's procedure guarantees.

The Veitch diagrams for this new flip-flop assignment are shown in Fig. 2.8, and the simplest equations are

$$(2.30) \qquad D_V = X,$$

$$(2.31) \qquad D_W = V,$$

$$(2.32) \qquad D_Y = X \, \bar{W} \, Y + Z_1 + \bar{X} \, \bar{V} \, W,$$

$$(2.33) \qquad Z_1 = X \, \bar{V} \, \bar{W} \, \bar{Y},$$

$$(2.34) \qquad Z_2 = X \, \bar{V} \, \bar{Y} + Y_1,$$

$$(2.35) \qquad Y_1 = X \, W \, Y + \bar{X} \, V \, \bar{W},$$

$$(2.36) \qquad Z_3 = Y_1 + \bar{X} \, \bar{V}.$$

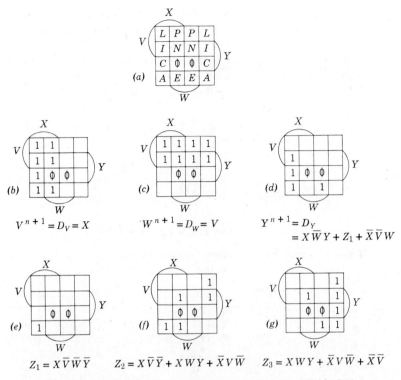

Fig. 2.8   Derivation of input equations for circuit of Table 2.11.

This set of gates has only 30 inputs, compared with the 47 of equations (2.22) to (2.29).

CHOOSING DECISION AND MEMORY ELEMENTS

2.6 MEMORY ELEMENTS OTHER THAN DELAY FLIP-FLOPS

Thus far we have very artificially limited our attention to circuits that use delay flip-flops as memory elements, and AND and OR gates as decision elements. This enabled us to concentrate on some of the various manipulations and rearrangements that have been devised for simplification, but we completely ignored the fact that computer designers have invented many other decision and memory elements with other logical properties.

In this section and the next, we shall show how these other circuit elements can be used to develop still other mechanizations of a given sequential circuit.

To begin with, let us continue to use the AND and OR decision elements, but allow ourselves the freedom of using some new type of memory element, different from the delay flip-flop, in building the circuit described by

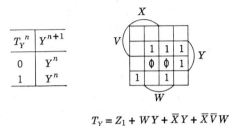

$T_Y{}^n$	Y^{n+1}
0	Y^n
1	Y^n

$$T_Y = Z_1 + WY + \overline{X}Y + \overline{X}\,\overline{V}W$$

Fig. 2.9   Trigger memory element for the flip-flop Y of Table 2.11.

$R_Y{}^n$	$S_Y{}^n$	Y^{n+1}
0	0	Y^n
0	1	1
1	0	0
1	1	?

(a) $R_Y = \overline{X}Y + Y_1$

(b) $S_Y = \overline{X}\,\overline{V}W + Z_1$

Fig. 2.10   Reset-set memory element for the flip-flop Y of Table 2.11.

(2.30) to (2.36). Since the logic (2.30), (2.31) for flip-flops V and W is about as simple as anything can be, let us see what can be done with the logic (2.32) of flip-flop Y.

Two of the most commonly used delay elements are the trigger and the reset-set (R-S) flip-flop; their truth tables are shown in Figs. 2.9 and 2.10, respectively. The trigger, like the delay element, has a single input. The trigger, however, has the property that when its input is 0, the flip-flop remains in whatever state is held previously; but when its input is 1, the circuit changes state, or "triggers." It can be shown algebraically, and should be evident from a quick look at the logic of the circuit, that the input T can be expressed as

$$(2.37) \qquad T_Y = \overline{Y}^n Y^{n+1} + Y^n \overline{Y}^{n+1};$$

that is, the trigger input must be 1 if and only if the circuit is in the 0 state and must be in the 1 state at the next bit time, or if it is in the 1 state and must be in the 0 state at the next bit time. To make flip-flop Y of (2.32) a trigger flip-flop, we need merely to substitute the equation for Y^{n+1}, plotted in Fig. 2.8d, into (2.37) and then simplify the resulting function. Alternatively, we can use a simple trick to plot T_Y directly on a Veitch diagram, so we can use previously described simplification techniques. To construct the Veitch diagram for T, we merely copy the 1's from the $\bar{Y}$ portion of Y^{n+1} onto the diagram, Fig. 2.9b, and the 0's from the Y portion of Y^{n+1}. The simplest resulting equation is

$$(2.38) \qquad T_Y = Z_1 + W\,Y + \bar{X}\,Y + \bar{X}\,\bar{V}\,W,$$

which requires eleven inputs and is clearly no improvement over the nine inputs required in (2.32) for the delay flip-flop.

The R-S flip-flop, whose truth table is given in Fig. 2.10, has two inputs. The R input $(R = 1, S = 0)$ *resets* the circuit to the 0 state, whatever its previous state has been. The S input $(R = 0, S = 1)$ *sets* the circuit to the 1 state. If both inputs are 0, the circuit remains in its preceding state. Finally, if both inputs are 1, the circuit designer shrugs his shoulder and says he is not sure what will happen—the flip-flop may remain in its previous state or may change states. This combination—both inputs 1—must therefore be avoided by the logical designer in his use of the R-S flip-flop.

Again, it is possible to derive algebraic expressions for R and S as a function of the current state and desired next state of a flip-flop. Thus we have

$$(2.39) \qquad R_Y = Y^n\,\bar{Y}^{n+1} + a\,\bar{Y}^n\,\bar{Y}^{n+1},$$

$$(2.40) \qquad S_Y = \bar{Y}^n\,Y^{n+1} + b\,Y^n\,Y^{n+1},$$

where a and b are arbitrary Boolean constants that may be chosen so as to obtain the simplest possible circuit mechanization. These equations, like (2.38), are simply explained. Equation (2.39), for example, states that the reset input *must* be 1 if the circuit is in the 1 state and must be in the 0 state at the next bit time, and that the input *may* be 1 if the circuit is in the 0 state and is to remain in that state until the next bit time.

Plots of R_Y and S_Y are shown in Fig. 2.10(a) and (b), and the resulting simplest input equations are

$$(2.41) \qquad R_Y = \bar{X}\,Y + Y_1,$$

$$(2.42) \qquad S_Y = \bar{X}\,\bar{V}\,W + Z_1.$$

These circuits have nine inputs, the same as the circuit (2.32) for the delay element, and therefore represent no improvement over the delay flip-flop.

Obviously, many additional types of memory elements are possible, at least from a logical point of view. If we consider two-input memory elements only, and restrict our discussion to those in which there are no forbidden input combinations, such as $R = S = 1$ in the R-S flip-flop, then we can describe each circuit with a table such as shown in Fig. 2.10. Since each entry in the Y^{n+1} column can have one of four different values— 0, 1, Y^n, and $\bar{Y}^n$—there are $4^4 = 256$ possible columns. Some of these columns represent trivial elements, some represent basically one-input memory elements, and some are duals of others. Wrzesinski [8] has shown

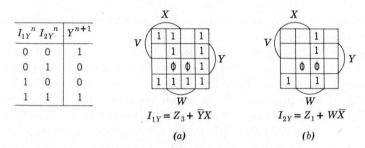

I_{1Y}^n I_{2Y}^n	Y^{n+1}
0 0	1
0 1	0
1 0	0
1 1	1

$I_{1Y} = Z_3 + \bar{Y}X$

(a)

$I_{2Y} = Z_1 + W\bar{X}$

(b)

Fig. 2.11 Wrzesinski's memory element for the flip-flop Y of Table 2.11.

that there are 108 basically distinct two-input circuits, and has made some progress on a solution to the interesting problem: Given a job to be done and a set of decision elements, which memory elements should be used so that the resulting logic is as simple as possible?

Wrzesinski's procedure is too complicated to be presented here. Application of his scheme led, however, to the choice of the two-input circuit having a truth table shown in Fig. 2.11. For this circuit, the output at time $n + 1$ is 1 only if the inputs are equal at time n; that is, only if I_1 and I_2 are both 0 or both 1. The generalized input equations for this circuit are

$$(2.43) \quad I_{1Y} = C_1 \bar{Y}^n \bar{Y}^{n+1} + C_2 Y^n \bar{Y}^{n+1} + K_1 \bar{Y}^n Y^{n+1} + K_2 Y^n Y^{n+1},$$

$$(2.44) \quad I_{2Y} = \bar{C}_1 \bar{Y}^n \bar{Y}^{n+1} + \bar{C}_2 Y^n \bar{Y}^{n+1} + K_1 \bar{Y}^n Y^{n+1} + K_2 Y^n Y^{n+1},$$

where C_1, C_2, K_1, and K_2 are arbitrary Boolean quantities. Proper choice of these arbitrary terms leads to the Veitch diagrams of Fig. 2.11a, b and to the solutions

$$(2.45) \quad I_{1Y} = Z_3 + \bar{Y} X,$$

$$(2.46) \quad I_{2Y} = Z_1 + W \bar{X}.$$

Since these circuits require only eight inputs compared to the ten of the delay circuit (2.25), they represent a slight improvement, assuming that we can persuade the circuit designer to devise a memory element having the I_1-I_2 logical properties.

2.7 DECISION ELEMENTS OTHER THAN AND AND OR GATES

Because the AND and OR gates have been cheap and easy to construct, and because they are easy for the designer to use, they were for many years the only decision elements employed to any great extent by the computer industry. The development of the transistor, however, has made a variety of other circuits practical and economical, and has given rise to many new and interesting problems for the mathematicians and engineers working in this field. Two of the most important new decision elements, the NOR circuit and the NAND circuit, are most widely used and are easy to treat because they can easily be derived from AND-OR circuits. A third, the threshold element, presents a much more difficult problem.

The NOR and NAND *decision elements* are easily specified as follows:

(2.47)   NOR: $A \downarrow B \downarrow C \downarrow \ldots \downarrow X = \overline{A + B + C + \cdots + X}$,

(2.48)   NAND: $A \mid B \mid C \mid \ldots \mid X = \overline{A \cdot B \cdot C \cdots X}$.

Because of the close relationship of these circuits to AND and OR circuits, as indicated by these definitions, it is very easy to transform a simplified AND-OR circuit into a NAND or NOR circuit. For example, from (2.35) and (2.36) we have

(2.35) $Y_1 = X\,W\,Y + \bar{X}\,V\,\bar{W}$

(2.49) $= (X \mid W \mid Y) \mid (\bar{X} \mid V \mid \bar{W})$,

and

(2.36) $Z_3 = Y_1 + \bar{X}\,\bar{V}$

             $= (Y_1 + \bar{X})(Y_1 + \bar{V})$

(2.50) $= (Y_1 \downarrow \bar{X}) \downarrow (Y_1 \downarrow \bar{V})$.

The reader can verify that (2.49) and (2.50) are correct by applying the definitions (2.47) and (2.48).

The *threshold decision element* [7] has attracted particular interest because it is said to be a mathematical model of the neuron, the decision element in living things. The element can be thought of as having three parts, as shown in Fig. 2.12: a set of weights applied to the various inputs, an algebraic adder, and a discriminator. Each of the circuit inputs, with value 0 or 1, is multiplied by a positive or negative weight. The resulting signals are added algebraically in the adder, which may also have a constant input or bias. The result is applied to the discriminator, with

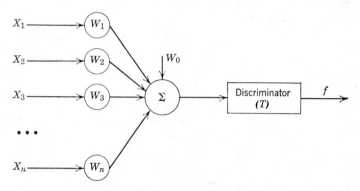

Fig. 2.12   Threshold decision element.

output 0 or 1, depending on whether its input is less than or greater than a threshold value assigned to the discriminator. If the two-valued inputs are $X_1, X_2, X_3, \ldots, X_n$, the weights are $W_1, W_2, W_3, \ldots, W_n$, the constant term is W_0, and the threshold is T, then the output Z will be 1 if and only if

$$(2.51) \qquad \sum_{i=1}^{n} W_i X_i + W_0 > T.$$

The obvious problem, then, is to find values for $W_0, W_1, \ldots, W_n$, and T appropriate to mechanize a given Boolean function.

It should be evident that a single threshold element of the kind described by (2.51) cannot realize every Boolean function. For example, consider the simple two-variable function

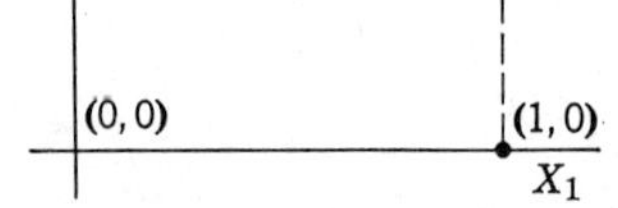

Fig. 2.13   Threshold-logic example.

$$(2.52) \qquad f = X_1 \bar{X}_2 + \bar{X}_1 X_2.$$

In Fig. 2.13 this function is plotted as two points on a rectangular coordinate system. The threshold element of (2.51) is a straight line in this coordinate system, and the slope and intercept of the line are determined by the weights and threshold. The *graphical* equivalent of the design problem is this: Can we draw a straight line in Fig. 2.13 in such a way that points (0. 1) and (1, 0) lie on one side of the line, and points (0, 0) and (1, 1) on the other? Obviously this is not possible, and so the function (2.52) is not realizable with a single threshold element.

As the number of variables in a function increases, the likelihood that it can be realized with a single threshold decision element becomes very

TABLE 2.12

Partitioning of Output Z_3 into Simpler Outputs Z_3^1 and Z_3^2

Input				Output		
V	W	Y	X	Z_3	Z_3^1	Z_3^2
0	0	0	0	1	1	0
0	0	0	1	0	0	0
0	0	1	0	1	1	0
0	0	1	1	0	0	0
0	1	0	0	1	1	0
0	1	0	1	0	0	0
0	1	1	0	0̸	1	0
0	1	1	1	0̸	0	1
1	0	0	0	1	1	0
1	0	0	1	0	0	0
1	0	1	0	1	1	0
1	0	1	1	0	0	0
1	1	0	0	0	0	0
1	1	0	1	0	0	0
1	1	1	0	0	0	0
1	1	1	1	1	0	1

small indeed, and so investigators have turned to the problem of designing circuits containing more than one such element. Minnick [4] and Stram [6] have published papers describing a design procedure, and we shall conclude this section by showing how a threshold circuit can be constructed for the function Z_3 of Fig. 2.8.

We begin by constructing a truth table for the function Z_3, as is shown in Table 2.12. The general procedure will be to find two or more functions Z_3^j that are simpler than Z_3, that can be mechanized with threshold elements, and that have Boolean sum equal to the desired function. We can then combine these subfunctions, or partitions, with a simple OR circuit, or with another threshold element, to form Z_3.

The two-variable circuits that are realizable by means of single threshold elements are easily found and tabulated, as in Table 2.13. Therefore we divide Table 2.12 into four groups of four terms each, as indicated by the dashed lines, where the four terms within a group differ only in the values of the last two variables Y and X. We now look for a function that is

TABLE 2.13
Two-Variable Circuits f_i That Are Realizable by Means of Single Threshold Elements, with Solutions

Input								Subscript i							
X_1	X_2	0	1	2	3	4	5	7	8	10	11	12	13	14	15
0	0	0	1	0	1	0	1	1	0	0	1	0	1	0	1
0	1	0	0	1	1	0	0	1	0	1	1	0	0	1	1
1	0	0	0	0	0	1	1	1	0	0	0	1	1	1	1
1	1	0	0	0	0	0	0	0	1	1	1	1	1	1	1
Parameter								Solution							
T		0	0	0	0	0	0	0	0	0	0	0	0	0	0
W_0		0	1	0	1	0	1	1	-1	0	1	0	1	0	1
W_1		0	-1	-1	-1	1	0	$-\frac{1}{2}$	1	0	-1	1	1	1	0
W_2		0	-1	1	0	-1	-1	$-\frac{1}{2}$	1	1	1	0	-1	1	0

included in Z_3 and that is easy to construct by means of the two-variable circuits of Table 2.13. The appearance of the function f_5 in the top three groups suggests that it might be used with good effect, and so we construct the function Z_3^1 of Table 2.12. This new function is equal to f_5 except for $V = W = 1$, and this in turn suggests the use of function f_7 of V and W since, in Boolean algebra,

$$(2.53) \qquad Z_3^1 = f_5(Y, X) \cdot f_7(V, W) = \bar{X} \cdot (\bar{V} + \bar{W}).$$

Referring to Table 2.13, we find how f_5 and f_7 can be constructed; see Table 2.14a, b. We read off the threshold mechanizations

$$f_5: 1 - X > 0, \qquad f_7: 1 - \tfrac{1}{2}V - \tfrac{1}{2}W > 0,$$

and then combine these two into a single one, obtaining the mechanization

$$(2.54) \qquad Z_3^1: 1 - X - \tfrac{1}{2}V - \tfrac{1}{2}W > 0.$$

The design of our circuit can be completed by adding to Table 2.12 a second column Z_3^2 having the property that

$$(2.55) \qquad\qquad Z_3 = Z_3^1 + Z_3^2.$$

This new function is easily constructed by using the threshold circuit

$$(2.56) \qquad\qquad Z_3^2 \colon W + Y + X - 2 > 0.$$

TABLE 2.14
Threshold Circuits f_i for the Output Z_3^1, with Solutions

(a) Circuit f_5			(b) Circuit f_7		
Input		i	Input		i
Y	X	5	V	W	7
0	0	1	0	0	1
0	1	0	0	1	1
1	0	1	1	0	1
1	1	0	1	1	0
Parameter	Solution		Parameter	Solution	
T	0		T	0	
W_0	1		W_0	1	
W_1	0		W_1	$-\frac{1}{2}$	
W_2	-1		W_2	$-\frac{1}{2}$	
Mechanization $1 - X > 0$			Mechanization $1 - \frac{1}{2}V - \frac{1}{2}W > 0$		

The finished circuit, which makes use of the function f_{14} in Table 2.13 to mechanize the logical sum, is shown in Fig. 2.14.

The procedure sketched does not guarantee a minimal circuit in any sense, and can be applied to many-variable logical circuits with only the help of computing devices. So far, threshold circuits have been used in existing computers in relatively few circumstances when their unique properties make it possible to construct a circuit that would be substantially more expensive if derived in any other way. The tools we have for determining such circumstances are thus far very primitive.

2.8 CONCLUSION

The sequential-circuit model devised by Huffman and others provides a very general approach to the design of any digital system. Starting with

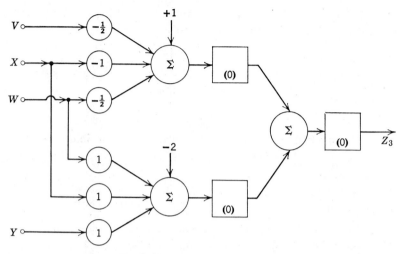

Fig. 2.14   Threshold circuit for Z_3.

a general verbal description of the functions such a system is to perform, we can draw up a table that specifies precisely how the circuit is to function, and can then manipulate the table to derive additional tables specifying other equivalent circuits.   Furthermore, going from such a table to a practical working circuit is a simple step for which adequate procedures are available.   The problem of finding the *simplest* such circuit, however, is one that has been tackled only in the most primitive fashion, and is a problem of considerable difficulty because of the variety and complexity of devices available.

MULTIPLE-CHOICE REVIEW PROBLEMS

1. Memory element is to flip-flop as decision element is to

(a) trigger,   (b) diode,   (c) AND gate,   (d) output.

2. The simplest AND-OR circuit for $f = A\overline{C} + AB + BC$ is

(a) $f = A\overline{C} + BC$,
(b) $f = AB + BC$,
(c) $f = A\overline{C} + AB + BC$,
(d) $f = \overline{C} + AB$.

3. The circuit described in the table, shown on the next page, can be built with no fewer than

(a) 3 flip-flops,   (b) 2 flip-flops,
(c) 1 flip-flop,   (d) no flip-flops.

Present state	Next state		Output	
	$X = 0$	$X = 1$	$X = 0$	$X = 1$
a	d	e	0	1
b	a	b	1	0
c	e	d	1	0
d	e	c	1	0
e	b	c	0	1

4. A majority logic circuit is described by the following algebraic inequality:

$$2 - 2A + B - C > 0 \quad \text{means } f = 1.$$

Which of the Boolean functions shown below correctly describes f?

(a) $\overline{A} + B\overline{C}$,

(b) $\overline{A}\overline{B} + AB\overline{C}$,

(c) $AC + \overline{A}\overline{B}\overline{C}$,

(d) $AC + A\overline{B}$.

5. The correct input logic for a delay flip-flop is

$$D_A = A\overline{B}\overline{C} + \overline{A}B.$$

If we want to substitute a trigger flip-flop for this delay flip-flop, its logic should be

(a) $T_A = A + B + C$,

(b) $T_A = A\overline{C} + \overline{A}C$,

(c) $T_A = \overline{A}\overline{B}\overline{C} + AB$,

(d) $T_A = B + AC$.

REFERENCES

1. Bartee, T. C., "Computer Design of Multiple-Output Logical Networks," *IRE Trans. on Electronic Computers* **EC-10** (1961), 21–30.
2. Hartmanis, J., "On the State Assignment Problem for Sequential Machines, I," *IRE Trans. on Electronic Computers* **EC-10** (1961), 157–164.
3. Huffman, D. A., "The Synthesis of Sequential Switching Circuits. I. II," *J. Franklin Inst.* **257** (1954), 161–190, 275–303.
4. Minnick, R. C., "Linear-Input Logic," *IRE Trans. on Electronic Computers* **EC-10** (1961), 6–16.
5. Phister, M., Jr., *Logical Design of Digital Computers*, John Wiley and Sons, New York, 1958.
6. Stram, O. B., "Arbitrary Boolean Functions of N Variables Realizable in Terms of Threshold Devices," *Proc. IRE* **49** (1961), 210–220.
7. Winder, R. O., "Some Recent Papers in Threshold Logic," *Proc. IRE* **49** (1961), 1100.
8. Wrzesinski, J. P., "Selection of a Computer Memory Element to Minimize Input Logic," M.S. Thesis, University of California, Los Angeles, 1961.

PART IB

COUNTING
AND ENUMERATION

Let me count the ways.

ELIZABETH BARRETT BROWNING
Sonnets from the Portuguese, XLIII

CHAPTER 3

Generating Functions

JOHN RIORDAN

3.1 INTRODUCTION

Generating functions were invented by Laplace. He gave them their name and their first systematic treatment in the course of his great work [10] on the theory of probability. The idea involved is simple: A sequence of numbers $a_0, a_1, a_2, \ldots$, which is a finitely or infinitely multiple entity in this separated form, is replaced by the function

$$(3.1) \qquad A(t) = a_0 + a_1 t + a_2 t^2 + \cdots + a_n t^n + \cdots,$$

which is called its *generating function;* inversely, the coefficient a_n as a function of n is called the *determined function.* The object of the replacement is simplification. Operations on the sequences correspond to simpler operations on their generating functions: The generating function becomes a tool in an algebra of sequences.

Several examples of such a simplification are given in this chapter.

Because of its origin, and because of its kinship with probability and classical combinatorial analysis, the generating function has great combinatorial significance. Indeed, it is implicit in the following informal definition given by De Morgan [6, p. 335]: *"Combinatorial analysis mainly consists in the analysis of complicated developments by means of a priori consideration and collection of the different combinations of terms which can enter the coefficients."* This statement, by the way, is one-sided, as noticed in the preface to the author's book [13]; it recognizes the fact that the coefficients of the terms of generating functions may be determined by combinatorial considerations, but ignores the reverse possibility that combinatorial problems may be solved by generating functions. As discussed later, even in the simplest part of the theory of combinations and permutations, generating functions are essential to presenting this theory in a simple general form.

For permutations, moreover, a form differing from (3.1) is convenient, namely,

$$(3.2) \qquad A(t) = a_0 + a_1 t + \frac{a_2 t^2}{2!} + \cdots + \frac{a_n t^n}{n!} + \cdots.$$

For obvious reasons this is called an *exponential* generating function.

Both (3.1) and (3.2) are instances of the general form

$$(3.3) \quad A(t) = a_0 f_0(t) + a_1 f_1(t) + a_2 f_2(t) + \cdots + a_n f_n(t) + \cdots.$$

In order that $f_n(t)$ should be associated uniquely with the coefficient a_n, it is required that the set of functions $\{f_0(t), f_1(t), \ldots\}$ be linearly independent.

Another instance of (3.3) is provided by

$$f_0(t) = 1, \qquad f_n(t) = t(t-1) \cdots (t-n+1), \qquad n = 1, 2, \ldots,$$

the *falling factorial*. The two forms (3.1) and (3.2), however, dominate both combinatorial analysis and probability.

It is clear that there is an easy extension of (3.3) to several variables. For concreteness, the extension to two variables is as follows:

$$(3.4) \qquad A(t, u) = a_{00} f_0(t) g_0(u) + \cdots + a_{nm} f_n(t) g_m(u) + \cdots.$$

A different kind of extension to many variables is MacMahon's introduction of symmetric functions, the basis of his original development of combinatory analysis [11]. The idea may be seen in the following very special case. A simple two-variable generating function with a finite number of terms is

$$A(t, u) = a_{00} + a_{10} t + a_{01} u + a_{20} t^2 + a_{11} tu + a_{02} u^2.$$

This is not symmetrical in t and u for arbitrary coefficients, but the sum $A(t, u) + A(u,t)$ clearly is symmetrical; in fact, we have,

$$A(t, u) + A(u, t) = 2a_{00} + (a_{10} + a_{01})(t + u)$$
$$+ (a_{20} + a_{02})(t^2 + u^2) + 2a_{11} tu.$$

This is MacMahon's form; it is mentioned here only for completeness. Because of its great generality, it is not adapted either to easy explanation or to easy manipulation for interesting results.

The course of development to be presented is as follows. The use of generating functions is examined first in the intuitively simple setting of combinations, and then in the setting of permutations. Next, elementary relations governing their uses are given and exemplified in a variety of results. Then, a number of kinds of generating function appearing in probability and statistics are described; consideration of their inter-relations provides further examples of the chameleonic character of the subject. Finally, to indicate how intricate and elaborate the use of generating functions may become, a brief résumé is given of a theorem due to George Pólya [12], which N. G. de Bruijn [4] has called the "fundamental theorem in enumerative combinatorial analysis," and which will be developed much more fully in Chapter 5. It is hoped that the examples given here will persuade the reader that, knowingly and unknowingly, he has been using generating functions all his life.

3.2 GENERATING FUNCTIONS FOR COMBINATIONS

Consider the combinations of three objects, say a, b, c. These may be shown as follows:

$$
\begin{array}{lll}
a & ab & abc \\
b & ac & \\
c & bc &
\end{array}
$$

This tabular presentation, although perhaps natural for engineers and scientists, is not a mathematical object. Its generating function, namely (unity is added by convention),

$$1 + (a + b + c)t + (ab + ac + bc)t^2 + abct^3,$$

has a closely similar look, expresses exactly the same information, and is a mathematical object, a familiar third-degree polynomial.

Now the function

$$C_3(t) = (1 + at)(1 + bt)(1 + ct)$$

is readily generalized, on labeling objects $o_1, o_2, \ldots, o_n$, to

$$
\begin{aligned}
(3.5) \qquad C_n(t) &= (1 + o_1 t)(1 + o_2 t) \cdots (1 + o_n t) \\
&= 1 + a_1 t + a_2 t^2 + \cdots + a_n t^n,
\end{aligned}
$$

where a_k is the kth *elementary symmetric function* of the n variables o_1 to o_n. Note that $C_n(t)$, after multiplication of its factors, contains the actual exhibition of combinations. If only the *number* of combinations is in question, the object labels may be ignored, and the generating function (now an enumerating generating function—or *enumerator*, for short) becomes

$$(3.6) \qquad C_n(t) = (1 + t)^n = \sum_{k=0}^{n} C(n, k) t^k,$$

with $C(n, k)$ the number of combinations of n things taken k at a time. It is clear from (3.6) that $C(n, k)$ is a binomial coefficient. Even at this early stage, the generating function can be interesting, as is shown by the following examples.

Example 3.1   Since

$$C_n(t) = (1 + t)^n = (1 + t)C_{n-1}(t),$$

it follows from (3.6) that

$$C(n, k) = C(n - 1, k) + C(n - 1, k - 1),$$

which is the basic recurrence relation for binomial coefficients.

Example 3.2   Since $C_n(1) = 2^n$, it follows that

$$2^n = \sum_{k=0}^{n} C(n, k) = \sum_{0}^{n} \binom{n}{k}.$$

In addition, we have

$$C_n(-1) = 0 = \sum_{k=0}^{n} (-1)^k C(n, k) = \sum_{0}^{n} (-1)^k \binom{n}{k}.$$

The last expression in each case employs the usual notation for the binomial coefficients.

Define the even function $C_n^e(t)$ by

$$2C_n^e(t) = (1 + t)^n + (1 - t)^n = 2 \sum_{k=0}^{[n/2]} C(n, 2k)t^{2k},$$

and the odd function $C_n^o(t)$ by

$$2C_n^o(t) = (1 + t)^n - (1 - t)^n = 2 \sum_{k=0}^{[(n-1)/2]} C(n, 2k + 1)t^{2k+1}.$$

Then we have

$$C_n^e(1) = C_n^o(1) = 2^{n-1} = \sum_{k=0}^{[n/2]} \binom{n}{2k} = \sum_{k=0}^{[(n-1)/2]} \binom{n}{2k + 1}.$$

Example 3.3   Write $n = n - m + m$. Then the relation

$$(1 + t)^n = (1 + t)^{n-m}(1 + t)^m$$

implies the Vandermonde identity

$$\binom{n}{k} = \sum_{j=0}^{k} \binom{n - m}{j} \binom{m}{k - j}.$$

Now, look again at (3.5), the generating function for combinations of n distinct objects. In its factored form, each object is represented by a binomial and each binomial spells out the fact that its object has two possibilities in any combination: Either it is absent (the term 1) or it is present (the term $o_i t$ for object o_i). Equation (3.5) is the generating function for combinations *without repetition,* a qualification previously unnecessary. It is clear that repetitions of any kind are accounted for by a corresponding generating function. For example, if an object may appear in a combination zero, one, or two times, then the function is the polynomial $1 + ot + o^2 t^2$; if the number of repetitions is unlimited, it is the function $1 + ot + o^2 t^2 + \cdots = (1 - ot)^{-1}$; if the number of repetitions must be even, it becomes $1 + o^2 t^2 + o^4 t^4 + \cdots$; if the number is specified as 1 or a prime, it is $ot + o^2 t^2 + o^3 t^3 + o^5 t^5 + \cdots$; etc. Moreover, the specification of repetitions may be made arbitrarily for each object; the generating function is a representation of this specification in its factored form as well as a representation of the corresponding combinations in its developed form. The possibilities are endless, and the following examples are merely for concreteness.

Example 3.4 For combinations of n objects with no restriction on the number of repetitions for any object, the enumerating generating function is $(1 + t + t^2 + \cdots)^n$, and we have

$$(1 + t + t^2 + \cdots)^n = (1 - t)^{-n}$$
$$= \sum_{k=0}^{\infty} \binom{-n}{k} (-t)^k$$
$$= \sum_{k=0}^{\infty} \binom{n+k-1}{k} t^k.$$

The number of combinations with repetition of n objects, k at a time, is the same as the number of combinations without repetition of $n + k - 1$ objects, k at a time.

Example 3.5 For the problem of Example 3.4, with the added specification that each object must appear at least once, the enumerator is $(t + t^2 + \cdots)^n$, and we have

$$(t + t^2 + \cdots)^n = t^n (1 - t)^{-n}$$
$$= \sum_{k=n}^{\infty} \binom{k-1}{k-n} t^k.$$

3.3 GENERATING FUNCTIONS FOR PERMUTATIONS

For permutations, the generating function is a little less inclusive; it is reduced to an enumerator. In the simplest case, for n distinct things and no repetitions, the number of permutations k at a time is $P(n, k)$, where

$$P(n, k) = k! C(n, k),$$

since the positions of the objects in a combination of k may be permuted in $k(k-1)\cdots 1 = k!$ ways. Hence we have

$$(3.7) \qquad (1+t)^n = \sum_{k=0}^{n} C(n,k)t^k = \sum_{k=0}^{n} \frac{P(n,k)t^k}{k!}.$$

The enumerator for permutations is an *exponential* generating function.

When repetitions are allowed, the enumerator for any object is a series containing a term $t^k/k!$ for each k in the specification of repetitions. If an object may appear zero, one, or two times, the enumerator is the polynomial $1 + t + t^2/2!$; if unlimited repetition is specified, it is $1 + t + t^2/2! + \cdots = e^t$; etc. The following examples illustrate a few of the possibilities.

Example 3.6 Consider permutations k at a time of n objects with repetition. The enumerator is $(e^t)^n = e^{nt}$, and

$$e^{nt} = \sum_{k=0}^{\infty} \frac{n^k t^k}{k!}.$$

The number of permutations in question is n^k, a result easily found by other means.

Example 3.7 Consider again the permutations of Example 3.6, with the added condition that each object must appear at least once. The enumerator is $(e^t - 1)^n$, and we have

$$(e^t - 1)^n = \sum_{j=0}^{n} \binom{n}{j} (-1)^j e^{(n-j)t}$$

$$= \sum_{k=0}^{\infty} \frac{t^k}{k!} \sum_{j=0}^{n} \binom{n}{j} (-1)^j (n-j)^k.$$

If E is the shift operator, $Ef(n) = f(n+1)$ and $\Delta = E - 1$, then the inner sum may be written as $\Delta^n 0^k$, which is equal to $n!S(k,n)$, with $S(k,n)$ a *Stirling number of the second kind*, a ubiquitous number in combinatorics. Thus

$$\frac{(e^t - 1)^n}{n!} = \sum_{k=0}^{\infty} \frac{S(k,n)t^k}{k!}$$

is the exponential generating function for Stirling numbers of the second kind.

3.4 ELEMENTARY RELATIONS: ORDINARY GENERATING FUNCTION

Considered as a tool in an algebra of sequences, a point of view made familiar in papers (see [1]) and a book by E. T. Bell [2], the ordinary generating function, namely the formal power series, is an element of a commutative ring, provided addition and multiplication are properly defined.

The definition of *addition* is the natural one: If $A(t)$ is the generating function of the sequence (a_n), $n = 0, 1, 2, \ldots$, and $B(t)$ is the generating function of the sequence (b_n), $n = 0, 1, 2, \ldots$, then we have

(3.8) $$C(t) = A(t) + B(t)$$

provided

$$c_n = a_n + b_n, \qquad n = 0, 1, 2, \ldots.$$

Similarly, *multiplication* is defined by stating that

(3.9) $$C(t) = A(t)B(t)$$

provided

$$c_n = a_n b_0 + a_{n-1}b_1 + \cdots + a_{n-k}b_k + \cdots + a_0 b_n, \qquad n = 0, 1, 2, \ldots.$$

The coefficient c_n is often called the *Cauchy product*, which is the same as the *convolution*, of the sequences (a_n) and (b_n).

The zero of the ring is of course $0 + 0x + \cdots$, whereas the unity is $1 + 0x + 0x^2 + \cdots$. The inverse $A'(t)$ corresponding to the sequence (a_n'), $n = 0, 1, 2, \ldots$, is defined by

(3.10) $$A(t)A'(t) = A'(t)A(t) = 1,$$

so that if $a_0 \neq 0$, then

$$a_0 a_0' = 1,$$
$$a_n a_0' + a_{n-1}a_1' + \cdots + a_1 a_{n-1}' + a_0 a_n' = 0, \qquad n = 1, 2, \ldots.$$

Solution of these equations shows that a_n' may be written as the determinant (given in [1])

$$a_n' = (-1)^n a_0^{-n-1} \begin{vmatrix} a_1 & a_2 & a_3 & \cdots & a_{n-2} & a_{n-1} & a_n \\ a_0 & a_1 & a_2 & & a_{n-3} & a_{n-2} & a_{n-1} \\ 0 & a_0 & a_1 & & a_{n-4} & a_{n-3} & a_{n-2} \\ \cdot & \cdot & \cdot & & \cdot & \cdot & \cdot \\ 0 & 0 & 0 & & a_3 & a_1 & a_2 \\ 0 & 0 & 0 & & 0 & a_0 & a_1 \end{vmatrix}.$$

In this ring, the *derivative* $DA(t)$, $D \equiv d/dt$, is defined by

$$DA(t) = a_1 + 2a_2 t + \cdots + na_n t^{n-1} + \cdots,$$

and its jth iterate is

$$D^j A(t) = j!a_j + \cdots + (n)_j a_n t^{n-j} + \cdots,$$

with $(n)_j = n(n-1) \cdots (n - j + 1)$, the *falling factorial with j factors*. These iterates correspond to term-by-term differentiation. When the series has a positive radius of convergence, the formal operations correspond to the same operations on the sum of the series. Results, however, obtained by equating coefficients after formal manipulations of generating functions are true apart from convergence, a fact that seems to have been a folk-theorem before the appearance in 1923 of the fine paper by E. T. Bell [1] to which we have already referred.

Some examples may be useful in giving substance to these formal statements.

Example 3.8 (a) Suppose $B(t) = 1 + t + t^2 + \cdots$. Then the function $C(t) = A(t)B(t)$ generates the sequence (c_n), where

$$c_n = a_n + a_{n-1} + \cdots + a_0, \qquad n = 0, 1, 2, \ldots,$$

and

$$c_n - c_{n-1} = a_n, \qquad n = 1, 2, \ldots.$$

Thus c_n is the cumulation of coefficients a_k, $k = 0, 1, \ldots, n$, and the difference formula corresponds to $(1 - t)C(t) = A(t)$, consistent with $B(t) = (1 - t)^{-1}$, the natural expectation.

(b) If c_n is the complementary cumulation given by

$$c_n = a_{n+1} + a_{n+2} + \cdots,$$

and if

$$A(1) = a_0 + a_1 + a_2 + \cdots,$$

then we have

$$(1 - t)C(t) = A(1) - A(t),$$

a form useful in probability, usually with $A(1) = 1$.

(c) If $A(t)$ is finite, $A(t) = a_0 + a_1 t + \cdots + a_n t^n$, and $B(t) = (1 - t)^{-1}$, as in part a, then we obtain

$$C(t) = A(t)(1 - t)^{-1}$$
$$= c_0 + c_1 t + \cdots + c_n t^n + A(1)t^{n+1}(1 - t)^{-1}.$$

We write

$$C^*(t) = c_0 + c_1 t + \cdots + c_n t^n$$
$$= [A(t) - A(1)t^{n+1}](1 - t)^{-1}.$$

In particular, let $A(t) = (1 + t)^n$, the generating function for combinations. Then

$$C_n(t) = [(1 + t)^n - 2^n t^{n+1}](1 - t)^{-1}$$
$$= (1 + t)C_{n-1}(t) + 2^{n-1}t^n.$$

Iterating the last result leads to

$$C_n(t) = (1 + t)^n + \sum_{k=1}^{n} 2^{k-1} t^k (1 + t)^{n-k}.$$

From the definition we have

$$c_k = \binom{n}{k} + \binom{n}{k-1} + \cdots + 1, \qquad k = 0, 1, \ldots, n,$$

and from the iterated result, on equating coefficients of t^k, we obtain

$$c_k = \binom{n}{k} + \sum_{j=1}^{k} 2^{j-1} \binom{n-j}{k-j}.$$

Equating the two expressions for c_k leads to the identity

$$c_{k-1} = \sum_{j=0}^{k-1} \binom{n}{k} = \sum_{j=1}^{k} 2^{j-1} \binom{n-j}{k-j}.$$

Moreover, we have

$$c_k - c_{k-1} = \binom{n}{k} = \sum_{j=0}^{k} 2^j \left[\binom{n-1-j}{k-j} - \binom{n-1-j}{k-1-j} \right]$$

$$= \sum_{j=0}^{k} 2^j \frac{n-2k+j}{n-k} \binom{n-1-j}{k-j}.$$

Example 3.9 The Fibonacci numbers may be defined by $f_0 = f_1 = 1$, $f_n = f_{n-1} + f_{n-2}$, $n = 2, 3, \ldots$. Hence their generating function is

$$F(t) = 1 + t + 2t^2 + 3t^3 + 5t^4 + 8t^5 + 13t^6 + 21t^7 + 34t^8 + \cdots.$$

The function $F(t)$, however, also satisfies the equation

$$F(t) = 1 + t + (f_1 + f_0)t^2 + \cdots + (f_{n-1} + f_{n-2})t^n + \cdots$$

$$= 1 + (t + t^2)F(t)$$

$$= (1 - t - t^2)^{-1}.$$

Thus, if $1 - t - t^2 = (1 - at)(1 - bt)$, so that $a + b = 1$, $ab = -1$, and

(3.11)

$$a = \frac{1 + \sqrt{5}}{2},$$

$$b = \frac{1 - \sqrt{5}}{2},$$

then the partial-fraction expansion

$$[(1 - at)(1 - bt)]^{-1} = (a - b)^{-1}[a(1 - at)^{-1} - b(1 - bt)^{-1}]$$

shows that

$$F(t) = \sum_{k=0}^{\infty} \frac{a^{k+1} - b^{k+1}}{a - b} t^k.$$

Accordingly, we have

$$(a - b)f_k = a^{k+1} - b^{k+1},$$

an explicit and well-known formula, with a, b given by (3.11). Incidentally, the procedure we have followed has a natural extension to the solution of linear recurrence relations of any order.

The formula provides an easy proof of a familiar identity. First, we have

$$(a - b)^2 f_k^2 = a^{2k+2} + b^{2k+2} - 2(ab)^{k+1},$$

from which we obtain

$$(a - b)^2 f_{k-1} f_{k+1} = a^{2k+2} + b^{2k+2} - (ab)^k(a^2 + b^2),$$

so that, since $ab = -1$,

$$f_k^2 - f_{k-1} f_{k+1} = (-1)^k, \qquad k = 0, 1, 2, \ldots.$$

Example 3.10   Continuing with Fibonacci sequences, we note that the generating function for f_{n+m}, $n = 0, 1, 2, \ldots$, $m > 0$ is

$$F(t; m) = \sum_{n=0}^{\infty} f_{n+m} t^n = [F(t) - (1 + t + \cdots + f_{m-1} t^{m-1})]t^{-m}$$

$$= [1 - (1 - t - t^2)(1 + t + \cdots + f_{m-1} t^{m-1})]t^{-m} F(t)$$

$$= (f_m + t f_{m-1}) F(t).$$

Equating coefficients of t^n leads to the recurrence relation

$$f_{n+m} = f_m f_n + f_{m-1} f_{n-1},$$

a result that can be verified by iteration of the basic recurrence relation for these numbers.

Example 3.11   Still another kind of relation for Fibonacci numbers is provided by the study of the equation $(1 - t)^{-m} F(t) = F_m(t)$. First, we have

$$F_1(t) = (1 - t)^{-1} F(t)$$

$$= (2 + t) F(t) - (1 - t)^{-1},$$

by partial-fraction expansion. Now notice that $2 + t = f_2 + t f_1$, and therefore equating the two forms yields

$$[(1 - t)(f_2 + t f_1) - f_0] F(t) = f_1.$$

This suggests looking for a generalization of the latter, which turns out to be

$$[(1 - t)(f_{2m} + tf_{2m-1}) - f_{2m-2} - tf_{2m-3}]F(t) = f_{2m-1},$$

an identity verified by use of the basic recurrence relation. Now we obtain

$$F_2(t) = (1 - t)^{-1}F_1(t)$$
$$= (2 + t)(1 - t)^{-1}F(t) - (1 - t)^{-2}$$
$$= (f_4 + tf_3)F(t) - f_3(1 - t)^{-1} - (1 - t)^{-2},$$

since, by the identity with $m = 2$, we have

$$[(f_4 + tf_3) - (1 - t)^{-1}(f_2 + tf_1)]F(t) = f_3(1 - t)^{-1}.$$

The general result is now evident, namely

$$F_m(t) = (f_{2m} + tf_{2m-1})F(t) - f_{2m-1}(1 - t)^{-1} - f_{2m-3}(1 - t)^{-2} - \cdots$$
$$- f_{2m+1-2k}(1 - t)^{-k} - \cdots - f_1(1 - t)^{-m};$$

this is easily proved by mathematical induction and the identity. Since Example 3.4 has shown that

$$(1 - t)^{-m} = \sum_{n=0}^{\infty} \binom{m + n - 1}{n} t^n,$$

and Example 3.10 that $F(t; m) = (f_m + tf_{m-1})F(t)$, the corresponding Fibonaccian identity is

$$\sum_{k=0}^{m} \binom{m + k - 1}{k} f_{n-k} = f_{n+2m} - \sum_{k=1}^{m} \binom{k + n - 1}{n} f_{2m+1-2k}.$$

Its first two instances are

$$f_n + f_{n-1} + \cdots + f_0 = f_{n+2} - 1,$$
$$f_n + 2f_{n-1} + \cdots + (n + 1)f_0 = f_{n+4} - f_3 - (n + 1) = f_{n+4} - (n + 4).$$

Example 3.12 While we are on the Fibonaccian theme, it is interesting to solve Problem 3, Chapter 2, of [13] in an alternative way. The problem asks for the generating functions of (a_n) and (b_n), where

$$(3.12) \quad a_n = \sum_{k=0}^{n} \binom{n + k}{2k}, \qquad b_n = \sum_{k=0}^{n-1} \binom{n + k}{2k + 1}, \qquad n = 1, 2, \ldots,$$

and $a_0 = 1$, $b_0 = 0$. By the simple recurrence relation for binomial coefficients (Example 3.1), it is found that

$$a_{n+1} = a_n + b_{n+1},$$
$$b_{n+1} = a_n + b_n,$$

so the first few values in the number sequences are as shown in Table 3.1. The table is seen to be a zigzag way of writing the Fibonacci sequence.

TABLE 3.I
The Number
Sequences (3.12)

n	a_n	b_n
0	1	0
1	2	1
2	5	3
3	13	8
4	34	21
5	89	55

Indeed, by inspection we have $a_n = f_{2n}$, $b_n = f_{2n-1}$. If we write $A_{2n} = a_n$, $A_{2n-1} = b_n$, then the two recurrence relations are instances of the single (Fibonacci) recurrence relation $A_n = A_{n-1} + A_{n-2}$, and we have

$$A(t) = \sum_{n=0}^{\infty} A_n t^n = (1 - t - t^2)^{-1} = F(t).$$

The generating functions asked for,

$$a(t) = \sum_{n=0}^{\infty} a_n t^n, \qquad b(t) = \sum_{n=0}^{\infty} b_n t^n,$$

are given by the relations

$$a(t^2) = \tfrac{1}{2}[A(t) + A(-t)] = (1 - t^2)(1 - 3t^2 + t^4)^{-1},$$

$$b(t^2) = \frac{t}{2}[A(t) - A(-t)] = t^2(1 - 3t^2 + t^4)^{-1}.$$

These expressions are in agreement with those obtained in the book [13], which, however, does not identify their Fibonaccian character. In a notation like that of Example 3.2, we have

$$a(t) = F_e(\sqrt{t}) \quad (e \text{ for even}), \qquad b(t) = t^{\frac{1}{2}}F_o(\sqrt{t}) \quad (o \text{ for odd}).$$

A further result of the identification is that the definitions of a_n and b_n specified in the problem coalesce into the single relation

$$f_n = \sum_{k=0}^{m} \binom{n-k}{k}, \qquad m = \left[\frac{n}{2}\right],$$

a result a trifle disguised in Problem 1, Chapter 1, of [13].

The relations for $a(t^2)$, $b(t^2)$ used previously are instances of a process that De Morgan [6] has called *multisection of series*. Briefly, if

$$A(t) = a_0 + a_1 t + a_2 t^2 + \cdots$$

is an ordinary power-series generating function, and α is a primitive nth root of unity, $\alpha = \exp(2\pi i/n)$, where $i = \sqrt{-1}$, then the kth n-sectional series,

$$A_k(t; n) = a_k t^k + a_{k+n} t^{k+n} + \cdots, \qquad k = 0, 1, \ldots, n-1,$$

is given by

$$A_k(t; n) = n^{-1} \sum_{j=1}^{n} \alpha^{-kj} A(\alpha^j t).$$

3.5 ELEMENTARY RELATIONS: EXPONENTIAL GENERATING FUNCTION

The chief difference of exponential generating functions from what was given in Sec. 3.4 flows from the definition of a product of generating functions, namely if

$$A(t) = a_0 + a_1 t + \cdots + \frac{a_k t^k}{k!} + \cdots,$$

and $B(t)$, $C(t)$ are the similar series with coefficients b_k, c_k, $k = 0, 1, 2, \ldots$, then we have

$$C(t) = A(t)B(t),$$

provided

$$c_k = a_k b_0 + k a_{k-1} b_1 + \cdots + \binom{k}{j} a_{k-j} b_j + \cdots + a_0 b_k.$$

Because of the presence of binomial coefficients in the last expression, it is more easily remembered and used in the symbolic form

$$c_k = (a + b)^k, \qquad a^k \equiv a_k, \qquad b^k \equiv b_k.$$

Actually this is a standard form of the *Blissard* or *umbral* calculus, or algebra, the basic structure of which is developed in [3]. For working purposes, it is sufficient to proceed as follows: In any sequence (a_n), replace a_k by a^k, $k = 0, 1, \ldots$, treating the exponent as a power for all formal operations; at the completion of this, restore the indexes. The procedure, of course, is suggested by the fact that it makes exponential generating functions look like and behave like exponential functions; thus we have

$$A(t) = \exp at, \qquad a^k \equiv a_k,$$

$$DA(t) = a \exp at = \sum_{n=0}^{\infty} \frac{a_{n+1} t^n}{n!},$$

and so on. As with the exponential function, for convenience we shall represent $A(t)$ interchangeably by exp at or e^{at}. It should be noted that $a^0 \equiv a_0$ is not necessarily unity as in ordinary algebra, and that

$$(a + a)^n = \sum_{k=0}^{n} \binom{n}{k} a_{n-k} a_k.$$

The temptation of the following sequence of operations must be resisted:

$$(a + a)^n = \sum_{k=0}^{n} \binom{n}{k} a^{n-k} a^k = \sum_{k=0}^{n} \binom{n}{k} a^n = \sum_{k=0}^{n} \binom{n}{k} a_n = 2^n a_n.$$

The unit generating function is $1 + 0t + 0t^2/2! + \cdots$, while the zero has all coefficients 0. The inverse function,

$$A'(t) = \exp a't, \qquad (a')^n \equiv a_n,$$

is defined by

(3.13) $$1 = A(t)A'(t) = e^{(a+a')t},$$

or by

$$a_0 a_0' = 1, \qquad (a + a')^n = 0, \qquad a^n \equiv a_n, \qquad (a')^n \equiv a_n' \qquad n = 1, 2, \ldots.$$

Hence we have

$$a_0' = \frac{1}{a_0}, \qquad a_2' = -\frac{a_2}{a_0^2} + \frac{2a_1^2}{a_0^3},$$

$$a_1' = -\frac{a_1}{a_0^2}, \qquad a_3' = -\frac{a_3}{a_0^2} + \frac{6a_2 a_1}{a_0^3} - \frac{6a_1^3}{a_0^4}.$$

A concise expression for a_n', using the multivariable polynomials Y_n of E. T. Bell, is given in Problem 24, Chapter 2, of [13]. These polynomials will be described later.

Example 3.13 Suppose that

$$B(t) = 1 + \alpha t + \cdots + \frac{\alpha^n t^n}{n!} + \cdots,$$

$$A(t) = e^{at}, a^n \equiv a_n;$$

then by definition we have

$$C(t) = A(t)B(t) = e^{ct}, \qquad c^n \equiv c_n,$$

provided

$$c_n = (a + \alpha)^n = \sum_{k=0}^{n} \binom{n}{k} a_{n-k} \alpha^k.$$

Note, however, that we also have

$$A(t) = C(t)B(-t) = e^{(c-\alpha)t}, \qquad c^n \equiv c_n,$$

and

$$a_n = (c - \alpha)^n = \sum_{k=0}^{n} \binom{n}{k} c_{n-k}(-\alpha)^k;$$

the latter is an inverse relation easily proved by substitution. Thus we obtain

$$\begin{aligned}
c_n &= \sum_{k=0}^{n} \binom{n}{k} \alpha^{n-k} a_k = \sum_{k=0}^{n} \binom{n}{k} \alpha^{n-k} \sum_{j=0}^{k} \binom{k}{j} (-\alpha)^{j-k} c_j \\
&= \sum_{j=0}^{n} c_j \alpha^{n-j} \sum_{k=j}^{n} \binom{n}{k} \binom{k}{j} (-1)^{k-j} \\
&= \sum_{j=0}^{n} c_j \alpha^{n-j} \sum_{k=j}^{n} \binom{n}{j} \binom{n-k}{k-j} (-1)^{k-j} \\
&= \sum_{j=0}^{n} c_j \alpha^{n-j} \binom{n}{j} (1-1)^{n-j}.
\end{aligned}$$

Example 3.14 The inverse relation for

$$c_n = (a + b)^n, \qquad a^n \equiv a_n, \qquad b^n \equiv b_n,$$

is

$$a_n = (b' + c)^n, \qquad (b')^n = b'_n, \qquad c^n \equiv c_n,$$

since for $f^n \equiv f_n, f = a, b, c$, we have

$$e^{ct} = e^{(a+b)t} = e^{at} e^{bt},$$

$$e^{(b'+c)t} = e^{b't} e^{ct} = e^{(b'+b)t} e^{at} = e^{at}.$$

In the particular case $b_n = \alpha^n$, we have $b'_n = (-\alpha)^n$, and the result is the same as that of Example 3.13.

Example 3.15 Suppose that $b_n = n!$; then $B(t) = (1 - t)^{-1}$, $B'(t) = 1 - t$, and the dual relations are

$$c_n = \sum_{k=0}^{n} \binom{n}{k} (n - k)! a_k,$$

$$a_n = c_n - n c_{n-1}.$$

The joint implication of these relations is readily verified.

Example 3.16 Make the substitution $t = \log(1 + u)$. Then operating formally, we obtain

$$\begin{aligned}
e^{at} = e^{a[\log(1+u)]} &= (1 + u)^a, \qquad a^n \equiv a_n, \\
&= \sum_{k=0}^{\infty} \binom{a}{k} u^k = \sum_{k=0}^{\infty} \frac{(a)_k u^k}{k!}, \\
&= e^{(a)u},
\end{aligned}$$

where the falling factorial $(a)_k$ is defined by

$$(a)_k = a(a-1)\cdots(a-k+1),$$
$$= \sum_{j=0}^{k} s(k,j)a_j, \qquad a^n \equiv a_n,$$

with $s(k,j)$ the *Stirling number of the first kind* defined by the relation just given. The formal operations are verified by

$$e^{a[\log(1+u)]} = \sum_{k=0}^{\infty} a_k \frac{[\log(1+u)]^k}{k!}$$
$$= \sum_{k=0}^{\infty} a_k \sum_{n=0}^{\infty} s(n,k) \frac{u^n}{n!}$$
$$= \sum_{n=0}^{\infty} \frac{u^n}{n!} \sum s(n,k)a_k.$$

The exponential generating function for the Stirling numbers used on the second line is well known; a sketch of its derivation appears as Problem 13a, Chapter 2, of [13].

Example 3.17 (a) Bernoulli numbers b_n, $n = 0, 1, 2, \ldots$, have the exponential generating function

$$e^{bt} = \frac{t}{e^t - 1}, \qquad b^n \equiv b_n.$$

Hence we have

$$(e^t - 1)\, e^{bt} = t,$$

or, equating coefficients,

$$(b+1)^n - b_n = \delta_{n1}, \qquad b^n \equiv b_n,$$

with δ_{n1} the Kronecker delta. Instances of this equation are

$$b_0 - b_0 = 0,$$
$$b_1 + b_0 - b_1 = 1,$$
$$b_2 + 2b_1 + b_0 - b_2 = 0,$$
$$3b_2 + 3b_1 + b_0 = 0,$$

so that $b_0 = 1$, $b_1 = -\frac{1}{2}$, $b_2 = \frac{1}{6}$. Except for b_1, all Bernoulli numbers of odd index are zero, as shown by

$$\tfrac{1}{2}[e^{bt} - e^{b(-t)}] = \frac{1}{2}\left(\frac{t}{e^t - 1} + \frac{t}{e^{-t} - 1}\right) = \frac{1}{2}\frac{t(1 - e^t)}{e^t - 1} = -\tfrac{1}{2}t,$$

or by

$$\tfrac{1}{2}[e^{bt} + e^{b(-t)}] = \frac{t}{2}\frac{1 + e^t}{e^t - 1} = \tfrac{1}{2}t + e^{bt}.$$

(b) Make the substitution $t = \log(1 + u)$ as in Example 3.16. Then we have

$$e^{(b)u} = \frac{\log(1 + u)}{u} = \sum_{n=0}^{\infty} \frac{(-u)^n}{n + 1}$$

and

$$(b)_n = \sum s(n, k)b_k = \frac{(-1)^n n!}{n + 1}.$$

Instances of this are

$$b_0 = 1,$$
$$b_1 = -\tfrac{1}{2},$$
$$b_2 - b_1 = \tfrac{2}{3},$$
$$b_3 - 3b_2 + 2b_1 = -\tfrac{6}{4}.$$

(c) Now use the exponential generating function for Stirling numbers of the second kind appearing in Example 3.7:

$$e^{bt} = \frac{t}{e^t - 1} = \frac{\log(1 + e^t - 1)}{e^t - 1}$$

$$= \sum_{n=0}^{\infty} \frac{(-1)^n}{n + 1} (e^t - 1)^n$$

$$= \sum_{n=0}^{\infty} \frac{(-1)^n n!}{(n + 1)} \sum_{k=0}^{\infty} S(k, n) \frac{t^k}{k!}.$$

Equating coefficients, we obtain

$$b_n = \sum_{k=0}^{n} \frac{(-1)^k k! S(n, k)}{k + 1}.$$

This relation is a dual to the one of part b of this example in the sense that each of the relations

$$a_n = \sum_{k=0}^{n} s(n, k)b_k \qquad \text{and} \qquad b_n = \sum_{k=0}^{n} S(n, k)a_k$$

implies the other. The duality is a consequence of the orthogonality relations

$$\sum_{k=0}^{n} s(n, k)S(k, m) = \sum_{k=0}^{n} S(n, k)s(k, m) = \delta_{nm},$$

with δ the Kronecker delta. In their turn these relations follow directly from the definitions of the Stirling numbers, which it may be useful to display together; they are

$$(x)_n = \sum_{k=0}^{n} s(n, k)x^k \qquad \text{and} \qquad x^n = \sum_{k=0}^{n} S(n, k)(x)_k.$$

3.6 GENERATING FUNCTIONS IN PROBABILITY AND STATISTICS

As is proper to their origin, generating functions are endemic in probability and statistics. The first form is the probability generating function

$$(3.14) \qquad P(t) = p_0 + p_1 t + p_2 t^2 + \cdots .$$

The coefficients p_n, $n = 0, 1, 2, \ldots$, as probabilities, are nonnegative and at most 1. Usually we take $P(1) = 1$.

The next form is the generating function of *ordinary moments*, a sequence of numbers m_n, $n = 0, 1, 2, \ldots$, with

$$(3.15) \qquad m_n = \sum_{k=0}^{\infty} k^n p_k.$$

Then we have

$$(3.16) \qquad m(t) = e^{mt} = \sum_{n=0}^{\infty} \frac{m_n t^n}{n!} = P(e^t).$$

The *central* moments M_n, $n = 0, 1, 2, \ldots$, are defined by

$$(3.17) \qquad M_n = (m - m_1)^n, \qquad m^k \equiv m_k.$$

Hence we have

$$(3.18) \qquad M(t) = e^{Mt} = e^{(m-m_1)t} = e^{-m_1 t} P(e^t).$$

The *factorial moments*, $(m)_n$, $n = 0, 1, 2, \ldots$, are defined by

$$(3.19) \qquad (m)_n = \sum_{k=0}^{\infty} s(n, k) m_k = \sum_{k=0}^{\infty} (k)_n p_k.$$

It follows from Example 3.16 that

$$(3.20) \qquad \exp (m)t = \exp [m \log (1 + t)] = P(1 + t).$$

Use of (3.14) gives the same result.

The closely related (and frequently more convenient) *binomial* moments are defined by

$$(3.21) \qquad B_n = \frac{(m)_n}{n!} = \sum_{k=0}^{\infty} \binom{k}{n} p_k, \qquad m^k \equiv m_k;$$

for them, we have

$$(3.22) \qquad B(t) = \sum_{n=0}^{\infty} B_n t^n = e^{(m)t} = P(1 + t).$$

Note that $B(t)$ is an ordinary generating function.

A further kind of moment, useful only in special circumstances, deserves notice for formal reasons; this is the *figurate moment*, defined by

$$(3.23) \qquad C_n = \sum_{k=0}^{\infty} \binom{k+n-1}{n} p_k.$$

The word *figurate* is intended to suggest the *figurate numbers*, with generating function $(1 - t)^{-n}$, just as binomial suggests binomial coefficients. We have

$$(3.24) \qquad C(t) = \sum_{n=0}^{\infty} C_n t^n = P[(1 - t)^{-1}].$$

The interrelations of the four generating functions, probability, ordinary moment, binomial moment, and figurate moment, are shown in Table 3.2.

TABLE 3.2
Interrelations of Generating Functions

Function	$P(\cdot)$	$m(\cdot)$	$B(\cdot)$	$C(\cdot)$
$P(t)$	t	$\log t$	$t - 1$	$(t - 1)t^{-1}$
$m(t)$	e^t	t	$e^t - 1$	$1 - e^{-t}$
$B(t)$	$1 + t$	$\log (1 + t)$	t	$t(t + 1)^{-1}$
$C(t)$	$(1 - t)^{-1}$	$\log (1 - t)^{-1}$	$t(1 - t)^{-1}$	t

The table hides interesting relations. In the first place, the relation in the third column of the first row,

$$P(t) = B(t - 1),$$

corresponds to a formula inverse to (3.21), namely

$$(3.25) \qquad p_n = \sum_{k=0}^{\infty} (-1)^k \binom{n+k}{k} B_{n+k}.$$

Next, the relation $m(t) = B(e^t - 1)$ implies the generating function for Stirling numbers of the second kind, just as the inverse relation

$$m[\log (1 + t)] = B(t)$$

implies that for Stirling numbers of the first kind as already developed in Example 3.16.

Finally, the relation

$$C(t) = B[t(1 - t)^{-1}]$$

is a disguised form of Euler's transformation of series. This follows from the fact that

$$B[t(1-t)^{-1}] = \sum_{k=0}^{\infty} B_k t^k (1-t)^{-k}$$

$$= \sum_{n=0}^{\infty} t^n \sum_{k=0}^{n-1} \binom{n-1}{k} B_{k+1}.$$

Hence we obtain

(3.26)

$$C_n = \sum_{k=0}^{n-1} \binom{n-1}{k} B_{k+1} = B(B+1)^{n-1} = (B+1)^n - (B+1)^{n-1}, \quad B^k \equiv B_k,$$

and

(3.27)　$C_0 + C_1 + C_2 + \cdots + C_n = (B+1)^n \equiv (E+1)^n B_0 = \beta_n,$

with E the *shift operator*: $EB_n = B_{n+1}$. The corresponding generating-function relation, with $\beta(t)$ written for the generating function of β_n, is

(3.28)　　　$\beta(t) = (1-t)^{-1}C(t) = (1-t)^{-1}B[t(1-t)^{-1}].$

Since by Example 3.13, $(B+1)^n = \beta_n$ implies

$$B_n = (\beta-1)^n = \Delta^n \beta_0, \qquad \beta^k \equiv \beta_k,$$

we have

$$\beta(t) = \beta_0 + \beta_1 t + \cdots + \beta_n t^n + \cdots$$

$$= \beta_0 (1-t)^{-1} + \Delta^n \beta_0 t^n (1-t)^{-n-1} + \cdots,$$

which is Euler's transformation of series. Note that the inverse to (3.28) is

$$(1+t)^{-1}\beta[t(1+t)^{-1}] = B(t).$$

Note also that iteration of (3.28) leads to

$$\beta_2(t) = (1-t)^{-1}\beta[t(1-t)^{-1}] = (1-2t)^{-1}B[t(1-2t)^{-1}],$$

(3.29)　$\beta_m(t) = (1-t)^{-1}\beta_{m-1}[t(1-t)^{-1}] = (1-mt)^{-1}B[t(1-mt)^{-1}]$

$$= \sum \beta_n(m)t^n.$$

Clearly we have

$$\beta_n(m) = (B+m)^n, \qquad B^k \equiv B_k.$$

This is the generalization given by Touchard [15].

Another important generating function is the *cumulant* or *semi-invariant* generating function. If λ_n is the nth cumulant, with $\lambda_0 = 0$, and $\lambda(t)$ is its exponential generating function, then the *cumulants* are defined by

(3.30)　　　　　$m(t) = e^{mt} = e^{\lambda(t)}, \qquad m^n \equiv m_n.$

Differentiation leads to

$$m'(t) = m\,e^{mt} = \lambda'(t)\,e^{mt},$$

so that

(3.31) $m_{n+1} = \lambda(m + \lambda)^n, \qquad m^n \equiv m_n, \qquad \lambda^n \equiv \lambda_n.$

Instances of (3.31) are

$$m_1 = \lambda_1, \qquad m_3 = \lambda_1 m_2 + 2\lambda_2 m_1 + \lambda_3,$$
$$m_2 = \lambda_1 m_1 + \lambda_2, \qquad m_4 = \lambda_1 m_3 + 3\lambda_2 m_2 + 3\lambda_3 m_1 + \lambda_4.$$

Hence we have

$$m_1 = \lambda_1, \qquad m_3 = \lambda_3 + 3\lambda_2\lambda_1 + \lambda_1^3,$$
$$m_2 = \lambda_2 + \lambda_1^2, \qquad m_4 = \lambda_4 + 4\lambda_3\lambda_1 + 3\lambda_2^2 + 6\lambda_2\lambda_1^2 + \lambda_1^4,$$

and

$$\lambda_1 = m_1, \qquad \lambda_3 = m_3 - 3m_2 m_1 + 2m_1^3,$$
$$\lambda_2 = m_2 - m_1^2, \qquad \lambda_4 = m_4 - 4m_3 m_1 - 3m_2^2 + 12m_2 m_1^2 - 6m_1^4.$$

If the *Bell polynomial* $Y_n(fy_1, fy_2, \ldots, fy_n)$ is defined by

(3.32) $$Y_n = \sum_{\substack{\text{partitions} \\ \text{of } n}} \frac{n!}{k_1! \cdots k_n!} \left(\frac{y_1}{1}\right)^{k_1} \cdots \left(\frac{y_n}{n!}\right)^{k_n} f_k,$$

where $k = k_1 + k_2 + \cdots + k_n$, $n = k_1 + 2k_2 + \cdots + nk_n$, and the summation extends over all solutions of the latter in nonnegative integers, that is, over all partitions of n, then

(3.33) $m_n = Y_n(\lambda_1, \lambda_2, \ldots, \lambda_n),$

(3.34) $\lambda_n = Y_n(fm_1, fm_2, \ldots, fm_n), \qquad f^k \equiv f_k = (-1)^{k-1}(k-1)!$

Extensions to many variables are made in a simple way, as exemplified by

$$B(t, u) = P(1 + t, 1 + u).$$

Example 3.18   (a) The simplest probability generating function is $q + pt, q + p = 1$, associated with the toss of a single coin. For n independent tosses, the generating function is that of the Bernoulli distribution, namely

$$P(t) = (q + pt)^n, \qquad q + p = 1.$$

The simplest moment-generating function is the binomial

$$B(t) = P(1 + t)$$
$$= (1 + pt)^n = \sum_{k=0}^{n} \binom{n}{k} p^k t^k.$$

The binomial moments themselves are

$$B_k = \binom{n}{k} p^k, \qquad k = 0, 1, \ldots, n.$$

The ordinary moments follow at once from

$$m_k = \sum_{j=0}^{n} S(k, j)(m)_j$$

$$= \sum_{j=0}^{n} S(k, j) j! B_j$$

$$= \sum_{j=0}^{n} S(k, j)(n)_j p^j,$$

with $(n)_j$ the falling factorial.

(b) Alternatively, ordinary moments may be found by recurrence. It is desirable to indicate the independent variables n and p explicitly by writing $m_k(n, p)$ for m_k and $e^{tm(n, p)}$ for the generating function. Then we have

$$e^{tm(n, p)} = \sum_{k=0}^{\infty} \frac{m_k(n, p) t^k}{k!} = (q + p\, e^t)^n.$$

With $D_p = d/dp$, $D_t = d/dt$, it may be verified that

$$pq D_p\, e^{tm(n, p)} = D_t\, e^{tm(n, p)} - np\, e^{tm(n, p)},$$

which corresponds to

$$m_{k+1}(n, p) = npm_k(n, p) + pq D_p m_k(n, p).$$

The first few instances are

$$m_1 = np,$$
$$m_2 = (np)^2 + npq,$$
$$m_3 = (np)^3 + 3n^2 p^2 q + npq(q - p).$$

Example 3.19 Consider the *geometric* or *Pascal* probability generating function, defined by

$$p_n = pq^n, \qquad p + q = 1, \qquad n = 0, 1, \ldots.$$

For this, we have

$$P(t) = p(1 + qt + q^2 t^2 + \cdots)$$
$$= p(1 - qt)^{-1},$$

and

$$B(t) = p(p - qt)^{-1} = \sum_{n=0}^{\infty} \frac{q^n t^n}{p^n},$$

$$m(t) = p(1 - qe^t)^{-1} = p + q(1 - q)(e^{-t} - q)^{-1}$$
$$= p + qH(q, -t),$$

where $H(x, t) = (1 - x)(e^t - x)^{-1}$ is a generating function used by Euler for polynomials with coefficients closely related to Eulerian (as distinguished from Euler) numbers. More exactly, if

$$H(x, t) = (1 - x)(e^t - x)^{-1} = \sum_{n=0}^{\infty} \frac{A_n(x)(x - 1)^{-n} t^n}{n!},$$

then the *Eulerian numbers* A_{nk} are given by

$$A_n(x) = \sum_{k=1}^{n} A_{nk} x^{k-1}.$$

A concise summary of properties of these numbers and polynomials appears in [5].

By definition, we have

$$m_n = \sum_{k=0}^{n} k^n p_k = p \sum_{k=0}^{n} k^n q^k,$$

and therefore, by the formula for $m(t)$, we get

$$m_n = q(-1)^n (q - 1)^{-n} A_n(q)$$
$$= q(1 - q)^{-n} A_n(q);$$

hence we obtain

$$qA_n(q) = (1 - q)^{n+1} \sum_{k=0}^{n} k^n q^k,$$

an identity found, working from the other end, in Problem 2, Chapter 2, of [13].

It is interesting to notice that

$$H(2, t) = (2 - e^t)^{-1} = \sum_{k=0}^{\infty} (e^t - 1)^k$$
$$= \sum_{k=0}^{\infty} \sum_{j=0}^{\infty} \frac{k! S(j, k) t^j}{j!}.$$

Thus we have

$$H_n(2) = A_n(2) = \sum_{k=0}^{n} k! S(n, k) = \sum_{k=1}^{n} A_{nk} 2^{k-1}.$$

Example 3.20   The mean of the Bernoulli distribution with parameters

n and p is np. If n increases while $np = \lambda$, a constant, the limiting distribution is Poisson, as is apparent from

$$\lim P(t; n, p) = \lim_{n \to \infty} \left[1 + \frac{\lambda(t - 1)}{n} \right]^n$$
$$= \exp \lambda(t - 1)$$
$$= \sum_{n=0}^{\infty} \frac{e^{-\lambda}(\lambda t)^n}{n!}.$$

The *Poisson binomial moment-generating function* is

$$B(t) = \exp \lambda t = \sum_{n=0}^{\infty} \frac{\lambda^n t^n}{n!},$$

from which we obtain

$$B_n = \frac{\lambda^n}{n!}$$

$$(m)_n = \lambda^n$$

$$m_n = \sum_{k=0}^{n} S(n, k)\lambda^k.$$

Then we have $m_n \equiv m_n(\lambda) = a_n(\lambda)$, where $a_n(\lambda)$ is the enumerator of permutations by number of ordered cycles.

3.7 PÓLYA'S ENUMERATION THEOREM

The theorem in question is that already described in Sec. 3.1 as the fundamental theorem in enumerative combinatorial analysis. It was discovered in 1937 by George Pólya in his study [12] of the enumeration of trees, linear graphs, and chemical structures. It is concerned with the enumeration of collections of objects, subject to prescribed order equivalence, according to properties inherent in the objects. A presentation, with a number of exemplifications, is given in the author's book [13], and also in a later expository paper [14]. Frank Harary has used the theorem on numerous occasions, notably in [9]; see also Chapter 6. It has been generalized in the elegant paper of de Bruijn [4], which will be reviewed in Chapter 5. Numerous variations of a lemma with almost the same content as the theorem appear in Golomb's paper [8]. Part of de Bruijn's generalization is developed from a specific context in Gilbert and Riordan [7].

Because an understanding of the theorem adequate for its intelligent use seems to require prolonged exposure, it is probable that the brief presentation given in this chapter will have low carrying power. It will not be wasted if it begins the exposure for some readers, or reinforces a dormant fascination in others. In any case, it will be thoroughly developed in Chapter 5.

The theorem is perhaps clearest in the intuitive setting given by Pólya himself. The objects are *figures*, and their totality s is called a *figure-store* (*Figurenvorrat*). The figures may be distinct or may be classifiable according to one or more properties. If these properties are measured by integers, so that the *weight* $(w_1, w_2, \ldots, w_p)$ represents an arbitrary class, when p properties are in question, having $N(w_1, w_2, \ldots, w_p)$ members, then the multivariable ordinary generating function (or one of its more general relatives),

$$(3.35) \qquad s(x_1, x_2, \ldots, x_p) = \sum_s N(w_1, w_2, \ldots, w_p) x_1^{w_1} x_2^{w_2} \cdots x_p^{w_p},$$

may be taken as the *store enumerator*.

A fixed set of n labeled points in space serves for the configuration, which is made by placing a figure from the store at each point. The figures are chosen independently, hence with replacement. The configurations are grouped into equivalence classes in the following way: For a *given* permutation group H, two configurations C_1 and C_2 are *equivalent*, or *equivalent* (H), if there is a permutation of the n labeled points P such that $PC_1 = C_2$. Two equivalent configurations must contain the same figures.

For the purposes of the theorem, the group H is specified by a *cycle index*, which is a multivariable generating function,

$$(3.36) \qquad H(t_1, t_2, \ldots, t_n) = h^{-1} \sum_H h_{k_1, k_2, \ldots,} t_1^{k_1} t_2^{k_2} \cdots,$$

with h the total number of permutations in the group, and $h_{k_1, k_2, \ldots}$ the number with k_1 cycles of length 1, k_2 of length 2, and so on. If H contains just one permutation, the identity, then we have $H = t_1^n$. At the other extreme, if H contains all $n!$ permutations of n elements, so that H is the symmetric group, then

$$(3.37) \quad H = S_n(t_1, t_2, \ldots, t_n) = \sum_{\substack{\text{partitions} \\ \text{of } n}} \frac{1}{k_1! \cdots k_n!} t_1^{k_1} \left(\frac{t_2}{2}\right)^{k_2} \cdots \left(\frac{t_n}{n}\right)^{k_n}$$

$$= \frac{Y_n(y_1, \ldots, y_n)}{n!}, \qquad y_k = (k-1)! t_k,$$

the last by comparison with (3.32). If $S_0 = 1$, then this cycle index has the generating function

$$(3.38) \qquad \sum_{n=0}^{\infty} S_n(t_1, t_2, \ldots, t_n) u^n = \exp\left(t_1 u + \frac{t_2 u^2}{2} + \frac{t_3 u^3}{3} + \cdots\right).$$

The weight of a configuration is the sum of the weights of its figures, in the sense that the sum of $(w_1, w_2, \ldots)$ and $(w_1', w_2', \ldots)$ is $(w_1 + w_1', w_2 + w_2', \ldots)$. This entails the result that if the cycle index is t_1'', so that there are no order equivalences, then the configuration enumerator $T_n(x_1, x_2, \ldots, x_p)$ is given by

$$(3.39) \qquad T_n(x_1, x_2, \ldots, x_p) = s^n(x_1, x_2, \ldots, x_p).$$

The store enumerator is multiplicative, as is usual for independent choices.

The theorem may now be stated. If, purely for convenience, the variables of the enumerator are limited to two, it is as follows.

Theorem (Pólya). If n figures are chosen independently from a store of figures having multiplicative enumerator $s(x, y)$, and if the order equivalence of choices is specified by a group of permutations with cycle index $H_n(t_1, \ldots, t_n)$, then the enumerator of inequivalent configurations is given by

$$T_n(x, y) = H_n(s_1, s_2, \ldots, s_n),$$

with

$$s_k = s(x^k, y^k).$$

For completeness, it is useful to have a statement of the corresponding lemma.

Lemma. If G is a finite group, of order g, of transformations operating on a finite set of objects, and if two objects are equivalent when one is transformed into the other by a transformation of G, then the number of inequivalent objects is

$$T = g^{-1} \sum_{t \in G} I(t),$$

where $I(t)$ is the number of objects left invariant by transformation t of G, and the sum is over all g members of G.

A single but rather nice use of the theorem is the following.

Example 3.21 Suppose that the store enumerator is a function of a single variable, namely

$$s(x) = 1 + x + x^2 + \cdots = (1 - x)^{-1}.$$

As noted in Sec. 3.2, this is the enumerator for any object in combinations with unlimited repetition. It is also the enumerator for parts of a partition, with zero parts permitted, where a *partition* of a given number is defined to be a collection of numbers $1, 2, \ldots$, with sum equal to the number, without regard to order. (When order is considered, MacMahon's term is *composition*.)

If $H = t_1^n$, so that there is no order equivalence, then by Pólya's theorem we have

$$T_n(x) = s_n(x) = (1 - x)^{-n},$$

as in Example 3.4. Note that $T_n(x)$ is the enumerator of compositions with n parts, zero parts permitted, or—what is the same thing—with *at most* n parts and zero parts excluded.

If $H = S_n(t_1, t_2, \ldots, t_n)$, so that all orders are alike, then

$$T_n(x) = S_n(s_1, s_2, \ldots, s_n), \qquad s_k = (1 - x^k)^{-1},$$

and, by the above remark, $T_n(x)$ is the enumerator of partitions with at most n parts. Now by an early theorem in the theory of partitions, the number of partitions with at most n parts equals the number with largest part n; its enumerator is $1/[(1 - x)(1 - x^2), \ldots, (1 - x^n)]$. Thus there is an identity, namely

$$S_n[(1 - x)^{-1}, \ldots, (1 - x^n)^{-1}] = (1 - x)^{-1} \cdots (1 - x^n)^{-1},$$

This is the formula actually developed, quite differently, by MacMahon for finding simple general formulas for the number of partitions with largest part n. The simplest nontrivial instance of the identity, namely the one with $n = 2$, is given by

$$\frac{2}{(1 - x)(1 - x^2)} = \frac{1}{(1 - x)^2} + \frac{1}{1 - x^2},$$

since $2S_2(t_1, t_2) = t_1^2 + t_2$. If we write $D(k; 1, 2, \ldots, n)$ for the *denumerant* of k into parts $1, 2, \ldots, n$, that is, the number of partitions of k with largest part n, then the identity shows that

$$2D(k; 1, 2) = k + 2 \qquad \text{for } k = 2j,$$

$$= k + 1 \qquad \text{for } k = 2j + 1.$$

The next instance may be written as

$$\frac{6}{(1 - x)(1 - x^2)(1 - x^3)} = \frac{1}{(1 - x)^3} + \frac{3}{(1 - x)(1 - x^2)} + \frac{2}{1 - x^3}$$

$$= \frac{1}{(1 - x)^3} + \frac{3}{2(1 - x)^2} + \frac{3}{2(1 - x^2)} + \frac{2}{1 - x^3}.$$

Equating coefficients of x^k shows that $D(k; 1, 2, 3)$ is the nearest integer to $(k + 3)^2/12$, a result due to De Morgan.

MULTIPLE-CHOICE REVIEW PROBLEMS

1. The combinatorial interest in generating functions lies in

   (a) their aid to mathematical rigor,
   (b) their use as a tool in the algebra of sequences,
   (c) their reproductive suggestiveness,
   (d) none of the above.

2. The ordinary generating function is related to

   (a) Cauchy products,
   (b) umbral calculus,
   (c) differential calculus,
   (d) none of the above.

3. The variable t of a generating function $A(t)$ is to be regarded as

   (a) a real variable,
   (b) a complex variable,
   (c) an abstract mark,
   (d) none of the above.

4. The generating function for Fibonacci numbers is

   (a) $(1 - t)^{-1}$,
   (b) $(1 - t - t^2)^{-1}$,
   (c) $(1 - t - t^3)^{-1}$,
   (d) none of the above.

5. The binomial-moment generating function $B(t)$ is related to the probability distribution function $P(t)$ by

   (a) $B(t) = P(1 + t)$,
   (b) $B(t) = P(e^t)$,
   (c) $B(t) = P[\log (1 + t)]$,
   (d) none of the above.

REFERENCES

1. Bell, E. T., "Euler Algebra," *Trans. Amer. Math. Soc.* **25** (1923), 135–154.
2. Bell, E. T., *Algebraic Arithmetic*, Amer. Math. Soc. Colloquium Publications **7**, American Mathematical Society, Providence, Rhode Island, 1927.
3. Bell, E. T., "Postulational Bases for the Umbral Calculus," *Amer. J. Math.* **62** (1940), 717–724.
4. de Bruijn, N. G., "Generalization of Pólya's Fundamental Theorem in Enumerative Combinatorial Analysis," *Nederl. Akad. Wetensch. Proc. Ser. A* **62** = *Indag. Math.* **21** (1959), 59–69.
5. Carlitz, L., "Eulerian Numbers and Polynomials," *Math. Mag.* **32** (1959), 247–260.
6. De Morgan, A., *Differential and Integral Calculus*, Kegan Paul, Trench, Trübner and Company, Ltd., London, 1842.

7. Gilbert, E. N., and J. Riordan, "Symmetry Types of Periodic Sequences," *Illinois J. Math.* **5** (1961), 657–665.

8. Golomb, S. W., "A Mathematical Theory of Discrete Classification," in C. Cherry (editor), *Information Theory*, Butterworths, London, 1961.

9. Harary, F. "The Number of Linear, Directed, Rooted and Connected Graphs," *Trans. Amer. Math. Soc.* **78** (1955), 445–463.

10. Laplace, P. S., *Théorie analytique des probabilitiés*, Mme. ve. Courcier, Paris, 1812.

11. MacMahon, P. A., *Combinatory Analysis*, Cambridge University Press, Vol. I, London, 1915; Vol. II, London, 1916.

12. Pólya, G., "Kombinatorische Anzahlbestimmungen für Gruppen, Graphen, und Chemische Verbindungen," *Acta Math.* **68** (1937), 145–253.

13. Riordan, J., *An Introduction To Combinatorial Analysis*, John Wiley and Sons, New York, 1958.

14. Riordan, J., "The Combinatorial Significance of a Theorem of Pólya," *J. Soc. Indust. Appl. Math.* **5** (1957), 225–237.

15. Touchard, J., "Sur certaines equations fonctionelles," *Proc. Int. Math. Congress*, Toronto, Canada, 1924, 465–472.

CHAPTER 4

Lattice Statistics

ELLIOTT W. MONTROLL

4.1 INTRODUCTION

This chapter is concerned with the counting of the number of ways certain events can occur on lattices. Approximation methods will be avoided; we shall be interested only in the exact solutions of the problems presented in the limit of large lattices. Much of the analysis will be valid only for one- and two-dimensional lattices. The difference between most two- and three-dimensional lattice problems seems to be enormous. Generating functions (see Chapter 3) will be exploited throughout this chapter.

A lattice is defined in the usual way. Consider a set of three non-coplanar unit vectors **i**, **j**, and **k**. The endpoints of the vectors

$$\mathbf{l} \equiv (l_1, l_2, l_3) = l_1\mathbf{i} + l_2\mathbf{j} + l_3\mathbf{k},$$

with all l's ranging through the integers $0, \pm 1, \pm 2, \pm 3, \ldots$, generate a space lattice. Lattices of arbitrary dimensionality can be defined in a similar manner.

A crystalline solid (see Chapter 16) is essentially a collection of atoms or molecules that have equilibrium positions at lattice points on some space lattice. Many of the problems in lattice statistics have their origin in solid-state physics. One can distinguish the devoted solid-state physicist from the devoted problem solver, however, by seeing how the quest for "exact solutions" is carried out. The true solid-state physicist is willing to make all kinds of approximations, even reckless ones if necessary, so that he can get to a qualitative answer that will help him understand certain experiments. The problem solver puts his emphasis on the boundary condition—the solution must be exact even though he loses sight of the physics of the problem and gets lost in a jungle of mathematics. Each has his strength and each has his weaknesses.

4.2 RANDOM WALKS ON LATTICES AND THE PÓLYA PROBLEM

Let us consider the problem of the random walker on a one-dimensional lattice. He starts at the origin (Fig. 4.1) and he takes a step either to the right or to the left with probability $\frac{1}{2}$ each. We ask for the probability

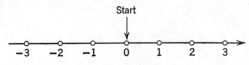

Fig. 4.1 Random walk.

that after t steps he is located at a lattice point l units from the origin. In the expression

$$\tfrac{1}{2}(e^{i\phi} + e^{-i\phi}),$$

the coefficient of $e^{i\phi}$, $\frac{1}{2}$, represents the probability of a step to the right, the probability that $l = 1$ after the first step, and the coefficient of $e^{-i\phi}$, $\frac{1}{2}$, that of a step to the left. The coefficient of $e^{2i\phi}$ in

$$[\tfrac{1}{2}(e^{i\phi} + e^{-i\phi})]^2,$$

$\frac{1}{4}$, represents the probability of the walker being at $l = 2$ after two steps, the coefficient $\frac{1}{2}$ of $e^{0i\phi}$ that he has returned to the origin, and the coefficient $\frac{1}{4}$ of $e^{-2i\phi}$ that he ends at $l = -2$ after two steps. Indeed, we have

(4.1)
$$P_t(l) = \text{prob. that walker is at } l \text{ after } t \text{ steps}$$
$$= \text{coefficient of } e^{i\phi l} \text{ in } [\tfrac{1}{2}(e^{i\phi} + e^{-i\phi})]^t.$$

An integral representation of this coefficient is immediately apparent from the identity

$$\frac{1}{2\pi} \int_{-\pi}^{\pi} e^{-i\phi m} \, d\phi = \delta_{m,0} = \begin{cases} 1 & \text{if } m = 0, \\ 0 & \text{if } m \text{ is a nonvanishing integer.} \end{cases}$$

If we operate with the integral operator

$$\frac{1}{2\pi} \int_{-\pi}^{\pi} e^{-i\phi l}(\cdot) \, d\phi$$

on the expression in (4.1), it filters out the coefficient of $e^{i\phi l}$ in (4.1), so that

$$P_t(l) = \frac{1}{2\pi} \int_{-\pi}^{\pi} [\tfrac{1}{2}(\underset{R}{e^{i\phi}} + \underset{L}{e^{-i\phi}})]^t e^{-i\phi l}\, d\phi$$

$$= \frac{1}{2\pi} \int_{-\pi}^{\pi} (\cos\phi)^t e^{-i\phi l}\, d\phi.$$

This argument can be applied immediately to the investigation of walks on a two-dimensional square lattice in which the walker can go to any one of his four nearest-neighbor points with probability $\tfrac{1}{4}$ at any step. Then the probability of going from the origin to lattice point $1 = (l_1, l_2)$ is

$$P_t(\mathbf{l}) = P_t(l_1, l_2)$$

$$= \frac{1}{(2\pi)^2} \int_{-\pi}^{\pi} \int_{-\pi}^{\pi} [\tfrac{1}{4}(\underset{R}{e^{i\phi_1}} + \underset{L}{e^{-i\phi_1}} + \underset{U}{e^{i\phi_2}} + \underset{D}{e^{-i\phi_2}})]^t e^{-i(\phi_1 l_1 + \phi_2 l_2)}\, d\phi_1\, d\phi_2.$$

Indeed, on an S-dimensional simple cubic lattice, with $2S$ possible steps from each lattice point to a nearest-neighbor lattice point, we have

$$P_t(\mathbf{l}) = P_t(l_1, l_2, \ldots, l_S)$$

$$= \frac{1}{(2\pi)^S} \int_{-\pi}^{\pi} \cdots \int_{-\pi}^{\pi} \left[\frac{1}{2S} \sum_{m=1}^{S} (e^{i\phi_m} + e^{-i\phi_m}) \right]^t e^{-i\,\boldsymbol{\phi}\cdot\mathbf{l}}\, d\phi_1 \cdots d\phi_S$$

$$= \frac{1}{(2\pi)^S} \int_{-\infty}^{\infty} \cdots \int_{-\infty}^{\infty} \left[\frac{1}{S}(c_1 + c_2 + \cdots + c_S) \right]^t e^{-i\,\boldsymbol{\phi}\cdot\mathbf{l}}\, d^S\boldsymbol{\phi},$$

where

$$c_j = \cos\phi_j = \tfrac{1}{2}(e^{i\phi_j} + e^{-i\phi_j}) \quad \text{and} \quad d^S\boldsymbol{\phi} \equiv d\phi_1\, d\phi_2 \cdots d\phi_S.$$

The generating function for all walks ending at $\mathbf{l}$, which involve successive jumps to nearest-neighbor points on an S-dimensional simple cubic lattice, is

(4.2)

$$U_S(z, \mathbf{l}) = \sum_{t=0}^{\infty} P_t(\mathbf{l}) z^t = \frac{1}{(2\pi)^S} \int_{-\pi}^{\pi} \cdots \int_{-\pi}^{\pi} \frac{e^{-i\,\boldsymbol{\phi}\cdot\mathbf{l}}\, d^S\boldsymbol{\phi}}{1 - (z/S)(c_1 + c_2 + \cdots + c_S)}.$$

Tables [17] of these functions exist for $S = 3$. In particular, the generating function for those walks that originate and terminate at the origin is

(4.3) $\quad U_S(z) = \dfrac{1}{(2\pi)^S} \displaystyle\int_{-\pi}^{\pi} \cdots \int_{-\pi}^{\pi} \dfrac{d^S\boldsymbol{\phi}}{1 - (z/S)(\cos\phi_1 + \cdots + \cos\phi_S)}.$

If we analyze a more exotic walk than to nearest-neighbor points, we easily gets dizzy in generating our steps, and a more systematic approach is advantageous. This is done in Sec. 4.3, where it is shown [19] that if

$$P(\mathbf{l}) = \text{prob. of steps from a point } \mathbf{l'} \text{ to } \mathbf{l''} = \mathbf{l'} + \mathbf{l},$$

then

(4.4) $$U_S(z, \mathbf{l}) = \frac{1}{(2\pi)^S} \int_{-\pi}^{\pi} \cdots \int_{-\pi}^{\pi} \frac{e^{-i\boldsymbol{\phi}\cdot\mathbf{l}} \, d^S\boldsymbol{\phi}}{1 - z\lambda(\phi_1, \phi_2, \ldots, \phi_S)},$$

with

$$\lambda(\boldsymbol{\phi}) \equiv \lambda(\phi_1, \phi_2, \ldots, \phi_S) = \sum_{\mathbf{l}} p(\mathbf{l}) e^{i\boldsymbol{\phi}\cdot\mathbf{l}}.$$

For a walk on a three-dimensional body-centered cubic lattice, we have

(4.5) $$\lambda(\boldsymbol{\phi}) = c_1 c_2 c_3,$$

whereas on a face-centered cubic lattice we have

(4.6) $$\lambda(\boldsymbol{\phi}) = \tfrac{1}{3}(c_1 c_2 + c_1 c_3 + c_2 c_3);$$

see Sec. 4.3.

An interesting question concerning random walks on lattices was asked many years ago by Pólya [26], and indeed to some extent answered by him. Does a random walker whose steps are from one lattice point to a nearest-neighbor lattice point eventually return to his starting point? If not, what is his escape probability for a given lattice?

To answer this question, we use some ideas of Feller [5] concerning the theory of recurrent events. Let ε be an event that might occur over and over again. Let

$f_j =$ probability ε occurs for the first time on the jth trial,

$u_j =$ probability ε occurs on the jth trial whether or not it has previously occurred.

Also define $u_0 \equiv 1$, and construct the generating functions of the sets $\{u_j\}$ and $\{f_j\}$,

(4.7) $$U(z) = \sum_{j=0}^{\infty} u_j z^j \quad \text{and} \quad F(z) = \sum_{j=1}^{\infty} f_j z^j.$$

Then we have

$$u_j = u_0 f_j + u_1 f_{j-1} + u_2 f_{j-2} + \cdots + u_{j-1} f_1.$$

Each term on the right-hand side represents an independent way in which ε occurs on the jth trial. Since $u_0 = 1$, $u_0 f_j$ is the probability that ε occurs for the first time on the jth trial, $u_1 f_{j-1}$ is the probability that it occurs on the $(j-1)$th trial for the first time and that it reoccurs again after one more trial, etc. Now multiply both sides by z^j and sum these equations from $j = 1$ to $j = \infty$. Then we get

$$\sum_{j=1}^{\infty} u_j z^j = \sum_{j=1}^{\infty} [u_0(f_j z^j) + (u_1 z)(f_{j-1} z^{j-1}) + \cdots + (u_{j-1} z^{j-1}) f_1 z],$$

or

$$U(z) - 1 = F(z)U(z),$$

so that

$$U(z) = \frac{1}{1 - F(z)},$$

and

$$F(z) = 1 - [U(z)]^{-1}.$$

The interpretation of $F(1)$ is quite clear, for the probability that ε ever occurs is

$$f_1 + f_2 + f_3 + \cdots = F(1),$$

since f_1, f_2, etc., represent the probabilities of the independent events that ε occurs for the *first time* on the first trial, f_2 for the *first time* on the second trial, etc. Hence we have

$$\text{prob. } \varepsilon \text{ ever occurs} = F(1) = 1 - [U(1)]^{-1}.$$

Two situations arise:

1. If $U(1) = \infty$, then $F(1) = 1$ and ε is certain to occur.

2. If $U(1) < \infty$, then $F(1) < 1$ so that a positive probability $[U(1)]^{-1}$ of nonoccurrence of ε exists.

In our random-walk problem, the walker is certain to return to the origin and indeed to return infinitely often, if the integral $U(1)$ of (4.3) [or (4.4)] diverges, since (4.3) [or (4.4)] is just the required generating function defined in (4.7). On the other hand, if (4.3) [or (4.4)] converges, then there is a positive probability for the escape of the walker without his ever returning to the origin.

Before proceeding to the exact calculation of probabilities of return to the origin on several lattices, let us examine the qualitative behavior of (4.3) when $z = 1$, namely the behavior of

$$U_S(1) = \frac{1}{(2\pi)^S} \int_{-\pi}^{\pi} \cdots \int_{-\pi}^{\pi} \frac{d^S \phi}{1 - (1/S)(\cos \phi_1 + \cdots + \cos \phi_S)}.$$

The convergence of this integral depends on the behavior of the denominator of the integrand, which, since

$$\cos \phi_j = 1 - \tfrac{1}{2}\phi_j^2 + \cdots,$$

behaves like

$$\tfrac{1}{2}(\phi_1^2 + \cdots + \phi_S^2) \equiv \tfrac{1}{2}r^2$$

as $(\phi_1, \phi_2, \ldots, \phi_S) \to 0$. We define r to be the radius variable in S-dimensional polar coordinates. The integral $U_S(1)$ can be expressed as the sum over two components, the first an integral over an S sphere of small radius a, with center at the origin, and the second the integral over the S-dimensional hypercube of volume $(2\pi)^S$, with the central sphere excluded. In a region not including the origin, the integrand is well behaved; no divergence can come from the second component of the integral. We calculate the contribution of the integral over a small spherical shell about the origin of the ϕ space in the neighborhood of $\phi = 0$, but we omit the contribution of the sphere of radius ϵ and then let $\epsilon \to 0$. As long as the exterior radius of the shell is small, the integrand depends only on r; hence we can use polar coordinates in the integration, and $d^S\phi$ is proportional to $r^{S-1}\,dr$. The required integral is then proportional [7, 19, 20] to

$$I(\epsilon) = \int_\epsilon^a \frac{r^{S-1}}{r^2}\,dr = \int_\epsilon^a r^{S-3}\,dr = \frac{1}{\epsilon} - \frac{1}{a} \qquad \text{for } S = 1,$$

$$= -\log \frac{\epsilon}{a} \qquad \text{for } S = 2,$$

$$= \frac{1}{S-2}(a^{S-2} - \epsilon^{S-2}) \qquad \text{for } S \geq 3.$$

As $\epsilon \to 0$, so that the neighborhood of the origin is included in the integration, we see that $I(\epsilon) \to \infty$ for $S = 1$ or 2, whereas $I(\epsilon) < \infty$ for $S \geq 3$. Hence $U_S(1)$ *diverges for S = 1, 2 and converges for S ≥ 3*. This is just Pólya's result: *A random walker who walks in the manner prescribed above is certain to return to his point of origin if he walks on a one- or two-dimensional lattice. A nonvanishing escape probability exists in n-dimensional lattices for n ≥ 3*.

We now go beyond Pólya to the exact calculation of $U_S(1)$ for the three three-dimensional cubic lattices: simple cubic, body-centered cubic, and face-centered cubic. By good fortune, G. N. Watson [30] has done all the work for us. He evaluated the three integrals

$$(4.8) \quad I_1 = \frac{1}{(2\pi)^3} \int_{-\pi}^{\pi} \int_{-\pi}^{\pi} \int_{-\pi}^{\pi} \frac{d\phi_1 \, d\phi_2 \, d\phi_3}{1 - \frac{1}{3}(\cos \phi_1 + \cos \phi_2 + \cos \phi_3)}$$

$$= \frac{4}{3\pi^2} [18 + 12\sqrt{2} - 10\sqrt{3} - 7\sqrt{6}]K^2[(2 - \sqrt{3})(\sqrt{3} - \sqrt{2})]$$

$$= 1.5163860591,$$

$$(4.9) \quad I_2 = \frac{1}{(2\pi)^3} \int_{-\pi}^{\pi} \int_{-\pi}^{\pi} \int_{-\pi}^{\pi} \frac{d\phi_1 \, d\phi_2 \, d\phi_3}{1 - \cos \phi_1 \cos \phi_2 \cos \phi_3}$$

$$= \frac{[\Gamma(\frac{1}{4})]^4}{4\pi^3} = 1.3932039297,$$

$$(4.10) \quad I_3 = \frac{1}{(2\pi)^3} \int_{-\pi}^{\pi} \int_{-\pi}^{\pi} \int_{-\pi}^{\pi} \frac{d\phi_1 \, d\phi_2 \, d\phi_3}{1 - \frac{1}{3}(c_1 c_2 + c_2 c_3 + c_3 c_1)}$$

$$= \frac{9[\Gamma(\frac{1}{3})]^6}{2^{14/3}\pi^4} = 1.3446610732,$$

where $K(k)$ is the complete elliptic integral of the second kind. At first glance, it is strange that anyone of his own free will would have investigated such integrals. Actually van Peype [25] encountered the above integrals in devising a theory of ferromagnetic anistropy based on spin-wave theory. Unable to evaluate them in closed form, he resorted to graphical integration. H. A. Kramers, van Peype's thesis supervisor, then put the problem of the evaluation of the three integrals (4.8), (4.9), and (4.10) to R. H. Fowler, who communicated it to G. H. Hardy; whereupon, to quote Watson [30], "the problem then became common knowledge in Cambridge and subsequently in Oxford, whence it made the journey to Birmingham without difficulty."

If we examine (4.2) for $S = 3$ and $z = 1$, and if we combine (4.5) and (4.6) with (4.4) for $z = 1$, we see that I_1, I_2, and I_3 are respectively the generating functions $U_3(1)$ for random walks on simple body-centered and face-centered cubic lattices. Then since, as we have shown, the probability of eventual return of a walker to his origin is given by $F(1) = 1 - [U(1)]^{-1}$, for the various cubic lattices [19] we have

prob. of return to origin $\doteq 0.340537330$ for simple cubic lattice,

$= 0.282229985$ for body-centered cubic lattice,

$= 0.256318237$ for face-centered cubic lattice.

The number of nearest neighbors to a lattice point in a simple cubic lattice is 6; in a body-centered cubic lattice, it is 8; and in a face-centered cubic lattice it is 12. It is not surprising that the return probability diminishes as the number of nearest neighbors increases, since more ways for escape exist. The escape probability also increases with dimensionality for a fixed number of nearest neighbors to each lattice point. For example, the probability of return in a four-dimensional simple cubic lattice, for which there are eight nearest neighbors to each lattice point, is 0.20 as compared with the 0.28 figure given on page 102. Earlier estimates of return probabilities on simple cubic lattices have been 0.35 by McCrea and Whipple [16] and 0.34054 by Domb [2].

4.3 MORE GENERAL RANDOM WALKS ON LATTICES

Consider a simple cubic lattice of infinite extent in S dimensions. Let l and l' be typical lattice vectors that characterize lattice points. The components of l and l' are integers:

$$l = (l_1, l_2, \ldots, l_s), \qquad \text{where } l_j = 0, \pm 1, \pm 2, \ldots,$$

with an analogous expression for l'. Let

$$p(l - l') = \text{prob. of step from } l' \text{ to } l$$

at any stage of the walk if the walker is at l'. Furthermore, let

$$P_t(l) = \text{prob. that the walker is at } l \text{ after } t \text{ steps.}$$

Then

(4.11) $$\sum_l p(l - l') = 1 \qquad \text{for all } l',$$

where l extends over the entire lattice, and similarly

$$\sum_l P_t(l) = 1.$$

The probability of a walker being at l after $t + 1$ steps is

(4.12) $$P_{t+1}(l) = \sum_{l'} p(l - l') P_t(l').$$

Let $\boldsymbol{\phi}$ be a vector of S components,

$$\boldsymbol{\phi} = (\phi_1, \phi_2, \ldots, \phi_S),$$

where the ϕ_j are real numbers, $j = 1, 2, \ldots, S$. If

$$\Pi_t(\boldsymbol{\phi}) = \sum P_t(l) e^{i\boldsymbol{\phi} \cdot l}$$

and

$$\lambda(\boldsymbol{\phi}) = \sum_l p(l) e^{i\boldsymbol{\phi} \cdot l},$$

then from (4.12) we obtain

$$\Pi_{t+1}(\boldsymbol{\phi}) = \Pi_t(\boldsymbol{\phi})\lambda(\boldsymbol{\phi}).$$

If, however, the walk starts at the origin when $t = 0$, then we have $\Pi_0(\boldsymbol{\phi}) = 1$, so that

$$\Pi_t(\boldsymbol{\phi}) = [\lambda(\boldsymbol{\phi})]^t.$$

This expression can be inverted to find $P_t(\mathbf{l})$ by using the well-known formula: If

$$g(\boldsymbol{\phi}) = \sum_{\text{all } l_j=-\infty}^{\infty} f(\mathbf{l})e^{i\mathbf{l}\cdot\boldsymbol{\phi}},$$

then

$$f(\mathbf{l}) = \frac{1}{(2\pi)^S} \int_0^{2\pi} \cdots \int_0^{2\pi} g(\boldsymbol{\phi})e^{-i\boldsymbol{\phi}\cdot\mathbf{l}} d^S\boldsymbol{\phi},$$

so that

$$P_t(\mathbf{l}) = \frac{1}{(2\pi)^S} \int_0^{2\pi} \cdots \int_0^{2\pi} [\lambda(\boldsymbol{\phi})]^t e^{-i\boldsymbol{\phi}\cdot\mathbf{l}} d^S\boldsymbol{\phi}.$$

The random-walk generating function is then

$$U_S(t,\mathbf{l}) = \sum_{t=0}^{\infty} z^t P_t(\mathbf{l})$$

$$= \frac{1}{(2\pi)^S} \int_0^{2\pi} \cdots \int_0^{2\pi} \frac{e^{-i\boldsymbol{\phi}\cdot\mathbf{l}} d^S\boldsymbol{\phi}}{1 - z\lambda(\phi_1, \phi_2, \ldots, \phi_s)},$$

the formula referred to in (4.4).

The random walk between nearest neighbors on a three-dimensional body-centered cubic lattice is equivalent to one on a simple cubic lattice of lattice spacing 1 on which a walker can take only steps that correspond to the eight displacement vectors

$$(\pm 1, \pm 1, \pm 1).$$

If each step can occur with the same probability 1/8, then we have

$$\lambda(\boldsymbol{\phi}) = \sum_l p(\mathbf{l})e^{i\boldsymbol{\phi}\cdot\mathbf{l}} = \cos\phi_1 \cos\phi_2 \cos\phi_3.$$

The random walk between nearest neighbors on a face-centered cubic lattice is equivalent to one on a single cubic lattice of lattice spacing 1 on which the walker can take only steps that correspond to the twelve displacements $(\pm 1, \pm 1, 0)$, $(\pm 1, 0, \pm 1)$, $(0, \pm 1, \pm 1)$, each with probability 1/12. Then we have

$$\lambda(\boldsymbol{\phi}) = \sum_l p(\mathbf{l})e^{i\boldsymbol{\phi}\cdot\mathbf{l}} = \tfrac{1}{3}(\cos\phi_1 \cos\phi_2 + \cos\phi_1 \cos\phi_3 + \cos\phi_2 \cos\phi_3).$$

4.4 THE PFAFFIAN AND THE DIMER PROBLEM

At the turn of the century a variety of algebraic objects similar to determinants were studied; but finding no place in applied mathematics nor having any connection with the foundations of mathematics, they sank into oblivion. One of these, the Pfaffian, has recently been revived by Caianiello [1] because of its natural appearance in quantum field theory, in which one is sometimes concerned with certain anticommuting operators.

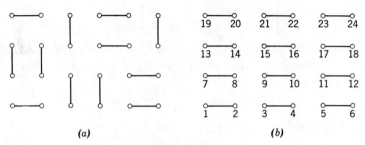

(a)   (b)

Fig. 4.2   Dimer covering configurations.

An interesting property of a Pfaffian is that it is just the right kind of quantity for enumerating the number of ways dimers can be placed on a two-dimensional lattice. This connection was discovered independently by H. N. V. Temperley and M. E. Fisher [29] and P. Kasteleyn [11]. The dimer problem had first been mentioned by Fowler and Rushbrooke [8] in 1937. It is remarkable that its solution on a two-dimensional lattice should have been found twice simultaneously twenty-five years later, and by essentially the same methods.

The dimer problem originated in the investigation of the adsorption of diatomic molecules on surfaces. It is the following: To find the number of ways diatomic molecules, which we henceforth call *dimers*, can cover a doubly periodic lattice, of lattice spacing equal to a dimer length, so that each dimer covers two adjacent lattice points and no lattice point remains uncovered. Two covering configurations of a 4 × 6 lattice are shown in Fig. 4.2. Since each dimer occupies two lattice points, covering configurations can exist only on lattices with an even number of lattice points. Thus a quadratic lattice with square unit cells must have an even number of lattice points in at least one direction to be coverable by dimers. We choose this to be the horizontal direction.

In order to connect the dimer problem with a Pfaffian, it is appropriate, of course, to define a Pfaffian. Consider a set of antisymmetric quantities

$\{a(p, p')\}$, with values given for all pairs p and p' that are selected from the integers $1, 2, 3, \ldots, 2N$. The antisymmetry of the a's implies that

(4.13)
$$a(p, p') = -a(p', p) \qquad \text{if } p \neq p',$$
$$= 0 \qquad \text{if } p = p'.$$

Indeed, the set $\{a(p, p')\}$ might be thought of as being composed of the elements of the antisymmetric matrix

$$A = \begin{bmatrix} a(1, 1) & a(1, 2) & \cdots & a(1, 2N) \\ a(2, 1) & a(2, 2) & \cdots & a(2, 2N) \\ \cdots\cdots\cdots\cdots\cdots\cdots\cdots\cdots\cdots \\ a(2N, 1) & a(2N, 2) & \cdots & a(2N, 2N) \end{bmatrix},$$

with the $a(p, p')$ satisfying (4.13). In one notation, the *Pfaffian* is exhibited as half a determinant—a thing that started out to be a determinant but did not quite succeed:

$$P\{a(p, p')\} = \begin{vmatrix} a(1, 2) & a(1, 3) & a(1, 4) & \cdots & a(1, 2N) \\ & a(2, 3) & a(2, 4) & \cdots & a(2, 2N) \\ & & \cdots\cdots\cdots \\ & & \cdots\cdots \\ & & & & a(2N - 1, 2N) \end{vmatrix}$$

It is defined, however, as

(4.14) $$P\{a(p, p')\} = \sum_P{}' \delta_P a(p_1 p_2) a(p_3 p_4) \cdots a(p_{2N-1}, p_{2N}),$$

the summation extending over all permutations $P = (p_1, p_2, \ldots, p_{2N})$ of the integers $(1, 2, 3, \ldots, 2N)$ such that

(4.15) $$p_1 < p_2, p_3 < p_4, p_5 < p_6, \ldots, p_{2N-1} < p_{2N},$$

(4.16) $$p_1 < p_3 < p_5 < \cdots < p_{2N-1}.$$

The factor δ_p is chosen to be $+1$ or -1 depending on whether P is an even or an odd permutation of $(1, 2, \ldots, 2N)$, an odd permutation being a sequence of integers that can be obtained by an odd number of interchanges of pairs in the set $(1, 2, \ldots, 2N)$. The sequence 1324 is an odd permutation of 1234, being the result of interchanging 2 and 3. The expansion of the Pfaffian [1] for $N = 2$ is

(4.17)
$$\begin{vmatrix} a(1, 2) & a(1, 3) & a(1, 4) \\ & a(2, 3) & a(2, 4) \\ & & a(3, 4) \end{vmatrix} = a(1, 2)a(3, 4) - a(1, 3)a(2, 4) + a(1, 4)a(2, 3).$$

We might examine a variety of planar lattices on which dimers could be placed. We have already exhibited a square lattice. Some other possibilities include the hexagonal lattice, Fig. 4.3, the triangular lattice, Fig. 4.4, and the bathroom-tile lattice, Fig. 4.5.

We identify a dimer configuration by a sequence of pairs of points, in which each member of the pair represents an end of a dimer,

$$C = \{p_1, p_2; p_3, p_4; p_4, p_5; \ldots\},$$

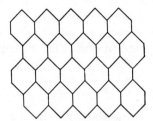

Fig. 4.3 Hexagonal lattice.

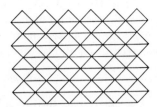

Fig. 4.4 Triangular lattice.

the p's in each pair being nearest neighbors on the lattice. Uniqueness of representation is ensured by ordering the p's in C so that

(4.18) $$p_1 < p_2, p_3 < p_4, p_5 < p_6, \ldots,$$

and

(4.19) $$p_1 < p_3 < p_5 < \cdots.$$

The configurations of Fig. 4.2a and b are respectively

$$\{1, 2; 3, 9; 4, 10; 5, 6; 7, 13; 8, 14; 11, 12; 15, 21; 16, 17;$$
$$18, 24; 19, 20; 22, 23\},$$

$$\{1, 2; 3, 4; 5, 6; 7, 8; 9, 10; 11, 12; 13, 14; 15, 16; 17, 18;$$
$$19, 20; 21, 22; 23, 24\}.$$

The latter configuration is an example of the "standard configuration," C_0, introduced by Kasteleyn. It is that configuration in which all dimers are horizontal on a square lattice.

Generally, we count from left to right, and bottom to top, in a diagram to obtain the scheme characterized by (4.18) and (4.19).

The inequalities (4.18) and (4.19) are

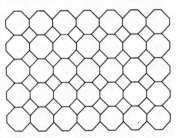

Fig. 4.5 Bathroom-tile lattice.

exactly those in (4.15) and (4.16). Therefore a one-to-one correspondence
exists between a dimer configuration and the terms in the expansion of
a Pfaffian. The elements $a(p, p')$ must be chosen to be 0 for those values
of p and p' that cannot be connected by a dimer. In view of this correspondence, we might hope to express the generating function for dimers
on a rectangular lattice,

$$(4.20) \qquad \Phi_{m,n}(z_1, z_2) = \sum_{N_1, N_2} g(N_1, N_2) z_1^{N_1} z_2^{N_2},$$

as a Pfaffian. Here $g(N_1, N_2)$ represents the number of dimer configurations with N_1 horizontal and N_2 vertical dimers in a rectangular lattice
with m horizontal lattice points and n vertical ones. The summation
extends over all N_1 and N_2 such that

$$\tfrac{1}{2}nm = N_1 + N_2.$$

There is a one-to-one correspondence between the terms in (4.20) and
those in a Pfaffian of order nm, determined by

$|a(p, p')| = z_1$ if p and p', with $p < p'$, are horizontal nearest neighbors,

$|a(p, p')| = z_2$ if p and p', with $p < p'$, are vertical nearest neighbors,

$a(p, p') = 0$ otherwise.

Actually, we would like to show that

$$(4.21) \qquad\qquad P\{A\} = \Phi_{mn}(z_1, z_2).$$

This, however, requires all terms in the expansion (4.14) to be positive.
The fundamental issue is then the following. Can phase factors of modulus
unity of $a(p, p')$ be found, such that all terms in the expansion of $P\{A\}$
can be made positive as is required?

Kasteleyn [12] has shown that signs can be given to $a(p, p')$ in such a
way that all terms in $P\{A\}$ are positive, provided that the lattice is
planar with no articulation points and no bridges; that is, no bonds can
cross over or bridge other bonds. We shall now review his ideas, which
he kindly conveyed to the author in several informative discussions.

Let us start with the identity permutation P_0 and its corresponding
dimer configuration, Fig. 4.2b, and assign the value z_1 to certain matrix
elements, as follows

$$a(1, 2) = a(3, 4) = a(5, 6) = \cdots = a(23, 24) = z_1;$$

that is, the matrix elements associated with all horizontal bonds that
connect an odd numbered point on the left with an even numbered point
on the right are given the value z_1. On a general square lattice with m

lattice points in the horizontal direction, this term in the Pfaffian has the value $z_1^{nm/2}$.

Any term in the Pfaffian can be expressed in a variety of equivalent ways. Consider the term associated with the permutation

$$(4.22) \qquad P: \delta_P a(p_1 p_2) a(p_3 p_4) \cdots a(p_{2N-1}, p_{2N}),$$

where the p's are arranged in the canonical Pfaffian order (4.15) and (4.16). This term remains invariant under a violation of the inequalities (4.16). For example, suppose we replace the pair (p_1, p_2) by (p_3, p_4) to obtain a new permutation P'. Then $\delta_P = \delta_{P'}$ since it involves an even number of interchanges. Also, the only change in the a's is associated with replacing $a(p_1 p_2) a(p_3 p_4) \cdots$ by $a(p_3 p_4) a(p_1 p_2) \cdots$; hence the value of the new representation of the term (4.22) is the same as that of (4.22). Furthermore, (4.22) is invariant to violations of the inequalities (4.15) and (4.16). Suppose the first is violated; then

$$a(p_1', p_2) \to a(p_2, p_1) = -a(p_1 p_2),$$

so that the sign of the product of the a's is changed. However, this permutation, which we denote by P'', involves one additional interchange of p's. Hence we have $\delta_P = -\delta_{P''}$. This -1 factor just compensates for the -1 that results from the antisymmetry of the a's, and the new representation of (4.21) has the same numerical value as (4.22). All the representations of P that result from (4.15) and (4.16) are associated with the same dimer configuration, because the same letters are connected in the a's. It is convenient to distinguish various representations by replacing dimer graphs such as those shown in Fig. 4.2 by graphs with directed bonds such that $a(p, p')$ is identified by a directed bond with arrow pointing from p to p'.

We have prescribed signs for one-fourth of the bonds that connect neighboring lattice points. The designation of the remaining three-fourths is made by comparing the sign of a term associated with an arbitrary permutation (4.22) and that of P_0, or indeed the signs of any two arbitrary permutations, say P_1 and P_2. Note that if the dimer configuration corresponding to P_1 is superimposed on that corresponding to P_2, the resulting graph is composed of unchanged dimers and *closed polygons*. Figure 4.6 is the superposition of Fig. 4.2b on Fig. 4.2a, the dimers of Fig. 4.2b being the dotted lines. Those dimers that are the same in P_1 and P_2 do not contribute any difference between the terms in the Pfaffian corresponding to P_1 and P_2. We have seen that all representations of P_1 and P_2 that correspond to various arrangements of *directed* bonds yield the same numerical terms as those that appear in the Pfaffian from P_1 and P_2. Let us then choose representations P_1' and P_2' such that all closed polygons are pointed in a clockwise direction and the indices representing lattice points are ordered in a cyclic fashion in all polygons. We call the polygons of the superposition graphs *superposition polygons*. Let P_3 be the permutation that converts P_1' to P_2'; that is, we have $P_2' = P_3 P_1'$. It is well known in the theory of permutations that $\delta_{P_2} = \delta_{P_3} \delta_{P_1'}$.

Now in each closed polygon the dotted-line dimers of P_2' can be converted into the solid-line dimers of P_1' by cyclic permutations, corresponding to each polygon being rotated clockwise through one bond length. The

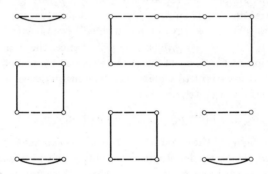

Fig. 4.6 Superposition of polygons of Fig. 4.2.

cyclic permutations associated with each polygon are independent of each other, no common indices appearing in different cycles. Hence δ_{P_3} is the product $\delta_1 \delta_2 \delta_3 \cdots \delta_j$, being $+1$ if the jth cyclic permutation is even and -1 if it is odd. To discover the sign of a given cyclic permutation, consider

$$\begin{bmatrix} p_1 & p_2 & p_3 & \cdots & p_{n-1} & p_n \\ p_2 & p_3 & p_4 & \cdots & p_n & p_1 \end{bmatrix}.$$

The final state $(p_2 p_3 \cdots p_1)$ is derived from $(p_1 p_2 \cdots p_n)$ by interchanging p_2 and p_1, then p_3 and p_1, then p_4 and p_1, etc., until p_1 appear on the right. Since each interchange yields a factor (-1), we have $\delta = (-1)^{n-1}$. Each superposition polygon in Fig. 4.6 spans an even number of lattice points, since every polygon must be an alternating one, the same number of sides coming from the P_1' as from the P_2' dimer configurations. Hence each δ_j is -1, and we have

$$\delta_{P_3} = (-1)^s,$$

s being the total number of polygons generated by the superposition.

Our requirement that all terms in the Pfaffian be positive can be achieved if we can make one term positive, which we have already done with the standard configuration on a square lattice, and find bond-sign prescriptions that give all terms the same sign. The sign of P_1 is the same as that of P_2 if in a given superposition polygon the sign of the factor associated with the terms in P_1, or P_1', is the same as that of the factor associated with the terms of P_2, or P_2'. If the factor associated with the dotted lines, that is, P_1', on the jth superposition polygon is

$$\delta_{P'_1}^{(j)} a(p_1 p_2) a(p_3 p_4) \cdots a(p_{2n_j-1} p_{2n_j}),$$

then that associated with the solid lines, that is, P_2', on the jth superposition polygon is

$$\delta_{P'_2}^{(j)} a(p_2 p_3) a(p_4 p_5) \cdots a(p_{2n_j}, p_1),$$

since each superposition polygon is an alternation of dotted and solid lines. The condition that the signs of these two terms are the same is

$$\delta_{P'_1}^{(j)} \operatorname{sign} a(p_1 p_2) \operatorname{sign} a(p_3 p_4) \cdots \operatorname{sign} a(p_{2n_j-1}, p_{2n_j})$$
$$= \delta_{P'_2}^{(j)} \operatorname{sign} a(p_2 p_3) \operatorname{sign} a(p_4 p_5) \cdots \operatorname{sign} a(p_{2n_j}, p_1).$$

Since all signs are $+1$ or -1, and since

$$\delta_{P_3}^{(j)} = \delta_{P'_2}^{(j)} \delta_{P'_1}^{(j)} = -1,$$

this condition is equivalent, with $p_{2n_j+1} \equiv p_1$, to

(4.23)
$$\prod_{k=1}^{2n_j} \operatorname{sign} a(p_k, p_{k+1}) = -1.$$

Thus the problem of the prescription of signs is solved if the product of the signs over any closed polygon that can be a superposition polygon is -1. This is the case if an odd number of -1's appears on any superposition polygon.

The appropriate sign prescription for a square lattice is given in Fig. 4.7. Then we have

$$a(p, p') = \begin{cases} z_1 \text{ for horizontal bonds with } p' > p, \\ (-1)^p z_2 \text{ for vertical bonds with } p' > p, \\ 0 \text{ if } p \text{ and } p' \text{ are not nearest neighbors}, \\ -a(p', p). \end{cases}$$

The product of signs over any of the three superposition polygons of Fig. 4.6 is -1, as required. The proof that this is a correct prescription, and indeed the proof that it is always possible to find an appropriate prescription on any planar lattice with no articulation points or bridges follows [12].

The sign of a given matrix element can be identified by an arrow such that its pointing from p to p' implies that $a(p, p') > 0$. For example, the arrow assignment of Fig. 4.7 is given in Fig. 4.8. Note that in the remainder of this chapter all arrow arguments are to be interpreted in this way; the discussion of arrows used in the derivation of (4.23) has been completed and is to be forgotten. Any polygon with an even number of clockwise sides is called *clockwise even*; one with an odd number of

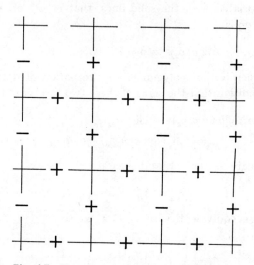

Fig. 4.7 Sign prescription for a square lattice.

clockwise sides is called *clockwise odd*. Since all superposition polygons contain an even number of sides, they are counterclockwise odd if they are clockwise odd. Now (4.23) is equivalent to the statement that every superposition polygon must be clockwise odd; that is, in a complete circulation around a superposition cycle an odd number of arrows must point from p_j to p_{j+1}.

It is possible to arrange arrows so that all lattice polygons that do not surround interior lattice points and that do not contain double points are clockwise odd. On the first such polygon selected, introduce an odd number of clockwise arrows, the remainder being counterclockwise. An adjacent polygon has a certain number of bonds in common with its neighbor, and at least one new unlabeled bond exists. Of the bonds that are not common, label all except one in an arbitrary manner. The last one can then be pointed in the required direction so that the polygon is clockwise odd. This process can be continued in such a way that at each stage we have a simply connected domain, until all polygons are labeled with no bonds remaining unlabeled. Arrow assignments will be given to the square, hexagonal, and triangular lattices at the end of this section.

With this assignment of arrows, it can now be proved [12] that the number of clockwise arrows on a closed polygon and the number of enclosed lattice points have opposite parity. This is done by induction. We have just proved it on the assumption that there are no enclosed lattice points, since 0 is even and the number of clockwise arrows is odd. Now consider a polygon Γ_n that encircles n polygons of the type discussed above, and suppose that our theorem is true for this polygon. We can construct a new polygon Γ_{n+1} that borders Γ_n and a single adjacent polygon Γ_1 with no enclosed lattice points; we must prove that the validity of our theorem for Γ_n implies its validity for Γ_{n+1}. Suppose Γ_n surrounds γ lattice points, contains α clockwise arrows, and has β bonds in common with Γ_1, which has α' (an odd number, as we saw) clockwise bonds. The number of lattice points interior to Γ_{n+1} is $\gamma + \beta - 1$. The number of clockwise bonds in Γ_{n+1} is the number in Γ_n, plus the number in Γ_1, minus the number lost to Γ_1 as a result of their becoming inside Γ_{n+1}, minus the number lost to Γ_n by their becoming inside Γ_{n+1}. Since clockwise bonds of Γ_n that are common to Γ_1 are counterclockwise on Γ_1, each common bond is clockwise to either Γ_n or Γ_1. We then find that the number of clockwise bonds in Γ_{n+1} is $\alpha + \alpha' - \beta$. The assumption that our theorem is valid for Γ_n implies that α and γ have opposite parity. Since α' was proved above to be odd, as is the -1 in $\gamma + \beta - 1$, it follows that $(\alpha + \alpha' - \beta)$ and $(\gamma + \beta - 1)$ have opposite parity, as was required, and our assertion is valid.

The problem of the signs will now be solved by showing that a superposition polygon necessarily encloses an even number of lattice points, for this will imply that all superposition polygons are clockwise odd and therefore that (4.23) holds for all such polygons. The superposition of two arbitrary dimer configurations yields only superposition polygons, which span an even number of lattice points and superimposed isolated dimers while leaving no lattice points uncovered. Hence a superposition polygon can surround other superposition polygons and dimers only in such a way that an even number of lattice points are surrounded. This is what we set out to prove.

The directed graphs in Figs. 4.8 and 4.9 give the sign prescriptions for square, hexagonal, and triangular lattices.

4.5 CYCLIC MATRICES

An nth-order matrix A, with elements

$$a(j, k) \equiv a(k - j)$$

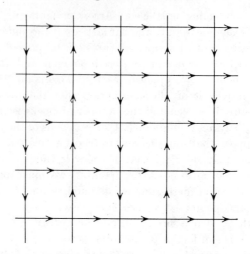

Fig. 4.8 Directed bonds on a square lattice.

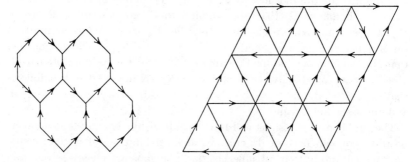

Fig. 4.9 Directed bonds on hexagonal and triangular lattices.

that depend only on $k - j$, and that are periodic so that

$$a(k + n) = a(k),$$

is called cyclic [22]. An example is

$$\begin{bmatrix} a(0) & a(1) & a(2) & a(3) \\ a(3) & a(0) & a(1) & a(2) \\ a(2) & a(3) & a(0) & a(1) \\ a(1) & a(2) & a(3) & a(0) \end{bmatrix}.$$

The matrix elements might be numbers or matrices. Suppose they are

mth-order matrices. Such cyclic matrices can easily be transformed to matrices in which the diagonal elements are mth-order matrices and the off-diagonal elements vanish. Let R be an orthogonal matrix in which the elements are mth-order matrices,

$$R(k, l) = n^{-\frac{1}{2}} I_m e^{2\pi i k l/n},$$

where I_m is the mth-order identity matrix. The elements of R^{-1} are

$$R^{-1}(k, l) = n^{-\frac{1}{2}} I_m e^{-2\pi i k l/n}.$$

The elements of

$$B \equiv R^{-1} A R$$

are

$$
\begin{aligned}
B(j, k) &= \sum_{l,s=1}^{n} R^{-1}(j, l) a(s - l) R(s, k) \\
&= \frac{1}{n} \sum_{l=1}^{n} e^{[2\pi i (k-j) l]/n} \sum_{s=1}^{n} a(s - l) e^{[2\pi i (s-l) k]/n} \\
&= \delta_{j,k} \lambda \frac{2\pi j}{(n)},
\end{aligned}
$$

which are $m \times m$ matrices that vanish unless $j = k$, so that

$$
B =
\begin{bmatrix}
\lambda\left(\dfrac{2\pi}{n}\right) & 0 & \cdots \\
0 & \lambda\left(\dfrac{4\pi}{n}\right) & 0 & \cdots \\
0 & 0 & \lambda\left(\dfrac{6\pi}{n}\right) & \cdots \\
& \cdot & \cdot & \cdot
\end{bmatrix}.
$$

The s summation is independent of l because of the periodicity of $a(k)$; indeed, we have

$$\lambda(\phi) = \sum_{s=1}^{n} a(s) e^{is\phi}.$$

Since the determinant of A is invariant under an orthogonal transformation, we obtain

$$\det A = \det B = \prod_{j=1}^{n} \det \lambda\left(\frac{2j\pi}{n}\right).$$

In the limit as $n \to \infty$, we let $\phi = 2\pi s/n$ so that $d\phi = 2\pi/n$. Then we have

$$\frac{1}{n} \log \det A = \frac{1}{n} \sum_{j=1}^{n} \log \det \lambda\left(\frac{2j\pi}{n}\right) \sim \frac{1}{2\pi} \int_{0}^{2\pi} \log \det \lambda(\phi) \, d\phi.$$

This result is easily generalized to the situation in which $\mathbf{j}$ and $\mathbf{k}$ are vectors with S components and the matrix elements satisfy

$$a(\mathbf{j}, \mathbf{k}) = \mathbf{a}(\mathbf{k} - \mathbf{j}),$$

$$a(\mathbf{k} + \mathbf{n}) = a(\mathbf{k}),$$

$$\mathbf{k} = (k_1, k_2, \ldots) \quad \text{and} \quad \mathbf{n} = (n_1, n_2, \ldots, n_s).$$

Then A is diagonalized by the orthogonal matrix

$$R(\mathbf{k}, \mathbf{l}) = I_m \prod_{\alpha=1}^{s} \frac{1}{n_\alpha^{1/2}} e^{2\pi i k_\alpha l_\alpha / n_\alpha},$$

so that

$$B(\mathbf{j}, \mathbf{k}) = \delta_{j_1, k_1} \, \delta_{j_2, k_2} \cdots \lambda \left(\frac{2\pi j_1}{n_1}, \frac{2\pi j_2}{n_2}, \cdots \right)$$

where

$$\lambda(\phi_1, \phi_2, \ldots) = \sum_u a(\mathbf{u}) e^{i\mathbf{u} \cdot \boldsymbol{\phi}}.$$

Then

(4.24) $\det A = \det B = \displaystyle\prod_{j_1=1}^{n_1} \prod_{j_2=1}^{n_2} \cdots \det \lambda \left(\frac{2\pi j_1}{n_1}, \frac{2\pi j_2}{n_2}, \ldots \right).$

In the limit as $n_1 \to \infty$, $n_2 \to \infty$, ..., we let $\phi_\alpha = 2\pi j_\alpha / n_\alpha$, so that $d\phi_\alpha = 2\pi / n_\alpha$. Then we have

(4.25)

$$\frac{1}{n_1 n_2 \cdots} \log \det A = \frac{1}{n_1 n_2 \cdots} \sum_{j_1=1}^{n_1} \sum_{j_2=1}^{n_2} \cdots \log \det \lambda \left(\frac{2\pi j_1}{n_1}, \frac{2\pi j_2}{n_2}, \ldots \right)$$

$$\sim \frac{1}{(2\pi)^S} \int_0^{2\pi} \cdots \int_0^{2\pi} \log \det \lambda(\phi_1, \phi_2, \ldots, \phi_S) \, d^S\phi.$$

4.6 EVALUATION OF DIMER PFAFFIAN

The most convenient algorithm [1] for the evaluation of the Pfaffian $P\{A\}$ is the relation

(4.26) $\det A = [P\{A\}]^2.$

Hence our dimer problem would be solved if $\det A$ could be evaluated. The lattice of directed bonds in Fig. 4.8 has lost the translational symmetry of a lattice of undirected bonds in that a translation one step to the right changes the direction of the vertically directed bonds. If, however, we introduce new unit cells containing a pair of horizontal lattice points, then the lattice is invariant under horizontal translations of an integral number of unit cells.

It is convenient to work with a new numbering system of lattice points. We represent the lattice point p by three indices: p_1, the x coordinate of the unit cell; p_2, the y coordinate of the unit cell; and Γ, which is L for the left-hand member of the unit cell and R for the right-hand member. Finally, we define the matrix $A(p; p')$, in which the elements connect lattice points of unit cell p with those of unit cell p', by

$$A(p_1, p_2; p_1', p_2') = \begin{bmatrix} a(p_1, p_2, L; p_1', p_2', L) & a(p_1, p_2, L; p_1', p_2', R) \\ a(p_1, p_2, R; p_1', p_2', L) & a(p_1, p_2, R; p_1', p_2', R) \end{bmatrix}.$$

All these matrix elements vanish except those associated with nearest-neighbor pairs. Hence we have

$$(4.27) \quad A(p_1, p_2; p_1', p_2') = 0 \quad \text{unless} \quad \begin{cases} p_1 = p_1' \pm 1, & p_2 = p_2', \\ p_2 = p_2' \pm 1, & p_1 = p_1', \quad \text{or} \\ p_1 = p_1', & p_2 = p_2'. \end{cases}$$

Notice that $A(p_1, p_2; p_1', p_2')$ depends only on $(p_1' - p_1)$ and $(p_2' - p_2)$; that is, every unit cell (p_1, p_2) is related to its neighbors in a manner that depends only on direction (except for end effects, which will be discussed later) and not explicitly on (p_1, p_2).

Consider $A(p_1, p_2; p_1, p_2)$. This corresponds to the bonding between two points in the same unit cell. The two nonvanishing elements are

$$a(p_1, p_2, L; p_1, p_2, R) = z_1 = -a(p_1, p_2, R; p_1, p_2, L),$$

since the first of these elements corresponds to a horizontal bond pointing from the left-hand member to the right-hand member of cell (p_1, p_2). Thus

$$(4.28) \qquad A(p_1, p_2; p_1, p_2) = \begin{array}{c} \\ L \\ R \end{array} \!\! \begin{array}{c} L \quad\;\; R \\ \begin{bmatrix} 0 & z_1 \\ -z_1 & 0 \end{bmatrix} \end{array} \equiv a(0, 0),$$

where $a(u, v)$ is defined by

$$(4.29) \qquad A(p_1, p_2; p_1', p_2') = a(p_1' - p_1, p_2' - p_2) = a(p' - p).$$

Next consider $A(p_1, p_2; p_1 + 1, p_2)$. This connects a given unit cell with one to its right. The right-hand member of the left-hand cell can be bonded by a horizontal dimer to the left-hand member of the right-hand cell, so that

$$a(p_1, p_2, R; p_1 + 1, p_2, L) = z_1.$$

.

Since no other bonds can exist between cells (p_1, p_2) and $(p_1 + 1, p_2)$, we have

$$(4.30) \qquad A(p_1, p_2; p_1 + 1, p_2) = \begin{array}{c} \\ L \\ R \end{array} \overset{\displaystyle \begin{array}{cc} L & R \end{array}}{\begin{bmatrix} 0 & 0 \\ z_1 & 0 \end{bmatrix}} \equiv a(1, 0).$$

Similarly, we obtain

$$(4.31) \qquad A(p_1, p_2; p_1 - 1, p_2) = \begin{bmatrix} 0 & -z_1 \\ 0 & 0 \end{bmatrix} \equiv a(-1, 0).$$

Now consider $a(p_1, p_2; p_1, p_2 + 1)$. This connects a given unit cell with the one above it by vertical bonds. Here

$$a(p_1, p_2, R; p_1, p_2 + 1, R) = z_2,$$

corresponding to an upwardly directed vertical bond, whereas

$$a(p_1, p_2; L; p_1, p_2 + 1, L) = -z_2,$$

corresponding to a downwardly directed bond in the vertical direction. The off-diagonal elements vanish in this case because no dimer can connect the left-hand member of the lower cell with the right-hand member of the upper one. Hence we have

$$(4.32) \qquad A(p_1, p_2; p_1, p_2 + 1) = \begin{bmatrix} -z_2 & 0 \\ 0 & z_2 \end{bmatrix} \equiv a(0, 1),$$

and similarly

$$(4.33) \qquad A(p_1, p_2; p_1, p_2 - 1) = \begin{bmatrix} z_2 & 0 \\ 0 & -z_2 \end{bmatrix} \equiv a(0, -1).$$

Now (4.27) is equivalent to

$$a(u, v) = 0,$$

unless either (1) $u = v = 0$, (2) $u = 0$ and $v = \pm 1$, or (3) $v = 0$ and $u = \pm 1$.

Suppose the lattice has m points in each horizontal row, m being even, and n in each vertical column. As the lattice becomes very large, with $m \to \infty$ and $n \to \infty$, one would expect the exact nature of boundary conditions to have little influence on the problem. We shall discuss this at length in Sec. 4.8. Here, however, for simplicity we choose periodic boundary conditions, so that

$$a(k_1 + \tfrac{1}{2}m, k_2 + n) = a(k_1, k_2),$$

our lattice having $m/2$ unit cells in the x direction and n in the y direction.

Then we can immediately apply (4.25) for the evaluation of det A:

$$(4.34) \qquad \frac{2}{mn} \log \det A \sim \frac{1}{(2\pi)^2} \int_0^{2\pi} \int_0^{2\pi} \log \det \lambda(\phi_1, \phi_2)\, d\phi_1\, d\phi_2,$$

where, from (4.24) and (4.28), (4.30), (4.31), (4.32), and (4.33), we have

$$(4.35) \qquad \begin{aligned} \lambda(\phi_1, \phi_2) &= \sum_{u_1, u_2} a(u_1, u_2) e^{i(u_1\phi_1 + u_2\phi_2)} \\ &= a(0, 0) + a(1, 0)e^{i\phi_1} + a(-1, 0)e^{-i\phi_1} \\ &\qquad\qquad\qquad + a(0, 1)e^{i\phi_2} + a(0, -1)e^{-i\phi_2} \\ &= \begin{bmatrix} -z_2 e^{i\phi_2} + z_2 e^{-i\phi_2} & z_1 - z_1 e^{-i\phi_1} \\ -z_1 + z_1 e^{i\phi_1} & z_2 e^{i\phi_2} - z_2 e^{-i\phi_2} \end{bmatrix}, \end{aligned}$$

so that

$$(4.36) \qquad \det \lambda(\phi_1, \phi_2) = 4(z_1^2 \sin^2 \tfrac{1}{2}\phi_1 + z_2^2 \sin^2 \phi_2)$$

and

$$\begin{aligned} \frac{2}{mn} \log \det A &\sim \frac{1}{(2\pi)^2} \int_0^{2\pi} \int_0^{2\pi} \log 4(z_1^2 \sin^2 \tfrac{1}{2}\phi_1 + z_2^2 \sin^2 \phi_2)\, d\phi_1\, d\phi_2 \\ &= \frac{1}{(2\pi)^2} \int_0^{2\pi} \int_0^{2\pi} \log 2[(z_1^2 + z_2^2) - z_1^2 \cos \theta_1 - z_2^2 \cos \theta_2]\, d\theta_1\, d\theta_2. \end{aligned}$$

Our Pfaffian (4.26) then has the form

$$(4.37)$$
$$P\{A\} \sim \exp\left\{\frac{mn}{(2\pi)^2} \int_0^\pi \int_0^\pi \log 2[(z_1^2 + z_2^2) - z_1^2 \cos \theta_1 - z_2^2 \cos \theta_2]\, d\theta_1\, d\theta_2\right\},$$

which, from (4.21), is just the generating function

$$\Phi_{mn}(z_1, z_2) = P\{A\}$$

for dimer configurations.

The integral in (4.37) is evaluated as follows. Let

$$F(X_1, X_2) = \frac{1}{\pi^2} \int_0^\pi \int_0^\pi \log (X_1 + X_2 - X_1 \cos \theta_1 - X_2 \cos \theta_2)\, d\theta_1\, d\theta_2.$$

Note that

$$\begin{aligned} F(X_1, 0) &= \frac{1}{\pi} \int_0^\pi \log [X_1(1 - \cos \theta_1)]\, d\theta_1 \\ &= \log \tfrac{1}{2}X_1, \end{aligned}$$

and similarly

$$F(0, X_2) = \log \tfrac{1}{2}X_2.$$

Also, we have

$$\frac{\partial F}{\partial X_1} = \frac{1}{\pi^2} \int_0^\pi \int_0^\pi \frac{(1 - \cos \theta_1)\, d\theta_1\, d\theta_2}{X_1 + X_2 - X_1 \cos \theta_1 - X_2 \cos \theta_2}.$$

The θ_2 integration leads to

$$\frac{\partial F}{\partial X_1} = \frac{1}{\pi} \int_0^\pi \frac{(1 - \cos \theta_1)\, d\theta_1}{[X_2 + X_1(1 - \cos \theta_1)]^2 - X_2^2}.$$

Let $y = 1 - \cos \theta_1$; then

$$\frac{\partial F}{\partial X_1} = \frac{1}{\pi} \int_0^2 \frac{dy}{[X_1(2 - y)(2X_2 + X_1 y)]^{1/2}}.$$

Since

$$-\frac{2}{X_1} \frac{d}{dy} \tan^{-1} \left[\frac{X_1(2 - y)}{2X_2 + yX_1} \right]^{1/2} = \frac{1}{[X_1(2 - y)(2X_2 + X_1 y)]^{1/2}},$$

it follows that

$$\frac{\partial F}{\partial X_1} = \frac{2}{\pi X_1} \tan^{-1} \left(\frac{X_1}{X_2} \right)^{1/2}$$

$$= \frac{2}{\pi X_1} \left[\left(\frac{X_1}{X_2} \right)^{1/2} - \frac{1}{3} \left(\frac{X_1}{X_2} \right)^{3/2} + \frac{1}{5} \left(\frac{X_1}{X_2} \right)^{5/2} - \cdots \right].$$

By integrating term by term and employing the boundary condition A, we find

$$F(X_1, X_2) = \log \tfrac{1}{2} X_1 + \frac{4}{\pi} \left[\left(\frac{X_1}{X_2} \right)^{1/2} - \frac{1}{3^2} \left(\frac{X_1}{X_2} \right)^{3/2} + \frac{1}{5^2} \left(\frac{X_1}{X_2} \right)^{5/2} - \cdots \right].$$

In particular $F(2, 2)$, as required in (4.37), where $z_1 = z_2 = 1$, has the value

$$F(2, 2) = \frac{4}{\pi} \left\{ 1 - \frac{1}{3^2} + \frac{1}{5^2} - \cdots \right\} = \frac{4G}{\pi}.$$

The sum, abbreviated here by G, is called *Catalan's constant*. A computation yields

$$G = 1 - 3^{-2} + 5^{-2} - 7^{-2} + \cdots \doteq 0.915965594.$$

We now find the number of dimer configurations [6, 11] to be

$$\Phi_{m, n}(1, 1) \sim \exp \frac{mn}{(2\pi)^2} \int_0^\pi \int_0^\pi \log 2(2 - \cos \theta_1 - \cos \theta_2)\, d\theta_1\, d\theta_2$$

$$= \frac{mn}{2} e^{2G/\pi},$$

where $mn/2$ is the number of dimers filling our $n \times m$ lattice. Numerically, we have

$$e^{2G/\pi} \doteq 1.791623.$$

This is close to the original estimate of Fowler and Rushbrooke [8] of the integral (4.37), which was obtained by extrapolating results from small-circumference infinite cylinders, with m ranging from 2 to 8 and $n \to \infty$.

The exact results for finite rectangles and toruses have also been obtained, by Fisher [6] and Kasteleyn [11]. One interesting numerical result due to Fisher is the number of ways a chess board (8×8 squares) can be covered by 32 dominoes, each domino covering two squares; it is

$$\Phi_{8,8}(1, 1) = 12,988,816 = 2^4 \times (901)^2.$$

Results also exist for several other types of lattices.

4.7 THE ISING PROBLEM

This section contains a discussion [3, 23] of the thermodynamic properties of the Ising model of ferromagnetism. In this primitive model of ferromagnetism, we associate an elementary magnet or electron spin with each lattice point, and we identify the state of the jth spin through a variable σ_j with the property that

$$\sigma_j = \begin{cases} +1 & \text{if the magnetic moment is "up," } \uparrow, \\ -1 & \text{if the magnetic moment is "down," } \downarrow. \end{cases}$$

Furthermore, each spin is postulated to interact with its neighboring spins so that the preferred spin orientation is the parallel one; that is, the state of lowest energy of a pair of spins, σ and σ', is that in which they are parallel. An interaction energy J is then introduced, so that

$$\left. \begin{array}{l} \text{energy of parallel state } \uparrow\uparrow \text{ or } \downarrow\downarrow = -J \\ \text{energy of antiparallel state } \uparrow\downarrow \text{ or } \downarrow\uparrow = J \end{array} \right\} = -J\sigma\sigma'.$$

The ferromagnetic state of a lattice is one with a nonvanishing residual magnetic moment, the nonmagnetic state that with a random distribution of spin orientations with no residual magnetic moment. If only nearest-neighbor pairs interact, and if the energy of interaction is additive, then the energy of a lattice spin state $\{\sigma_1, \sigma_2, \ldots\}$ is

$$E\{\sigma_1, \sigma_2, \ldots, \sigma_N\} = -J \sum_{n.n.} \sigma_j\sigma_k,$$

where the sum is taken over all pairs j and k that are nearest neighbors ($n.\,n.$) in the lattice. Thermodynamic properties of the lattice are obtained [3] from the partition function

$$Z = \sum_{\sigma_1 = \pm 1} \cdots \sum_{\sigma_N = \pm 1} \exp\left(\tfrac{1}{2} K \sum_{j,k=1}^{N} a_{jk} \sigma_j \sigma_k\right)$$

$$= \sum_{\sigma_1 = \pm 1} \cdots \sum_{\sigma_N = \pm 1} \prod_{n.n.} e^{K \sigma_j \sigma_k},$$

where

$$a_{jk} = 1 \qquad \text{if } j \text{ and } k \text{ are nearest neighbors,}$$
$$= 0 \qquad \text{otherwise,}$$

and

$$K = \frac{J}{kT},$$

in which k is Boltzmann's constant and T the absolute temperature. If Z is considered to be a function of the temperature, then the internal energy of the Ising ferromagnet is given by

$$U = kT^2 \frac{\partial}{\partial T} \ln Z,$$

and the heat capacity is

$$C = \frac{\partial U}{\partial T}.$$

Other thermodynamic properties are obtained in the standard way from Z.

Let c be the number of nearest neighbors to a given lattice point. Then since $\sigma^{2n} = 1$ and $\sigma^{2n+1} = \sigma$, we have

$$(4.38) \qquad Z = (\cosh K)^{cN/2} \sum_{\{\sigma\}} \prod_{n.n.} (1 + z \sigma_i \sigma_j),$$

where $cN/2$ is the total number of nearest-neighbor pairs and

$$z = \tanh K.$$

On a square lattice, and on a simple cubic lattice, we have $c = 4$, and $c = 6$, respectively.

Let us now expand the product in (4.38):

$$(4.39) \quad Z = (\cosh K)^{Nc/2} \sum_{\sigma_1 = \pm 1} \cdots \sum_{\sigma_N = \pm 1} Q(\sigma_1, \sigma_2, \cdots, \sigma_N),$$

where

$$Q(\sigma_1, \sigma_2, \cdots, \sigma_N) = 1 + z \sum_{n.n.} (\sigma_i \sigma_j) + z^2 \sum_{n.n.} (\sigma_i \sigma_j) \sum_{n.n.} (\sigma_k \sigma_l) + \cdots.$$

The coefficient of z^s is the sum of all products of $2s$ σ's. The σ's occur in pairs corresponding to nearest-neighbor pairs, and no pair occurs more than once in the same product.

A diagram can be associated with any product of σ's by introducing a bond to connect each pair of points i and j that go with each $(\sigma_i\sigma_j)$. Since each nearest-neighbor σ pair occurs at most once in a given product, no bond appears more than once in a given graph. In Fig. 4.10, we have constructed the graph associated with the term

$$(\sigma_1\sigma_6)(\sigma_6\sigma_7)(\sigma_7\sigma_8)(\sigma_8\sigma_3)(\sigma_9\sigma_{14})(\sigma_{12}\sigma_{13})(\sigma_9\sigma_{10})(\sigma_{10}\sigma_{15})(\sigma_{15}\sigma_{14}),$$

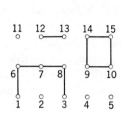

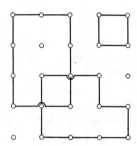

Fig. 4.10 Typical graph in expansion of Ising partition function.

Fig. 4.11 Typical non-vanishing (closed) graph in expansion of Ising partition function.

which is typical of the terms that occur in (4.39). Since

$$\sum_{\sigma=\pm1} \sigma = 0 \quad \text{and} \quad \sum_{\sigma=\pm1} \sigma^2 = 2,$$

the only terms of (4.39) that contribute to Z are those in which each σ_j appears an even number of times. Since a bond can appear only once in a term, σ_j can occur at most c times (for example, four times on a square lattice). This also means that the only graphs of interest for a square net are those in which each lattice point is attached to an even number of bonds (0, 2, or 4). Then all contributing graphs must be a superposition of simple closed polygons (polygons that do not cross themselves) that have no common sides but that can meet at vertices. Conversely, there is a nonzero term in Z for each superposition of simple closed polygons. Polygons that are the superposition of simple polygons may cross each other provided no "double" or higher-order bonds occur. For simplicity, we call such graphs *closed graphs* (Fig. 4.11). Each closed graph of m bonds contributes a term $z^m 2^N$ to Z after we sum over $\sigma_1 \cdots \sigma_N$. Hence the partition function can be written as

(4.40) $$Z = 2^N (\cosh K)^{Nc/2} \sum_r n(r) z^r,$$

where $n(r)$ is the number of closed graphs of r bonds that can be constructed on the lattice. We have $n(0) \equiv 1$, and on a square or simple cubic lattice we have $n(r) \equiv 0$ unless r is even.

We can generalize (4.40) to include an anisotropic interaction constant [23]. If, for example, K represents the interaction between lattice points of a square lattice that lie on the same row, and K' the interaction between pairs in the same column, then we have

(4.41) $$Z = (\cosh K \cosh K')^N 2^N \sum_{r,s} n(r,\,s) z_1^r z_2^s,$$

$$z_1 = \cosh K, \qquad z_2 = \cosh K',$$

where $n(r,\,s)$ is the number of closed graphs with $r + s$ bonds, r in the horizontal and s in the vertical direction.

We might first apply (4.40) to the calculation of the partition function of a ring of N points. Clearly, the only graph that can be drawn on a ring, without introducing double bonds, is one with N bonds that encircles the ring completely. Such a graph can be constructed in only one way. Since $c = 2$ on a ring, we have

$$Z = 2^N (\cosh K)^N [n(0) + n(N) \tanh^N K]$$
$$= (2 \cosh K)^N + (2 \sinh K)^N \sim (2 \cosh K)^N \qquad \text{as} \quad N \to \infty.$$

The thermodynamics of the two-dimensional Ising model was first determined by L. Onsager [13, 24]. This calculation was very important because it was the first example clearly indicating that statistical mechanics is capable of describing a phase transition, here from the ordered ferromagnetic to a disordered nonferromagnetic state. Kac and Ward expressed the partition function as a certain determinant and obtained the Onsager results in an alternative manner [10, 23, 27]. Some gaps existed in their analysis. These were recently filled in by S. Sherman [28]. Hurst and Green [9, 4] re-expressed the Kac-Ward theory in terms of Pfaffians in a paper that inspired the work of Fisher and Kasteleyn on dimers. Finally, Kasteleyn [12] has recently shown the connection between the Ising problem and the dimer problem. We now derive Onsager's results by using this connection.

Consider dimers on a "bathroom-tile" lattice and the closed graphs that are significant for the Ising problem on a square lattice. We allow one more degree of freedom of bond arrangement than previously. We permit bonds to appear as diagonals in the small squares, and allow one diagonal bond to bridge another; but as usual no lattice point can be occupied by two bonds. (We might consider the small squares as cities and the horizontal and vertical bonds connecting small squares as roads connecting cities.) For each closed polygon in the Ising problem, one can find a dimer configuration. For example, compare Figs. 4.12 and 4.13.

Each lattice point on a square Ising lattice must be the intersection of an even number of bonds, 0, 2, or 4; no lattice point may be a meeting

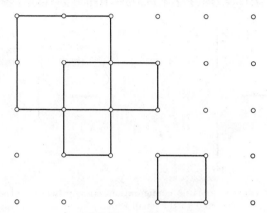

Fig. 4.12 Ising graph

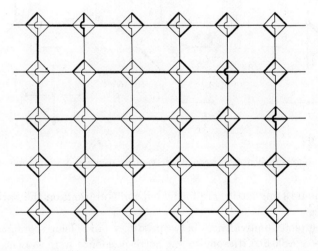

Fig. 4.13 A bathroom-tile dimer graph corresponding to Fig. 4.12.

place of an odd number of bonds. A one-to-one correspondence exists between horizontal and vertical dimers that connect the small squares on the bathroom-tile lattice and the horizontal and vertical bonds that connect nearest-neighbor lattice points on the square Ising lattice.

No dimer configuration exists on the bathroom-tile lattice such that only two vertical dimers and one horizontal dimer, or vice versa, come together, for then point 1 (or its analogue) of Fig. 4.14a would be left uncovered. Similarly, no dimer configuration can exist with only one horizontal (or vertical) dimer being connected to a small square of the bathroom-tile lattice (see Fig. 4.14a, b). Hence forbidden configurations

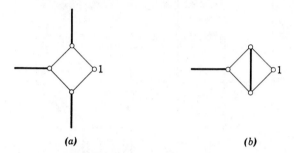

(a)   (b)

Fig. 4.14 Demonstration that no configuration exists in which an odd number of external bonds are connected to the small square, for then one point on the small square remains uncovered.

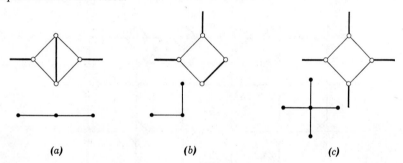

(a)   (b)   (c)

Fig. 4.15 Correspondence between dimer configurations and Ising configurations.

on the square Ising lattice are forbidden configurations on the bathroom-tile lattice.

The complete equivalence between dimer and Ising configurations would be established if a one-to-one correspondence were shown to exist among all configurations at lattice points on an Ising lattice and dimer configurations at small squares on the bathroom-tile lattice. Each lattice point on a quadratic Ising lattice must be in one of four possible states, with four bonds attached, with two in the same direction attached, with two at right angles attached, or with none attached. The meeting of two horizontal bonds is in one-to-one correspondence with the dimer configuration of Fig. 4.15. The meeting of a vertical bond and a horizontal bond is in one-to-one correspondence with a dimer configuration (see Fig. 4.15b.) The meeting of four bonds on the square Ising lattice are in one-to-one correspondence with the configuration of Fig. 4.15c in the bathroom-tile lattice.

Only one situation remains to be examined, that in which no bonds are attached to an Ising lattice point. If the one-to-one correspondence could be made in this case, then that between all square Ising configurations and all bathroom-tile configurations would be complete. But here difficulty arises. It is clear from Fig. 4.16 that three dimer configurations exist in

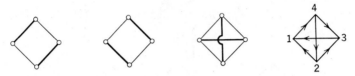

Fig. 4.16 Three dimer configurations in which a small square is not connected to its neighbor.

which a small square is not connected to its neighbor. This gives a three-to-one correspondence. Fortunately, one ray of hope remains. We are still free to prescribe signs to our various dimer configurations. If the elements of the Pfaffian that we use to generate dimer configurations are chosen so that configurations of Figs. 4.16a, b are positive while that of Fig. 4.16c is negative, then the contributions of two of the configurations cancel and only one is left, as is required. The remaining problem is to give a prescription of signs that makes all terms positive in the expansion of the Pfaffian except those that have bridges, which must be negative.

It can be immediately verified that the array of arrows given in Fig. 4.17 satisfy our requirements. First consider all directed bonds except those that are crossed in the interior of the small squares. Each small square, such as $\{1, 2, 3, 4\}$, and each octagon, such as $\{4, 3, 5, 6, 7, 8, 9, 10\}$, is clockwise odd, so that all superposition polygons that do not have sides in the interior of small squares are also clockwise odd and indeed contribute positively to the Pfaffian associated with dimers on this lattice. Now introduce the horizontal bonds directed to the left in each small square. The triangles into which they dissect the squares are still clockwise odd, as are those that result from the downwardly directed vertical bonds. Hence all superposition polygons that do not contain bridges are clockwise odd, and all dimer configurations that are in one-to-one correspondence with Ising configurations occur with a positive sign in the expansion of the dimer Pfaffian.

Now consider in detail the contribution of each of the configurations of

Fig. 4.16. Suppose all lattice points other than those of a single small square are covered by some dimer configuration. Such a configuration occurs in three terms of the dimer-generating function, one for each of the arrangements of Fig. 4.16 of the specified square. Each term factors into two parts, one factor being associated with the small square and the other with contributions from all other lattice points, the latter being the same

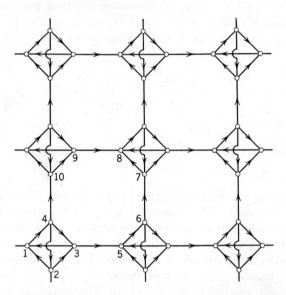

Fig. 4.17 Array of arrows satisfying prescribed requirements.

in all three terms. The sum of the three also has two factors, one being the Pfaffian associated with the small square given by (4.17), and the other being the contribution of the remaining dimers. The first term in (4.17) corresponds to Fig. 4.16a, the second to Fig. 4.16b, and the third to Fig. 4.16c.

We prescribe the values

(4.42) $a(i, j) = 1$

to each bond directed from i to j on a small square, so that $a(j, i) = -1$ in view of the antisymmetry of the a's. From the bond directions assigned in Fig. 4.16, we note that (4.17) has the value

$$(-1)(-1) - (-1)(-1) + (1)(1) = 1.$$

Hence, with the assignment (4.42), our small square—and indeed any small square covered by two dimers—contributes a factor 1 to a given over-all lattice configuration.

If we assign the following values to horizontal and vertical bonds of the large octagons on the bathroom-tile lattice,

$$(4.43) \quad a(i,j) = z_1 \quad \text{for each horizontal bond directed from } i \text{ to } j,$$
$$\qquad\qquad = z_2 \quad \text{for each vertical bond directed from } i \text{ to } j,$$

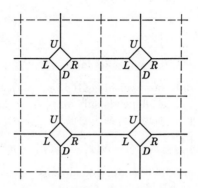

Fig. 4.18 Dotted squares corresponding to unit cell in bathroom-tile lattice.

then we can show that the Ising partition function [see (4.41)] is

$$(4.44) \qquad\qquad Z = (2 \cosh K \cosh K')^N P(A),$$

where $P(A)$ is the dimer-generating function with the preceding weights. Consider a term in the expansion of the Pfaffian. It corresponds to a dimer configuration in which a certain number of small squares are covered by internal dimer configurations without being connected to other small squares. Such squares merely contribute a factor 1 to the term. Other small squares are connected by either vertical or horizontal bonds with respective weights z_2 and z_1, as required in the Ising problem. These dimers make up a pattern that corresponds to some closed polygon configuration of the Ising partition function. From our earlier discussion, each term in (4.44) corresponds to a term in (4.41) and vice versa; the sign of each term in (4.42) has been made positive, as required, through our sign prescription.

We now proceed with the evaluation of the Pfaffian $P(A)$ by calculating its associated antisymmetrical determinant (4.26), with value $[P(A)]^2$. Our bathroom-tile lattice has the symmetry of a square lattice if each unit cell is chosen to contain a small square of four lattice points (see Fig. 4.18). A typical unit cell is characterized by (p_1, p_2), as was done in

Sec. 4.6. Here, however, the index Γ takes on four values, L, R, U, D, instead of two, since we now have four instead of two points per unit cell; L represents the left, R the right, U the top, and D the bottom point in each small square. The 2×2 matrices $A(p, p')$ of the dimer problem on a single square lattice must be replaced by 4×4 matrices of the form

$$
A(p_1, p_2; p_1', p_2') =
\begin{array}{c}
\\ R \\ L \\ U \\ D
\end{array}
\overset{\begin{array}{cccc} R & L & U & D \end{array}}{
\left[
\begin{array}{cccc}
a(p_1, p_2, R : p_1', p_2', R) & \cdot & \cdot & \cdot & \cdot \\
\cdot & \cdot & \cdot & \cdot & \cdot & \cdot \\
\cdot & \cdot & \cdot & \cdot & \cdot & \cdot \\
a(p_1, p_2, D : p_1', p_2', R) & \cdot & \cdot & \cdot & \cdot
\end{array}
\right]}.
$$

As in Sec. 4.6, the only nonvanishing $A(p_1, p_2; p_1', p_2')$ are those listed in (4.27), and $A(p_1, p_2; p_1', p_2')$ depends only on $p_1' - p_1$, and $p_2' - p_2$, that is, on the vector $\mathbf{p}' - \mathbf{p}$, so that, repeating (4.29), we abbreviate

$$
A(p_1, p_2; p_1', p_2') = a(p_1' - p_1, p_2' - p_2).
$$

We again employ periodic boundary conditions. First we find

$$
A(p_1 p_2; p_1 p_2) =
\begin{array}{c}
\\ R \\ L \\ U \\ D
\end{array}
\overset{\begin{array}{cccc} R & L & U & D \end{array}}{
\left[
\begin{array}{cccc}
0 & 1 & -1 & -1 \\
-1 & 0 & 1 & -1 \\
1 & -1 & 0 & 1 \\
1 & 1 & -1 & 0
\end{array}
\right]} = a(0, 0).
$$

The top row corresponds to connections between the left-hand corner of a small square and various other corners. Since a given point cannot connect with itself, the (R, R) element is zero. The (R, L) element corresponds to a dimer bond from the right-hand to the left-hand point. It is given the value $+1$ because it is directed from R to L in Fig. 4.17. The (R, U) and (R, D) elements are negative because the bonds are directed toward R. Generally, the element in the Γth row and Γ'th column is positive if the bond conecting Γ and Γ' is directed toward Γ', and negative if it is directed toward Γ.

Next consider $A(p_1, p_2; p_1 + 1, p_2)$. This connects a given unit cell with one to its right. The R corner of the left-hand cell can be bonded to the L corner of the right-hand cell. No other bonding can take place. Since this is a horizontal bond between two small squares, it is weighted by z_1 [see (4.43)]; thus we have

$$
a(p_1, p_2, R; p_1 + 1, p_2, L) = z_1
$$

and

$$A(p_1, p_2; p_1 + 1, p_2) = \begin{array}{c} \\ R \\ L \\ U \\ D \end{array}\begin{array}{cccc} R & L & U & D \\ \left[\begin{array}{cccc} 0 & z_1 & 0 & 0 \\ 0 & 0 & 0 & 0 \\ 0 & 0 & 0 & 0 \\ 0 & 0 & 0 & 0 \end{array}\right] \end{array} = a(1, 0).$$

Similarly, we obtain

$$A(p_1; p_2; p_1 - 1, p_2) = \left[\begin{array}{cccc} 0 & 0 & 0 & 0 \\ -z_1 & 0 & 0 & 0 \\ 0 & 0 & 0 & 0 \\ 0 & 0 & 0 & 0 \end{array}\right] = a(-1, 0),$$

$$A(p_1, p_2; p_1, p_2 + 1) = \left[\begin{array}{cccc} 0 & 0 & 0 & 0 \\ 0 & 0 & 0 & 0 \\ 0 & 0 & 0 & z_2 \\ 0 & 0 & 0 & 0 \end{array}\right] = a(1, 0),$$

$$A(p_1, p_2; p_1, p_2 - 1) = \left[\begin{array}{cccc} 0 & 0 & 0 & 0 \\ 0 & 0 & 0 & 0 \\ 0 & 0 & 0 & 0 \\ 0 & 0 & -z_2 & 0 \end{array}\right] = a(0, -1).$$

All other $a(u, v)$ vanish. In a lattice with $n \times n = N$ unit cells, corresponding to an $n \times n$ Ising lattice, the value of $[P(A)]^2 = \det A$ is given by (4.25) with $n_1 = n_2 = n$, so that

$$(4.45) \qquad [P(A)]^2 \sim \exp\left[\frac{N}{(2\pi)^2}\int_0^{2\pi}\int_0^{2\pi} \log \det \lambda(\phi_1, \phi_2)\, d\phi_1\, d\phi_2\right],$$

where

$$\lambda(\phi_1, \phi_2) = a(0, 0) + a(1, 0)e^{i\phi_1} + a(-1, 0)e^{-i\phi_1}$$
$$+ a(0, 1)e^{i\phi_2} + a(0, -1)e^{-i\phi_2}$$

$$= \left[\begin{array}{cccc} 0 & 1 + z_1 e^{i\phi_1} & -1 & -1 \\ -(1 + z_1 e^{-i\phi_1}) & 0 & 1 & -1 \\ 1 & -1 & 0 & 1 + z_2 e^{i\phi_2} \\ 1 & 1 & -(1 + z_2 e^{-i\phi_2}) & 0 \end{array}\right].$$

From this we obtain

(4.46) $\det \lambda(\phi_1, \phi_2) = (1 + z_1^2)(1 + z_2^2)$
$$- 2z_1(1 - z_2^2) \cos \phi_1 - 2z_2(1 - z_1^2) \cos \phi_2.$$

Combining (4.44), (4.45), (4.46), and (4.41), we find the celebrated equation first derived by Onsager:

$$\frac{1}{N} \log Z = \log 2 + \frac{1}{(2\pi)^2} \int_0^{2\pi} \int_0^{2\pi} Q \, d\phi_1 \, d\phi_2,$$

where

$$Q = \log \left(\cosh 2K_1 \cosh 2K_2 - \sinh 2K_1 \cos \phi_1 \sinh 2K_2 \cos \phi_2 \right) d\phi_1 \, d\phi_2.$$

In the symmetrical case, we have $K_2 = K_1 = K = J/kT$, and the internal energy per lattice point is found to be

$$E = kT^2 \frac{\partial (N^{-1} \log Z)}{\partial T}$$
$$= -J \coth 2K[1 + (2\pi)^{-1}(2 \tanh^2 2K - 1)K_1(k_1)],$$

where

$$k_1 = 2 \sinh 2K \cosh^{-2} 2K,$$

and $K_1(k_1)$ is the complete elliptic integral of the first kind, namely,

$$K_1(k_1) = \int_0^{\pi/2} (1 - k_1^2 \sin^2 \phi_1)^{-1} \, d\phi_1.$$

A critical point exists at $2 \tanh^2 2K = 1$, at which the system changes its state; the state is ordered (ferromagnetic) at lower temperatures and disordered at higher. There is no latent heat.

The heat capacity is $C = \partial E/\partial T$. Since one of the terms of E is proportional to $|T - T_c| \log |T - T_c|$ near $T = T_c$, there is a term in C proportional to $\log |T - T_c|$, the heat capacity thus having a logarithmic singularity, a so-called λ-point, at T_c.

The Ising-model partition function in a magnetic field is the generating function for any lattice problem in which a two-state variable is defined at each lattice point and some interaction exists between nearest-neighbor lattice points.

4.8 SOME REMARKS ON PERIODIC BOUNDARY CONDITIONS

The calculations of Secs. 4.6 and 4.7 were made by using periodic boundary conditions. It is of interest to outline some justification for this

simplification. The central problem of both those sections was the calculation of the determinant of an antisymmetric matrix, the determinant
being the square of a desired Pfaffian. The elements of the antisymmetric
matrix A, of course, satisfy

$$a(p, p') = -a(p', p) = e^{i\pi}a(p', p).$$

The eigenvalue problem

(4.47) $$\sum_{p'} a(p, p')x(p') = \lambda x(p)$$

can be converted to a new eigenvalue problem

(4.48) $$\sum_{p'} b(p, p')\phi(p') = \mu\phi(p),$$

the $b(p, p')$ being elements of a Hermitian matrix B.

Let

$$b(p, p') = e^{i\pi/2}a(p, p'),$$

where $a(p, p')$ is real, as was the case in Secs. 4.6 and 4.7. Then

$$b(p', p) = e^{i\pi/2}a(p', p) = e^{-i\pi/2}a(p, p')$$
$$= b^*(p, p'),$$

b^* being the complex conjugate of b. If we let

$$\phi(p) = e^{-i\pi/2}x(p), \qquad \mu = e^{i\pi/2}\lambda,$$

then (4.47) implies (4.48). Let $\{\lambda_j\}$ and $\{u_j\}$ be the set of eigenvalues of A
and of B, respectively. Then

$$\det A = \prod\lambda_j = e^{-i\pi N/2}\prod\mu_j$$
$$= e^{-i\pi N/2}\det B$$

if A, and therefore B, are $N \times N$ matrices.

Clearly, we have

$$\log (e^{i\pi N/2} \det A) = \log \det B$$
$$= \sum \log \mu_j$$
$$= \int g(\mu) \log \mu \, d\mu,$$

where $g(\mu)$ is the distribution function of the eigenvalues μ_j of B; that is,
$g(\mu) \, d\mu$ is the number of eigenvalues of B with values between μ and $\mu +
d\mu$. Since B is Hermitian, all μ are real.

The basic theorem needed to justify the replacement of A by another
matrix A' with periodic elements is the following:

Ledermann's Theorem [14]. If in a Hermitian matrix the elements of r rows and their corresponding columns are modified in any way whatever, provided that the matrix elements remain Hermitian, then the number of eigenvalues that lie in any given interval cannot increase or decrease by more than $2r$.

Let B' be the matrix with periodic elements that replaces B. The eigenvalues of B' occur in pairs,

$$\mu_{\pm}(\phi_1, \phi_2) = \pm 2(z_1^2 \sin^2 \tfrac{1}{2}\phi_1 + z_2^2 \sin^2 \phi_2)^{\frac{1}{2}},$$

where $\phi_1 = 2\pi j/n$ and $\phi_2 = 2\pi k/m$, with $N = nm$; see (4.35) and (4.36). As $N \to \infty$, the distribution of μ's becomes everywhere dense in the interval

$$[-2(z_1^2 + z_2^2)^{\frac{1}{2}}, \; 2(z_1^2 + z_2^2)^{\frac{1}{2}}].$$

Hence any small subinterval $\Delta\mu$ of this interval contains the order of N eigenvalues. In the employment of periodic boundary conditions, however, matrix elements from all elementary cells on the boundary of the lattice have been artificially changed. Since the number of boundary cells is

$$2(n + m) = O(\sqrt{N}),$$

Ledermann's theorem implies that at most $O(\sqrt{N})$ of the N eigenvalues move in or out of the subinterval $\Delta\mu$ as a result of the change in boundary conditions. Therefore $g(\mu)$, which is $O(N)$, changes only by $O(\sqrt{N})$, which is negligible in the limit as $N \to \infty$. Thus one is justified in using periodic boundary conditions.

4.9 LATTICE STATISTICS OF SLIGHTLY DEFECTIVE LATTICES

Up to this point we have been concerned with perfectly periodic lattices Sometimes we wish to understand the influence of a small number of defective points or bonds in a lattice. A variety of important properties of solids results from a small number of defects such as impurities or vacancies. Much of semiconductor technology is based on such properties. Localized vibrations of crystals often exist around defects. Many mathematical techniques have been developed for investigating such problems. Some of them will be reviewed here through their application to lattice statistics. Because of space limitations, only very simple examples will be considered.

The result of a few defect points on random walks is discussed first. In accordance with Secs. 4.2 and 4.3, the probability that a random walker is at a point l after t steps, $P_t(l)$, is determined from a set of equations

(4.49)
$$P_{t+1}(l) = \sum_{l'} \not p(l, l') P_t(l'),$$

where $\not p(l, l')$ is the probability of a step from l' to l by a walker known to be at l'. In a perfect lattice, $\not p(l, l')$ was postulated to be a function of only $l - l'$. Since in an imperfect lattice some points differ from others, this is no longer the case. We express $\not p((l, l')$ as

(4.50)
$$\not p(l, l') = p(l - l') + q(l, l'),$$

the component $p(l - l')$ being that of the perfect lattice and $q(l, l')$ the perturbation. After any step a walker is certain to be somewhere; hence

$$\sum_l \not p(l, l') = 1 \qquad \text{for all } l'.$$

This corresponds to the conservation of walkers. For a perfect lattice, it was expressed in (4.11) as

$$\sum_l p(l - l') = 1.$$

Hence we take

$$\sum_l q(l, l') = 0 \qquad \text{for all } l'.$$

Let $G(l, z)$ be the generating function for $\{P_t(l)\}$:

$$G(l, z) = \sum_{t=0}^{\infty} z^t P_t(l).$$

On the supposition that at $t = 0$, before the first step, the walker is at l_0, we have

(4.51)
$$P_0(l) = \delta_{l, l_0}.$$

Multiplication of (4.49) by z^{t+1}, summation from $t = 0$ to ∞, and incorporation of (4.50) and (4.51) yields

$$G(l, z) - \delta_{l, l_0} = z \sum_{l'} \not p(l, l') G(l', z),$$

or

$$G(l, z) - z \sum_{l'} p(l - l') G(l', z) = \delta_{l, l_0} + z \sum_{l'} q(l, l') G(l', z).$$

This set of linear equations can be solved in terms of the Green's function $U(l, z)$, which is the solution of

(4.52)
$$U(l, z) - z \sum_{l'} p(l - l') U(l', z) = \delta_{l, 0}.$$

The exact formula for $U(l, z)$ has been given for nearest-neighbor walks on various cubic lattices in Sec. 4.2 and for general lattice walks in Sec. 4.3. The inhomogeneous equation

$$(4.53) \qquad G(l, z) - z \sum_{l'} p(l - l')G(l', z) = F(l)$$

has the solution

$$(4.54) \qquad G(l, z) = \sum_{l'} U(l - l', z)F(l'),$$

as can be verified by substituting this expression into the left-hand side of (4.53) and employing (4.52) to obtain the right-hand side of (4.53). Now if we let

$$F(l) = \delta_{l, l_0} + z \sum_{l'} q(l, l')G(l', z)$$

in (4.53), then (4.54) becomes

$$G(l, z) = \sum_{l'} U(l - l', z)\left[\delta_{l', l_0} + z \sum_{l''} q(l', l'')G(l'', z) \right],$$

or

$$(4.55) \qquad G(l, z) = U(l - l_0, z) + z \sum_{l', l''} U(l - l', z)q(l', l'')G(l'', z).$$

Generally, this equation can be solved by iteration. Under the special condition, however, that $q(l, l')$ vanish except for a few points, say

$$l' = l'_1, l'_2, \ldots, l'_{n'} \qquad \text{and} \qquad l = l_1, l_2, \ldots, l_n,$$

a closed-form solution can be derived. Let the various l's run through their set of possible values. Then for $j = 1, 2, \ldots, n'$, we obtain

$$G(l_j, z) = U(l_j - l_0, z) + z \sum_{k=1}^{n} \sum_{m=1}^{n'} U(l_j - l_k, z)q(l_k, l_m)G(l_m, z).$$

This set of n equations can be solved for the n unknowns $G(l_j, z)$. When they are substituted into (4.55), we find

$$(4.56) \quad G(l, z) = U(l - l_0, z) + z \sum_{k=1}^{n} \sum_{m=1}^{n'} U(l - l_k, z)q(l_k, l_m)G(l_j, z).$$

To demonstrate an application of this result, we derive a formula for the result of defects on the solution of the Pólya problem. Let a walker start from the origin. With $l_0 = 0$, the probability that he never returns was shown in Sec. 4.2 to be

$$(4.57) \qquad\qquad\qquad \frac{1}{G(0, 1)}.$$

In a perfect lattice, this is $1/U(0, 1)$; but in the defective lattice, the generating function $G(l, z)$ plays the same role as $U(l, z)$ does in the perfect lattice. Hence the probability of never returning to origin is

$$\left[U(0, 1) + \sum_{k=1}^{n} \sum_{m=1}^{n'} U(-l_k, 1)q(l_k, l_m)G(l_m, 1) \right]^{-1}.$$

We now apply this formula to a special example. Let the nonvanishing q's be

$$q(l_1, l_1) = \epsilon$$

and

$$q(l, l_1) = -\frac{\epsilon}{2S} \qquad \text{if } l \text{ is a nearest neighbor to } l_1$$

in an S-dimensional simple cubic lattice in which a walker jumps at each step to any nearest-neighbor point with equal probability. If $\epsilon = 1$, then the point l_1 is an absorbing trap for walkers. If $\epsilon < 1$, a walker at l_1 remains there with a probability ϵ at the moment that he might normally take a step.

Then from (4.56), with $l_0 = 0$, we have

$$G(l, z) = U(l, z) + z\epsilon G(l_1, z)\left[U(l - l_1, z) - \frac{1}{2S} \sum_{l_k} U(l - l_k, z) \right],$$

where $\{l_k\}$ represents the set of points that are nearest neighbors of l_1. From (4.52), however, as applied to this special case, we obtain

$$\frac{z}{2S} \sum_{l_k} U(l - l_k, z) = U(l - l_1, z) - \delta_{l-l_1,0},$$

so that

$$G(l, z) = U(l, z) + \epsilon[(z - 1)U(l - l_1, z) + \delta_{l-l_1,0}]G(l_1, z).$$

An expression can be found for $G(l_1, z)$ by setting $l = l_1$. Then we have

$$G(l_1, z) = \frac{U(l_1, z)}{(1 - \epsilon) + \epsilon(1 - z)U(0, z)},$$

so that, if $l \neq l_1$,

$$G(l, z) = U(l, z) + \frac{\epsilon(z - 1)U(l - l_1, z)U(l_1, z)}{(1 - \epsilon) + \epsilon(1 - z)U(0, z)}.$$

The central quantity in the Pólya problem of the probability of escape from a starting position is $G(0, 1)$, which has the form

$$G(0, 1) = \lim_{z \to 1} \left[U(0, z) + \frac{\epsilon(z - 1)\,|U(l_1, z)|^2}{(1 - \epsilon) + \epsilon(1 - z)U(0, z)} \right].$$

As would be expected, the dimensionality plays an important role in a discussion of the influence of the defect on the unperturbed solution $U(0, 1)$.

The function $|U(\mathbf{l}, z)|$ is bounded in a three-dimensional lattice, and

$$G(0, 1) = U(0, 1) \qquad\qquad \text{if } \epsilon < 1,$$

$$= U(0, 1) - \frac{|U(\mathbf{l}_1, 1)|^2}{U(0, 1)} \qquad \text{if } \epsilon = 1.$$

Hence, if $\epsilon < 1$, then the perturbation has no effect on the probability of return to the origin. On the other hand, if the point $\mathbf{l}_1$ is a trap from which a particle never escapes ($\epsilon = 1$), then there is a correction to $U(0, 1)$. Let $|\mathbf{l}_1|$ represent the distance of the trap from the origin. Then as $|\mathbf{l}_1| \to \infty$, we have

$$U(\mathbf{l}_1, 1) \sim \frac{3}{2\pi l_1} \; ;$$

this asymptotic formula as well as other properties and tables of $U(\mathbf{l}, z)$ are given in [18]. From Sec. 4.1, however, we have

$$U(0, 1) = 1.51638 \ldots \equiv u.$$

Then as $|\mathbf{l}_1| \to \infty$, the escape probability from the origin becomes

$$\frac{1}{u}\left(1 + \frac{9}{4\pi^2 u^2 l_1^2} + \cdots\right),$$

so that the probability of never returning to the origin is augmented by an amount inversely proportional to the square of the distance of the trapping defect from the origin.

In a one-dimensional lattice, we have

$$U(l, z) = \frac{1}{\pi} \int_0^\pi \frac{\cos lx \, dx}{1 - z \cos x} = \frac{(-1 + \sqrt{1 - z^2})^l}{z^l \sqrt{1 - z^2}}.$$

Hence we obtain

$$G(0, z) = (1 - z^2)^{-\frac{1}{2}} - \frac{\epsilon(1 - z)[1 - \sqrt{1 - z^2}]^{2l}(1 - z^2)^{-\frac{1}{2}}}{z^{2l}[(1 - \epsilon)\sqrt{1 - z^2} + \epsilon(1 - z)]}.$$

As $z \to 1$, we get

$$G(0, z) \sim (1 - z^2)^{-\frac{1}{2}} - \frac{\epsilon}{2(1 - \epsilon)} \qquad \text{if } \epsilon < 1,$$

$$\sim 2l_1 \qquad\qquad\qquad\qquad \text{if } \epsilon = 1.$$

Then from (4.57), the probability of a walker never returning to the origin is $1/(2l_1)$ for $\epsilon = 1$.

Certain functions that appear in lattice statistics can be expressed as additive functions of the eigenvalues of appropriate matrices. Let $\{\lambda_j\}$ be the set of eigenvalues of a matrix M. Then such an additive function F has the form

$$F = \sum_j f(\lambda_j).$$

An example is the characteristic function of the distribution of the eigenvalues of the matrix

$$N \langle \exp i\alpha\lambda \rangle_{\mathrm{av}} = \sum_j e^{i\alpha\lambda_j}.$$

The Fourier transform of this average (α being the Fourier-transform variable) is just the distribution function of the λ's. The determinant of the matrix M is

$$\det M = \prod_j \lambda_j,$$

so that

$$\log \det M = \sum \log \lambda_j$$

is also an additive function of the λ's. The generating function for dimer configurations is such a function [see (4.34)], as is the Ising partition function.

The detailed calculations of the influence of defects on the number of dimer configurations and on properties of the Ising model are rather lengthy and will not be discussed here. Instead, a formalism for the investigation of sums of additive functions will be outlined. It has been used in the theory of the result of defects on lattice vibrations as well as in the meson-pair theory of nucleon interactions.

Let the λ_j be roots of a characteristic equation,

$$g(\lambda) = 0.$$

If

$$g(\lambda) = \prod_j (\lambda - \lambda_j),$$

then

$$\frac{d \log g}{d\lambda} = \sum \frac{1}{\lambda - \lambda_j},$$

and

$$\frac{1}{2\pi i} \int_C f(z) \frac{d \log g(z)}{dz} \, dz = \frac{1}{2\pi i} \int_C f(z) \sum \frac{1}{z - \lambda_j} \, dz$$
$$= \sum f(\lambda_j) = F.$$

The contour C is to be taken in a clockwise direction around a domain that includes all λ_j. It has been assumed that $f(z)$ has no poles in this domain.

Let $\det M_0(\lambda) = g(\lambda)$ for a perfect lattice. Suppose that the corresponding matrix for a defective lattice is

$$M = M_0 + \Delta.$$

Then we have

$$M = M_0(I + M_0^{-1}\Delta),$$

$$\det M = (\det M_0)[\det(I + M_0^{-1}\Delta)],$$

$$\log \det M = \log \det M_0 + \log \det(I + M_0^{-1}\Delta).$$

If F_0 is the additive function of interest for a perfect lattice, namely

$$F_0 = \frac{1}{2\pi i} \int_C f(z) \frac{d \log \det M_0(z)}{dz} dz,$$

then the variation in F due to the existence of defects is

$$\Delta F = F - F_0 = \frac{1}{2\pi i} \int_C f(z) \frac{d \log \det(I + M_0^{-1}\Delta)}{dz} dz.$$

The elements of M_0^{-1} are essentially the Green's functions discussed at the beginning of this section and are assumed to be known. Then, although $\det M_0$ is a determinant with a large number of rows and columns, $\det(I + M_0^{-1}\Delta)$ can be reduced to one of a small number of rows and columns if only a small number of defects exists, that is, if Δ has only a small number of nonvanishing elements.

The simplest example is

$$\Delta_{kk} = \epsilon, \quad \text{and} \quad \text{all other} \quad \Delta_{ij} = 0.$$

Then we have

$$I + M_0^{-1}\Delta =
\begin{bmatrix}
1 & 0 & 0 & \cdot & \cdot & \cdot \\
0 & 1 & 0 & \cdot & \cdot & \cdot \\
0 & 0 & 1 & \cdot & \cdot & \cdot \\
\cdot & \cdot & \cdot & \cdot & \cdot & \cdot \\
\cdot & \cdot & \cdot & \cdot & \cdot & \cdot
\end{bmatrix}$$

$$+
\begin{bmatrix}
M_{11}^{-1} & M_{12}^{-1} & \cdot & \cdot & \cdot \\
M_{21}^{-1} & M_{22}^{-1} & \cdot & \cdot & \cdot \\
\cdot & \cdot & \cdot & \cdot & \cdot
\end{bmatrix}
\begin{bmatrix}
0 & 0 & \cdot & \cdot & \cdot & \cdot \\
0 & 0 & \cdot & \cdot & \cdot & \cdot \\
\cdot & \cdot & \cdot & \cdot & \cdot & \cdot \\
\cdot & 0 & \Delta_{kk} & 0 & \cdot & \cdot \\
\cdot & \cdot & \cdot & \cdot & \cdot & \cdot
\end{bmatrix}$$

$$=
\begin{bmatrix}
1 & 0 & 0 & \cdot & \cdot & 0 & M_{1k}^{-1}\Delta_{kk} & 0 & \cdot & \cdot & \cdot & \cdot & \cdot & \cdot \\
0 & 1 & 0 & \cdot & \cdot & 0 & M_{2k}^{-1}\Delta_{kk} & 0 & \cdot & \cdot & \cdot & \cdot & \cdot & \cdot \\
0 & 0 & 1 & \cdot & \cdot & 0 & M_{3k}^{-1}\Delta_{kk} & 0 & \cdot & \cdot & \cdot & \cdot & \cdot & \cdot \\
\cdot & \cdot & \cdot & \cdot & \cdot & \cdot & \cdot & \cdot & \cdot & \cdot & \cdot & \cdot & \cdot & \cdot \\
\cdot & \cdot & \cdot & \cdot & \cdot & \cdot & \cdot & \cdot & \cdot & \cdot & \cdot & \cdot & 1 & 0 \\
\cdot & \cdot & \cdot & \cdot & \cdot & \cdot & \cdot & \cdot & \cdot & \cdot & \cdot & \cdot & 0 & 1
\end{bmatrix}$$

The value of det $(I + M_0^{-1}\Delta)$ is found by expanding by columns to obtain

$$\det (I + M_0^{-1}\Delta) = I + M_{kk}^{-1}\Delta_{kk} = 1 + \epsilon M_{kk}^{-1},$$

so that

$$\Delta F = \frac{1}{2\pi i} \int_C f(z) \frac{d \log [1 + \epsilon M_{kk}^{-1}(z)]}{dz} dz.$$

A variety of applications of formulas such as these has been made to the theory of lattice vibrations. The functions M_{kk}^{-1} for simple cubic lattices have recently been tabulated.

MULTIPLE-CHOICE REVIEW PROBLEMS

1. A dimer is

    (a) a method for covering a doubly periodic lattice,
    (b) a diatomic molecule,
    (c) a ten-cent article,
    (d) none of the above.

2. A Pfaffian is a particular kind of

    (a) matrix,
    (b) determinant,
    (c) quadratic form,
    (d) none of the above.

3. A certain Hermitian matrix (a_{ij}) of order 5 has eigenvalues $-3, 0, 3, 5, 8$. If the value a_{33} is increased by 7, then the resulting matrix necessarily has, in the interval $(4, 9)$,

    (a) at least one eigenvalue,
    (b) exactly two eigenvalues,
    (c) at most four eigenvalues,
    (d) none of the above.

4. The value of the Pfaffian

$$\begin{vmatrix} 3 & 4 & 2 \\ & 1 & 5 \\ & & 7 \end{vmatrix}$$

is

    (a) 3,
    (b) $4(3 - \lambda)^2 - 2(1 - \lambda)^2 + 5(7 - \lambda)^2$,
    (c) the same as the value of the corresponding skew symmetric determinant,
    (d) none of the above.

5. A superposition polygon

    (a) necessarily is clockwise odd,
    (b) necessarily is clockwise even,
    (c) might be either clockwise odd or clockwise even,
    (d) cannot logically have clockwise parity assigned to it.

REFERENCES

1. Caianiello, E. R., "Theory of Coupled Quantized Fields," *Nuovo Cimento* (10) **14** (1959), *Supplemento*, 177–191.
2. Domb, C., "On Multiple Returns in the Random-Walk Problem," *Proc. Cambridge Philos. Soc.* **50** (1954), 586–591.
3. Domb, C., "On the Theory of Cooperative Phenomena in Crystals," *Advances in Phys.* **9** (1960), 149–244.
4. Dyhne, A. M., and Ju. B. Rumer, "Thermodynamics of a Plane Ising-Onsager Dipole Lattice," *Soviet Physics Uspekhi* **4** (1962), 698–705.
5. Feller, W., *An Introduction to Probability Theory and Its Applications*, John Wiley and Sons, New York, 1951.
6. Fisher, M. E., "Statistical Mechanics of Dimers on a Plane Lattice," *Phys. Rev.* **124** (1961), 1664–1672.
7. Foster, F. G., and I. J. Good, "On a Generalization of Pólya's Random-Walk Theorem," *Quart. J. Math.* (2) **4** (1953), 120–126.
8. Fowler, R. H., and G. S. Rushbrooke, "Statistical Theory of Perfect Solutions," *Trans. Faraday Soc.* **33** (1937), 1272–1294.
9. Hurst, C. A., and H. S. Green, "New Solution of the Ising Problem for a Rectangular Lattice," *J. Chem. Phys.* **33** (1960), 1059–1062.
10. Kac, M., and J. C. Ward, "A Combinatorial Solution of the Two-dimensional Ising Model," *Phys. Rev.* **88** (1952), 1332–1337.
11. Kasteleyn, P. W., "The Statistics of Dimers on a Lattice," *Physica* **27** (1961), 1209–1225.
12. Kasteleyn, P. W., "Dimer Statistics and Place Transitions," *J. Math. Phys.* (1963), 287–293.
13. Kaufman, B., and L. Onsager, "Crystal Statistics. II. Partition Function Evaluated by Spinor Analysis," *Phys. Rev.* **76** (1949), 1232–1252.
14. Ledermann, W., "Asymptotic Formulae Relating to the Physical Theory of Crystals," *Proc. Roy. Soc. London, Series A* **182** (1943–1944), 362–377.
15. Lifshitz, I. M., "Some Problems of the Dynamic Theory of Non-ideal Crystal Lattices," *Nuovo Cimento* (10) **3** (1956), *Supplemento*, 716–734.
16. McCrea, W. H., and T. F. W. Whipple, "Random Paths in Two and Three Dimensions," *Proc. Roy. Soc. Edinburgh* **60** (1940), 281–298.
17. Maradudin, A. A., P. Mazur, E. W. Montroll, and G. H. Weiss, "Remarks on the Vibrations of Diatomic Lattices," *Rev. Mod. Phys.* **30** (1958), 175–196.
18. Maradudin, A. A., E. W. Montroll, G. H. Weiss, R. Herman, and H. W. Milnes, "Green's Functions for Monatomic Simple Cubic Lattices," *Acad. Roy. Belg. Cl. Sci. Mem. Coll. in-4°* (2), **14** (1960), No. 7, 1–176.
19. Montroll, E. W., "Random Walks in Multidimensional Spaces, Especially on Periodic Lattices," *J. Soc. Indust. Appl. Math.* **4** (1956), 241–260.
20. Montroll, E. W., "Topics on Statistical Mechanics of Interacting Particles," in Cécile de Witt (editor), *La théorie des gaz neutres et ionisés: Le problème des n corps à température non nulle*, John Wiley and Sons, New York, 1960.

21. Montroll, E. W., and R. B. Potts, "Effect of Defects on Lattice Vibrations," *Phys. Rev.* **100** (1955), 525–543.

22. Montroll, E. W., and R. B. Potts, "Effect of Defects on Lattice Vibrations: Interaction of Defects and an Analogy with Meson Pair Theory," *Phys. Rev.* **102** (1956), 72–84.

23. Newell, G. F., and E. W. Montroll, "On the Theory of the Ising Model of Ferromagnetism," *Rev. Mod. Phys.* **25** (1953), 353–389.

24. Onsager, L., "Crystal Statistics. I. A Two-Dimensional Model with an Order-Disorder Transition," *Phys. Rev.* **65** (1944), 117–149.

25. Peype, W. F. van, "Zur Theorie der magnetischen Anistropie kubisher Kristalle beim absoluten Nullpunkt," *Physica* **5** (1938), 465–483.

26. Pólya, G., "Über eine Aufgabe der Wahrscheinlichkeitsrechnung betreffend die Irrfahrt im Strassennetz," *Math. Ann.* **84** (1921), 149–160.

27. Potts, R. B., and J. C. Ward, "The Combinatorial Method and the Two-Dimensional Ising Model," *Progr. Theoret. Phys.* **13** (1955), 38–46.

28. Sherman, S., "Combinatorial Aspects of the Ising Model for Ferromagnetism I. A Conjecture of Feynman on Paths and Graphs," *J. Math. Phys.* **1** (1960) 202–217.

29. Temperley, H. N. V., and M. E. Fisher, "Dimer Problem in Statistical Mechanics —An Exact Result," *Phil. Mag.* **6** (1961), 1061–1063.

30. Watson, G. N., "Three Triple Integrals," *Quart. J. Math. Oxford Ser.* **10** (1939) 266–276.

31. Wentzel, G., "Pair Theory of Nuclear Forces," *Helvetic Phys. Acta* **15** (1942) 111–126.

CHAPTER 5

Pólya's Theory of Counting

N. G. DE BRUIJN

5.1 INTRODUCTION

A large part of combinatorial analysis is concerned with counting: counting numbers of possibilities, numbers of solutions, numbers of sets of a given type, etc.

Frequently the technical difficulty of finding general formulas for the required numbers can be overcome by the method of generating functions, which is a tool, in the sense that it does the work for us after a small amount of organization on our part, as indicated in Chapters 3 and 4.

Many of the difficulties in combinatorial analysis, however, are of a conceptual rather than of a technical nature. This happens when counting becomes confused because of different objects that have to be considered as being equal. In modern language this means that we have an equivalence relation in a set of objects, and that we are no longer counting the number of objects themselves but the number of equivalence classes. For example, if we say that there are exactly five regular solids, we tacitly use equivalence induced by geometric similarity. Sometimes the equivalence relation is given by a group of permutations of a finite set of objects, and then we have a simple result (Lemma 5.1, p. 150): The number of classes is equal to the average, taken over the group, of the number of elements that are left invariant by a group element.

A third type of difficulty in counting is that we are not always counting all objects with the same weight. For instance, if we say that every algebraic equation of degree n has exactly n roots, we have provided each root with a weight equal to its multiplicity.

These three aspects of counting—generating functions, equivalences introduced by groups, and weights—meet each other in a very elegant theorem due to Pólya [10]. Since it has this central position in counting, we have called it the *fundamental theorem* in enumerative combinatorial analysis. Sections 5.2 to 5.8 are devoted to this theorem, and most of the remaining sections deal with results obtained in the same spirit as Pólya's and lead to formulas of similar structure.

A large and important part of Pólya's paper [10] dealt with the counting of trees. These questions will not be discussed here, for the possible notions of equivalence of trees are all quite complicated; these questions are reserved for Chapter 6. In this chapter, we shall aim at a complete description and an elementary presentation of combinatorial concepts, replacing primitive intuitive notions by concrete things such as sets and mappings. In questions about equivalence classes of trees, however, it seems hard to carry out such a program in a first introduction; what is worse, it might detract from the elegance of a subject in which intuition plays such a great part. We refer the reader to Chapter 6 and to some of the references cited in Sec. 5.14.

Little knowledge of group theory is expected of the reader, and on the whole we have tried to present the theory as simply as possible, fortified by numerous examples.

We follow the usual modern conventions concerning notation of mappings. If f is a mapping of a set S into a set T, then for each $s \in S$ we denote the image of s under this mapping by $f(s)$. Thus $f(s)$ is an element of T, and the symbol $f(s)$ will never be used for the mapping itself. If f is a mapping of S into T, and if g is a mapping of T into U, then gf is the composite mapping defined by the property that, for each $s \in S$, it maps s onto $g[f(s)]$. If, moreover, h is a mapping of U into a set V, then we have $h(gf) = (hg)f$. Accordingly, we can omit parentheses completely, if we so desire, and we can even write $hgfs$ for the image of the element $s \in S$ under the mapping hgf.

The set of *all* mappings of S into T is denoted by T^s.

If S is a finite set, then $|S|$ denotes the number of its elements. If G is a group, then $|G|$ is called the *order* of the group. If we have a group of permutations of a set S, then $|S|$ is called the *degree* of the permutation group.

5.2 CYCLE INDEX OF A PERMUTATION GROUP

Let S be a finite set. A *permutation* of S is a one-to-one mapping of S onto itself. If π is a permutation, and s is any element of S, then πs denotes the element onto which s is mapped by π. If π_1 and π_2 are permutations, then the product $\pi_1\pi_2$ is defined as the composite mapping obtained by applying first π_2, then π_1; thus for all $s \in S$ we have

$$(\pi_1\pi_2)s = \pi_1(\pi_2 s).$$

It is well known that if π is given, then we can split S in a unique way into *cycles*, that is, subsets of S that are cyclically permuted by π. If l is the length of such a cycle, and if s is any element of that cycle, then the cycle consists of

$$s, \pi s, \pi^2 s, \ldots, \pi^{l-1}s,$$

where $\pi^2 = \pi\pi$, etc.

If S splits into b_1 cycles of length 1, b_2 cycles of length 2, etc., then we say that π is of *type* $\{b_1, b_2, b_3, \ldots\}$. Obviously, $b_i = 0$ for all but at most a finite number of i's; certainly, $b_i = 0$ for $i > m$, where m stands for the number of elements of S. Furthermore, we clearly have

$$b_1 + 2b_2 + 3b_3 + \cdots = m,$$

the sum of the lengths of the cycles being the total number of elements in S.

We now define the cycle index of a permutation group. Letting G be a group whose elements are the permutations of S, the group operation being the multiplication introduced previously, we define a special polynomial in m variables $x_1, \ldots, x_m$, with nonnegative coefficients, as follows. For each $g \in G$ we form the product $x_1^{b_1} x_2^{b_2}, \ldots, x_m^{b_m}$, if $\{b_1, b_2, b_3, \ldots\}$ is the type of g. Taking the sum of these terms and dividing by the number of elements of G, we get the polynomial

$$P_G(x_1, x_2, \ldots, x_m) = |G|^{-1} \sum_{g \in G} x_1^{b_1} x_2^{b_2} \cdots x_m^{b_m},$$

which we call the *cycle index* of G.

Example 5.1 If G consists of the identity permutation only, then we have $P_G = x_1{}^m$, since this identity permutation is of type $\{m, 0, 0, \ldots\}$. Since here P_G depends on m, this simple example illustrates the fact that the cycle index of G depends not only on the structure of G as an abstract group but also on the interpretation of its elements as permutations of a set S.

Example 5.2 Let S be the set of *vertices* of a cube, so that $m = 8$, and let G be the set of all those permutations of S that can be produced by rotations of the cube. There are $6 \times 4 = 24$ such rotations. These can be divided into the following five categories.

(a) The identity.

(b) Three $180°$ rotations around lines connecting the centers of opposite faces.

(c) Six $90°$ rotations around lines connecting the centers of opposite faces.

(d) Six $180°$ rotations around lines connecting the midpoints of opposite edges.

(e) Eight $120°$ rotations around lines connecting opposite vertices.

Since $1 + 3 + 6 + 6 + 8 = 24$, this list is exhaustive.

It is easy to visualize the cycles of S in each case. In (a) there are eight cycles of length 1; a permutation of category (b) produces four cycles of length 2; in (c) there are two cycles of length 4; (d) contains four cycles of length 2; and in (e) there are two cycles of length 1 and two of length 3. Therefore, the cycle index is

$$P_G = \tfrac{1}{24}(x_1^8 + 9x_2^4 + 6x_4^2 + 8x_1^2x_3^2).$$

Example 5.3 Let S be the set of *edges* of a cube, so that $m = 12$, and let G be the set of all 24 permutations of S that can be produced by rotations of the cube.

The rotations are the same as in Example 5.2. It is now necessary to see what the rotations do with the edges. The identity produces a permutation of type $\{12, 0, 0, \ldots\}$. A rotation of category (b) produces the type $\{0, 6, 0, 0, \ldots\}$, (c) gives $\{0, 0, 0, 3, 0, \ldots\}$. (d) gives $\{2, 5, 0, 0, \ldots\}$, and (e) gives $\{0, 0, 4, 0, \ldots\}$. Therefore we have

$$P_G = \tfrac{1}{24}(x_1^{12} + 3x_2^6 + 6x_4^3 + 6x_1^2x_2^5 + 8x_3^4).$$

Example 5.4 Again we take rotations of the cube, but now S is the set of all *faces*. Our five categories now produce the types $\{6, 0, 0, \ldots\}$, $\{2, 2, 0, 0, \ldots\}$, $\{2, 0, 0, 1, 0, \ldots\}$, $\{0, 3, 0, 0, \ldots\}$, $\{0, 0, 2, 0, \ldots\}$, respectively, and therefore

$$(5.1) \qquad P_G = \tfrac{1}{24}(x_1^6 + 3x_1^2x_2^2 + 6x_1^2x_4 + 6x_2^3 + 8x_3^2).$$

Example 5.5 Let S be a set of m elements, and let G be the group of *all* permutations of S. That is, G is the so-called *symmetric group* of degree m. Its cycle index turns out to be equal to the coefficient of z^m in the development of

$$(5.2) \qquad \exp\left(zx_1 + \frac{z^2x_2}{2} + \frac{z^3x_3}{3} + \cdots\right)$$

as a power series in z.

We shall show in Example 5.17 that this is a consequence of Pólya's theorem. It is also possible to give a direct proof, a sketch of which follows. The expression (5.2) can be written as

$$\sum_{b_1=0}^{\infty} \frac{z^{b_1}x_1^{b_1}}{b_1!} \sum_{b_2=0}^{\infty} \frac{z^{2b_2}x_2^{b_2}}{b_2!\,2^{b_2}} \cdots.$$

The coefficient of z^m is obtained by summing the expression

$$(5.3) \qquad x_1^{b_1} x_2^{b_2} \cdots (b_1! b_2! 2^{b_2} b_3! 3^{b_3} \cdots)^{-1}$$

over all possible $b_1, b_2, \ldots$, satisfying $b_1 + 2b_2 + 3b_3 + \cdots = m$. Apart from a factor $m!$, the coefficient in (5.3) is equal to the number of permutations of type $\{b_1, b_2, \ldots\}$. This proves our statement.

Example 5.6   Let S be any finite group, and again let $|S| = m$. If a is a fixed element of S, then the mapping $s \to as$ is easily seen to be a permutation of S. Denoting this permutation by g_a, we observe that if a runs through S, then the g_a's form a permutation group G. We have $g_a g_b = g_{ab}$, and thus this G is isomorphic to S itself, the isomorphism being defined by $g_a \leftrightarrow a$. The group G is called the *Cayley representation* of the group S. We are interested in its cycle index.

If $a \in S$, then the *order* of a is the least positive integer k such that $a^k = e$, where e stands for the unit element of S. Denote this order by $k(a)$. The permutation g_a splits S into cycles that are all of length $k(a)$: If s is any element of S, then it belongs to the cycle obtained by

$$s \to as \to a^2 s \to \cdots \to a^{k(a)} s = s.$$

It follows that $k(a)$ divides m, and that there are $m/k(a)$ cycles of length $k(a)$. Thus we get the cycle index

$$(5.4) \qquad P_G = \frac{1}{m} \sum_{a \in S} [x_{k(a)}]^{m/k(a)}.$$

This sum can also be written as

$$(5.5) \qquad P_G = \frac{1}{m} \sum_{d \mid m} \nu(d) (x_d)^{m/d},$$

where d runs through the divisors of m, and $\nu(d)$ represents the number of elements $a \in S$ with order $k(a) = d$.

Example 5.7   As a special case of Example 5.6, we take the following cyclic permutation group. Let S be the group of all mth roots of unity, $e^{2\pi i j/m}$, where $j = 1, \ldots, m$, and i is the imaginary unit. Then for each $a \in S$, the mapping $s \to as$ is a permutation of S, and the group of these permutations is a cyclic group.

If $a = e^{2\pi i j/m}$, then the order of a is $k(a) = m/(m, j)$, where (m, j) stands for the greatest common divisor of m and j. Thus by (5.4) the cycle index is

$$P_G = \frac{1}{m} \sum_{j=1}^{m} (x_{m/(m,j)})^{(m,j)}.$$

A second expression is obtained from (5.5):

$$P_G = \frac{1}{m} \sum_{d \mid m} \phi(d)(x_d)^{m/d},$$

where ϕ is Euler's function; that is, $\phi(d)$ is the number of integers n with $1 \le n \le d$, $(n, d) = 1$.

Example 5.8 Let G be a group of permutations of a set S, and let H be a group of permutations of a set T. Assume that S and T have no elements in common, and let U be their union. To each choice of $g \in G$ and $h \in H$ there corresponds a permutation of U, defined by

$$u \to gu \quad \text{if} \quad u \in S, \qquad u \to hu \quad \text{if} \quad u \in T.$$

Denote this permutation of U by $g \times h$. It is easy to see that these permutations form a group whose order is the product of the orders of G and H. It is called the *direct product* of G and H, and is denoted by $G \times H$.

If $g \in G$ and $h \in H$, and if g is of type $\{b_1, b_2, \ldots\}$ and h is of type $\{c_1, c_2, \ldots\}$, then $g \times h$ is of type $\{b_1 + c_1, b_2 + c_2, \ldots\}$, since each cycle in U lies either entirely in S or entirely in T. Hence the term of the cycle index of $G \times H$ corresponding to the element $g \times h$ is equal to the product of the term in P_G corresponding to g and the term in P_H corresponding to h. Applying this to all terms of P_G and all terms of P_H, we obtain the product formula

$$P_{G \times H} = P_G \cdot P_H.$$

Remark

The type of a permutation g tells something about the permutation and about its powers $g^2, g^3, \ldots$, but we can say very little about the type of a product $g_1 g_2$ from the types of the factors. Accordingly, although the cycle index may give information about combinatorial questions concerning the permutation group, it does not reveal a great deal about the multiplicative structure of the group. Actually, Pólya ([10], p. 176) gives an example of two nonisomorphic permutation groups with identical cycle indexes. Therefore the cycle index does not always determine the group uniquely. Pólya takes two nonisomorphic groups of order p^3, where p is a prime > 2, which both have the property that each element has order p, apart from the unit elements. The effect is that the expression (5.5) for the cycle index of the Cayley representation gives the same result in the two cases.

5.3 THE MAIN LEMMA

An essential part of Pólya's theory consists of a simple lemma, presumably published for the first time by Burnside ([3], Sec. 145, Theorem VII). For convenience we give here a more general lemma; it is, however, a trivial consequence of the original one. It does not deal with a group of permutations, but with a group having elements that *act* as permutations.

Let G be a finite group. Assume that the elements of G act as permutations on S; that is, there is a mapping of G into the symmetric group of S. In other words, to each $g \in G$ we have attached a permutation of S. We denote that permutation by π_g. Assume that the mapping is homomorphic, that is,

$$(5.6) \qquad\qquad \pi_{gg'} = \pi_g \pi_{g'}$$

for all $g \in G$, $g' \in G$. Notice that different elements of G need not correspond to different permutations.

In this situation, G introduces an equivalence relation on the elements of S. Two elements s_1, s_2 of S are called *equivalent*, written $s_1 \sim s_2$, if there exists a group element g with $\pi_g s_1 = s_2$. It is easy to prove these results:

1. $s \sim s$ for all $s \in S$ [since (5.6) shows that if e is the unit element of G, then π_e is the identity permutation, that is, the one that leaves S pointwise fixed].

2. If $s_1 \sim s_2$, then $s_2 \sim s_1$ [since (5.6) shows that if $g' = g^{-1}$, then we have $\pi_{g'} = (\pi_g)^{-1}$].

3. If $s_1 \sim s_2$ and $s_2 \sim s_3$, then $s_1 \sim s_3$ [for if $\pi_g s_1 = s_2$, $\pi_{g'} s_2 = s_3$, then we have $\pi_{(g'g)} s_1 = \pi_{g'}(\pi_g s_1) = \pi_{g'} s_2 = s_3$].

It is well known that (1), (2), and (3) together imply that $\sim$ is an *equivalence relation*, by virtue of which S can be split into a number of *equivalence classes*. Two elements are equivalent if and only if they belong to the same class.

In this special situation the equivalence classes are called *transitive sets*. We can now formulate the lemma.

Lemma 5.1. The number of transitive sets equals

$$\frac{1}{|G|} \sum_{g \in G} \psi(g),$$

where $|G|$ denotes the number of elements of G, and, for each g, $\psi(g)$ denotes the number of elements of S that are invariant under π_g, that is, the number of $s \in S$ for which $\pi_g s = s$.

PROOF. We consider all pairs (g, s) with $g \in G$, $s \in S$, $\pi_g s = s$. The number n of these pairs can be counted in two ways. First, for each fixed g, we can count the number of s satisfying $\pi_g s = s$, and therefore the number of pairs is

$$n = \sum_{g \in G} \psi(g).$$

On the other hand, for each $s \in S$, we can count the number of g with $\pi_g s = s$. Denoting this number by $\eta(s)$, we infer that

(5.7) $$\sum_{s \in S} \eta(s) = \sum_{g \in G} \psi(g).$$

For fixed s, the elements of G with the property $\pi_g s = s$ form a subgroup of G, which we shall denote by G_s. Its order $|G_s|$ equals $\eta(s)$.

If s_1 is equivalent to s, then the number of g with $\pi_g s = s_1$ is equal to $|G_s|$. This follows from the fact that there is an element $h \in G$ with $\pi_h s_1 = s$, and now $\pi_g s = s_1$ means the same thing as $hg \in G_s$; thus if s_1 and s are fixed, then the number of possibilities for g is exactly equal to the number of elements of G_s.

Accordingly, G can be divided into subsets, each having $|G_s|$ elements and each corresponding to just one element of the equivalence class to which s belongs. It follows that this equivalence class has $|G|/|G_s|$ elements. Hence we have

$$\eta(s) = \frac{|G|}{\text{number of elements in equivalence class of } s}.$$

Summing with respect to s, we find that the sum of the $\eta(s)$, for all s belonging to one and the same equivalence class, equals $|G|$. Therefore the sum of *all* $\eta(s)$ equals $|G|$ times the number of equivalence classes, and now (5.7) proves the lemma.

5.4 FUNCTIONS AND PATTERNS

Let D and R be finite sets. We consider functions defined on D, with values in R; in other words, we consider mappings of D into R. The set D is called the *domain*, and R is called the *range*. The set of all such functions is denoted by R^D. The number of elements of R^D is $|R|^{|D|}$, since if we want to construct a function f we have, for each element $d \in D$, $|R|$ possible choices for $f(d)$, and these choices are independent.

Furthermore, we suppose that we are given a group G of permutations of D. This group introduces an equivalence relation in R^D: Two functions f_1, f_2 (both in R^D) are called *equivalent* (denoted $f_1 \sim f_2$) if there exists an element $g \in G$ such that

(5.8) $$f_1(gd) = f_2(d) \qquad \text{for all } d \in D.$$

Equation (5.8) can be abbreviated, of course, to $f_1 g = f_2$, since $f_1 g$ is the notation for the composite mapping "first g, then f_1." It is easy to check the usual equivalence conditions: (1) $f \sim f$; (2) if $f_1 \sim f_2$, then $f_2 \sim f_1$; (3) if $f_1 \sim f_2$ and $f_2 \sim f_3$, then $f_1 \sim f_3$. The first condition follows from the fact that the identity permutation belongs to G, the second from the fact that if g belongs to G, then also the inverse permutation g^{-1} belongs to G, and the third from the fact that if $g_1 \in G$, $g_2 \in G$, then the composite mapping $g_1 g_2$ belongs to G.

Thus $\sim$ is an equivalence relation, by virtue of which the set R^D splits into equivalence classes. These equivalence classes will be called *patterns*.

Example 5.9   Let D consist of the set of all six faces of a cube, and let G be the group of all permutations of D that can be produced by rotations of the cube (see Example 5.4). Let R consist of the two words *red* and *blue*. An element $f \in R^D$ can be considered as a way of painting the cube so that each face is painted either red or blue. This can be done in 2^6 ways. If two such cubes, placed in parallel positions, are colored in different ways, it may happen that one of them can be rotated in such a fashion that we can no longer see their differences. In that case they belong to the same pattern.

There are ten patterns, which can be described as follows (in parentheses we give the number of functions in each pattern): (*a*) all faces red (1); (*b*) five faces red, one face blue (6); (*c*) two opposite faces blue, the other four faces red (3); (*d*) two adjacent faces blue, the other four faces red (12); (*e*) three faces, meeting at one vertex, red, the three meeting at the opposite vertex blue (8); (*f*) two opposite faces, and one of the remaining faces, red, the three others blue (12); (*g*), (*h*), (*i*), (*j*) obtained from (*d*), (*c*), (*b*), (*a*) on interchanging the words red and blue. As a check, note that

$$1 + 6 + 3 + 12 + 8 + 12 + 12 + 3 + 6 + 1 = 2^6.$$

Example 5.10   Let D consist of the three elements $\{1, 2, 3\}$, let G be the symmetric group of D (that is, the group of all six permutations of D), and let R consist of two elements x and y. There are eight functions, but only four patterns. These can be denoted by the symbols x^3, $x^2 y$, xy^2, y^3, respectively. For example, $x^2 y$ represents the class of mappings f such that two of the values $f(1), f(2), f(3)$ are equal to x, and the remaining one is y. The pattern x^3 consists of one function only, the one defined by $f(1) = f(2) = f(3) = x$.

There may be some advantage in considering x and y as independent variables and in attaching to each f the product $f(1) f(2) f(3)$, which does not depend on the order of the factors. In other words, since the symmetric group is involved, saying that two functions f_1, f_2 are equivalent means the same thing as saying that the products $f_1(1)f_1(2)f_1(3)$ and $f_2(1)f_2(2)f_2(3)$ are identical. Therefore the patterns are characterized by the possible values of the product, namely, the monomials x^3, x^2y, xy^2, y^3.

5.5 WEIGHT OF A FUNCTION; WEIGHT OF A PATTERN

We again take finite sets D and R, and a permutation group G of D. To each element of R we assign a *weight*. This weight may be a number, or a variable, or, more generally, an element of a commutative ring containing the rational numbers. Thus we can form sums and products of weights, and rational multiples of weights, and these operations satisfy the usual associative, commutative, and distributive laws. The weight assigned to element $r \in R$ will be called $w(r)$.

Once these weights have been chosen, we can define the weight $W(f)$ of a function $f \in R^D$ as the product

$$(5.9) \qquad W(f) = \prod_{d \in D} w[f(d)].$$

If f_1 and f_2 are equivalent in the sense of Sec. 5.4, that is, if they belong to the same pattern, then they have the same weight. This follows from the fact that if $f_1 g = f_2$, $g \in G$ [see (5.8)], then we have

$$\prod_{d \in D} w[f_1(d)] = \prod_{d \in D} w[f_1(gd)] = \prod_{d \in D} w[f_2(d)],$$

since the first and the second product have the same factors, in a different order only, and since multiplication of weights is commutative.

Since all functions belonging to one and the same pattern have the same weight, we may define the weight of the pattern as this common value. Thus if F denotes a pattern, we shall denote the weight of F by $W(F)$; using the same symbol W both for weights of functions and for weights of patterns can hardly cause confusion.

Example 5.11 We take the cube-coloring case of Example 5.9, and form the ring of all polynomials in two variables x and y, with rational coefficients. The set R has the elements red and blue, to which we assign as weights the values x and y, respectively. The ten patterns (a), ..., (j) now get as weights

$$x^6,\ x^5y, x^4y^2,\ x^4y^2,\ x^3y^3,\ x^3y^3,\ x^2y^4,\ x^2y^4,\ xy^5,\ x^6,$$

respectively. From this it can be seen that different patterns need not have different weights.

Example 5.12 In Example 5.10, R had two elements, x and y. If we consider x and y as variables, there is no objection to assigning the element x the weight x and the element y the weight y. Now the symbols x^3, x^2y, xy^2, y^2 actually become pattern weights. In this case, the weight characterizes the pattern: Different patterns have different weights.

Example 5.13 If we take $w(r) = 1$ for all $r \in R$, then we have $W(f) = 1$ for all functions and $W(F) = 1$ for all patterns.

5.6 STORE AND INVENTORY

As before, we have sets D and R, and each $r \in R$ has a weight. Thinking of R as a set from which we have to choose function values, we call R the *store*. Since the weights can be added, a weight sum exists; this sum is called the *store enumerator*, or the *inventory* of R:

$$(5.10) \qquad \text{inventory of } R = \sum_{r \in R} w(r).$$

Example 5.14 The terminology suggests that the inventory gives a reasonably accurate description of what R contains, but this might be only partially true. Let R be a set consisting of three boxes of soap (called s_1, s_2, s_3), two packages of tea (called t_1, t_2), and four bottles of wine (called w_1, w_2, w_3, w_4). When we take nine variables s_1', s_2', s_3', t_1', t_2', w_1', w_2', w_3', w_4', and give s_1 the weight s_1', etc., the inventory becomes

$$s_1' + s_2' + s_3' + t_1' + t_2' + w_1' + w_2' + w_3' + w_4',$$

and the value of this sum gives complete information about the store. A storekeeper will usually apply a simpler system, for he is not much interested in fine distinctions between completely equivalent objects. He defines symbols s, t, w, standing for the abstract notions "box of soap," "package of tea," "bottle of wine"; he gives s_1, s_2, s_3 each the weight s; t_1, t_2 each the weight t; w_1, w_2 the weight w; and w_3, w_4 the weight $\frac{1}{2}w$ (because w_3 and w_4 are half bottles, a distinction thus far irrelevant). His inventory is $3s + 2t + 3w$. Sometimes, however, the storekeeper, or the tax inspector, has a keen interest in the value of the store in dollars. If he estimates a box of soap at 3, a package of tea at 1, a bottle of wine at 2, and a half bottle at 1, then his inventory becomes $9 + 2 + 4 + 2 = 17$. Now the inventory is just a number; it gives no information about what the store consists of, apart from the fact that its total value is $17. Finally, there is the possibility that he is instructing his clerk in counting; then, he gives a weight 1 to each object, and his inventory becomes just the number of objects, that is, 9.

5.7 INVENTORY OF A FUNCTION

We have finite sets D and R, and we want to consider the set R^D of all mappings of D into R. Each $r \in R$ has a weight $w(r)$; hence, each $f \in R^D$ has a weight

$$W(f) = \prod_{d \in D} w[f(d)].$$

Now the inventory of R^D is just a power of the inventory of R, the exponent being the number of elements of D:

$$(5.11) \qquad \text{inventory of } R^D = \sum_f W(f) = \left[\sum_{r \in R} w(r) \right]^{|D|}.$$

This can be seen as follows. The $|D|$th power can be written as the product of $|D|$ factors. If in each factor we select a single term, and if we take the product of these terms, then we get one term of the full expansion of the product—which consists of $|R|^{|D|}$ terms, according to the number of choices that can be made. We take some one-to-one correspondence between the factors of (5.11) and the elements of D; by virtue of that correspondence, we can say that the selection of a term from each factor can be described by a mapping f of D into R. Now to f there corresponds the term

$$\prod_{d \in D} w[f(d)]$$

of the full product expansion. Since this term is exactly $W(f)$, we notice that the full product equals the sum of all $W(f)$, which is the inventory of R^D.

We next evaluate the inventory of a certain subset S of R^D. Let D be dissected into a number of disjoint components $D_1, \ldots, D_k$, whence

$$|D| = |D_1| + \cdots + |D_k|.$$

We want to consider the set S of all functions f with the property that f is constant on each component; f may be, but need not be, different on different components. These f can be considered as composite functions $f = \phi\psi$, where ψ and ϕ are defined as follows: ψ is the function that maps d onto the index of the component to which d belongs, so that we always have $d \in D_{\psi(d)}$, and ϕ is a mapping of the set $\{1, \ldots, k\}$ into R. Notice that ψ is a fixed function, and that there are $|R|^k$ possibilities for ϕ.

We have the following result:

$$(5.12) \qquad \text{inventory of } S = \prod_{i=1}^{k} \sum_{r \in R} [w(r)]^{|D_i|}.$$

This can be seen again by investigating the full expansion of the product.

A term of that expansion is obtained by selecting a term in each factor of (5.12), and this means selecting a mapping ϕ of the set $\{1, \ldots, k\}$ into R. Thus this ϕ produces the term

$$\{w[\phi(1)]\}^{|D_1|} \cdots \{w[\phi(k)]\}^{|D_k|} = \prod_{i=1}^{k} \{w[\phi(i)]\}^{|D_i|}.$$

If $\phi\psi = f$, then this term is exactly $W(f)$, since obviously

$$\{w[\phi(i)]\}^{|D_i|} = \prod_{d \in D_i} w\{\phi[\psi(d)]\},$$

and

$$\prod_{i=1}^{k} \prod_{d \in D_i} w[f(d)] = W(f).$$

In this way, each $f \in S$ is obtained exactly once. Thus the sum of $W(f)$ for all $f \in S$ is equal to the sum of all terms of the expansion of the product in (5.12), and this proves (5.12).

Example 5.15   We want to distribute m counters over three persons P_1, P_2, and P_3, with the condition that P_1 obtains the same number as P_2. In how many ways is this possible? We are not interested in the individual counters, but only in the number of counters each person gets. That is to say, we want to have functions f defined on the set $D = \{P_1, P_2, P_3\}$, with range $R = \{0, 1, 2, \ldots\}$, and with the restrictions $f(P_1) = f(P_2)$ and $f(P_1) + f(P_2) + f(P_3) = m$. Put $\{P_1, P_2\} = D_1$, $\{P_3\} = D_2$.

We take a variable x, and we assign to the elements $0, 1, 2, \ldots$ of R the weights $1, x, x^2, x^3, \ldots$, respectively. Thus the functions we are interested in have weight x^m.

By (5.12), the inventory of all functions that are constant on each D_i equals

(5.13) $(1 + x^2 + x^4 + \cdots)(1 + x + x^2 + \cdots)$.

The number we were seeking is the coefficient of x^m in this expansion. Since

$$(1 - x^2)^{-1}(1 - x)^{-1} = \tfrac{1}{4}(1 + x)^{-1} + \tfrac{1}{2}(1 - x)^{-2} + \tfrac{1}{4}(1 - x)^{-1},$$

for the required number of functions we obtain

$$\tfrac{1}{2}(m + 1) + \tfrac{1}{4}[1 + (-1)^m],$$

that is, $\tfrac{1}{2}m + 1$ if m is even, and $\tfrac{1}{2}m + \tfrac{1}{2}$ if m is odd. It is very easy to verify this result directly. It might be remarked that the required number can also be interpreted as the number of partitions of m into parts 1 and 2.

The fact that the store is an infinite set, and that the inventory is the sum of an infinite series, need not greatly disturb us. We can just cut off the store, replacing it by $\{0, 1, \ldots, m\}$; the remaining elements cannot play any role in our problem because of the restriction $f(P_1) + f(P_2) + f(P_3) = m$. Moreover, the coefficient of x^m in (5.13) is identical to the coefficient of x^m in

$$(1 + x^2 + x^4 + \cdots + x^{2m})(1 + x + \cdots + x^m).$$

5.8 THE PATTERN INVENTORY; PÓLYA'S THEOREM

Again we take finite sets D and R, and we have a permutation group G of D. The elements of R have weights $w(r)$, and according to (5.9) the functions $f \in R^D$ and the patterns F have weights $W(f)$ and $W(F)$. Instead of the function inventory $\sum_f W(f)$ determined in (5.11), we now ask for the inventory $\sum_F W(F)$ of the set of all patterns. This inventory is given by the following result.

Theorem 5.1 (Pólya's Fundamental Theorem). The pattern inventory is

$$(5.14) \qquad \sum_F W(F) = P_G\left\{\sum_{r \in R} w(r), \sum_{r \in R} [w(r)]^2, \sum_{r \in R} [w(r)]^3, \ldots\right\},$$

where P_G is the cycle index. In particular, if all weights are chosen to be equal to unity, then we obtain

$$(5.15) \qquad \text{the number of patterns} = P_G(|R|, |R|, |R|, \ldots).$$

PROOF. Let ω be one of the possible values that the weight of a function may have. Let S be the set of all functions $f, f \in R^D$, satisfying $W(f) = \omega$.

If $g \in G$, and $f_1 = f_2 g$, then f_1 and f_2 have the same weight (Sec. 5.5). Hence if $f_1 \in S$, then $f_1 g^{-1} \in S$. Thus to each $g \in G$ there corresponds a mapping π_g of S into itself, defined by

$$\pi_g f = fg^{-1};$$

here π_g is a permutation, since it has an inverse, $\pi_{g^{-1}}$.

The mapping $g \to \pi_g$ satisfies the homomorphy condition (5.6). Thus if $g \in G$, $g' \in G$, then (5.6) holds since for each $f \in S$ we have

$$\pi_{gg'}f = f(gg')^{-1}, \qquad \pi_g(\pi_{g'}f) = \pi_g(fg'^{-1}) = fg'^{-1}g^{-1}, \qquad (gg')^{-1} = g'^{-1}g^{-1}.$$

Two elements f_1 and f_2 of S are equivalent in the sense of Sec. 5.3 if and only if they are equivalent in the sense of Sec. 5.4: The existence of a group element g with $\pi_g f_2 = f_1$ means the same thing as the existence of a group element g with $f_2 = f_1 g$. Accordingly, the patterns, as far as they are contained in S, are the equivalence classes of Sec. 5.3. Now Lemma 5.1

shows that the number of patterns, as far as they are contained in S, equals

(5.16)
$$\frac{1}{|G|} \sum_{g \in G} \psi_\omega(g),$$

where $\psi_\omega(g)$ denotes the number of functions f with $W(f) = \omega$, $fg^{-1} = f$ (or, what amounts to the same thing, $f = fg$).

The patterns contained in S all have weight ω; therefore, if we multiply (5.16) by ω and sum over all possible values of ω, we obtain the pattern inventory

$$\sum W(F) = \frac{1}{|G|} \sum_\omega \sum_{g \in G} \psi_\omega(g)\omega.$$

Obviously, we have

$$\sum_\omega \psi_\omega(g)\omega = \overset{(g)}{\underset{f}{\sum}} W(f),$$

where $\overset{(g)}{\underset{f}{\sum}}$ means that summation is taken over all $f \in R^D$ that satisfy $f = fg$. It follows that

(5.17)
$$\sum W(F) = \frac{1}{|G|} \sum_{g \in G} \overset{(g)}{\underset{f}{\sum}} W(f).$$

In order to evaluate $\overset{(g)}{\underset{f}{\sum}} W(f)$, we remark that g is a permutation of D, and therefore D splits into cycles that are cyclically permuted by g. The condition $f = fg$ means that

$$f(d) = f(gd) = f(g^2 d) = \cdots,$$

whence f is constant on each cycle of D. Conversely, each f that is constant on every cycle automatically satisfies $fg = f$, since $g(d)$ always belongs to the same cycle as d itself. Thus if the cycles are $D_1, D_2, \ldots, D_k$, then the sum $\overset{(g)}{\underset{f}{\sum}} W(f)$ is just the inventory evaluated in Sec. 5.7 and expressed by (5.12).

Let $\{b_1, b_2, \ldots\}$ be the type of g. This means that, among the numbers $|D_1|, \ldots, |D_k|$, the number 1 occurs b_1 times, the number 2 occurs b_2 times, etc. Consequently, we have

(5.18)
$$\overset{(g)}{\underset{f}{\sum}} W(f) = \left\{ \sum_{r \in R} w(r) \right\}^{b_1} \left\{ \sum_{r \in R} [w(r)]^2 \right\}^{b_2} \cdots .$$

The number of factors is finite, but we do not bother to invent a notation for the last one; after all, we may take the point of view that all b_i are zero from a certain i onward, that powers with exponent 0 are equal to 1, and that the product of infinitely many factors 1 is still equal to 1.

The expression (5.18) can be obtained by substitution of

$$x_1 = \sum_{r \in R} w(r), \qquad x_2 = \sum_{r \in R} [w(r)]^2, \qquad x_3 = \sum_{r \in R} [w(r)]^3, \ldots,$$

into the product $x_1^{b_1} x_2^{b_2} x_3^{b_3} \ldots$, which is the term corresponding to g in $|G| \, P_G$ (see Sec. 5.2). Summing with respect to g and dividing by $|G|$, we infer that the value of (5.17) is obtained by making the preceding substitution into $P_G(x_1, x_2, x_3, \ldots)$, and this proves Pólya's theorem.

Example 5.16   We consider Example 5.9. Here D is the set of faces of a cube, G the rotational substitution group, R the set of two colors, red and blue. According to (5.15), the number of color schemes is equal to $P_G(2, 2, 2, \ldots)$, and P_G was given in Example 5.4. We obtain

$$\tfrac{1}{24}(2^6 + 3.2^4 + 6.2^3 + 6.2^3 + 8.2^2) = 10,$$

the same number found, by inspection, in Example 5.9.

We next propose the question: How many color patterns show four red faces and two blue faces? To this end, we give weight x to red and y to blue and we ask for the pattern inventory. By (5.14) and (5.1), this inventory is

$$\tfrac{1}{24}[(x + y)^6 + 3(x + y)^2(x^2 + y^2)^2 + 6(x + y)^2(x^4 + y^4)$$
$$+ \ 6(x^2 + y^2)^3 + 8(x^3 + y^3)^2].$$

The coefficient of $x^4 y^2$ is

$$\tfrac{1}{24}(15 + 9 + 6 + 18 + 0) = 2.$$

Indeed, there are just two patterns with four red faces [(c) and (d) in Example 5.9]. For the complete pattern inventory, we easily obtain

$$x^6 + x^5 y + 2x^4 y^2 + 2x^3 y^3 + 2x^2 y^4 + xy^5 + y^6,$$

and this is in accordance with Example 5.11.

Example 5.17   Let D have m elements, and let G be the symmetric group of D. Let R have n elements, $R = \{1, \ldots, n\}$, and let these elements have weights $u_1, \ldots, u_n$, respectively. (The special case $m = 3$, $n = 2$ was considered in Example 5.10.) The pattern inventory is

$$(5.19) \qquad P_G(u_1 + u_2 \cdots + u_n, u_1^2 + \cdots + u_n^2, \ldots, u_1^m + \cdots + u_n^m).$$

It is our purpose to determine P_G by comparing (5.19) with an expression for the inventory that can be found independently.

Two functions f_1 and f_2 of R^D are equivalent if and only if each $r \in R$ occurs as often as a value of $f_1(d)$ as it occurs as a value of $f_2(d)$. In other words, they are equivalent if and only if they have the same weight.

Therefore, a pattern F can be adequately described as a function η defined on R, with values in $N = \{0, 1, 2, \ldots\}$, under the restriction that

$$\sum_{r \in R} \eta(r) = m.$$

The weight of F is

$$W(F) = u_1^{\eta(1)} \cdots u_n^{\eta(n)},$$

and thus it is easy to see that the pattern inventory $\Sigma\, W(F)$ is equal to the coefficient of z^m in the expansion of

$$\prod_{i=1}^{n} (1 + zu_i + z^2 u_i^2 + \cdots).$$

This expression is

$$\prod_{i=1}^{n} \exp\,(zu_i + \tfrac{1}{2} z^2 u_i^2 + \tfrac{1}{3} z^3 u_i^3 + \cdots)$$

$$= \exp\left(z \sum_{i=1}^{n} u_i + \tfrac{1}{2} z^2 \sum_{i=1}^{n} u_i^2 + \cdots \right),$$

which can be described as the result of the substitution

(5.20) $$x_1 = \sum_{i=1}^{n} u_i, \qquad x_2 = \sum_{i=1}^{n} u_i^2, \ldots$$

into

$$\exp\,(zx_1 + \tfrac{1}{2} z^2 x_2 + \tfrac{1}{3} z^3 x_3 + \cdots).$$

The coefficient of z^m in the power series expansion of this expression is a polynomial $Q(x_1, x_2, \ldots)$. Comparing our result with (5.19), we infer that P_G and Q have the same value under the substitution (5.20). By taking $n = m$, we can now show that P_G and Q, as polynomials in $x_1, \ldots, x_m$, are identical; this follows from the fact that if $x_1, \ldots, x_m$ are given complex numbers, then we can find $u_1, \ldots, u_m$ such that (5.20) holds with $n = m$. Thus we have proved the result indicated in Example 5.5, that is, that (5.2) is the generating function for the cycle index of the symmetric group.

Example 5.18   Let D be a finite set, and let G be a permutation group of D. Two subsets D_1 and D_2 are called *equivalent* if, for some $g \in G$, we have $gD_1 = D_2$. This means that D_2 is the set of all elements gd obtained by letting d run through D_1. As usual, we can form classes of equivalent subsets, and we ask for the number of classes.

The subsets can be brought into one-to-one correspondence with functions. Let R consist of two elements, "yes" and "no," and assign the weight 1 to each. If f is a mapping of D into R, then to f we let correspond the subset of all $d \in D$ with $f(d) =$ yes. We evidently get all subsets in this way, and equivalence of functions corresponds to equivalence of subsets. Each function has weight 1, so the number of subset classes equals the pattern inventory. By Pólya's theorem, it equals $P_G(2, 2, 2, \ldots)$.

If we take as weights $w(\text{no}) = 1$, $w(\text{yes}) = w$, where w is a variable, then the subsets of k elements correspond to functions f with $W(f) = w^k$. Thus the number of classes that consist of subsets with k elements each equals the coefficient of w^k in the pattern inventory, which is $P_G(1 + w, 1 + w^2, 1 + w^3, \ldots)$ in this case. Summing over all k, we obtain $P_G(2, 2, 2, \ldots)$ again, since the sum of the coefficients of a polynomial $p(w)$ equals $p(1)$.

A detailed example is provided by our painted cube (see Example 5.16). The subsets correspond, in an obvious way, to colorings of the cube. In each coloring, the red faces form a subset; and from each subset we can make a coloring by painting it red and its complement blue.

5.9 GENERALIZATION OF PÓLYA'S THEOREM

In the preceding sections we considered mappings of D into R, with an equivalence notion introduced by a permutation group G of D. We shall now treat a more general situation, in which there is also given a permutation group H of R, and in which equivalence of mappings is defined with the aid of both groups. Two mappings, $f_1 \in R^D$ and $f_2 \in R^D$, will be called *equivalent*, $f_1 \sim f_2$, if there exist elements $g \in G$ and $h \in H$ such that $f_1 g = h f_2$, that is,

$$(5.21) \qquad f_1(gd) = h f_2(d) \qquad \text{for all } d \in D.$$

We shall show that this is an equivalence relation.

1. The relation $f_1 \sim f_1$ is shown by taking for g the unit element of G, and for h the unit element of H.

2. If $f_1 \sim f_2$ then, for some $g \in G$, $h \in H$, we have $f_1[g(g^{-1}d)] = h f_2(g^{-1}d)$ for all $d \in D$, since $g^{-1}d \in D$ for all $d \in D$. It follows that $f_2(g^{-1}d) = h^{-1}f_1(d)$ for all $d \in D$; since $g^{-1} \in G$, $h^{-1} \in H$, this implies that $f_2 \sim f_1$.

3. If $f_1 \sim f_2$, $f_2 \sim f_3$, then there exist g, g', both $\in G$, and h, h', both $\in H$, such that

$$f_1(g\,d') = h f_2(d'), \qquad f_2(g'\,d) = h' f_3(d)$$

for all $d \in D$, $d' \in D$. Taking $d' = g'd$, we infer that

$$f_1(gg'\,d) = h f_2(g'\,d) = h h' f_3(d)$$

for all $d \in D$. Since $gg' \in G$, $hh' \in H$, it follows that $f_1 \sim f_3$.

By virtue of these properties of the equivalence defined by (5.21), we can say that R^D splits into equivalence classes, and again these equivalence classes will be called *patterns*.

Next we assume that each $f \in R^D$ has a certain *weight* $W(f)$, and that all these weights are elements of one and the same commutative ring. At present we do not assume, as we did in the earlier sections, that these $W(f)$ are obtained from certain weights of the elements of R by a formula of the type (5.9); but we shall still make a quite strong assumption, that equivalent functions have the same weight:

$$(5.22) \qquad f_1 \sim f_2 \text{ implies } W(f_1) = W(f_2).$$

If F denotes a pattern, we define its weight $W(F)$ as the common value of all $W(f)$ with $f \in F$, just as we did in Sec. 5.5. Again, the problem is to evaluate the pattern inventory, that is, the sum of the weights of all patterns. In analogy to (5.17), we shall establish the following result.

Lemma 5.2. The pattern inventory is

$$(5.23) \qquad \Sigma W(F) = |G|^{-1} |H|^{-1} \sum_{g \in G} \sum_{h \in H} \overset{(g,h)}{\underset{f}{\sum}} W(f),$$

where $\overset{(g,h)}{\underset{f}{\sum}} W(f)$ means the sum of $W(f)$ extended over all f that satisfy $fg = hf$.

PROOF. The proof is a straightforward extension of the proof of (5.17). Let ω be a possible value of W, and let S be the set of all $f \in R^D$ with $W(f) = \omega$. Furthermore, we consider the direct product $G \times H$ (see Example 5.8), consisting of all products $g \times h$ for $g \in G$, $h \in H$. We have the multiplication rule

$$(g \times h)(g' \times h') = (gg') \times (hh').$$

To each $g \times h \in G \times H$ we let correspond a permutation $\pi_{g \times h}$ of S, defined by the assertion that

$$\pi_{g \times h} f_1 = f_2 \text{ means that } f_2 = hf_1 g^{-1}.$$

If $f_2 = hf_1 g^{-1}$, then $W(f_1) = W(f_2)$, by (5.15); consequently, $\pi_{g \times h}$ transforms S into itself. Furthermore, $\pi_{g \times h}$ has an inverse, since

$$h^{-1}(hf_1 g^{-1})(g^{-1})^{-1} = f_1.$$

Hence $\pi_{g \times h}$ is a permutation of S.

The mapping $g \times h \rightarrow \pi_{g \times h}$ satisfies the homomorphy condition (5.6). This follows from the fact that if $g \in G$, $g' \in G$, $h \in H$, $h' \in H$, then for each $f \in S$ we have

$$\pi_{(g \times h)(g' \times h')} f = \pi_{gg' \times hh'} f = (hh') f (gg')^{-1}$$

and

$$\pi_{g \times h}(\pi_{g' \times h'} f) = \pi_{g \times h}(h' f g'^{-1}) = h(h' f g'^{-1}) g^{-1},$$

and therefore

$$\pi_{(g \times h)(g' \times h')} = \pi_{g \times h} \pi_{g' \times h'}.$$

Furthermore, that two elements f_1 and f_2 are equivalent in the sense of (5.21) means that there exist g and h such that $\pi_{g \times h} f_2 = f_1$. Therefore, by application of Lemma 5.1 to the group $G \times H$, the number of patterns belonging to S turns out to be

$$(5.24) \qquad |G \times H|^{-1} \sum_{g \in G} \sum_{h \in H} \psi_\omega(g, h),$$

where $\psi_\omega(g, h)$ is the number of f with $W(f) = \omega$, $\pi_{g \times h} f = f$. The latter condition means $hf = fg$.

Finally, when we multiply (5.24) by ω, sum over all possible ω, and use

$$\sum_\omega \psi_\omega(g, h) \omega = \sum_f^{(g,h)} W(f),$$

we have completed our proof.

Example 5.19 We shall give a fairly general example of attaching a weight $W(f)$ to each $f \in R^D$ in such a way that the invariance property (5.22) holds.

Assume that D is divided into disjoint subsets $D_1, \ldots, D_l$ such that each D_i is invariant under G, that is, that for all $g \in G, d \in D_i, i = 1, \ldots, l$, we have $g(d) \in D_i$. Similarly, assume that R is divided into pieces $R_1, \ldots, R_k$ such that each $R_j, j = 1, \ldots, k$ is invariant under H.

Let $\psi(j; n_1, \ldots, n_l)$ be a function of the integer variables $j, n_1, \ldots, n_l$, where $1 \le j \le k, 0 \le n_1 < \infty, \ldots, 0 \le n_l < \infty$. The function values are supposed to lie in a commutative ring.

If $f \in R^D, 1 \le i \le l, r \in R$, then $n_i(f, r)$ denotes the number of elements $d \in D_i$ that satisfy $f(d) = r$. Now we define as the weight of f the value

$$(5.25) \qquad W(f) = \prod_{j=1}^k \prod_{r \in R_j} \psi[j; n_1(f, r), \ldots, n_l(f, r)].$$

In order to establish (5.22), we first remark that, for all $i, g, h,$ and r, $i = 1, \ldots, l; g \in G; h \in H; r \in R$, we have

$$(5.26) \qquad n_i(hfg^{-1}, r) = n_i(f, h^{-1}r),$$

since the number of $d \in D_i$ with $hfg^{-1}(d) = r$ equals the number of $d \in D_i$ with $hf(d) = r$, D_i being permuted by g, and this equals the number of $d \in D_i$ with $f(d) = h^{-1}r$ (the d's in the two cases are the same).

It follows from (5.26) that if in (5.25) we replace f by hfg^{-1}, then the factors of the product

$$\prod_{r \in R_j} \psi[j; n_1(f, r), \ldots, n_l(f, r)]$$

are only permuted, for if r runs through R_j, then $h^{-1}r$ runs through R_j. Thus replacing f by hfg^{-1} does not affect the product. This proves that $W(hfg^{-1}) = W(f)$; that is, W satisfies (5.22).

In [2], a special case of (5.25) was considered; it was assumed that $l = 1$ and that H is the direct product $H_1 \times \cdots \times H_k$ of permutation groups of $R_1, \ldots, R_k$, respectively. Under these assumptions the result (5.23) could still be expressed in terms of the cycle indexes of the groups involved. We shall not treat this here; we only remark that by further specialization we can obtain Pólya's fundamental theorem and two other results that will be considered in Secs. 5.10 and 5.12. Pólya's theorem is obtained by taking $R_1, \ldots, R_k$ to be one-element sets, so that $k = |R|$, and by taking for each H_i the group consisting of the unit element only, with

$$\psi[j; n(f, r)] = (w_j)^{n(f,r)}.$$

Here $n(f, r)$ is the number of $d \in D$ with $f(d) = r$; w_j is some weight assigned to R_j, but as each R_j has only one element, this means the same thing as assigning weights to the elements of R.

The weights used in Secs. 5.10 and 5.12 can be considered as special cases of (5.25). Both for Sec. 5.10 and for Sec. 5.12 we have $k = l = 1$, and thus ψ depends only on one variable n, and

$$W(f) = \prod_{r \in R} \psi[n(f, r)];$$

$n(f, r)$ is the number of $d \in D$ with $f(d) = r$. In order to get the case of Sec. 5.10, we have to take $\psi(0) = 1$, $\psi(1) = 1$, $\psi(2) = \psi(3) = \cdots = 0$, where $W(f) = 1$ if f is one-to-one, and $W(f) = 0$ otherwise. This case of Sec. 5.12 can be obtained by taking simply $\psi(n) = 1$ for all n, and therefore $W(f) = 1$ for all f.

5.10 PATTERNS OF ONE-TO-ONE MAPPINGS

As in Sec. 5.9, we have finite sets D and R, subject to permutation groups G and H. We now define the weight $W(f)$ of any $f \in R^D$ by

$$W(f) = 1 \qquad \text{if } f \text{ is a one-to-one mapping},$$
$$= 0 \qquad \text{otherwise}.$$

Obviously, if $g \in G$, $h \in H$, then the mapping hfg^{-1} is one-to-one if and only if f is one-to-one, and thus W satisfies (5.22). The inventory $\Sigma W(F)$

is just the number of patterns of one-to-one functions (pattern being defined as in Sec. 5.9 with the aid of equivalence induced by G and H). In order to determine this number, we shall apply (5.23), and accordingly we must first evaluate $\sum\limits_{f}^{(g,h)} W(f)$.

Let g and h be fixed for the time being, $g \in G$, $h \in H$. Let g be of type $\{b_1, b_2, \ldots\}$, where b_i indicates the number of cycles of length i; see Sec. 5.2. Furthermore, let h be of type $\{c_1, c_2, c_3, \ldots\}$. We want to find the number of one-to-one mappings f of R into D that satisfy $fg = hf$.

Let f be such a mapping, and let d be some element of D, belonging to a cycle of length j. This cycle consists of the elements

(5.27) $$d, gd, g^2d, \ldots, g^{j-1}d,$$

and we have $g^jd = d$. We notice that $fg = hf$ implies

$$fg^2 = fgg = hfg = hhf = h^2f, \qquad fg^3 = h^3f,$$

etc. Hence f maps the elements (5.27) onto

(5.28) $$hfd, h^2fd, \ldots, h^{j-1}fd,$$

and we have $h^jfd = fg^jd = fd$. It follows that the length of the cycle of R to which fd belongs is a divisor of j.

As yet we have not used the condition that f is one-to-one, and therefore this first part of the argument can also be used in Sec. 5.12. In this situation, however, we do assume that f is one-to-one. It follows that there are no repetitions in the sequence (5.28), and therefore the length of the cycle to which fd belongs is equal to j itself. Thus a cycle of D of length j is mapped onto a cycle of R of length j. Different cycles of D are mapped onto different cycles of R, since f is one-to-one.

It is now easy to get a survey over all one-to-one mappings f that satisfy $fg = hf$. In order to construct such an f, for each cycle of D we can select a cycle of R having the same length, taking care that each cycle of R is selected at most once. Each time a selection has been made, if j is the length of the cycles under consideration, then there are still j possibilities for fixing the correspondence between the j elements of the D-cycle and the j elements of the R-cycle. This follows from the fact that if d is an element of the D-cycle, then we can map it onto any element r of the R-cycle, provided we map gd onto hr, g^2d onto h^2r, etc.

The number of one-to-one mappings of a set of b_j elements into a set of c_j elements equals

(5.29) $$c_j(c_j - 1)(c_j - 2) \cdots (c_j - b_j + 1),$$

which is zero if $c_j < b_j$. Therefore for the number of one-to-one mappings f of D into R subject to $fg = hf$, we have

$$(5.30) \qquad \overset{(g,h)}{\underset{f}{\sum}} W(f) = \prod_j j^{b_j} c_j (c_j - 1) \cdots (c_j - b_j + 1).$$

The product runs over all j for which $b_j > 0$; but if we interpret (5.29) as unity if $b_j = 0$, we can take the product over all values $j = 1, 2, 3, \ldots$ as well.

We can write $j^b c(c-1) \cdots (c-b+1)$ as the bth derivative of $(1+jz)^c$ at the point $z = 0$. Accordingly, (5.30) can be expressed as the result of a number of partial differentiations with respect to variables $z_1, z_2, \ldots$ at the point $z_1 = z_2 = \cdots = 0$:

$$(5.31) \qquad \left(\frac{\partial}{\partial z_1}\right)^{b_1} \left(\frac{\partial}{\partial z_2}\right)^{b_2} \left(\frac{\partial}{\partial z_3}\right)^{b_3} \cdots (1 + z_1)^{c_1} (1 + 2z_2)^{c_2} (1 + 3z_3)^{c_3} \cdots.$$

Thus far g and h have been fixed. Now summing with respect to g and h and dividing by the orders $|G|$ and $|H|$, we obtain the pattern inventory (5.23), which is, in this case, the number of patterns of one-to-one mappings. The differential operator in (5.31) is obtained from a term of the cycle index $P_G(x_1, x_2, \ldots)$ of G upon substitution of $x_1 = \partial/\partial z_1$, $x_2 = \partial/\partial z_2, \ldots$, and the operand in (5.31) is obtained upon substitution of $x_1 = 1 + z_1, x_2 = 1 + 2z_2, \ldots$ into a term of the cycle index $P_H(x_1, x_2, \ldots)$ of H. Thus by summation we obtain the following result.

Theorem 5.2. The number of patterns of one-to-one mappings of D into R [pattern being defined by equivalence induced by permutation groups G of D and H of R according to (5.21)] equals

$$(5.32) \qquad P_G\left(\frac{\partial}{\partial z_1}, \frac{\partial}{\partial z_2}, \frac{\partial}{\partial z_3}, \ldots\right) P_H(1 + z_1, 1 + 2z_2, 1 + 3z_3, \ldots),$$

evaluated at $z_1 = z_2 = z_3 \cdots = 0$.

The expression given in Theorem 5.2 can be simplified slightly if $|R| = |D|$. (If $|R| < |D|$, the number of one-to-one mappings is zero, of course.) Then we always have

$$\sum_j b_j = \sum_j c_j,$$

so that in (5.31) we have either $b_1 = c_1$, $b_2 = c_2, \ldots$, or at least once $b_j > c_j$. In the latter case, (5.31) vanishes. Thus it is easy to see that (5.31) is always equal to

$$\left(\frac{\partial}{\partial z_1}\right)^{b_1} \left(\frac{\partial}{\partial z_2}\right)^{b_2} \left(\frac{\partial}{\partial z_3}\right)^{b_3} \cdots z_1^{c_1} (2z_2)^{c_2} (3z_3)^{c_3} \cdots$$

evaluated at $z_1 = z_2 = z_3 = \cdots = 0$. Hence we have established the following result.

Theorem 5.3. If the assumptions of Theorem 5.2 hold, and if moreover $|R| = |D|$, then the number of patterns is equal to

$$(5.33) \qquad P_G\left(\frac{\partial}{\partial z_1}, \frac{\partial}{\partial z_2}, \frac{\partial}{\partial z_3}, \ldots\right) P_H(z_1, 2z_2, 3z_3, \ldots),$$

evaluated at $z_1 = z_2 = z_3 = \cdots = 0$.

In this case the situation is entirely symmetrical; if $|R| = |D|$, then the one-to-one mappings of D *into* R are mappings of D *onto* R, and their inverse mappings are one-to-one mappings of R onto D. Indeed, one can verify directly that at the point $z_1 = z_2 = z_3 = \cdots = 0$, the expression (5.33) is equal to

$$(5.34) \qquad P_H\left(\frac{\partial}{\partial z_1}, \frac{\partial}{\partial z_2}, \frac{\partial}{\partial z_3}, \ldots\right) P_G(z_1, 2z_2, 3z_3, \ldots).$$

Example 5.20 In how many geometrically different ways can we arrange the faces of a cube in cyclic order? This means that we want to know the number of patterns of one-to-one mappings in the following situation: D is the set of faces of a cube and G is the group of permutations induced by rotations (Example 5.4); R is the set of the 6th roots of unity, and again H is the group of permutations induced by rotations in the complex plane (Example 5.7, with $m = 6$).

The cycle indexes are

$$P_G = \tfrac{1}{24}(x_1^6 + 6x_2^3 + 8x_3^2 + 3x_1^2x_2^2 + 6x_1^2x_4),$$

$$P_H = \tfrac{1}{6}(x_1^6 + x_2^3 + 2x_3^2 + 2x_6),$$

and thus (5.33) gives

$$\tfrac{1}{6} \cdot \tfrac{1}{24}(6! + 6 \cdot 2^3 \cdot 3! + 16 \cdot 3^2 \cdot 2!) = 9$$

as the number of patterns.

Example 5.21 We take a finite group S, and we consider all permutations of S. Two permutations π_1 and π_2 are said to be *equivalent* if there are group elements $a, b \in S$ such that $a\pi_1 b = \pi_2$. Here the left-hand side $a\pi_1 b$ is the permutation that turns any $s \in S$ into $a\pi_1(bs)$; notice that bs is the product of the group elements b and s and that $a\pi_1(bs)$ is the product of the group elements a and $\pi_1(bs)$.

We ask for the number of equivalence classes. It is easily verified that this means just the number of patterns of one-to-one mappings in the following situation: $D = R = S$, $G = H =$ the Cayley representation of the group S. By (5.5), we have

$$P_G = P_H = m^{-1} \sum_{k|m} \nu(k)(x_k)^{m/k},$$

where $m = |S|$ and, for each divisor k of m, the symbol $\nu(k)$ represents the number of elements $s \in S$ having order k. From (5.33), we obtain

$$m^{-2} \sum_{k|m} [\nu(k)]^2 k^{m/k} \left(\frac{m}{k}\right)!$$

as the number of patterns.

5.11 LABELING AND DELABELING

At the very beginning of combinatorial analysis we find situations of the type studied in the preceding sections. Sometimes the formulations in old-fashioned combinatorial analysis are obscure, and often we can clarify them only by making careful statements about the groups and equivalences under consideration. For example, in all situations where expressions such as "the order of the elements is not taken into account" are used, this means that in a certain set of objects we introduce an equivalence relation defined by a permutation group (often the symmetric group) and that we pass from the original set of objects to the set of equivalence classes. Similarly, if it is stated that some objects are identified, or considered as equal, or if it is stated that some objects are different but indistinguishable, we again are concerned with the procedure of introducing equivalence classes. In this context, the words *labeling* and *delabeling* are useful. Vaguely speaking, labeling gives means of identification in a set where the elements are too much alike, and delabeling is the converse procedure of simplification by means of neglecting unimportant differences. Furthermore, there are intermediate cases in which the elements of a set bear labels, but in which some of the labels are confusingly alike.

Let D and R be finite sets. Any one-to-one mapping f of D into R is called a *labeled subset* of R. If S is the subset of R onto which f is mapped, then S is the subset that has been labeled, and D is called the *set of labels*. Notice that a labeled subset of R is not a subset of R. If one wishes, it can be considered as a set, the set of pairs $[d, f(d)]$, where d runs through D. Indeed, it might be better to speak of a "subset with a labeling" instead of a "labeled subset." If, in particular, D is the set $\{1, 2, \ldots, m\}$, then the labeled subsets are often called *variations* or *m-permutations*.

If we take a permutation group H of R, then we can define equivalence classes of labeled subsets, and these classes are just the patterns of one-to-one mappings studied in Sec. 5.10, with the specialization that the permutation group of D consists of the unit element only. Thus we have $P_G = x_1^m$, if m is the number of elements of D, and by Theorem 5.2 we obtain, for the number of patterns,

(5.35)
$$\left[\left(\frac{d}{dz}\right)^m P_H(1 + z, 1, 1, \ldots) \right]_{z=0}.$$

If, in addition, H consists of the unit element only, then, if R has n elements, we have

$$\left[\left(\frac{d}{dz}\right)^m (1 + z)^n \right]_{z=0} = n(n - 1)(n - 2) \cdots (n - m + 1),$$

which is the well-known result for the number of variations of m elements chosen from a set of n elements.

In the special case $m = n$, by virtue of Theorem 5.3 we may replace (5.35) by

$$\left[\left(\frac{d}{dz}\right)^m P_H(z, 0, 0, \ldots) \right]_{z=0},$$

and this is easily seen to be equal to $|H|^{-1}m!$. We can verify this result directly as follows. From each labeling f we can obtain the equivalence class to which f belongs. It is the set of all hf, where h runs through H. As all these hf, with f fixed, are different, the equivalence class contains exactly $|H|$ elements. Thus there are $|H|^{-1}m!$ equivalence classes.

An example is the number of possible dice. Here R is the set of faces of a cube, H is the group of permutations induced by rotations, $D = \{1, \ldots, 6\}$; the number of patterns is $6!/24 = 30$.

We next discuss *complete delabeling*; that is, we introduce the symmetric group in the set of labels and we form patterns with respect to this group—and, at the same time, with respect to the group H on R. It does not matter which labels are attached to elements of R, only which elements of R do get labels and which do not. Accordingly, the number of patterns is just the number of subset patterns, restricted to subsets of m elements. According to Example 5.18, this is equal to the coefficient of w^m in the polynomial

$$P_H(1 + w, 1 + w^2, 1 + w^3, \ldots).$$

As a check, we show that the same result can be obtained from Theorem 5.2. Now G is the symmetric group of degree m, so $P_G(x_1, x_2, \ldots)$ equals the coefficient of w^m in the expansion of

$$\exp\left(wx_1 + \tfrac{1}{2}w^2x_2 + \tfrac{1}{3}w^3x_3 + \cdots\right);$$

see Example 5.5. In order to evaluate the effect of the differential operator in (5.32), we notice that, by Taylor's series,

$$\left[\exp\left(a\frac{d}{dz}\right)f(z)\right]_{z=0} = f(0) + af'(0) + \frac{a^2}{2!}f''(0) + \cdots = f(a),$$

provided f is a polynomial, for then there is no trouble about the convergence. Thus, in our case, (5.32) becomes the coefficient of w^m in

$$P_H[1 + w, 1 + 2(\tfrac{1}{2}w^2), 1 + 3(\tfrac{1}{3}w^3), \ldots],$$

and this is exactly what we obtained previously.

In the special case that the group H consists of the identity only, we have

$$P_H(x_1, x_2, \ldots) = x_1^m,$$

whence the number of patterns becomes the coefficient of w^m in the expansion of $(1 + w)^n$. This is a well-known result, for here the patterns are just the (unlabeled) subsets of m elements taken from R. In combinatorial analysis one usually calls these things "combinations"—a rather superfluous expression nowadays since they can be called just "subsets."

We next discuss a single example of *incomplete delabeling*.

Example 5.22 We have six labels, $d_1, d_2, d_3, d_4, d_5, d_6$, to be pasted on the faces of a cube, one on each face. The labels have colors. Label d_1 is yellow and d_2 is black. The labels d_3 and d_4 are both violet and cannot be distinguished from each other since the lettering has become illegible. The same thing holds for d_5 and d_6, which are both purple. To make it more difficult, the whole experiment is conducted by someone who does not know what is violet and what is purple. He is not colorblind; although he does see the difference, he just does not know which color is which. Furthermore, he is not interested, since the people who are supposed to admire his color pattern are in the same state. Of course, labelings that can be obtained from each other by rotations of the cube are to be identified. We ask for the number of patterns.

Now D is the set of 6 labels, and R is the set of 6 faces. The group G consists of 8 permutations, characterized by the conditions that d_1 and d_2 are fixed and that the subset $\{d_3, d_4\}$ is mapped either onto itself or onto the set $\{d_5, d_6\}$. The cycle index of this group is

$$P_G(x_1, x_2, \ldots) = \tfrac{1}{8}(x_1^6 + 2x_1^4x_2 + 3x_1^2x_2^2 + 2x_1^2x_4),$$

and P_H is given by the right-hand side of (5.1).

The number of patterns is given by (5.34). This is easily evaluated since, for example, the operator $(\partial/\partial z_1)^4(\partial/\partial z_2)$ has a nonzero effect only on the term in $z_1^4 z_2$. For the number of patterns, we obtain

$$\tfrac{1}{8} \cdot \tfrac{1}{24}\{6! + 3 \cdot 3 \cdot 2!2^2 \cdot 2! + 6 \cdot 2 \cdot 2!4 \cdot 1!\} = 5.$$

We can check this by describing the patterns as follows. (1) Yellow and black are on opposite faces, and of the remaining faces opposites have equal colors. (2) Yellow and black are opposite, and of the remaining faces opposites have different colors. (3) Yellow and black are on adjacent faces and the one opposite yellow has the same color as the one opposite black. (4) The bottom face is black, the front face is yellow, and the top face and the face on the left have the same color. (5) The bottom face is black, the front face is yellow, and the top face and the face on the right have the same color.

5.12 THE TOTAL NUMBER OF PATTERNS

We again have finite sets D and R, and permutation groups G and H of D and R, respectively. These groups define an equivalence notion [see (5.21)] in the set R^D of all mappings of D into R, and we ask for the number of patterns. Thus the question is the same as in Sec. 5.10, except that we no longer restrict our attention to one-to-one mappings. In other words, we are asking for the pattern inventory, based on the weight $W(f)$ defined to be 1 for all $f \in R^D$.

We again apply (5.23), and in order to evaluate $\overset{(g,h)}{\underset{f}{\sum}} W(f)$, we use the first part of Sec. 5.10. This shows the following: If f satisfies $fg = hf$, then each cycle of D is mapped onto a cycle of R with length equal to a divisor of the length of the first-mentioned cycle of D. Moreover, within the cycles the correspondence is cyclic in the following sense. If d is an element of the cycle of D, and if $f(d) = r$, then the mapping is given by $f(gd) = hr$, $f(g^2d) = h^2r$, etc.

Conversely, every function f with this property satisfies $fg = hf$. Therefore we can easily compute the number of possibilities we have for f. In each cycle of D we select an element, called the *selected element*. The number of possibilities for the element of R onto which a selected element can be mapped by an f is

$$(5.36) \qquad\qquad \sum_{j|i} jc_j,$$

where i is the length of the cycle of D to which the selected element belongs, the c's are taken from the type $\{c_1, c_2, \ldots\}$ of h, and the sum is taken over all divisors j of i. Because the choices of function values for the various selected elements are independent, and because these choices determine f completely, the number of f equals the product of (5.36) taken over all selected elements. Since there are b_i cycles of length i, we obtain

$$(5.37) \quad \overset{(g,h)}{\underset{f}{\sum}} W(f) = \prod_i \left(\sum_{j|i} jc_j \right)^{b_i}$$

$$= (c_1)^{b_1}(c_1 + 2c_2)^{b_2}(c_1 + 3c_3)^{b_3}(c_1 + 2c_2 + 4c_4)^{b_4}(c_1 + 5c_5)^{b_5} \cdots .$$

Notice that a power with exponent 0 has to be interpreted as 1 in this context even if the base is zero. Therefore there is no difficulty about the interpretation of the infinite product.

As in Sec. 5.10, we can interpret (5.37) as a derivative. A power a^b can be written as the bth derivative of e^{az} at $z = 0$; if $a = b = 0$, this still gives the desired value $0^0 = 1$. Accordingly, the expression (5.37) can be written as

$$\left(\frac{\partial}{\partial z_1} \right)^{b_1} \left(\frac{\partial}{\partial z_2} \right)^{b_2} \left(\frac{\partial}{\partial z_3} \right)^{b_3} \cdots \exp \left(\sum_i z_i \sum_{j|i} jc_j \right),$$

evaluated at $z_1 = z_2 = z_3 = \cdots = 0$. The exponent can be expressed as

$$\sum_i z_i \sum_{j|i} jc_j = \sum_j jc_j(z_j + z_{2j} + z_{3j} + \cdots).$$

Now notice that the differential operator is obtained from $x_1^{b_1} x_2^{b_2} x_3^{b_3} \cdots$ on substitution of $x_1 = \partial/\partial z_1$, $x_2 = \partial/\partial z_2, \ldots$, and that the operand is obtained from $x_1^{c_1} x_2^{c_2} x_3^{c_3} \cdots$ on substitution of

$$x_1 = e^{z_1 + z_2 + z_3 + \cdots}, \, x_2 = e^{2(z_2 + z_4 + z_6 + \cdots)}, \, x_3 = e^{3(z_3 + z_6 + z_9 + \cdots)}, \ldots .$$

Thus, summing with respect to g and h, and using (5.23), we obtain the following result.

Theorem 5.4. The total number of patterns of mappings of D into R ["patterns" being defined by equivalence induced by permutation groups G of D and H of R according to (5.21)] equals

$$P_G\left(\frac{\partial}{\partial z_1}, \frac{\partial}{\partial z_2}, \frac{\partial}{\partial z_3}, \ldots \right) P_H[e^{z_1 + z_2 + z_3 + \cdots}, e^{2(z_2 + z_4 + z_6 + \cdots)}, e^{3(z_3 + z_6 + z_9 + \cdots)}, \ldots],$$

evaluated at $z_1 = z_2 = z_3 = \cdots = 0$.

There is a second expression for the number of patterns, which sometimes may turn out to be simpler to handle. Notice that (5.37) is obtained by substituting $x_1 = c_1$, $x_2 = c_1 + 2c_2, \ldots$ into $x_1^{b_1} x_2^{b_2} \cdots$. Summing for g and dividing by $|G|$, we obtain

$$P_G(c_1, c_1 + 2c_2, c_1 + 3c_3, c_1 + 2c_2 + 4c_4, \ldots),$$

in which the ith argument is

$$\sum_{j \mid i} jc_j.$$

Thus, again using (5.23), for the number of patterns we have

$$(5.38) \qquad |H|^{-1} \sum_{h \in H} P_G(c_1, c_1 + 2c_2, c_1 + 3c_3, c_1 + 2c_2 + 4c_4, \ldots).$$

Here it has to be remembered that $\{c_1, c_2, c_3, \ldots\}$ is the type of h.

We shall discuss some examples. In all of them, m and n stand for the numbers of elements of D and R, respectively:

$$m = |D|, \qquad n = |R|.$$

Example 5.23   We specialize by taking for G the group consisting of the identity only, where $P_G = x_1^m$. Then the patterns can be called *patterns of variation with repetition*, for if we write D as a sequence $\{1, \ldots, m\}$, then for each $f \in R^D$ the sequence $\{f(1), \ldots, f(m)\}$ may show repetitions, since our functions f are not required to be one-to-one. For the number of patterns, Theorem 5.4 gives the expression

$$(5.39) \qquad \left[\left(\frac{d}{dz} \right)^m P_H(e^z, 1, 1, \ldots) \right]_{z=0}.$$

If we make a further specialization by requiring also that H consists of the unit element only, then the patterns will be nothing other than functions $f \in R^D$. Indeed, (5.39) reduces to n^m in this case, for $P_H(e^z, 1, 1, \ldots)$ becomes e^{nz}.

A second specialization of (5.39) is obtained by keeping R and H general but requiring that D has only one element d_1. Now two functions f_1 and f_2 are equivalent if and only if $f_1(d_1)$ and $f_2(d_1)$ belong to the same transitive set of R, that is, if and only if $f_1(d_1)$ can be mapped onto $f_2(d_1)$ by some permutation of H. Consequently, the number of patterns is just the number of transitive sets. Indeed, if we take $m = 1$ in (5.39), it specializes to our main lemma, Lemma 5.1, since any term

$$|H|^{-1} x_1^{c_1} x_2^{c_2} x_3^{c_3} \cdots$$

produced by a permutation h of the type $\{c_1, c_2, \ldots\}$ gives as its contribution to (5.39) nothing but $|H|^{-1} c_1$, and c_1 is the number of elements of R that are invariant under the permutation h.

Example 5.24   We specialize by taking for G the symmetric group of D, but H is not specialized. The effect of the symmetric group is, roughly speaking, that we are interested only in the *number* of elements of D that a function f maps onto a given element of R and not in the elements themselves. Thus our patterns can be described as patterns of mappings ϕ of R into the set $N = \{0, 1, 2, \ldots\}$, with the restriction

$$\sum_{r \in R} \phi(r) = m,$$

whereas patterns are formed with respect to the group H acting on the domain, which is R in this case. Thus our number of patterns can be computed either from Theorem 5.4 or from Theorem 5.1, and we shall show that the results are the same in the two cases. Theorem 5.1 can be applied by providing N with weights $1, w^1, w^2, w^3, \ldots$ and asking for patterns with weight w^m; the result is the coefficient of w^m in

$$P_H(1 + w + w^2 + w^3 + \cdots, 1 + w^2 + w^4 + \cdots,$$
$$1 + w^3 + w^6 + \cdots, \ldots).$$

If we apply Theorem 5.4, we have as differential operator the coefficient of w^m in

$$\exp\left(w\frac{\partial}{\partial z_1} + \tfrac{1}{2}w^2\frac{\partial}{\partial z_2} + \cdots\right);$$

see Example 5.5. The effect of this operator on a function $\phi(z_1, z_2, \ldots)$ at $z_1 = z_2 \cdots = 0$ is

$$\phi(w, \tfrac{1}{2}w^2, \tfrac{1}{3}w^3, \ldots).$$

Therefore substitution of $z_1 = w$, $z_2 = \tfrac{1}{2}w^2$, $z_3 = \tfrac{1}{3}w^3, \ldots$, into the arguments of P_H in Theorem 5.4 is easily seen to produce

$$\exp(z_1 + z_2 + z_3 + \cdots) = \exp(w + \tfrac{1}{2}w^2 + \tfrac{1}{3}w^3 + \cdots)$$
$$= (1 - w)^{-1},$$
$$\exp\{2(z_2 + z_4 + z_6 + \cdots)\} = \exp(w^2 + \tfrac{1}{2}w^4 + \tfrac{1}{3}w^6 + \cdots)$$
$$= (1 - w^2)^{-1},$$

etc. Thus we obtain $P_H[(1 - w)^{-1}, (1 - w^2)^{-1}, (1 - w^3)^{-1}, \ldots]$; that is, we have the same result as was previously obtained.

Strictly speaking, the application of Taylor's series should require somewhat more care than in Sec. 5.11, for here we are not applying it to a polynomial, but to a sum of exponential functions.

Example 5.25   Next we specialize by taking for H the symmetric group of all permutations of R, and for G we take the identity only. Now the patterns turn out to be class partitions of D into at most m parts, where a *class partition* is defined as a set of disjoint subsets (classes) of D whose union is D; we use the term "class partition" instead of the simpler "partition" because we shall employ that word later with a different meaning. The foregoing result follows from the fact that if $f \in R^D$, then f defines a class partition by putting into one class all d that are mapped by f onto one and the same element of R.

According to (5.39) and Example 5.5, we find that the number of patterns is the coefficient of w^n in

$$\left[\left(\frac{d}{dz} \right)^m \exp \left(we^z + \tfrac{1}{2}w^2 + \tfrac{1}{3}w^3 + \cdots \right) \right]_{z=0},$$

and this is easily seen to be $m!$ times the coefficient of $z^m w^n$ in the expansion of $(1 - w)^{-1} e^{w(e^z - 1)}$.

From this result on the number of class partitions into at most n parts, it is not difficult to deduce that the number of class partitions into *exactly* n parts equals $m!$ times the coefficient of $z^m w^n$ in $e^{w(e^z - 1)}$, and that the *total* number of class partitions equals $m!$ times the coefficient of z^m in $e^{e^z - 1}$. This is a special case of a result due to Hadwiger [7] for the number of class partitions of n objects into m subsets of at most k each.

Example 5.26   If we take for H the symmetric group, but now without any specialization of G, we get *patterns* of class partitions of D. Unfortunately, it seems impossible to simplify the result of Theorem 5.4 [or (5.38)] essentially in this situation.

If we specialize G to the symmetric group of D, the patterns become class partitions of a set of unidentifiable objects. That is, the only thing that is of concern now is the size of the classes, not the contents. Thus our patterns can be brought into one-to-one correspondence with the partitions of the number m into at most n parts. A partition of m is a solution $\{b_1, b_2, \ldots\}$ of the equation

$$b_1 + 2b_2 + 3b_3 + \cdots = m,$$

in nonnegative integers $b_1, b_2, b_3, \ldots$. We say that m has been partitioned into b_1 1's, b_2 2's, b_3 3's, $\ldots$, and accordingly, $b_1 + b_2 + b_3 + \cdots$ is called the *number of parts* of the partition.

The number of patterns can be obtained from Example 5.24 by specializing H to be the symmetric group. We obtain the coefficient of z^m in

$$P_H[(1 - z)^{-1}, (1 - z^2)^{-1}, (1 - z^3)^{-1}, \ldots],$$

which becomes, in this case, the coefficient of $z^m w^n$ in

$$\exp \left[w(1 - z)^{-1} + \tfrac{1}{2}w^2(1 - z^2)^{-1} + \tfrac{1}{3}w^3(1 - z^3)^{-1} + \cdots \right],$$

and this expression can be reduced to

$$\exp \left[\log (1 - w)^{-1} + \log (1 - wz)^{-1} + \log (1 - wz^2)^{-1} + \cdots \right]$$

$$= \prod_{k=1}^{\infty} \frac{1}{1 - wz^k}.$$

This is the well-known result for the generating function of the number of partitions of a given number into a given number of components (see [13], page 112).

Example 5.27 We again consider the situation at the beginning of Example 5.26, but we now specialize by taking $n = 2$. Thus we get patterns of class partitions into at most two parts. By (5.38), the number of these is

$$\tfrac{1}{2} P_G(2, 2, 2, \ldots) + \tfrac{1}{2} P_G(0, 2, 0, 2, \ldots),$$

since $P_H = \tfrac{1}{2} x_1^2 + \tfrac{1}{2} x_2$.

If we compare this to the number of patterns of partitions into two *labeled* classes (see Example 5.18), $P_G(2, 2, 2, \ldots)$, we observe that the term $P_G(0, 2, 0, 2, \ldots)$ represents the number of *symmetric* partitions of D into two classes D_1 and D_2. That is, $P_G(0, 2, 0, 2, \ldots)$ represents the number of patterns of subsets D_1 that are equivalent to their complement, where *equivalence* is defined by the permutations of G and *patterns* are defined by this equivalence.

As a particular example, we take for D the set of the 10th roots of unity, and for G the group of 10 rotations. The cycle index (see Example 5.7) is

$$P_G = \tfrac{1}{10}(x_1^{10} + x_2^5 + 4x_5^2 + 4x_{10}),$$

and therefore

$$P_G(0, 2, 0, 2, \ldots) = \tfrac{1}{10}(2^5 + 4 \cdot 2) = 4.$$

Letting ω denote a primitive 10th root of unity, we describe the patterns by indicating a representative from each pattern.

$$\{1, \omega, \omega^2, \omega^3, \omega^4\}, \{1, \omega^2, \omega^4, \omega^6, \omega^8\}, \{1, \omega, \omega^4, \omega^7, \omega^8\}, \{1, \omega, \omega^2, \omega^4, \omega^8\}.$$

5.13 THE KRANZ GROUP

In this section, which will be independent of Secs. 5.8, 5.9, 5.10, 5.11, and for the larger part independent of Secs. 5.2 and 5.7, we shall derive Pólya's result for the cycle index of his so-called Kranz group.

Let S and T be finite sets, and let G and H be permutation groups of S and T, respectively. We consider the cartesian product $S \times T$, that is, the set of all pairs (s, t), $s \in S$, $t \in T$. We can construct special permutations of $S \times T$ as follows: Choose an element $g \in G$, and to each $s \in S$ choose an element $h_s \in H$. These elements determine a permutation of $S \times T$, defined by

$$(s, t) \to (gs, h_s t), \qquad s \in S, t \in T.$$

There are $|G|\,|H|^{|S|}$ such permutations, and they form a group, which Pólya [10] called the *Kranz* $G[H]$. Its cycle index can be expressed very elegantly in terms of the cycle indexes of G and H.

Theorem 5.5 (Pólya [10]). We have

(5.40)
$$P_{G[H]}(x_1, x_2, \ldots) = P_G[P_H(x_1, x_2, x_3, \ldots), P_H(x_2, x_4, x_6, \ldots), \ldots],$$

where the right-hand side is obtained on substitution of

$$y_k = P_H(x_k, x_{2k}, x_{3k}, \ldots) \quad \text{into} \quad P_G(y_1, y_2, y_3, \ldots).$$

PROOF. We fix an element γ of $G[H]$ by selecting $g \in G$, $h_1 \in H$, ..., $h_m \in H$. We want to find its type. Let $s_1, \ldots, s_k$ be a cycle of g, with length k. That is, we have

$$s_1 \in S, \ldots, s_k \in S, gs_1 = s_2, g^2 s_1 = s_3, \ldots, g^{k-1} s_1 = s_k, g^k s_1 = s_1.$$

It follows that the set of all pairs (s_i, t), for $i = 1, \ldots, k$; $t \in T$, is transformed onto itself by γ. We call this set a *block* and we want to find in what cycles this block splits under the influence of γ. We claim that this depends only on the product

$$h^* = h_{s_k} h_{s_{k-1}} \cdots h_{s_2} h_{s_1},$$

in the following way: Let h^* be of type $\{c_1, c_2, c_3, \ldots\}$; then the block splits into c_1 cycles of length k, c_2 cycles of length $2k$, etc. This is shown by investigating what γ and its powers do to an element (s_1, t). Successive applications of γ give

$$(s_1, t) \to (s_2, h_{s_1} t) \to (s_3, h_{s_2} h_{s_1} t) \to \cdots \to (s_1, h^* t).$$

Thus (s_1, t) is reproduced for the first time after kl applications of γ, if l is the first exponent for which $(h^*)^l t = t$. It follows that (s_1, t) generates a cycle of length kl. Noting that (s_1, t) and (s_1, t') generate the same cycle under application of γ if and only if t and t' generate the same cycle under application of h^*, we infer that our block contains exactly c_l cycles of length kl, and this holds for each l.

Accordingly, we attach the product $|H|^{-k} x_k^{c_1} x_{2k}^{c_2} \cdots$ to this block. If we sum this for $h_{s_1}, \ldots, h_{s_k}$, where $g, s_1, \ldots, s_k$ are fixed, we obtain $P_H(x_k, x_{2k}, x_{3k}, \ldots)$. This follows from the fact that if $h_{s_1}, \ldots, h_{s_k}$ all run through H, then the product h^* runs $|H|^{k-1}$ times through H.

Next we consider all elements γ of $G[H]$ arising from a single g of type $\{b_1, b_2, \ldots\}$. It is not difficult to show, by a technique analogous to the one of Sec. 5.7, that the contribution of these elements to the cycle index

$P_{G[H]}$ is

$$[P_H(x_1, x_2, \ldots)]^{b_1} [P_H(x_2, x_4, \ldots)]^{b_2} [P_H(x_3, x_6, \ldots)]^{b_3} \cdots .$$

Finally, by summing over all $g \in G$, we complete the proof of the theorem.

Pólya's original proof of Theorem 5.5 consisted of a very elegant, although not very easy, application of his fundamental theorem, Theorem 5.1. A brief sketch follows.

Let R be some finite set, and provide its elements r with weights $w(r)$. We consider mappings of T into R, and we form patterns λ by means of the equivalence defined by H. Let Λ be the set of these patterns. Its inventory is

$$(5.41) \qquad \sum_{\lambda \in \Lambda} W(\lambda) = P_H\{\sum w(r), \sum [w(r)]^2, \ldots\}.$$

Next consider mappings of S into Λ, and form patterns ψ by means of the equivalence defined by G. Defining weights $W^*(\psi)$ in the obvious way, for the inventory of the ψ's we obtain

$$(5.42) \qquad \sum_{\psi} P_G\{\sum W(\lambda), \sum]W(\lambda)]^2, \sum [W(\lambda)]^3, \ldots\}.$$

The sums

$$\sum [W(\lambda)]^2, \sum [W(\lambda)]^3, \ldots,$$

can be obtained by application of (5.41) to new weights, $w^2, w^3, \ldots$. Therefore (5.42) equals the expression we obtain by substituting

$$(5.43) \qquad x_1 = \sum w(r), \; x_2 = \sum [w(r)]^2, \ldots,$$

into the right-hand side of (5.40).

The next step is to construct a one-to-one correspondence between the set of ψ's and that of patterns that arise from the equivalence introduced by $G[H]$ in the set $R^{(S \times T)}$. Once this has been done, we can infer that our ψ inventory is also equal to the effect of (5.43) on the left-hand side of (5.40). Finally, (5.40) can be deduced from the fact that R and w are arbitrary in a way similar to that in Example 5.17.

Example 5.28 In n-dimensional vector space we consider the set R consisting of the $2n$ points

$$(1, 0, 0, \ldots, 0), (-1, 0, \ldots, 0), (0, 1, 0, \ldots, 0), (0, -1, 0, \ldots, 0), \ldots,$$
$$(0, \ldots, 0, 1), (0, \ldots, 0, -1).$$

Let G be the group of all linear homogeneous transformations of the space that map R onto itself. We can consider R as the cartesian product of a set $\{1, \ldots, n\}$ of n objects and the set $\{-1, 1\}$, by making a pair (k, ε) for $1 \leq k \leq n$, $\varepsilon = \pm 1$ correspond to the point $(x_1, \ldots, x_n)$ defined by

$x_k = \varepsilon$ and $x_i = 0$ if $i \neq k$. Furthermore, the group G, considered as a permutation group of R, is easily seen to be $S_n[S_2]$, where S_n is the symmetric group of degree n and S_2 is the one of degree 2.

By Theorem 5.5 and Example 5.5, we observe that the cycle index of G equals the coefficient of w^n in the power series development of

$$\exp\left[\tfrac{1}{2}w(x_1^2 + x_2) + \tfrac{1}{4}w^2(x_2^2 + x_4) + \tfrac{1}{6}w^3(x_3^2 + x_6) + \cdots\right].$$

Example 5.29 We have n cubes, whose faces we want to color red or blue. We ask for the number of ways to do this when equivalences are defined by permutations of the set of cubes and rotations of the separate cubes.

The group under consideration is $S_n[G]$, where S_n is the symmetric group of degree n and G is the cube-face group of Example 5.4. By (5.15) we have to substitute $x_1 = x_2 = \cdots = 2$ into its cycle index in order to find the required number. If we make this substitution in any of the polynomials

$$(5.44) \quad P_G(x_1, x_2, x_3, \ldots), P_G(x_2, x_4, x_6, \ldots), P_G(x_3, x_6, x_9, \ldots), \ldots,$$

we always get $P_G(2, 2, 2, \ldots)$, and this equals 10 (see Example 5.16). Thus the answer to our question is $S_n(10, 10, 10, \ldots)$. This number equals the coefficient of w^n in the development of

$$\exp\left(10w + \tfrac{1}{2} \cdot 10w^2 + \tfrac{1}{3} \cdot 10w^3 + \cdots\right) = (1 - w)^{-10},$$

and therefore

$$S_n(10, 10, 10, \ldots) = \frac{(n + 9)!}{n!\,9!}.$$

We ask a second question, connected with Example 5.27. How many of the preceding patterns have the property that they do not change if we interchange the colors? According to Example 5.27, this number is found by substituting

$$x_1 = x_3 = x_5 = \cdots = 0, \, x_2 = x_4 = x_6 = \cdots = 2$$

into the cycle index. Under this substitution, the polynomials (5.44) become $P_G(0, 2, 0, 2, \ldots)$ and $P_G(2, 2, 2, \ldots)$, alternately. As

$$P_G(0, 2, 0, 2, \ldots) = 2,$$

we obtain

$$P_{S_n[G]}(0, 2, 0, 2, \ldots) = P_{S_n}(2, 10, 2, 10, \ldots).$$

This is the coefficient of w^n in

$$\exp\left(2w + \tfrac{1}{2} \cdot 10w^2 + \tfrac{1}{3} \cdot 2w^3 + \tfrac{1}{4} \cdot 10w^4 + \cdots\right)$$
$$= \exp\left[2 \log (1 - w)^{-1} + 8 \log (1 - w^2)^{-\frac{1}{2}}\right] = (1 - w)^{-2}(1 - w^2)^{-4}$$
$$= (1 + 2w + w^2)(1 - w^2)^{-6}$$
$$= 1 + 2w + 7w^2 + 12w^3 + 27w^4 + 42w^5 + 77w^6$$
$$+ 112w^7 + 182w^8 + 252w^9 + \cdots.$$

For example, if $n = 5$, the required number of patterns is 42.

5.14 EPILOGUE

We shall make some comments about the material presented in this chapter and about related subjects.

It was pointed out by Harary ([9], Addendum 1; see also Sec. 6.11) that many important aspects of Pólya's theory were anticipated by Redfield [12] in a slightly different terminology. Redfield seems to have been the first to introduce the cycle index of a group, which he called the "group reduction function."

The main lemma (Lemma 5.1) has, of course, been rediscovered quite often. It is basic for all questions about numbers of equivalence classes. A discussion of this lemma, with a number of examples, has been given by Golomb [6]; the cycle index has been used in most of these examples, although not explicitly.

Pólya's fundamental theorem was presented by Pólya [10], by Riordan [13] (see also [14] for applications), by Uhlenbeck and Ford [16], and, in a more sophisticated form, by Riguet (Appendix V in [1]). For applications to trees and graphs, initiated by Pólya [10], we refer to the survey by Harary [9] and to the references given there; see also Uhlenbeck and Ford [16] and Chapter 6 of this book. Presently we shall discuss only applications and generalizations that do not refer to the counting of trees and graphs.

First we give some general considerations. If we have one or more permutation groups, operating on sets of objects, then we can often consider new sets of objects that are, in a natural way, subject to permutations induced by the original permutations. For this new situation, we can try to find the number of transitive sets, and we can also ask for the cycle index. A typical example is Sec. 5.12, where D and R are the original sets, with groups G and H, whereas the new objects are mappings of D into R, and the group elements g and h induce the permutation given by $f \rightarrow hfg^{-1}$. These permutations form a group, and the transitive sets are what we called the patterns.

As a further example, we mention the permutations induced by G and H on the cartesian product of D and R, where $g \in G$, $h \in H$ induce the mapping $(d, r) \to (gd, hr)$. This permutation group is what Harary [8] calls the *direct product* of G and H (he uses the term *direct sum* for the group of our Example 5.8). This group induces a group of permutations of the subsets of $D \times R$. If, instead of all subsets, we consider only subsets that are graphs of mappings of D into R, we are back to the case of Sec. 5.12.

Similarly, the Kranz $G[H]$ of Sec. 5.13 generates a group that Harary [8] calls the exponentiation H^G. Any element of $G[H]$ permutes the cartesian product $S \times T$, and therefore it permutes the subsets of $S \times T$. A subset that is a graph of a mapping of S into T is transformed into another subset of that type. Now these graphs form the new objects, and their permutation group is H^G. It does not seem easy to get a simple expression for the cycle index of H^G, although the way the elements of $G[H]$ have been grouped together in the first proof of Theorem 5.5 can be of some help.

A relatively simple example is the following. Let G be a permutation group of R, and let Φ be the set of mappings of R into itself. Any $g \in G$ induces a permutation of Φ by $\phi \to g\phi g^{-1}$; see [4] and, for one-to-one mappings, [2].

An essential aspect of Pólya's theory is the use of weights, of such a general nature that we can take them to be variables. A result of this is that we can ask questions about the numbers of patterns subject to some extra conditions, such as the condition about four red faces and two blue faces in Example 5.16. That is, the use of variable weights enables us to get generating functions. This aspect was lacking in Secs. 5.10 and 5.12, but these are special cases of more general theorems presented in [2], where generating functions do occur.

A further application of Pólya's theory has been given by Pólya [11] and Slepian [15]. Their papers consider symmetry types of Boolean functions of n Boolean variables. A function of n Boolean variables is a function defined on a set of 2^n elements, and on this set we can define a permutation group of order $n!2^n$ and of degree 2^n by taking all permutations of the variables and all possible ways of replacing a number of variables by their negation. (Using Harary's exponentiation, we can denote this group by $S_2^{S_n}$, where S_2 and S_n are the symmetric groups of degrees 2 and n, respectively.) The "symmetry types" are just the patterns of mappings of the domain set of 2^n elements into a range consisting of 2 elements, where patterns are taken with respect to equivalences introduced by the permutation group $S_2^{S_n}$ of the domain. Needless to say, we can also determine the number of patterns in case the equivalences are extended by taking each Boolean function to be equivalent to its negation.

This refers to the symmetric group in the two-element range, and the answer is given by Example 5.27.

Finally, we mention an application of Theorem 5.4 (see Example 5.26) to the case where G is a cyclic or dihedral permutation group and H is a symmetric group, by Gilbert and Riordan [5]. These authors also give applications to musical chords.

Note added in proof. After this chaper was written, the author published a new generalization of Pólya's theory in a paper that has meanwhile appeared ["Enumerative Combinatorial Problems concerning Structures," *Nieuw Archief Wiskunde* (3) **11,** (1963) 142–161]. It gives a large number of applications of a method based on attaching a polynomial $U(y_1, y_2, \ldots)$ to a given class of structures provided with a permutation group. (The reader may think of the class of all labeled graphs with n nodes and k edges, and may take for the permutation group the group of all permutations of the labels.) An element of the permutation group either transforms a structure into an equivalent structure or leaves the structure invariant. In the latter case the permutation belongs to the automorphism group of the structure. We select a complete set of nonequivalent structures from our class, and for each one of these we determine P_Z, that is, the cycle index of the automorphism group. The sum of all these P_Z is the polynominal U. This polynomial gives the solution to many combinatorial problems concerning our structures, in a way that is completely analogous to the way in which P_G gives the solution for a single set of objects permuted by a permutation group.

For example, considering the special case of graphs previously mentioned, we can ask for the number of essentially different colored graphs with n nodes and k edges, where the nodes are to be colored with colors from a store containing r colors. The answer is simply $U(r, r, r, \ldots)$, which is analogous to (5.15). In this example, and in many other cases, the polynomial U can be determined explicitly.

MULTIPLE-CHOICE REVIEW PROBLEMS

1. Let m be a fixed integer, and let $n = 2m + 1$. Consider the 10^n numbers of n digits (a number might start with one or more zeros), each one being printed on a slip of paper. Two slips are considered to be the same if one of them can be transformed into the other by putting it upside down (thus 0698161 is identified with 1918690, as the up-side-down 0, 6, 9, 8, 1 cannot be distinguished from 0, 9, 6, 8, 1, respectively). The number of different slips is

    (a) $10^n - \frac{1}{2} \cdot 5^n + \frac{3}{2} \cdot 5^m$,

    (b) $5^n + 3 \cdot 5^m$,

    (c) $10^n + \frac{1}{2}(5^{m+1} - 5^n)$,

    (d) none of the above.

2. Let S be a finite set, and let G be a permutation group of S. Consider the set T of all ordered pairs (a, b) with $a \in S$, $b \in S$. To any permutation $g \in G$, let there correspond a permutation h of T, defined by $h(a, b) = (ga, gb)$. These h's form a permutation group H of T. Its cycle index P_H is

    (a) exactly the same as P_G,

    (b) entirely determined by P_G, but not by the structure of G as an abstract group,

    (c) not properly defined by the preceding statements since the structure of T depends on the structure of S and not only on the structure of G,

    (d) entirely determined by the structure of G as an abstract group.

3. Consider a hollow cube with thin walls, which we want to paint both on the inside and on the outside. Thus twelve faces have to be painted. The amount of paint available suffices for seven red and five blue faces. If P_G denotes the cycle index, the number of patterns (with respect to the orthogonal group) equals the coefficient of $r^7 b^5$ in

    (a) $[P_G(r + b, r^2 + b^2, r^3 + b^3, \ldots)]^2$,

    (b) $P_G[(r + b)^2, (r^2 + b^2)^2, (r^3 + b^3)^2, \ldots]$,

    (c) $P_G(r^2 + br + b^2, r^4 + b^2 r^2 + b^4, r^6 + b^3 r^3 + b^6, \ldots)$,

    (d) $P_G(r^2 + b^2, r^4 + b^4, r^6 + b^6, \ldots)$.

4. Consider patterns of mappings of a set S into the set of residue classes mod 6, with equivalence induced by a permutation group G of S. Assume that $|G| > 1$. The weight of a function is a residue class mod 6, simply defined as the product of the function values $\Pi_{s \in S} f(s)$. Now we conjecture that the sum of the weights of the patterns is mod 6 congruent to $P_G(3, 1, 3, 1, \ldots)$, since the sum of the kth powers of the numbers 1, 2, 3, 4, 5, 6 is mod 6 congruent to 3 if k is odd, and to 1 if k is even. This is

    (a) a consequence of Pólya's fundamental Theorem 5.1,

    (b) true, but not a consequence of Theorem 5.1, because of the restrictions made in that theorem on the ring to which the weights belong,

    (c) nonsense, since the residue classes are function values, and cannot be used as weights at the same time,

    (d) nonsense, since we have not defined division in the ring of residue classes mod 6.

5. In the ring of polynomials in the six variables $z_1, z_2, z_3, z_4, z_5, z_6$, with integer coefficients, we consider the polynomial $\phi = z_1 z_2 z_3 + z_4 z_5 z_6$. Furthermore, we consider a permutation group G of the six symbols 1, 2, 3, 4, 5, 6, consisting of all permutations π that satisfy

$$\phi(z_{\pi(1)}, z_{\pi(2)}, z_{\pi(3)}, z_{\pi(4)}, z_{\pi(5)}, z_{\pi(6)}) = \phi(z_1, z_2, z_3, z_4, z_5, z_6)$$

as an identity in $z_1, \ldots, z_6$. This group is

    (a) the direct product $S_2 \times S_2 \times S_2$,

    (b) the direct product $S_3 \times S_3$,

    (c) the Kranz $S_3[S_2]$,

    (d) the Kranz $S_2[S_3]$.

REFERENCES

1. Berge, C., *Théorie des graphes et ses applications*, Collection Universitaire de Mathematiques 2, Dunod, Paris, 1958. Translation by Alison Doig, *The Theory of Graphs and Its Applications*, John Wiley and Sons, New York, 1962.

2. de Bruijn, N. G., "Generalization of Pólya's Fundamental Theorem in Enumerative Combinatorial Analysis," *Nederl. Akad. Wetensch. Proc. Ser. A* **62** = *Indag. Math.* **21** (1959), 59–69.

3. Burnside, W., *Theory of Groups of Finite Order*, 2nd edition, Cambridge University Press, Cambridge, England, 1911; Dover Publications, New York, 1955.

4. Davis, R. L., "The Number of Structures of Finite Relations," *Proc. Amer. Math. Soc.* **4** (1953), 486–495.

5. Gilbert, E. N., and Riordan, J., "Symmetry Types of Periodic Sequences," *Illinois J. Math.* **5** (1961), 657–665.

6. Golomb, S. W., "A Mathematical Theory of Discrete Classification," in C. Cherry (editor), *Information Theory*, Fourth London Symposium, Butterworths, London, 1961.

7. Hadwiger, H., "Gruppierung mit Nebenbedingungen," *Mitt. Verein. Schweiz. Versich.-Math.* **43** (1943), 113–122.

8. Harary, F., "Exponentiation of Permutation Groups," *Amer. Math. Monthly* **66** (1959), 572–575.

9. Harary, F., "Unsolved Problems in the Enumeration of Graphs," *Magyar Tud. Akad. Mat. Kutató Int. Közl.* = *Publ. Math. Inst. Hungarian Acad. Sci.* **5** (1960), 63–95.

10. Pólya, G., "Kombinatorische Anzahlbestimmungen für Gruppen, Graphen und chemische Verbindungen," *Acta Math.* **68** (1937), 145–254.

11. Pólya, G., "Sur les types des propositions composées," *J. Symbolic Logic* **5** (1940), 98–103.

12. Redfield, J. H., "The Theory of Group-Reduced Distributions," *Amer. J. Math.* **49** (1927), 433–455.

13. Riordan, J., *An Introduction to Combinatorial Analysis*, John Wiley and Sons, New York, 1958.

14. Riordan, J., "The Combinatorial Significance of a Theorem of Pólya," *J. Soc. Indust. Appl. Math.* **5** (1957), 225–237.

15. Slepian, D., "On the Number of Symmetry Types of Boolean Functions of n Variables," *Canadian J. Math.* **5** (1953), 185–193.

16. Uhlenbeck, G. E., and G. W. Ford, "Theory of Linear Graphs with Application to the Theory of the Virial Development of the Properties of Gases," in J. de Boer and G. E. Uhlenbeck (editors), *Studies in Statistical Mechanics*, Vol. 1, 123–211, North Holland Publishing Company, Amsterdam, 1962.

CHAPTER 6

Combinatorial Problems in Graphical Enumeration

FRANK HARARY

6.1 INTRODUCTION

Our object in this chapter is to present several unsolved problems in the enumeration of graphs in the hope of stimulating additional active interest in this area. It is not likely that all these problems will be settled in the near future, for included among their solutions there would in particular be enough information to settle the four-color conjecture.

We first illustrate what is meant by a graphical enumeration problem, using graphs and directed graphs. We then develop the preliminary concepts concerning graphs in order to be able to state the unsolved problems concisely. Statements (without proofs) of several methods that have been used in the enumeration of graphs are given. The most important tool in this area is provided by the elegant and powerful enumeration method of Pólya. This method or a variation thereof has been used in most known solutions to such problems. We compare problems involving the number of trees of various kinds with analogous problems for graphs. Lists of 27 solved problems and 27 unsolved problems are presented. The importance of the unsolved problems and the nature of their essential difficulties are indicated. A comprehensive bibliography concludes this chapter.

6.2 GRAPHICAL PRELIMINARIES

A *graph* G consists of a finite set V of *points* $v_1, v_2, \ldots, v_p$, together with a prescribed set X of unordered pairs of distinct points of V. Each such pair of points u and v is a *line* $x = uv$ of the graph G. We then say that points u and v are *adjacent* and that the point u and the line x are *incident* with each other. Note that by definition a graph has no line joining a point with itself, nor does it have two distinct lines joining the same pair of points. If the definition of a graph is generalized to permit *parallel lines*, that is, lines joining the same pair of distinct points, the result is called a *multigraph*. If, in addition to parallel lines, we further allow the presence of *loops*, that is, lines joining points with themselves, then we have a *general graph*.

A *subgraph* of a graph G consists of a subset of V and a subset of X that themselves form a graph. A *spanning* subgraph of G has the same set V of points as G.

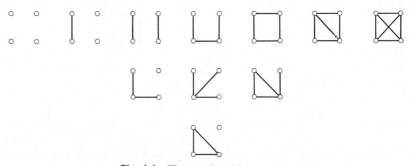

Fig. 6.1   The graphs of four points.

Two graphs are *isomorphic* if there exists a one-to-one correspondence between their point sets that preserves adjacency. In Fig. 6.1 we show all the graphs (up to isomorphism) of four points.

For a more extensive discussion of graph theory, see König [38], Berge [1], and Ore [45].

A *directed graph* (or more briefly a *digraph*) consists of a finite set V of points together with a prescribed collection of *ordered pairs* of distinct points of V. Each such ordered pair (u, v) is called a *directed line* (or a *line* when the meaning is clear by context), and is denoted $\overrightarrow{uv}$ (or more briefly uv when the direction is clear). The definition of isomorphism of digraphs is analogous to that of graphs. In Fig. 6.2 are shown all digraphs of three points. For an extensive treatment of digraphs *per se*, see Harary, Norman, and Cartwright [31].

Let g_{pq} be the number of graphs with p points and q lines. Enumerating the graphs of p points means finding an expression for the generating function, or counting series (polynomial),

$$g_p(x) = g_{p0} + g_{p1}x + g_{p2}x^2 + \cdots ,$$

in which the highest power of x is $p(p-1)/2$.

Denote by $\bar{g}_{pq}$ the number of digraphs with p points and q lines. Then

$$\bar{g}_p(x) = \bar{g}_{p0} + \bar{g}_{p1}x + \bar{g}_{p2}x^2 + \cdots$$

is the counting series for digraphs with p points, and here the highest power of x is $p(p-1)$. From Figs. 6.1 and 6.2 we see that the counting series for the graphs of four points and the digraphs of three points are respectively

$$g_4(x) = 1 + x + 2x^2 + 3x^3 + 2x^4 + x^5 + x^6,$$

$$\bar{g}_3(x) = 1 + x + 4x^2 + 4x^3 + 4x^4 + x^5 + x^6.$$

A *walk* of a graph is a sequence, beginning with a point and ending with a point, in which points and lines alternate and each line is incident with

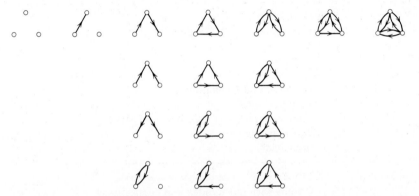

Fig. 6.2   The digraphs of three points.

the points immediately preceding and following it. A walk of the form $v_0, x_1, v_1, x_2, \ldots, v_n$ is said to *join* v_0 with v_n. The *length* of a walk is the number of lines in it. A *trajectory* is a walk in which all lines are distinct; a *path* is a walk in which all points are distinct. An *open walk* has distinct first and last points; a *closed walk* is not open. A *cycle* is a closed walk in which all points are distinct, except the first and last, which coincide. A *complete walk* contains all the points of G. A graph is *connected* if each pair of its points are joined by a path. A *tree* is a connected graph that has no cycles. The *distance* between two points is the length of a shortest path joining them. The *degree* of a point is the number of lines incident with it; the *partition* of a graph is the sequence of degrees of its points.

If each pair of points of a graph is joined by a line, the graph is called *complete* and is denoted by K_p.

There are corresponding definitions for digraphs. For example, a *directed path from v_1 to v_n* consists of a sequence of directed lines v_1v_2, $v_2v_3, \ldots, v_{n-1}v_n$, together with the n distinct points. If there is a directed path from u to v, then v is *reachable* from u. The *outdegree* of a point v is the number of directed lines from v; the *indegree* is the number of directed lines to v. The *partition* of a digraph is the sequence of ordered pairs consisting of the outdegree and indegree of each point.

TABLE 6.I

Unsolved Problems in the Enumeration of Graphs

Category	Problem	Number
Digraph	Strong digraphs	1
	Unilateral digraphs	2
	Digraphs with a source	3
Partition	Graphs with a given partition	4
	Homeomorphically irreducible graphs	5
	Regular graphs	6
	Eulerian graphs	7
Topological	k-colored and k-colorable graphs	8
	Planar graphs	9
	Planar graphs with additional properties	10
	Biplanar graphs	11
	Simplicial complexes	12
Connectivity	Graphs with given diameter and girths	13
	Graphs with given index and connectivity	14
	Nonseparable graphs	15
	Line graphs	16
Group	Symmetric graphs	17
	Identity graphs	18
	Graphs with a given group	19
	Latin squares	20
Electrical	Types of complete cycles in an n-cube	21
	Finite automata	22
	Indecomposable two-terminal networks	23
Physical	Ising problem of dimension n	24
	Ising problem with magnetic field	25
	Paving problem	26
	Cell-growth problem	27

In Secs. 6.4 to 6.10, we shall describe these 27 problems in more detail.

6.3 SOME UNSOLVED PROBLEMS

We regard a solution of each of the unsolved problems of Table 6.1 as a generating function in closed form for the number of graphs of each given kind with a given number p of points and a given number q of lines (or directed lines for digraphs). These problems are partitioned into seven categories combining related problems, as shown in Table 6.1.

6.4 PROBLEMS INVOLVING DIRECTED GRAPHS

A digraph D is *strongly connected* or *strong* if each point is reachable from each other point; D is *unilaterally connected* or *unilateral* if for any

Fig. 6.3   Digraphs with various kinds of connectedness.

two points, at least one is reachable from the other; D is *weakly connected* or *weak* if, for any division of its set of points into two nonempty subsets, there is a line from a point of one subset to a point of the other. Finally, D is *disconnected* if it is not weak. (See Fig. 6.3.) A *source* of a digraph is a point from which all other points are reachable. Consider the following categories.

Strong digraphs

Unilateral digraphs

Digraphs with a source

For each of these, we obtain from Fig. 6.2 the counting series for digraphs of three points. For strong digraphs this is

$$x^3 + 2x^4 + x^5 + x^6,$$

since, for example, there are two strong digraphs with three points and four lines. For unilateral digraphs, this counting series is

$$x^2 + 4x^3 + 4x^4 + x^5 + x^6,$$

and for those with a source it is

$$2x^2 + 4x^3 + 4x^4 + x^5 + x^6.$$

6.5 PROBLEMS INVOLVING PARTITIONS

Graphs with a given partition

From Fig. 6.1, we see that each graph of four points has a different partition. For example, the graph consisting of a single cycle of length 4 has partition $(2, 2, 2, 2)$ and is the only graph with this partition. Starting with graphs of five points, however, there exist partitions that belong to

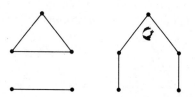

Fig. 6.4   Two graphs with the same partition.

more than one graph. An example is given by the two graphs shown in Fig. 6.4, each of which has the partition $(1, 1, 2, 2, 2)$.

Homeomorphically irreducible graphs

A graph is *homeomorphically irreducible* if it has no point of degree 2. Inspection of Fig. 6.1 shows that the counting series for homeomorphically irreducible graphs of four points is

$$1 + x + x^2 + x^3 + x^6.$$

Regular Graphs

A *regular* graph is one in which all points have the same degree; it is called *cubic* if each point has degree 3. Regular graphs are an interesting special case of graphs with a given partition.

Every regular graph of degree 1 has an even number $2n$ of points, which are joined by n lines to form n connected components. Every regular graph of degree 2 has a cycle for each of its components. The first interesting case of regular graphs is given by cubic graphs. The only cubic graph of four points is the complete graph shown in Fig. 6.1; hence the counting series for cubic graphs of four points is simply x^6.

Eulerian graphs and digraphs

Euler himself proved that a graph has a complete closed trajectory containing all its lines if and only if it is connected and every point has even degree. For this reason such graphs are called *Eulerian*. From Euler's result it follows that Eulerian graphs are subsumed in the category of

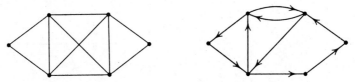

Fig. 6.5   An Eulerian graph and an Eulerian digraph.

graphs with a given partition. A digraph is Eulerian if and only if it is strong and each point has equal indegree and outdegree. In Fig. 6.5, an Eulerian graph and an Eulerian digraph are shown.

6.6   TOPOLOGICAL PROBLEMS

k-colored and *k*-colorable graphs

A graph is *k-colored* if each of its points has been assigned one of k colors in such a way that no two points of the same color are adjacent and all k colors have been used. Two *k*-colored graphs are regarded as

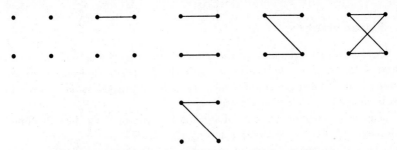

Fig. 6.6   The bicolored graphs with two points of each color.

isomorphic if there is an isomorphism between them as graphs such that two points of the first graph have the same color if and only if their image points have the same color. A graph is *k*-colorable if it can be *k*-colored. For example, the bicolored graphs with two points of each color are shown in Fig. 6.6, in which the two points to the left in each graph are regarded as colored with the first color, whereas the two points to the right are colored with the second color.

Only the numbers of bicolored and bicolorable graphs have been found [25, 33]. It is clear that there is a one-to-one correspondence between connected bicolored graphs and connected bicolorable graphs, but there are more disconnected bicolored graphs than disconnected bicolorable graphs. By a theorem of König [38], a graph is bicolorable if and only if all

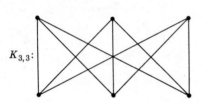
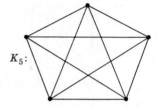

Fig. 6.7 Two nonplanar graphs.

its cycles have even length. Thus we see from Fig. 6.1 that the number of bicolorable graphs of four points is given by the series

$$1 + x + 2x^2 + 2x^3 + x^4.$$

Just as K_p denotes the complete graph with p points, we let $K_{m,n}$ denote the complete bicolored graph having m points of one color and n points of the other, in which two points are adjacent if and only if they are of different colors. In general, we let $K_{n_1, n_2, \ldots, n_k}$ denote the complete k-colored graph, with n_i points of the ith color.

Planar graphs

A *plane graph* is a subset of the plane containing a set of *nodes* (vertices) and a set of *closed arcs* (edges), with the property that any two arcs *intersect* in at most one point, which is a node. A graph is called *planar* if it is isomorphic with a plane graph.

The *subdivision of a line uv* of a graph is accomplished when this line is replaced by two lines uw and wv together with a new point w. A *subdivision of a graph* is the result of successive subdivisions of its lines. Two graphs are called *homeomorphic* if they have isomorphic subdivisions. Kuratowski has shown that a graph is planar if and only if it contains no subgraph homeomorphic to either of the graphs $K_{3,3}$ or K_5 shown in Fig. 6.7. For a very clear proof of this theorem, see Dirac and Schuster [10]. Another criterion for a graph to be planar was discovered by Whitney [69].

Since every graph of four points is planar, the counting series for the planar graphs of five points is obtained from that of all graphs of five points by subtracting x^{10}.

Planar graphs with additional properties

These problems include planar cubic graphs, planar Eulerian graphs, and planar k-colorable graphs. They are of interest mainly because of their connection with the four-color conjecture.

Biplanar graphs

A graph is *biplanar* if it is the union of two spanning planar subgraphs. Thus the graph K_5 shown in Fig. 6.7 is biplanar since it is the union of two planar subgraphs, for example, two cycles of length 5. Of course, every planar graph is biplanar, but not conversely. There is no known combinatorial criterion for a graph to be biplanar.

Simplicial complexes

An (abstract) *simplicial complex* consists of a set P of *points* and a collection S of subsets of P called *simplexes*, which satisfy the following two conditions:

1. Each point is a simplex.
2. Each nonempty subset of a simplex is a simplex.

The *dimension* of a simplex is one less than the number of points it contains. The problem is to find the number of isomorphism types of simplicial complexes with a given number of simplexes of each dimension. We illustrate this by using Fig. 6.1 to write the counting series for the simplicial complexes with four points, and a given number of 1-simplexes (lines) and 2-simplexes (triangles). Letting x and y be the variables representing the 1-simplexes and 2-simplexes, respectively, we find that this series is

$$1 + x + 2x^2 + 3x^3 + x^3y + 2x^4 + x^4y + x^5 + x^5y$$
$$+ x^5y^2 + x^6 + x^6y + x^6y^2 + x^6y^3 + x^6y^4.$$

6.7 PROBLEMS INVOLVING CONNECTIVITY

Graphs with given diameter and given girths

The *diameter* of a connected graph is the maximum distance between any two of its points. The *lower girth* is the length of any shortest cycle, and the *upper girth* is the length of a longest cycle. From Fig. 6.1 we see that the counting series for connected graphs of four points with diameter 2 is $x^3 + 2x^4 + x^5$, that exactly three of these connected graphs have lower girth 3, and also that there is the same number with upper girth 4.

Graphs of given index and given connectivity

The *removal of a point v*, and of the lines incident with v, from a graph G results in the graph $G - v$, which is the maximal subgraph of G not containing v. Similarly, the graph $G - x$ obtained on the *removal of a line* x from G is the maximal subgraph of G not containing x. The *connectivity* of G is the smallest number of points whose removal results in a graph that is disconnected or consists of a single point. The *index* of a connected graph is the smallest number of lines whose removal results in a tree.

From Fig. 6.1, we find that the counting series for connected graphs of four points with connectivity 1 is $2x^3 + x^4$. Among the connected graphs of four points, there are two graphs of index 1, one of index 2, and one of index 3.

Nonseparable graphs

A *cut point* of a connected graph is a point whose removal results in a disconnected graph. A *block of a graph* is a maximal connected subgraph with no cut points.

A *nonseparable graph* has exactly one block. Since a connected graph has connectivity 1 if and only if it has a cut point, it follows at once that nonseparable graphs have connectivity greater than 1. From Fig. 6.1, the counting series for nonseparable graphs of four points is $x^4 + x^5 + x^6$.

Line graphs

Two lines of a graph G are called *adjacent* if they both are incident with a common point of G. The *line graph of G* is the graph with points corresponding to the lines of G, in which two points are adjacent whenever the corresponding lines of G are adjacent (see Fig. 6.8). A graph is called a *line graph* if it is the line graph of some graph. It might be noted that here different authors employ different terminology, with some more descriptive than others. Thus the line graph of G is called its *interchange graph* by Ore [45] and its *derivative graph* by Sabidussi [61].

The concept of the line graph of a graph was introduced by Whitney [68], who showed that a line graph H is the line graph of only one graph unless $H = K_3$. Krausz [39] has obtained the following elegant characterization of line graphs: A graph G is a line graph if and only if there exists a partition of the set of lines of G into complete subgraphs such that no point of G lies in more than two of these subgraphs.

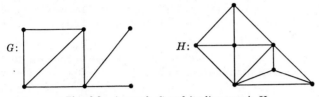

$G:$ $H:$

Fig. 6.8   A graph G and its line graph H.

The present problem is to find the number of line graphs with a given number of points and lines.

6.8   PROBLEMS INVOLVING GROUPS

Symmetric graphs

An *automorphism* of a graph is an isomorphism with itself. Two points of a graph are *similar* if there is an automorphism that maps one of the points onto the other; similarity of two lines is defined analogously.

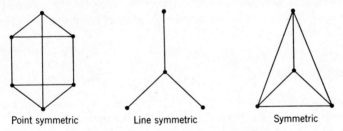

Point symmetric   Line symmetric   Symmetric

Fig. 6.9   Symmetric graphs.

A graph is *point-symmetric* if all its points are similar; it is *line-symmetric* if all its lines are similar; and it is *symmetric* if it is both point-symmetric and line-symmetric. In Fig. 6.9 are graphs of each type. The problem is to enumerate each of these three kinds of graphs having a given number of points and lines.

Identity graphs

In an *identity graph* the only automorphism is the identity mapping. In Fig. 6.10 are shown the smallest identity graphs respectively without and with cycles.

Graphs with a given group

The *group of a graph* is the collection of all its automorphisms; it is a permutation group acting on the set of points. It is known [13] that every finite group is abstractly isomorphic to the group of some graph, but it is not known generally whether a given permutation group is a graph group. The general problem, which includes this one, is to find the number of graphs with a given permutation group.

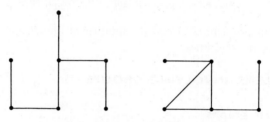

Fig. 6.10   Identity graphs.

The *line group* of G is the permutation group whose objects are the lines of G and whose permutations are induced by those of the group of automorphisms. We may also ask for the number of graphs with a given line group.

Latin squares

A *Latin square* of order n is a square matrix of order n in which each row and each column is a permutation of the integers $1, 2, \ldots, n$. Let L_n be the number of Latin squares in which the first row and first column are in the standard order $1, 2, \ldots, n$. The only known terms of the counting series for Latin squares are the first seven:

$$x + x^2 + x^3 + 4x^4 + 56x^5 + 9408x^6 + 16{,}942{,}080x^7 + \cdots.$$

Every Latin square may be regarded as a bicolored graph $K_{n,n}$ in which the lines are also colored. The points u_i of the first color correspond to the rows of the Latin square, whereas the points v_i correspond to the columns. Each line of $K_{n,n}$ is colored with one of n colors so that each point is incident with exactly one line of each color. The matrix interpretation of such a graph is that the color of the line $u_i v_j$ is the (i, j) entry of the matrix.

6.9 ELECTRICAL PROBLEMS

Dissimilar complete cycles in an n-cube

An n-cube is a graph of 2^n points, each of which is a binary sequence of n digits, in which two points are adjacent provided they differ in exactly one digit. In Fig. 6.11 are shown the 2-cube and 3-cube. Two complete cycles of an n-cube are *similar* if there is an automorphism of the cube

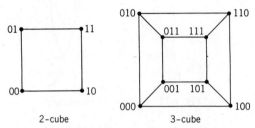

2-cube   3-cube

Fig. 6.11   Two cubes.

that maps one cycle onto the other. It is quite easy to convince oneself that there is exactly one similar type of complete cycle in a 3-cube. It has been shown by Gilbert [15] that the counting series for this problem begins $x^2 + x^3 + 9x^4$, where the coefficient of x^n is the number of dissimilar complete cycles in an n-cube. This coefficient is not known even for x^5. The total number of complete cycles also is not known for $n > 4$.

Finite automata

A *finite automaton*, or a *sequential machine* with two inputs 0, 1 (and a finite number of states), may be defined as follows. Consider a general digraph (that is, one permitting directed loops and parallel-directed lines), and call its points *states*, one of which is rooted and called the *initial state*. Each point has two lines from it, one labeled 0 and the other 1. The initial state is called a *source* (see Fig. 6.12). The two labels on the lines of the digraph are called *inputs* and serve to determine the next state of the machine when a particular given state and input are known.

Indecomposable two-terminal networks

A *two-terminal network* is a connected multigraph in which two points are marked u and v; they are called the *first terminal* and the *second terminal*. The *product* or *series connection* $N = N_1 N_2$ of two two-terminal networks N_1 and N_2 is the network obtained on identifying the points v_1 and u_2. The *sum* or *parallel connection* $N = N_1 + N_2$ is obtained on identifying u_1 with u_2 and also v_1 with v_2. These two operations on networks are illustrated in Fig. 6.13.

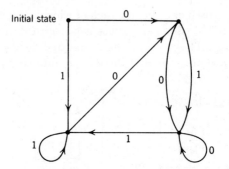

Fig. 6.12 A finite automaton.

A two-terminal network is *series-parallel* if it can be constructed from a finite succession of series and parallel connections starting with the network having exactly two adjacent points u and v. It is well known [57] that a two-terminal network is series-parallel if and only if it is *unidirectional;*

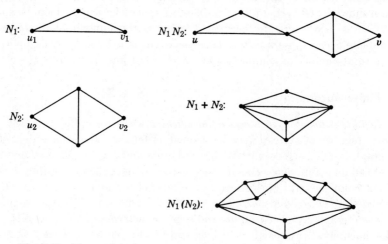

Fig. 6.13 The product, sum, and composition of two-terminal networks.

that is, no two paths from u to v contain any two points w_1 and w_2 in opposite orders.

The *composition* $N = N_1(N_2)$, where N_1 is series-parallel, is obtained on replacing each line of N_1, using unidirectionality, by the network N_2. In

Fig. 6.13, the composition of two networks, the first of which is series-parallel, is also illustrated.

A network N is *indecomposable* if it is not possible to write it in the form $N = N_1(N_2)$. Vetuchnovsky [67] has obtained upper and lower bounds for the number of indecomposable two-terminal series-parallel networks with a given number of points. The exact number is not known; its determination constitutes the present problem.

6.10 PHYSICAL PROBLEMS

The Ising problem of dimension *n*

In a *labeled graph* each point is colored with a different color so that it will be distinguished from all others. In a *two-dimensional lattice*

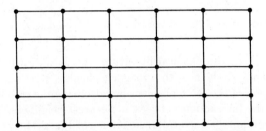

Fig. 6.14   A two-dimensional lattice graph.

graph, the points are ordered pairs (i, j), $i = 1, 2, \ldots, m$; $j = 1, 2, \ldots, n$. Two points are *adjacent* if the Euclidean distance between them is 1. An example is shown in Fig. 6.14; we note that in physical applications this would usually be considered as drawn on a torus, that is, with both pairs of opposite sides identified. An n-dimensional lattice graph is defined similarly.

Consider a labeled graph that is an n-dimensional lattice. A subgraph of this lattice is called *admissible* if and only if each of its points has even degree. Let A_q be the number of different labeled admissible subgraphs with q lines. The n-dimensional *Ising problem* is to find a generating function for the quantity A_q. This problem was proposed by Ising in [36]. For $n = 2$, Onsager [44] obtained a solution that did not use combinatorial methods. His procedure has not been generalized to higher dimensions. Recently, a purely combinatorial solution for $n = 2$ was derived by Sherman [63]. No real beginning has been made toward the solution of the problem for $n \geq 3$.

The Ising problem with magnetic field

By the *area* of an admissible labeled subgraph of a two-dimensional lattice we mean the minimum area enclosed by disjoint cycles constituting this subgraph. Let $A_{q,r}$ be the number of admissible labeled subgraphs with q lines and area r. The problem is to find a generating function for the

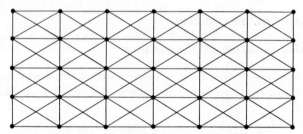

Fig. 6.15 A two-dimensional lattice graph with diagonals added.

quantities $A_{q,r}$. In the physical literature, this is shown to be the two-dimensional Ising problem with a magnetic field.

As a variation of the Ising problem, we also have the case known in the literature as the interaction between nonnearest neighbors. In Fig. 6.15 is shown the graph obtained from the lattice of Fig. 6.14 by joining those pairs of nonadjacent points that are nearest to each other. The problem is to determine the counting series for admissible subgraphs of such graphs.

A paving problem

Begin with a two-dimensional lattice having N squares, and consider n_1 squares and n_2 double squares (like dominoes) such that $n_1 + n_2 = N$. In how many ways can the labeled lattice be paved by these?

The cell-growth problem

Consider a one-celled animal that has a square shape and that can grow in the plane by adding a square cell of the same size to any of its sides. How many connected animals A_r with r cells are there, up to isomorphism? The animals are assumed to be simply connected in that they have no holes. In Fig. 6.16 are shown all such animals with up to five cells.

Recently, Read [54] has ingeniously extended the known results, which are now as shown in Table 6.2.

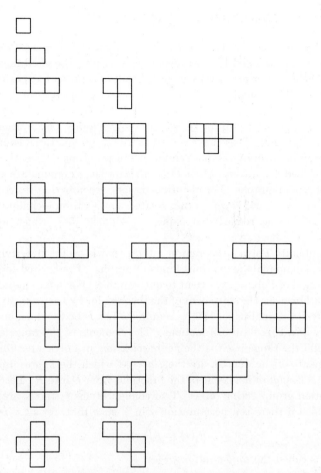

Fig. 6.16   The animals with at most five cells.

TABLE 6.2
The Number A_n of Simply Connected Animals of Area n

n	1	2	3	4	5	6	7	8	9	10
A_n	1	1	2	5	12	35	107	363	1248	4271

6.11 GRAPH-COUNTING METHODS

In the preceding chapter, de Bruijn [4] discussed Pólya's theory of counting and developed his own generalization [3]. We shall now state Pólya's theorem without proof in order that the present exposition be self-contained, and shall then mention a special case derived by Davis [8] and Slepian [64].

A theorem of Otter [46] for trees has been generalized by Norman [43] to arbitrary graphs. These have served as useful combinatorial lemmas in the solution of several graphical enumeration problems. Recently, Read [51] developed his *superposition theorem* in order to enumerate certain classes of general graphs. For historical reasons, we note that the counting theorem of Pólya and several other related enumeration techniques were anticipated in the remarkable paper of Redfield [56], which appears largely to have been overlooked.

We shall state Pólya's theorem in a form that is useful in deriving the counting polynomials for various kinds of graphs. The desired form is a specialization of Pólya's statement to one variable. For some enumeration problems, however, the statement of the theorem for two or more variables, as in Pólya's original paper [47], is required. Let D be the domain and R the range of a collection of functions. The elements of the range are the *figures*, and the range itself is the *figure collection*, in Pólya's terminology. The elements of the domain are the *places* at which the figures are to be located. A *configuration* is one of the functions from D into R. Let Y be a permutation group acting on D. Two configurations f_1 and f_2 are called *Y-equivalent* if there is a permutation α in Y such that, for all x in D,

$$f_1(x) = f_2(\alpha x).$$

Thus Y is called the *configuration group*.

To each figure there is assigned a nonnegative integer called its *content*. Let a_k denote the number of different figures of content k. Then the *figure-counting series* $a(x)$ is defined by

(6.1) $$a(x) = \sum_{k=0}^{\infty} a_k x^k.$$

Let Y be a permutation group of degree s and order h. Thus a configuration is a sequence of s figures. The *content of a configuration* is the sum of the contents of its figures. Let F_k denote the number of Y-inequivalent configurations of content k. The configuration-counting series $F(x)$ is defined by

(6.2) $$F(x) = \sum_{k=0}^{\infty} F_k x^k.$$

The object of Pólya's theorem is to express $F(x)$ in terms of $a(x)$ and Y. This is accomplished by means of the cycle index of Y, defined as follows. Let $h(j)$ denote the number of elements of Y of type $(j) = (j_1, j_2, \ldots, j_s)$, that is, having j_k cycles of length k, for $k = 1, 2, \ldots, s$, so that

$$(6.3) \qquad j_1 + 2j_2 + \cdots + sj_s = s.$$

Let $y_1, y_2, \ldots, y_s$ be s indeterminates. Then $Z(Y)$, the *cycle index* of Y, is defined as

$$(6.4) \qquad Z(Y) = \frac{1}{h} \sum_{(j)} h(j) y_1^{j_1} y_2^{j_2} \cdots y_s^{j_s},$$

where the sum is taken over all partitions (j) of s satisfying (6.3). For any function $f(x)$, let $Z[Y, f(x)]$ denote the function obtained from $Z(Y)$ by replacing each indeterminate y_k by $f(x^k)$.

Pólya's Theorem. The configuration series is obtained by substituting the figure-counting series into the cycle index of the configuration group. Symbolically, we have

$$(6.5) \qquad F(x) = Z[Y, a(x)].$$

This theorem reduces the problem of finding the configuration-counting series to the determination of the figure-counting series and the cycle index of the configuration group.

The following special case of Pólya's theorem has been independently discovered by Davis and Slepian. Very simply stated, the special case is obtained from Pólya's theorem (6.5) by substituting $x = 1$. Formally, this gives $F(1) = Z[Y, a(1)]$. From (6.2) we have $F(1) = \Sigma F_k$, and from (6.1) we obtain $a(1) = \Sigma a_k$; but $F(1)$ is the total number of inequivalent configurations, without regard to content, and similarly $a(1)$ is the total number of figures, without regard to content. Hence the substitution of $x = 1$ in (6.5) results in the following formula for the total number of configurations, in terms of the total number of figures and the configuration group. Using the notation of [20], let $B = F(1)$ and $b = a(1)$. Then (6.5) becomes

$$(6.6) \qquad B = \frac{1}{h} \sum_{(j)} h(j) b^{\Sigma j_k}.$$

Thus B is obtained at once from the cycle index of the configuration group.

Pólya [49] has written a beautiful and clear exposition of picture writing which gives an aid to intuition in thinking about graphical-enumeration problems.

The generalization of Pólya's theorem by de Bruijn considers the situation in which there is a permutation group acting on the range R, as well as the group Y acting on D.

Otter's dissimilarity characteristic theorem for trees, given in (6.7), was used as a lemma in his elegant enumeration [46] of trees in terms of *rooted trees*, that is, trees in which one point is distinguished from all others. A generalization (6.8) of this theorem by Norman [43] enabled him to solve the more general enumeration problem of finding the number of graphs with given blocks. These theorems and other formulas are derived in [30].

By the *number of dissimilar points* of a graph we mean the number of equivalence classes of similar points; analogous definitions may be given for lines, blocks, etc.

Dissimilarity characteristic theorem for trees

Let p' and q' be the numbers of dissimilar points and lines of a tree, and let q'' be the number of lines having points that are similar. Then we have

(6.7) $$p' - (q' - q'') = 1.$$

Dissimilarity characteristic theorem for graphs

Let G be a connected graph with k dissimilar blocks. Let p' be the total number of dissimilar points in G, and let p'_i be the number of dissimilar points in the ith dissimilar block of G. Then we have

(6.8) $$\sum_{i=1}^{k} (p'_i - 1) = p' - 1.$$

The application of equations (6.7) and (6.8) to graph-counting problems is made by summing each of these equations over the collection of all graphs to be enumerated. The term 1 when summed over all graphs obviously gives the total number of graphs, whereas the term p' becomes the number of rooted graphs under consideration. Combinatorial devices then serve to yield formulas for the summation of the remaining terms in these formulas.

Read [51] has developed his superposition theorem to treat an interesting class of enumeration problems that do not appear to be solvable by the preceding methods. The *superposition* of a collection of graphs all on the same set of points is the union of their sets of lines, including multiplicity. For example, in Fig. 6.17 is shown the multigraph G, which is the result of the superposition of graphs G_1, G_2, and G_3. One interpretation of superposition is that the lines of G_i have color i which is different from color j

for $j \neq i$, and these colors are preserved in G. If the points of any of the graphs G_i are labeled differently, the resulting superposed graph need not be isomorphic to the given one. Thus one can ask the question: Given n graphs $G_1, G_2, \ldots, G_n$, how many nonisomorphic superposed graphs can be formed from them? This number, it turns out, depends only on the automorphism groups $Y_1, Y_2, \ldots, Y_n$ of the graphs; it is given by an expression that we denote by $N(Y_1, Y_2, \ldots, Y_n)$. Let h_i be the order of

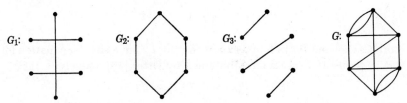

Fig. 6.17   Three graphs and their superposed multigraph G.

the group Y_i, and let $h_i(j)$ be the number of permutations of type (j) in the group Y_i as defined by (6.4).

Superposition Theorem. For n graphs $G_1, G_2, \ldots, G_n$, with automorphism groups $Y_1, Y_2, \ldots, Y_n$, let h_i be the order of the group Y_i, and let $h_i(j)$ be the number of permutations of type (j) in the group Y_i. Then the number of nonisomorphic superposed graphs that can be formed from $G_1, G_2, \ldots, G_n$ is given by

$$(6.9) \qquad N(Y_1, Y_2, \ldots, Y_n) = \frac{Q}{h_1 h_2 \cdots h_n},$$

where

$$Q = \sum_{(j)} h_1(j) h_2(j) \cdots h_n(j) (1^{j_1} 2^{j_2} \cdots s^{j_s} j_1! j_2! \cdots j_s!)^{n-1}.$$

6.12 TREE-COUNTING METHODS

We now present some of the results that have been obtained relative to the counting series for trees. Let T_p be the number of rooted trees with p points; then the counting series for rooted trees is denoted by

$$T(x) = \sum_{p=1}^{\infty} T_p x^p.$$

We define t_p and $t(x)$ similarly for unrooted trees. The well-known combinatorial identity [47],

$$(6.10) \qquad \sum_{n=0}^{\infty} Z[S_n, f(x)] = \exp \sum_{r=1}^{\infty} \frac{f(x^r)}{r},$$

where by definition $Z[S_0, f(x)] = 1$, is useful in obtaining some of the following formulas as well as in solving other counting problems.

Pólya showed that $T(x)$ satisfies the functional equation

(6.11) $$T(x) = x \exp \sum_{r=1}^{\infty} \frac{T(x^r)}{r} .$$

This equation is also readily derived from the earlier result of Cayley [6] that

(6.12) $$T(x) = x \prod_{r=1}^{\infty} (1 - x^r)^{-T_r}.$$

Both Cayley and Pólya found expressions for t_p, but a simpler equation for $t(x)$ in terms of $T(x)$ was first obtained by Otter [46], namely

(6.13) $$t(x) = T(x) - \tfrac{1}{2}[T^2(x) - T(x^2)].$$

In [19] it is shown that this equation may be derived from the formulas of Pólya by straightforward manipulations. Explicitly, we find

$$T(x) = x + x^2 + 2x^3 + 4x^4 + 9x^5 + 20x^6 + 48x^7 + 115x^8$$
$$+ 286x^9 + 719x^{10} + 1842x^{11} + 4766x^{12} + \cdots,$$

and

$$t(x) = x + x^2 + x^3 + 2x^4 + 3x^5 + 6x^6 + 11x^7 + 23x^8$$
$$+ 47x^9 + 106x^{10} + 235x^{11} + 551x^{12} + \cdots.$$

The methods of Pólya and Otter can be extended to count other kinds of trees. Let $h(x)$, $H(x)$, and $\bar{H}(x)$ be the counting series for homeomorphically irreducible trees, rooted trees, and planted trees, respectively, where a *planted* tree is a rooted tree in which the root has degree 1. Harary and Prins [32] show that $\bar{H}(x)$ satisfies the functional equation

(6.14) $$\bar{H}(x) = \frac{x^2}{1 + x} \exp \sum_{r=1}^{\infty} \frac{\bar{H}(x^r)}{rx^r} .$$

They next obtain the following expression for $H(x)$ in terms of $\bar{H}(x)$:

(6.15) $$H(x) = \frac{1 + x}{x} \bar{H}(x) - \frac{1}{2x}[\bar{H}^2(x) + \bar{H}(x^2)].$$

Otter's Theorem, (6.7), then yields

(6.16) $$h(x) = H(x) - \frac{1}{x^2}[\bar{H}^2(x) - \bar{H}(x^2)].$$

Explicitly, we have

$$h(x) = x + x^2 + x^4 + x^5 + 2x^6 + 2x^7 + 4x^8$$
$$+ 5x^9 + 10x^{10} + 14x^{11} + 26x^{12} + \cdots.$$

By an abuse of notation, let

$$Z(A_n - S_n) = Z(A_n) - Z(S_n),$$

where A_n denotes the alternating group of degree n. Riordan has established the following formula, which is related to (6.10):

(6.17) $$\sum_{n=0}^{\infty} Z[A_n - S_n, f(x)] = \exp \sum_{r=1}^{\infty} \frac{(-1)^{r+1}f(x^r)}{r}.$$

Let $u(x)$ and $U(x)$ be the counting series for trees and rooted trees for which the automorphism group is the identity group. Using (6.17), Harary and Prins [32] show that $U(x)$ satisfies the functional equation

(6.18) $$U(x) = x \exp \sum_{r=1}^{\infty} \frac{(-1)^{r+1}U(x^r)}{r},$$

and that $u(x)$ may be expressed in terms of $U(x)$ as follows:

(6.19) $$u(x) = U(x) - \tfrac{1}{2}[U^2(x) + U(x^2)].$$

Explicitly, we have

$$u(x) = x + x^7 + x^8 + 3x^9 + 6x^{10} + 15x^{11} + 29x^{12} + \cdots.$$

Riordan [58, 60] has counted some other kinds of trees, including various types of labeled and colored trees. In addition to homeomorphically irreducible trees and identity trees, Harary and Prins [32] have enumerated the following:

Trees with a given partition.
Trees with a given diameter.
Directed trees.
Oriented trees.
Signed trees.
Trees of given strength.
Trees of given type.

A *directed tree* is obtained from a tree by assigning to each line either one or both directions. An *oriented graph* is one in which each line is assigned a unique direction. A *signed graph* is one in which the lines are designated as either positive or negative. A *graph of strength n* is a multigraph in which no more than n lines join any pair of points. A *graph of type n* is obtained from a graph of strength n by assigning colors to its lines in such a way that any two distinct lines joining the same pair of points have different colors.

There is an interesting extension of trees that has also been enumerated. A *cactus* is a connected graph in which no line is in more than one cycle.

Thus each block of a cactus is a cycle or a single line. These graphs were formerly called *Husimi trees*. The cacti in which every block is a triangle were enumerated first in [29] and [34]; the general problem was solved by Norman [43], and its solution also was given by Ford, Norman, and Uhlenbeck [11–II].

6.13 COMPARISON OF SOLVED AND UNSOLVED PROBLEMS

We now present a list of solved enumeration problems, omitting the tree solutions mentioned previously (See Table 6.3). The rest of this section is devoted to an indication of the methods that have been used in solving these problems, and to a comparison of these solved problems with the unsolved problems listed earlier in Table 6.1.

Pólya's theorem is used to obtain the number of graphs of p points as follows. The pairs of distinct points among the p given points are taken as the figures, and the content of a figure is taken as 0 or 1, corresponding to nonadjacency or adjacency of these two points. Thus the figure series is $1 + x$. The configuration group that serves to count graphs is next obtained from the symmetric group of degree p by considering as the objects to be permuted the pairs of distinct points. The cycle index of the resulting group is then readily found, and Pólya's theorem gives the counting polynomial for the number of graphs with p points and a given number of lines.

The counting of rooted graphs is now an easy modification; it is obtained by taking any one of the objects permuted in the symmetric group as fixed before forming its *pair group*. The number of digraphs also is readily obtained from the number of graphs by constructing the *ordered pair group* analogous to the pair group.

Connected graphs are enumerated in terms of the total number of graphs by a combinatorial method that is exactly parallel to the enumeration of rooted trees in terms of themselves, as derived by Pólya. The procedure makes use of the identity (6.10). This result turns out to be particularly important because of its wide applicability. In general, it serves to count the number of connected graphs, or other configurations, having a given property when the total number, both connected and disconnected, is known. If desired, the formula can also serve to give the total number of graphs of a given kind in terms of the number of connected graphs of this kind.

The enumeration of signed graphs offers no difficulty and is obtained immediately, from the formula for the number of graphs, by a modification of the figure-counting series to $1 + x + y$, where the terms 1, x, and y indicate respectively no line, a positive line, and a negative line joining two points.

Using the line group of a graph as the configuration group and $1 + x$ as the figure-counting series, we immediately obtain the number of dissimilar

TABLE 6.3
Solved Enumeration Problems

Category	Problem	Reference
Graph	Graphs	[18]
	Rooted graphs	[18]
	Connected graphs	[18]
	Graphs of given strength	[18]
	Graphs of given type	[18]
	Signed graphs	[17]
	Subgraphs of a given graph	[21]
	Supergraphs of a given graph	[22]
	Graphs with given blocks	[43], [11-II]
Digraph	Directed graphs	[18]
	Weak digraphs	[18]
	Oriented graphs	[23]
	Tournaments	[9]
	Transitive digraphs	[68]
	Functional digraphs	[26], [53]
Partition	General graphs with a given partition	[51]
Group	Tree groups	[50]
Topological	Bicolored and bicolorable graphs	[25], [33]
	Self-complementary graphs	[55]
	Rooted planar triangulations	[65], [66]
Electrical	Two-terminal series-parallel networks	[57], [59]
	Types of Boolean functions	[48], [64]
	Spanning trees of a given graph	[37]
Labeled	Labeled graphs	[14], [11-I]
	Labeled series-parallel networks	[5]
	Labeled graphs with a given partition	[51]
	Labeled nonseparable graphs	[11-III]

subgraphs of a given graph. Analogous formulas for the number of dissimilar supergraphs of a given graph, and in general for the number of types of graphs between a given graph-subgraph pair, are readily formulated [24].

The methods described also enumerate weak digraphs, although unsolved problems 1 and 2 of Table 6.1, which ask for the number of strong and unilateral digraphs, have not been found amenable to this approach. Problem 3, which asks for the number of digraphs with a source, can be regarded as a generalization of problem 1, since any point of a strong digraph is a source. Problem 22, which asks for the number of finite automata, was recently (1963) solved by M. Harrison, but the enumeration of the important class of automata that are strongly connected remains unsolved.

The determination of the number of oriented graphs is analogous to the determination of the number of digraphs, but involves a modification of both the configuration group and the figure-counting series in order to take account of the condition that each line of an oriented graph has exactly one of two possible directions. Again, a figure is a pair of distinct points that either are nonadjacent or are joined by a line in exactly one direction. Hence the figure-counting series is $1 + 2x$, where the content of a figure is the number of lines it contains.

There are other special cases of digraphs that have been counted. *Tournaments* are complete oriented graphs, and were counted by Davis [9]. The result is also readily obtained as a special case of the formula given by Harary [23] for the number of oriented graphs. Moon [40] has observed that the counting series for strong tournaments can be expressed in terms of the counting series for all tournaments, and Moser [70] has shown that the total number of transitive oriented graphs with p points is

$$\frac{1}{p+1} \binom{2p}{p}.$$

A *functional digraph* is a digraph in which each point has outdegree 1. The number of functional digraphs has been found [26] by means of the characterization that each weak component of a finite functional digraph contains exactly one directed cycle together with rooted trees located at each point of the cycle. It follows that the configuration group for this problem is the cyclic group and that the figure-counting series is the known generating function for rooted trees. A simplification of this formula was obtained by Read [53].

We have already noted that the number of trees with a given partition is known. Read [51] has also determined the number of general graphs with a given partition, using his superposition theorem. His method does not appear to be applicable, however, when loops and parallel lines are not permitted. Thus there are solutions to two variations to Problem 4 of Table 6.1, which itself remains unsolved. Read's results also serve to enumerate general graphs that are homeomorphically irreducible, regular, or Eulerian, corresponding to unsolved Problems 5, 6, and 7.

The problem of enumerating bicolored and bicolorable graphs has recently been handled by the construction of a new binary operation on permutation groups, called *exponentiation*. An elementary exposition of the algebraic interaction between this operation and other already known operations on permutation groups is given in the note [25].

An intuitive indication of the difficulties involved in the enumeration of planar graphs is the relationship to the four-color conjecture. If the generating functions for the number of planar graphs and four-colorable planar graphs were obtained and shown to be equal, the four-color conjecture would be proved true. On the other hand, if these numbers were shown to be unequal, the conjecture would be disproved.

Tutte [65, 66] has recently attacked with enthusiasm and vigor the problem of enumerating planar graphs. A triangulation of the sphere is called *simple* if every triangle is a face. Tutte has counted simple triangulations of the sphere in which the exterior triangle is rooted, by distinguishing it from the other faces.

The *complement* $\bar{G}$ of a graph G has the same set of points as G, and two points are adjacent in $\bar{G}$ if and only if they are nonadjacent in G. A *self-complementary* graph is isomorphic to its complement. The complement of a digraph is defined similarly, as is a self-complementary digraph. Read [55] has recently counted self-complementary graphs and digraphs, solving what was one of the unsolved problems listed earlier, in [28].

By means of equation (6.9), Norman [43] has derived a formula for the number of connected graphs with given blocks. Nevertheless, no one has succeeded in deriving a formula for the number of nonseparable graphs with a given number of points and lines, Problem 15 of Table 6.1. The enumeration of graphs with given index and connectivity, Problem 14, is conceptually similar. There is a rather complete set of theorems involving the index of a graph and its connectivity, but these have not proven helpful in finding the kind of permutation-group characterizations of such graphs that would be useful in counting them.

Prins [50] has characterized those groups that are permutation groups of trees. In addition, he found the number of trees belonging to each such group. A corresponding criterion for graph groups is unknown. Such a criterion would provide a partial answer to Problem 19, which is that of finding the number of graphs with a given permutation group.

In general, it is conceptually simpler to enumerate labeled graphs of various kinds than to enumerate the corresponding unlabeled graphs. Of course, there are also very difficult problems for labeled graphs, for example the Ising problem. Some enumeration problems for labeled graphs are so simple, however, that they can be done instantly. For example, the counting series for labeled graphs with p points is

$$(1 + x)^{\binom{p}{2}}$$

since each distinct line of K_p is either in or not in a particular spanning subgraph.

The enumeration, by Gilbert [14], of connected labeled graphs in terms of all labeled graphs involves exactly the same combinatorial lemma used in [18] to count connected graphs in terms of all graphs.

As in [18], let $\gamma(x, y)$ count all the graphs with a given property, and let $\kappa(x, y)$ count the connected graphs among these; then we have

$$(6.20) \qquad 1 + \gamma(x, y) = \exp\left[\sum_{n=1}^{\infty} \frac{\kappa(x^n, y^n)}{n}\right].$$

The number of labeled trees with p points is p^{p-2}, a fact apparently first discovered by Cayley [6] and subsequently independently rediscovered many times. Recently, very elementary proofs have been given by Husimi [35], Clarke [7], and Moon [41].

Although labeled graphs with a given partition have been counted, by Read [51], the enumeration of unlabeled graphs in terms of this interesting and important parameter remains untouched. Similarly, Read [52] has obtained the enumeration of labeled k-colored graphs; the unlabeled case, Problem 8 of Table 6.1, is unsolved.

Asymptotic formulas are often of interest in enumeration problems. Such formulas have been derived for the number of trees by Pólya [47] and Otter [46], and for graphs and labeled graphs by Ford and Uhlenbeck [11–IV]. Several other scientists have derived asymptotic formulas for various kinds of graphs, mainly in order to apply the results to problems in theoretical physics and chemistry.

6.14 SOME APPLICATIONS OF ENUMERATION PROBLEMS TO OTHER FIELDS

The Ising problems

Newell and Montroll [42] give a very clear exposition of Ising problems, showing how they can be regarded as graphical-enumeration problems, and accordingly we shall follow their formulation of these problems—problems that have been found so compulsively interesting by many theoretical physicists as to give rise to a malady humourously called "The Ising Disease." The corresponding disease for mathematicians is probably "The Four Color Disease."

Criteria are known for a partition to be graphical. This subject was studied by Senior [62] in an attempt to classify chemical compounds by means of their configurations, that is, multigraphs in which the points are the atoms of a molecule and the number of lines joining a pair of points is the number of electrons shared by the corresponding two atoms. With saturated hydrocarbons, C_nH_{2n+2}, each configuration is a tree in which each point has degree 1 or degree 4. The problem of finding the number of such isomers, that is, different compounds with the same atomic constituents, was solved by Cayley [6], and in a more powerful way by Pólya [47]. Each of them began by generalizing the problem to finding the number t_n of unrooted trees. In this context, the number of such planted trees is equal to the number of monosubstituted saturated hydrocarbons.

Problem 27 of Table 6.1, the cell-growth problem, was proposed to the author by G. E. Uhlenbeck, who wanted to determine the number of different shapes of paving blocks for application to Problem 26, and through an intermediary by an anonymous biologist, who was curious concerning the number of different shapes of such animals with a given number of cells. Very recently, a computer was programmed to count the total number of such animals, multiply connected as well as simply connected. The resulting counting series, as far as it is known, is

$$x + x^2 + 2x^3 + 5x^4 + 12x^5 + 35x^6 + 108x^7$$

$$+ 369x^8 + 1285x^9 + 4655x^{10} + 17073x^{11} + 63600x^{12}.$$

The problem has also been studied by Golomb [16] under the name of *polyominoes*, or generalized dominoes, in order to handle the version of the paving problem in which each paving stone has the same area.

Combinatorial considerations play an important part in the statistical design of experiments. Two Latin squares (see Chapter 13) S and S' of order n are called *orthogonal* if their superposition contains each of the n^2 ordered pairs (i, j'), $i, j' = 1, 2, \ldots, n$, exactly once. In former days, one of two orthogonal Latin squares was written with Latin letters, the other with Greek letters, and their superposition was called a *Graeco-Latin square*.

A closed formula for the number of distinct Latin squares of arbitrary order n, Problem 23 of Table 6.1, appears to be extremely difficult. The numbers have been found through $n = 7$ by exhaustive methods.

Further enumeration problems are suggested by the recent work of Bose, Parker, and Shrikhande [2], in which they disproved Euler's long-standing conjecture that there are no orthogonal Latin squares or order $4n + 2$ for $n > 1$, by providing examples for each positive integer n. Their work will be discussed in greater detail in Chapter 13.

Although the author of this chapter has tried to solve each of the open problems of Table 6.1, without success thus far, he feels that several of them are not intrinsically impossible. It is his hope and expectation that at least a few of them will be solved in the near future.

MULTIPLE-CHOICE REVIEW PROBLEMS

1. The graph G shown in the accompanying figure is

    (a) nonplanar because it contains a subgraph homeomorphic to K_5,
    (b) nonplanar because it contains a subgraph homeomorphic to $K_{3,3}$,
    (c) nonplanar because it contains too many lines for its ten points,
    (d) planar.

G: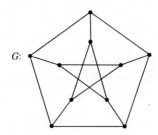

2. The connectivity of the graph G shown in the accompanying figure is

    (a) 1,   (b) 2,   (c) 3,   (d) 4.

3. The number of trees with seven points is

    (a) 48,   (b) 12,   (c) 11,   (d) 16,807.

4. The following partition is graphical:

    (a) $(1, 2, 3, 3, 4)$,   (b) $(0, 2, 2, 2, 4)$,   (c) $(1, 1, 2, 4, 4)$,   (d) $(2, 2, 3, 3, 4)$.

5. The cycle index of the alternating group of degree 4 is

    (a) $\frac{1}{12}(t_1^6 + 3t_1^2t_2^2 + 4t_2^3 + 2t_3^2 + 2t_6)$,
    (b) $\frac{1}{12}(t_1^4 + 3t_1t_3 + 8t_2^2)$,
    (c) $\frac{1}{12}(t_1^4 + 8t_1t_3 + 3t_2^2)$,
    (d) $\frac{1}{24}(t_1^4 + 6t_1^2t_2 + 8t_1t_3 + 3t_2^2 + 6t_4)$.

REFERENCES

1. Berge, C., *Théorie des graphes et ses applications*, Collection Universitaire de Mathematiques, **2**, Dunod, Paris, 1958. Translation by Alison Doig, *The Theory of Graphs and Its Applications*, John Wiley and Sons, New York, 1962.
2. Bose, R. C., E. V. Parker, and S. S. Shrikhande, "Further Results on the Construction of Mutually Orthogonal Latin Squares and the Falsity of Euler's Conjecture," *Canadian J. Math.* **12** (1960), 189–203.

3. de Bruijn, N. G., "Generalization of Pólya's Fundamental Theorem in Enumerative Combinatorial Analysis," *Nederl. Akad. Wetensch, Proc. Ser. A* **62** = *Indag. Math.* **21** (1959), 59–69.

4. de Bruijn, N. G., "Pólya's Theory of Counting," Chapter 5 in E. F. Beckenbach (editor), *Applied Combinatorial Mathematics*, John Wiley and Sons, New York, 1964.

5. Carlitz, L., and J. Riordan, "The Number of Labeled Two-Terminal Series-Parallel Networks," *Duke Math. J.* **23** (1956), 435–446.

6. Cayley, A., *Collected Mathematical Papers*, Cambridge University Press, Cambridge, England, 1889–1897; **3**, 242–246; **9**, 202–204, 427–460; **11**, 365–367; **13**, 26–28.

7. Clarke, L. E., "On Cayley's Formula for Counting Trees," *J. London Math. Soc.* **33** (1958), 471–474.

8. Davis, R. L., "The Number of Structures of Finite Relations," *Proc. Amer. Math. Soc.* **4** (1953), 486–495.

9. Davis, R. L., "Structures of Dominance Relations," *Bull. Math. Biophysics* **16** (1954), 131–140.

10. Dirac, G. A., and S. Schuster, "A Theorem of Kuratowski," *Nederl. Akad. Wetensch. Proc. Ser. A* **57** = *Indag. Math.* **16** (1954), 343–348.

11. Ford, G. W., and G. E. Uhlenbeck, "Combinatorial Problems in the Theory of Graphs. I. II. III. IV," *Proc. Nat. Acad. Sci. U. S. A.* (R. Z. Norman is the third co-author of Paper II) **42** (1956), 122–128, 203–208, 529–535; **43** (1957), 163–167.

12. Erdös, P., and T. Gallai, "Graphen mit Punkten vorgeschriebenen Grades," *Mat. Lapok* **11** (1960), 264–274.

13. Frucht, R., "Graphs of Degree 3 with a Given Abstract Group," *Canadian J. Math.* **1** (1949), 365–378.

14. Gilbert, E. N., "Enumeration of Labeled Graphs," *Canadian J. Math.* **8** (1956), 405–411.

15. Gilbert, E. N., "Gray Codes and Paths on the *n*-cube," *Bell System Tech. J.* **37** (1958), 815–826.

16. Golomb, S. W., "Checker Boards and Polyominoes," *Amer. Math. Monthly* **61** (1954), 675–682.

17. Harary, F., "On the Notion of Balance of a Signed Graph," *Michigan Math. J.* **2** (1953–54), 143–146.

18. Harary, F., "The Number of Linear, Directed, Rooted, and Connected Graphs," *Trans. Amer. Math. Soc.* **78** (1955), 445–463.

19. Harary, F., "Note on the Pólya and Otter Formulas for Enumerating Trees," *Michigan Math. J.* **3** (1955–56), 109–112.

20. Harary, F., "Note on an Enumeration Theorem of Davis and Slepian," *Michigan Math. J.* **3** (1955–56), 149–153.

21. Harary, F., "On the Number of Dissimilar Line-subgraphs of a Given Graph," *Pacific J. Math.* **6** (1956), 57–64.

22. Harary, F., "The Number of Dissimilar Supergraphs of a Linear Graph," *Pacific J. Math.* **7** (1957), 903–911.

23. Harary, F., "The Number of Oriented Graphs," *Michigan Math. J.* **4** (1957), 221–224.

24. Harary, F., "On the Number of Dissimilar Graphs between a Given Graph-Subgraph Pair," *Canadian J. Math.* **10** (1958), 513–516.

25. Harary, F., "On the Number of Bicolored Graphs," *Pacific J. Math.* **8** (1958), 743–755.

26. Harary, F., "The Number of Functional Digraphs," *Math. Annalen.* **138** (1959), 203–210.

27. Harary, F., "The Exponentiation of Permutation Groups," *Amer. Math. Monthly* **66** (1959), 572–575.

28. Harary, F., "Unsolved Problems in the Enumeration of Graphs," *Magyar Tud. Akad. Mat. Kutató Int. Közl* = *Publ. Math. Inst. Hungar. Acad. Sci.* **5** (1960), 63–95.

29. Harary, F., and R. Z. Norman, "The Dissimilarity Characteristic of Husimi Trees," *Ann. of Math.* **58** (1953), 134–141.

30. Harary, F., and R. Z. Norman, "Dissimilarity Characteristic Theorems for Graphs," *Proc. Amer. Math. Soc.* **11** (1960), 332–324.

31. Harary, F., R. Z. Norman, and D. Cartwright, *Structural Models: An Introduction to the Theory of Directed Graphs*, John Wiley and Sons, New York, 1965.

32. Harary, F., and G. Prins, "The Number of Homeomorphically Irreducible Trees, and Other Species," *Acta Math.* **101** (1959), 141–162.

33. Harary, F., and G. Prins, "Enumeration of Bicolourable Graphs," *Canadian. J. Math.* **15** (1963), 237–248.

34. Harary, F., and G. E. Uhlenbeck, "On the Number of Husimi Trees. I," *Proc. Nat. Acad. Sci., U.S.A.* **39** (1953), 315–322.

35. Husimi, K., "Note on Mayer's Theory of Cluster Integrals," *J. Chem. Phys.* **18** (1950), 682–684.

36. Ising, E., "Beitrag zur Theorie des Ferromagnetismus," *Z. Physik* **31** (1925), 253–258.

37. Kirchhoff, G. "Über die Auflösung der Gleichungen, auf welche man bei der Untersuchung der linearen Verteilung galvanischen Ströme geführt wird," *Ann. Phys. Chem.* **72** (1847), 497–508.

38. König, D., *Theorie der endlichen und unendlichen Graphen. Kombinatorische Topologie der Streckenkomplexe.* Mathematik in Monographien **16**, Akademische Verlagsgesellschaft, Leipzig, 1936; reprinted by Chelsea Publishing Company, New York, 1950.

39. Krausz, J., "Démonstration nouvelle d'une théorème de Whitney sur les resaux," (Hungarian), *Mat. Fiz. Lapok* **50** (1943), 75–85.

40. Moon, J. W., "On Some Combinatorial and Probabilistic Aspects of Bipartite Graphs," Doctoral Dissertation, University of Alberta, 1962.

41. Moon, J. W., "Another Proof of Cayley's Formula for Counting Trees," *Amer. Math. Monthly* **70** (1963), 846–847.

42. Newell, G. F., and E. W. Montroll, "On the Theory of the Ising Model of Ferromagnetism," *Rev. Modern Phys.* **25** (1953), 353–389.

43. Norman, R. Z., "On the Number of Linear Graphs with Given Blocks," Doctoral Dissertation, University of Michigan, 1954.

44. Onsager, L., "Crystal Statistics. I. A Two-Dimensional Model with an Order—Disorder Transition," *Phys. Rev.* **65** (1944), 117–149.

45. Ore, O., *Theory of Graphs*, Amer. Math. Soc. Colloq. Publs. **38**, American Mathematical Society, Providence, Rhode Island, 1962.

46. Otter, R., "The Number of Trees," *Ann. of Math.* **49** (1948), 583–599.

47. Pólya, G., "Kombinatorische Anzahlbestimmungen für Gruppen, Graphen, und chemische Verbingungen," *Acta Math.* **68** (1937), 145–254.

48. Pólya, G., "Sur les Types des Propositions Composées," *J. Symbolic Logic* **5** (1940), 98–103.

49. Pólya, G., "On Picture-Writing," *Amer. Math. Monthly* **63** (1956), 689–697.
50. Prins, G., "The Automorphism Group of a Tree," Doctoral Dissertation, University of Michigan, 1957.
51. Read, R. C., "The Enumeration of Locally Restricted Graphs. I. II," *J. London Math. Soc.* **34** (1959), 417–436, and **35** (1960), 344–351.
52. Read, R. C., "The Number of k-Colored Graphs on Labeled Nodes," *Canadian J. Math.* **12** (1960), 409–413.
53. Read, R. C., "A Note on the Number of Functional Digraphs." *Math. Ann.* **143** (1961), 109–110.
54. Read, R. C., "Contributions to the Cell-Growth Problem," *Canadian J. Math.* **14** (1962), 1–20.
55. Read, R. C., "On the Number of Self-complementary Graphs and Digraphs," *J. London Math. Soc.* **38** (1963), 99–104.
56. Redfield, J. H., "The Theory of Group-reduced Distributions," *Amer. J. Math.* **49** (1927), 433–455.
57. Riordan, J., and C. E. Shannon, "The Number of Two-Terminal Series-parallel Networks," *J. Math. Phys.* **21** (1942), 83–93.
58. Riordan, J., "The Numbers of Labeled Colored and Chromatic Trees," *Acta Math.* **97** (1957), 211–225.
59. Riordan, J., *An Introduction to Combinatorial Analysis*, John Wiley and Sons, New York, 1958.
60. Riordan, J., "The Enumeration of Trees by Height and Diameter," *IBM J. Res. Dev.* **4** (1960), 473–478.
61. Sabidussi, G., "Graph Derivatives," *Math. Z.* **76** (1961), 385–401.
62. Senior, J. K., "Partitions and Their Representative Graphs," *Amer. J. Math.* **73** (1951), 663–689.
63. Sherman, S., "Combinatorial Aspects of the Ising Model for Ferromagnetism, I, A Conjecture of Feynman on Paths and Graphs," *J. Math. Phys.* **1** (1960), 202–217.
64. Slepian, D., "On the Number of Symmetry Types of Boolean Functions of n Variables," *Canadian J. Math.* **5** (1953), 185–193.
65. Tutte, W. T., "A Census of Planar Triangulations," *Canadian J. Math.* **14** (1962), 21–38.
66. Tuttle, W. T., "A New Branch of Enumerative Graph Theory," *Bull. Amer. Math. Soc.* **68** (1962), 500–504.
67. Vetuchnovsky, F. Y., "On the Number of Indecomposable Nets and Some of Their Properties," *Doklady Akad. Nauk, U.S.S.R.* **123** (1958), 391–394.
68. Whitney, H., "Congruent Graphs and the Connectivity of Graphs," *Amer. J. Math.* **54** (1932), 150–168.
69. Whitney, H., "Non-separable and Planar Graphs," *Trans. Amer. Math. Soc.* **34** (1932), 339–362.
70. Wine, R. L., and J. E. Freund, "On the Enumeration of Decision Patterns Involving n Means," *Ann. Math. Stat.* **28** (1957), 256–259.

PART 2

CONTROL AND EXTREMIZATION

Fundamentally, improvements in control are really improvements in communicating information within an organization or mechanism.

JOHN VON NEUMANN
Fortune, June, 1955, p. 108

Dynamic Programming and Markovian Decision Processes, with Particular Application to Baseball and Chess

RICHARD BELLMAN

7.1 INTRODUCTION

The introduction of big-league baseball to California in 1960 revived, at least in the minds of *aficionados*, a long-standing question, "Are managers *really* necessary?" Do they exist solely to enforce curfew and to provide copy for sportswriters, or do they possess a crystal ball and a dowsing bat that enable them to make the right decisions at the right time?

Since baseball is a multistage decision process of stochastic type, as will be explained later, we can go a long way toward answering this question with the help of the theory of dynamic programming and digital computers.

We shall describe various formulations in some detail and then examine some of the analytic and computational aspects, none of which are trivial. Then we shall briefly point out that the same mathematical apparatus can be used to treat inventory, repair, and replacement processes, of interest in the areas of operations research and economics, and turn to a brief discussion of chess as a multistage decision process that illustrates the limitations of our methods.

7.2 BASEBALL AS A MULTISTAGE DECISION PROCESS; STATE VARIABLES

To simplify the discussion, let us begin with an analysis of the problems confronting the manager of the team at bat. Later, we shall consider the full problem, in which attention must be paid to the characteristics of the team in the field, to the particular abilities of the pitcher, to the arrangement of the fielders, etc. Initially, let us average over different pitchers, different defensive arrangements, etc., and assume therefore that we are facing an average pitcher, who delivers a strike with a certain probability p and a ball with a certain probability $1 - p$, and an average team.

With this as the background, what data does the manager require to carry out his tactical and strategic maneuvering? The full information pattern is the following:

1. The score.
2. The inning.
3. The batter.
4. The number of outs.
5. The count on the batter, that is, the number of balls and strikes.
6. The men on base and their location.

Let us first ignore the inning, which is to say we are initially contemplating the early stages of the game, and also let us ignore the score. We shall discuss these assumptions in more detail in Sec. 7.4.

We now proceed to enumerate the possible situations that can arise in a particular inning as far as the team at bat is concerned.

There are three possibilities for the number of outs, four possibilities for the number of balls, three possibilities for the number of strikes, eight possible ways in which there can be 0, 1, 2, or 3 men on base, nine possible batters (supposing, as we shall, that pinch hitters will not be used in the early innings). There are then

$$3 \times 4 \times 3 \times 8 \times 9 = 2592$$

different possible situations that can occur within an inning. Were we to take into account the inning, nine additional possibilities, this would increase the number to at least 23,328, whereas variations in scores would increase this total to a number in excess of 10^5, a respectably large number.

Let us for the moment denote these possibilities, in no particular order, by the symbols i, $i = 1, 2, \ldots, 2592$, and call them the *states* of the system. Subsequently, we shall point out that for conceptual and computational purposes, some arrangements are far preferable to others. The preferential labelings arise very naturally from the structure of the game, as is to be expected.

7.3 DECISIONS

We shall assume that the manager is responsible for every action taken by the batter and the base runners. In each of the foregoing situations, we suppose that he signals the players, and that they then follow his instructions as best they can.

In each state, the fundamental initiating action is the delivery of a pitch by the pitcher. The batter has the prerogative of taking the pitch, without swinging at it, or of swinging and attempting to hit the ball (we shall not distinguish between a full swing and a bunt). The men on base have the prerogative of attempting to steal, or not.

The manager then has the responsibility of deciding among the following alternatives:

1. The batter accepts the next pitch.
2. The batter swings at the next pitch if it is a strike.
3. The batter swings at the next pitch in any event, that is, he carries out a hit-and-run play.
4. Men on base are instructed to steal.

As a result of these decisions, certain events can take place:

1. The batter has a ball or strike added to his count, which for three balls or two strikes may mean respectively a walk or a strikeout.
2. The batter swings and misses—a strike and possibly an out.
3. The batter swings and hits a foul—a strike or not, depending on the count.
4. The batter swings and hits a fair ball; this may be a single, double, or triple out, and it may be a hit—a single, double, triple, or homer.

Let us enumerate the possible decisions D in some fashion and denote the set of all decisions by $\{D\}$.

In each of these eventualities, a certain number of runs may be scored, depending on the type of hit or out and the location of the runners. Observe, however, that whatever happens, we return to one of the 2592 possible situations enumerated in the previous section—or else the inning is over as a consequence of three outs.

A characteristic feature of baseball is that it is a stochastic process. Starting in a given state and having made a particular decision, we cannot possibly predict exactly what the next state is going to be. We can, however, determine the set of possible states that can result from a given decision in a particular state.

7.4 CRITERION FUNCTION

Let us now examine various ways of evaluating the decisions that are made. Our over-all objective, of course, is to win the game. Since this is a stochastic event, we must employ an averaging technique. The two most immediate measures are the probability of winning the game, and the probability of maximizing the expected number of runs scored. Clearly these are not equivalent criteria.

Both objectives possess important invariance properties that greatly aid the mathematical determination of optimal strategies. With either criterion, regardless of what has transpired in the past, one continues from the present state according to the same criterion. Strategy based on the probability of winning requires the additional state variables of the score and the inning, whereas that based on the expected number of runs does not require this information, an advantage that we shall exploit to reduce the number of state variables to a manageable size.

It is a reasonable approximation to suppose that in the early innings a team will play so as to maximize the number of runs that it scores in an inning, whereas in later innings it will play to either maximize the probability of scoring a run (protecting a lead) or maximize the probability of at least tying the score.

7.5 POLICIES, OPTIMAL POLICIES, AND COMBINATORICS

In any particular situation, specified by a state i, $i = 1, 2, \ldots, 2592$, we have a choice of any of a number of possible decisions. Since the decision D that is made is clearly dependent on the state i, it is a function of i,

Fig. 7.1 Schematic representation of baseball states.

which we shall denote by $D(i)$. Any function of this type, which maps the set of states onto the set of decisions, is called a *policy*. A policy that maximizes the criterion function is called an *optimal policy*.

The determination of winning strategies is equivalent to the determination of optimal policies.

The problem of obtaining optimal policies can in turn be conceived of as a combinatorial question along the following lines. Each sequence of events in an inning can be represented by a set of integers $(i_1, i_2, \ldots, i_k)$. Let us arrange the states in a linear order, as indicated in Fig. 7.1. Then this sequence of events can be represented by a graph, as shown in Fig. 7.2. This is a generalized random-walk process in which the transition probabilities from state to state change over time in accordance with the decisions that are made.

A priori, since each decision has a number of possible consequences, we have a stochastic graph, as indicated in Fig. 7.3.

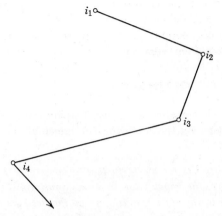

Fig. 7.2   Representation of a deterministic sequence of events.

It is theoretically possible to obtain policies by tracing out the consequences of various policies and averaging over the outcomes of these alternatives. This would be a laborious process, inelegant at best, and impossible to execute computationally. The number of possible paths increases in an alarming way.

Generally speaking, we cannot handle combinatorial processes on a computer in a routine fashion. To encompass the enormous numbers of cases that arise in dealing with even the simplest problems (see [6] for a discussion), we must consider the structure of the problem; that is, we must introduce sophisticated mathematical techniques.

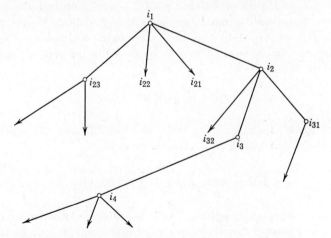

Fig. 7.3   Representation of a stochastic sequence of events.

One systematic technique is to transform combinatorial problems into analytic problems that are amenable to the classical methods of analysis. Can we perform this metamorphosis here in order to obtain optimal policies without actual enumeration?

7.6 PRINCIPLE OF OPTIMALITY

To accomplish this transformation, we employ the theory of dynamic programming, relying on the following fundamental characterization of optimal policies.

An optimal policy has the property that whatever the initial state and initial decisions are, the remaining decisions must constitute an optimal policy with regard to the state resulting from the decisions already made.

This simple and intuitive statement leads to a functional equation that permits us to determine optimal policies by analytic means.

7.7 FUNCTIONAL EQUATIONS

Let us begin with the obvious remark that the expected number of runs scored starting in a particular state i depends on this state. Define therefore the function

$$f(i) = \text{expected number of runs scored in an inning,}$$
$$\text{starting in state } i \text{ and employing an optimal policy.}$$

A decision D in state i results directly in an expected number of runs scored $r_i(D)$ and, with probability $p_{ij}(D)$, a transition into state j. The principle of optimality, cited in the foregoing section, asserts that if decision D is made in state i, and thereafter an optimal policy is followed, then the expected number of runs to be scored in the rest of the inning is given by the expression

$$(7.2) \qquad\qquad r_i(D) + \sum_j p_{ij}(D) f(j).$$

Since this holds for any choice of D, clearly in an optimal policy D is chosen so as to maximize (7.2). Hence, we have

$$(7.3) \qquad\qquad f(i) = \max_D \left[r_i(D) + \sum_j p_{ij}(D) f(j) \right],$$

an equation determining optimal baseball strategy. Although we hardly expect Casey Stengel to solve equations of this nature, it may be that we can obtain significant results with the aid of digital computers. As we shall see, this is the case as far as baseball is concerned, but not as far as chess is concerned.

7.8 EXISTENCE AND UNIQUENESS

Before we devote time to the analytic and computational solution of equations of this type, it is essential that we make sure that the equation has a solution, and that either it has a unique solution or we know how to focus on the solution that corresponds to the multistage decision process we are examining.

The general study of existence and uniqueness of equations of this nature leads to some very difficult problems; see [4, 5, 8, 9, 11, 12] and Chapter 8. In this case, however, using the analytic device discussed in the following section, we readily obtain the desired results. They are simple consequences of the fact that any policy has nonzero probability of leading to the "trap" state—three outs in this case.

7.9 ANALYTIC ASPECTS

Consider now the question of obtaining an exact analytic expression for $f(i)$, and thus a direct determination of the optimal policy. To illustrate the basic idea, let us start with the one-dimensional version, the scalar equation

$$(7.4) \qquad u = \max_q \, [a(q) + b(q)u],$$

where to avoid extraneous questions of continuity we suppose that q ranges over a discrete set S of values, that $a(q)$ is finite on this set, and that $0 \le b(q) < 1$ on this set. Provided that (7.4) has a solution, we have the relation

$$(7.5) \qquad u \ge a(q) + b(q)u$$

for each $q \in S$, and therefore

$$(7.6) \qquad u - b(q)u \ge a(q),$$

or

$$(7.7) \qquad u \ge \frac{a(q)}{1 - b(q)}.$$

Since equality holds for one q, we have

$$(7.8) \qquad u = \max_q \left[\frac{a(q)}{1 - b(q)} \right].$$

On the other hand, the method of successive approximations [2], or a fixed-point method [5], establishes the existence of a solution. As the reader can verify, one can go directly from (7.8) to (7.4).

Can we extend this explicit representation to the vector-matrix case? Suppose that

$$(7.9) \qquad \mathbf{x} = \max_q \, [\mathbf{a}(q) + B(q)\mathbf{x}],$$

where q again ranges over a discrete set, $\mathbf{x}$ is an n-dimensional vector, as is $\mathbf{a}(q)$, and $B(q)$ an $n \times n$ matrix with nonnegative elements. Then, as previously, we have

$$(7.10) \qquad \mathbf{x} \geq \mathbf{a}(q) + B(q)\mathbf{x}$$

for all $q \in S$, and

$$(7.11) \qquad [I - B(q)]\mathbf{x} \geq \mathbf{a}(q),$$

where I is the $n \times n$ identity matrix. The problem thus reduces to an examination of the conditions under which (7.11) implies

$$(7.12) \qquad \mathbf{x} \geq [I - B(q)]^{-1}\mathbf{a}(q).$$

This question is part of the general theory of positive operators; see [2, 3, 13, 15] for extensive discussions and many additional references. Matrices possessing this positivity property arise naturally in probability theory and mathematical economics and have been the subject of much research.

It is clear that (7.12) is a consequence of the nonnegativity of the elements of the matrix $[I - B(q)]^{-1}$. A simple and meaningful condition that guarantees this is

$$(7.13) \qquad \sum_j b_{ij}(q) < 1, \qquad i = 1, 2, \ldots, n,$$

with $b_{ij}(q) \geq 0$ for all i and j. Condition (7.13) is not directly satisfied in our case, but there is no difficulty in circumventing this obstacle.

Under the foregoing hypothesis, we can write the solution of (7.9) in the form

$$(7.14) \qquad \mathbf{x} = \max_q \, [I - B(q)]^{-1}\mathbf{a}(q),$$

a result first observed by Shapley [15]. Particular cases, such as (7.8), were given in [4].

In general, the explicit solution (7.14) is as useful as the explicit solution by Cramer's rule for solving systems of linear equations. As in this case, approximate methods bypassing matrix inverses are more useful and, as we shall see, more significant relative to the actual process.

7.10 DIRECT COMPUTATIONAL APPROACH

Let us now examine a direct computational approach to the solution of (7.2) based on the classical method of successive approximations, which will be discussed further in Chapter 8. Write

(7.15)
$$f_0(i) = \max_D r_i(D),$$
$$f_{n+1}(i) = \max_D \left(r_i(D) + \sum_j p_{ij}(D) f_n(j) \right), \qquad n = 0, 1, 2, \ldots$$

The interpretation of these equations, as far as the original process is concerned, is quite direct. The first function $f_0(i)$ represents the maximum average number of runs scored on a particular play, whereas $f_n(i)$ represents the maximum expected number of runs scored under the constraint that at most $n + 1$ additional men will come to bat. Since the probability of going through the batting order in a particular inning is quite small, we see that $f_8(i)$ should be an extremely good approximation to $f(i)$, and probably even $f_5(i)$ will also be in excellent agreement.

The foregoing approach avoids discussions of existence and uniqueness of solution. If, however, one is interested in obtaining $f(i)$ rather than $f_8(i)$, it is worth noting that the convergence is monotone and geometric; that is, we have

(7.16) $$f(i) - f_n(i) \le c_1 r^n, \qquad 0 < r < 1, \qquad c_1 > 0.$$

Now let us examine the storage aspects. To compute $f_{n+1}(i)$, we must store $f_n(i)$ and the quantities $r_i(D)$ and $p_{ij}(D)$. Since $1 \le i \le 2592$, we see that each matrix $[p_{ij}(D)]$ involves $(2592)^2$ entries, a number in excess of 6,250,000, which is a sizable number insofar as rapid access storage is concerned. What saves the situation is the fact that most of the $p_{ij}(D)$ are zero, regardless of the choice of D. In other words, from any particular state i one can go to only a small number of other states.

For calculations of this nature in connection with inventory and replacement processes, see [10] and [8].

7.11 STRATIFICATION

In many processes of this nature, some parts of the problem can be solved independently of others. There are very natural decompositions that enable us to determine the optimal policy in pieces. In the present case, one such decomposition is clear. First we can determine what to do when there are two out. The optimal policy for any state with two out is independent of any decision made for a smaller number of outs, since no action within an inning can decrease the number of outs.

Having calculated those values of $f(i)$ involving two outs in the state variable, we can then calculate the values of $f(i)$ involving one out, using the previously computed values. Once this is done, we turn to the calculation of those values of $f(i)$ involving no outs. The computational advantage of doing this is great, since sometimes we can reduce an involved problem, for which the storage requirements when considered in a straightforward fashion exceed modern computers, to one that is quite tractable.

The theoretical importance of stratification is that we can often use it to transform a formidable problem, in which a condition of the type

$$(7.17) \qquad \sum_j p_{ij}(D) < 1$$

does *not* hold for all i and all D, into a sequence of subproblems within each of which a condition of this nature—which, as pointed out, ensures nonnegativity of $[I - P(D)]^{-1}$—does hold.

This is the case here, since with two out all "swing" decisions involve a nonzero probability of an out and thus an end to the inning. Some of the "take" decisions, however, do not immediately involve the probability of an out. By means of successive elimination of these states, and use of (7.5) to (7.8), we can reduce all remaining equations to those for which (7.17) holds.

7.12 APPROXIMATION IN POLICY SPACE

In what has preceded, we have employed the classical method of successive approximations with some small modifications. Let us now turn to a method of successive approximations that has no counterpart in classical analysis, the technique of approximation in policy space. Returning to (7.2), let us begin by guessing an initial policy $D_0(i)$ and compute the return function $f_0(i)$ obtained on the basis of this policy. The return function $f_0(i)$ satisfies the functional equation

$$(7.18) \qquad f_0(i) = r_i(D_0) + \sum_j p_{ij}(D_0)f_0(j),$$

an equation that we can solve by direct iteration—or perhaps, if we wish to avoid some storage difficulties, by Monte Carlo methods. We now determine a superior policy $D_1(i)$ by carrying out the maximization

$$(7.19) \qquad \max_D \left[r_i(D) + \sum_j p_{ij}(D)f_0(j) \right], \qquad \text{for } i = 1, 2, \ldots.$$

Having obtained $D_1(i)$ in this fashion, we compute the new return function $f_1(i)$ by means of the equation

(7.20) $$f_1(i) = r_i(D_1) + \sum_j p_{ij}(D_1)f_1(j).$$

It is not difficult to establish rigorously the inequalities

(7.21) $$f_0(i) \leq f_1(i) \leq \cdots ,$$

for all i. Approximation in policy space yields monotone convergence, a most important property. Discussions of the general idea will be found in [4, 6, 8], and comments pertinent to processes of the present type are given in [8, Chapter 11], [5], and [11]. For the connection between this concept and the powerful technique of *quasilinearization*, see [13].

7.13 UTILIZATION OF EXPERIENCE AND INTUITION

A most important attribute of approximation in policy space is that it allows us to take maximum advantage of actual experience in carrying out the process. Consider, for example, the game of baseball. In many of the 2592 situations that might arise, the optimal decision is obvious. Consequently, we start immediately with the correct value of $D(i)$, the optimal policy, for many values of i. This excellent first guess speeds up the calculation of the solution to a vast extent.

Furthermore, it should be emphasized that human beings think automatically in terms of policies, and not immediately in terms of return functions. Actually, in most significant processes in the economic, industrial, and military spheres, the return functions do not exist, but policies are still meaningful. The concept of optimal policy, however, must then be modified.

Let us point out, finally, that very seldom is the return function, or indeed the optimal policy, of major significance. In many situations, it is impossible, or prohibitively expensive, to change from one policy to a policy of quite different type in a short space of time. What is often desired is a technique for changing the policy slowly while improving it uniformly. The technique of approximation in policy space, which permits one to maximize for some of the i's in (7.19) while using the old policy for the others, possesses the most important property of gradualism.

How many fanatics and would-be reformers deduce oversimplified and incorrect policies from incomplete models, which neglect the costs of rapid transition from state to state and ignore the dangers of transient effects—effects that can easily destroy the entire system!

7.14 TWO-TEAM VERSION OF BASEBALL

The actual game of baseball involves two teams that are constantly maneuvering and countermaneuvering to gain an advantage. This means that the determination of optimal baseball strategy requires the analysis and computational solution of multistage games. There is no difficulty in applying dynamic programming to the analytic formulation of processes of this nature [4]. The computing time per stage goes up a little but not drastically. Since there is no difficulty in principle, we shall bypass this excursion, however inviting, and turn to some other matters.

7.15 PROBABILITY OF TYING OR WINNING

Suppose that we are interested in determining optimal play in connection with tying the score. We then introduce the functions

(7.22) $f_k(i) = $ probability of scoring at least k runs,
            starting in state i and using an optimal policy,
            $k = 1, 2, \ldots.$

The outcome of a decision D will, with probability $p_{ijr}(D)$, be a new state j and a certain number r of runs scored. Hence the equation for $f_k(i)$, obtained from the principle of optimality, is

(7.23) $$f_k(i) = \max \sum_{0 \le r \le k} \sum_j p_{ijr}(D)f_{k-r}(j),$$

with $f_0(i) \equiv 1$. These equations are quite easy to solve computationally, starting with those for $f_1(i)$, as an application of the stratification technique mentioned in Sec. 7.11.

Equations of this nature will be particularly useful in analyzing strategy in late-inning and extra-inning play.

7.16 INVENTORY AND REPLACEMENT PROCESSES AND ASYMPTOTIC BEHAVIOR

In a number of important processes carried out daily on the economic and industrial scene, we meet equations of the type

(7.24) $$f_n(i) = \max_D \left[r(i, D) + \sum_j p_{ij}(D)f_{n-1}(j) \right], \qquad n \ge 1,$$

$$f_0(i) = \max_D [r(i, D)],$$

in which the functions $f_n(i)$, representing the minimum cost of carrying out a process, increase without bound as $n \to \infty$. Here n is not a variable introduced to keep account of the degree of approximation, as previously; rather, it is a significant state variable equal to the number of stages through which the process has yet to pass. Frequently, if a process is carried out repeatedly day after day, year after year, a certain pattern emerges. In other words, for large n we expect a *steady-state* behavior, as opposed to the *transient* behavior observed when n is small.

We suspect then that in many cases we can establish asymptotic relations of the type

$$(7.25) \qquad f_n(i) \sim nf(i)$$

as $n \to \infty$, and, in other cases, of the more precise type

$$(7.26) \qquad f_n(i) \sim nc_1 + f(i),$$

where c_1 is *independent* of i. As in ordinary Markovian processes, if there is enough opportunity for mixing, the asymptotic behavior is independent of the initial state.

Results of this sort have been established under various assumptions [5, 12]. The asymptotic form, the quantity c_1 and the $f(i)$, can be determined by means of linear-programming techniques, which sometimes may be quicker or more convenient. This fact has been noted and used by a number of authors, including Beckmann, Howard, Manne, and Dantzig-Wolfe [14].

For the computational solution of inventory, repair, and replacement problems using (7.1) directly, see [8, 10, 11]. In many cases, the analytic structure of the optimal policy can be obtained from (7.24); see [4] and the extensive work of Arrow, Karlin, Scarf, and others in [1].

7.17 CHESS AS A MULTISTAGE DECISION PROCESS

Let us conclude with a brief discussion of the game of kings as a multi-decision process. It turns out that it is much simpler to describe than the game of baseball.

The state of the system is the position of the pieces and pawns, with some slight additional data concerning castling and *en passant*. A move by either side constitutes a transformation of the position, and the criterion is simple: Win (or at least draw).

Let p denote the position at any time, and introduce the function $f(p)$ defined by the relations

$$(7.27) \qquad \begin{aligned} f(p) &= \quad 1, \quad \text{if } p \text{ is a winning position,} \\ &= -1, \quad \text{if } p \text{ is a losing position,} \\ &= \quad 0, \quad \text{if } p \text{ is a draw position.} \end{aligned}$$

Assuming that a move by white and a move by black constitute a stage, and using the principle of optimality, we can write the equation

$$(7.28) \qquad f(p) = \max_{W} \min_{B} f[T_{WB}(p)].$$

Here $T_{WB}(p)$ denotes the new position resulting from a move by white (W), and then a move by black (B). To avoid any difficulties caused by uncertainties concerning the length of the game, we can introduce a constraint on the number of moves and define the sequence of functions

$(7.29) \quad f_n(p) = 1,$ \qquad if p is a winning position within at most n moves,

etc. Then, in place of (7.28), we have the relation

$$(7.30) \qquad f_n(p) = \max_{W} \min_{B} f_{n-1}[T_{WB}(p)], \qquad n \geq 1,$$

with $f_0(p)$ determined by inspection of the position p.

This then is a complete and mathematically precise formulation of chess. Is it of any value?

7.18 COMPUTATIONAL ASPECTS

The question in the foregoing section pertains to the use of (7.30) for computational purposes. From the beginning, because of the inherent instability of chess, manifesting itself in the rather wild behavior of $f(p)$ (a function' that is an excellent approximation to a continuous non-differentiable function), we can reject any hope of an analytic solution of (7.28) or (7.30).

The question of analytic or computational solution should not be rejected immediately because of the very large (10^{60}?) number of possible games. After all, the function $f(x) = x(1 - x)$ has a very large (∞) number of possible values for $0 \leq x \leq 1$, and yet we can determine its maximum value quite easily. It is not the *number* of possibilities, but the *structure* of the process that matters.

By using dynamic programming, we concentrate not on the number of possible games, but on the number of possible states. This latter is a much smaller quantity, but still incredibly huge. Any perfunctory count will make evident a lower bound for the order of magnitude of the number of possible positions.

We can nevertheless use the stratification technique of Sec. 7.11 to obtain the solution to a large number of interesting chess problems. Most likely, we can solve all king-pawn endings using the foregoing equations. Let us discuss this briefly.

Aside from the possibility of a pawn queening, or transforming into another piece, the number of pieces on the board decreases over time, as does the number of pawns. Furthermore, pawns move down the board. We see then that the situation is ideally suited for the employment of the stratification method.

Combining this approach with the reduction of dimensionality presented in [7], we feel that we now possess a straightforward approach to the mathematical treatment of much of the end game of chess. The number of possible states in the middle game prevents us, however, from applying the foregoing approach successfully. As far as the full game of chess is concerned, we are as far away from any kind of precise mathematical analysis as we ever were.

MULTIPLE-CHOICE REVIEW PROBLEMS

1. In the study of multistage decision processes occurring in the real world,

    (a) any process can be accurately treated by mathematics,
    (b) it is only a question of how much we are willing to spend on computations to obtain a complete solution,
    (c) dynamic-programming techniques will always yield a feasible computational algorithm,
    (d) there are often formidable conceptual difficulties that prevent a quantitative approach.

2. The computational feasibility of a dynamic-programming solution hinges on

    (a) a combination of analytic formulation and digital computers,
    (b) nothing but the availability of a digital computer with a fast memory,
    (c) the use of ingenious approximation techniques,
    (d) the use of linear-programming techniques.

3. An analytic treatment of a multistage decision process is often facilitated by

    (a) the use of a digital computer,
    (b) a study of the asymptotic or steady-state behavior,
    (c) a formulation as an optimization process,
    (d) sheer inspiration.

4. Baseball

    (a) can be completely analyzed as a one-person multistage decision process by means of dynamic programming,
    (b) can be completely analyzed as a two-person multistage decision process by means of dynamic programming,
    (c) contains psychological features that make a complete analysis difficult,
    (d) has none of the above properties.

5. Chess

    (a) can be completely analyzed, given computers 10^6 times as fast as those we now have,

    (b) can be completely analyzed, given computers 10^6 times as big as those we now have,

    (c) is a multistage decision process devoid of psychological aspects,

    (d) has been the subject of much Sunday Supplement discussion.

REFERENCES

1. Arrow, K. J., S. Karlin, and H. Scarf, *Studies in the Mathematical Theory of Inventory and Production*, Stanford University Press, Stanford, Calif., 1958.

2. Beckenbach, E. F., and R. Bellman, *Inequalities*, Ergebnisse der Math., J. Springer-Verlag, Berlin, 1961.

3. Bellman, R., "Functional Equations in the Theory of Dynamic Programming—V: Positivity and Quasi-linearity," *Proc. Nat. Acad. Sci. U.S.A.* **41** (1955), 743–746.

4. Bellman, R., *Dynamic Programming*, Princeton University Press, Princeton, New Jersey, 1957.

5. Bellman, R., "A Markovian Decision Process," *J. Math. Mech.* **6** (1957), 679–684.

6. Bellman, R., *Adaptive Control Processes: A Guided Tour*, Princeton University Press, Princeton, New Jersey, 1961.

7. Bellman, R., "On the Reduction of Dimensionality for Classes of Dynamic Programming Processes," *J. Math. Anal. Appl.* **3** (1961), 358–360.

8. Bellman, R., and S. Dreyfus, *Applied Dynamic Programming*, Princeton University Press, Princeton, New Jersey, 1962.

9. Blackwell, D., "On the Functional Equation of Dynamic Programming," *J. Math. Anal. Appl.* **2** (1961), 273–276.

10. Dreyfus, S., "A Note on an Industrial Replacement Process," *Operational Res. Quart.* **8** (1957), 190–193.

11. Howard, R., *Dynamic Programming and Markov Processes*, John Wiley and Sons, New York, 1960.

12. Inglehart, D., "Dynamic Programming and Stationary Analyses of Inventory Problems," Doctoral Dissertation, Stanford University, Stanford, Calif., 1961.

13. Kalaba, R., "On Nonlinear Differential Equations, the Maximum Operation, and Monotone Convergence," *J. Math. Mech.* **8** (1959), 519–574.

14. Manne, A. S., "Linear Programming and Sequential Decisions," *Management Sci.* **6** (1960), 259–267.

15. Shapley, L. S., "Stochastic Games," *Proc. Nat. Acad. Sci. U.S.A.* **39** (1953), 1095–1100.

CHAPTER 8

Graph Theory and Automatic Control

ROBERT KALABA

8.1 INTRODUCTION

A problem of central importance in the theory of automatic control involves the transforming of a system from an initial state into a desired terminal state in the most efficient fashion. If, for example, a gust of wind causes an aircraft to begin rolling, then the automatic pilot is supposed to send control signals to the control surfaces in an effort to restore the craft as rapidly as possible to the horizontal position with no angular velocity about its longitudinal axis. Again, consider a spacecraft that is about to re-enter the earth's atmosphere. It is desired to fly along a trajectory that will bring the craft to the surface of the earth while keeping the maximum temperature to which the surface of the craft is exposed as low as possible. One might also wish to select a path that will minimize the maximum deceleration during descent.

From the mathematical viewpoint, it seems natural to consider such problems as belonging to the calculus of variations. Toward this end we introduce a state vector $\mathbf{x}(t)$, with $\mathbf{x}(0) = \mathbf{c}$, a control vector $\mathbf{y}(t)$, and a dynamical equation

$$(8.1) \qquad \dot{\mathbf{x}} = \mathbf{f}(\mathbf{x}, \mathbf{y}).$$

We wish to determine the control vector $\mathbf{y}(t)$ in such a manner that the system is transformed from the initial state $\mathbf{c}$ to some desired terminal state, say $\mathbf{x} = \hat{\mathbf{0}}$, as rapidly as possible. An extensive literature now exists concerning such problems [4, 17], especially for the case in which the function $\mathbf{f}$ is linear in $\mathbf{x}$ and $\mathbf{y}$, and the components of the control vector $\mathbf{y}$ are subject to certain constraints. Alternatively, we might wish

237

to determine the control vector $\mathbf{y}(t)$ in such a way that we minimize the maximum value of some function $g[\mathbf{x}(t), \mathbf{y}(t)]$ during the course of transforming the system from the initial state to the terminal state. There is a less extensive literature associated with such problems [1, 2].

If we keep in mind that the treatment of significant problems in this area will ultimately involve the use of digital computers, where all variables are rendered discrete, and that some control systems operate in a discrete fashion by design, it becomes of interest to consider a discrete formulation and treatment of an automatic control process as opposed to a continuous one. The aim of this chapter is to give an indication of the current state of affairs in this program.

8.2 TIME-OPTIMAL CONTROL AND THE NATURE OF FEEDBACK

Let us consider a system S that may be found in any of a finite number N of possible states. We represent these states by the nodes of a graph [9, 15] and number them $1, 2, \ldots, N$. In the course of time, the system changes its state; that is, it undergoes a sequence of transformations, which we call a *process*. When the system is in state i, we assume that a control decision can be made, the result of which is that the system is transformed into some new state. In general, when the system is in state i, only certain new states may be attained as the result of making a control decision; see Chapter 7.

In the passage from state i to state j, a certain amount of a resource such as energy must be consumed during the transformation. This assigns a number t_{ij} to the directed arc from i to j Let us consider that we wish to control the system S in such a way that if the system is disturbed from its equilibrium (or most desirable) state, say the state N, then we shall return it to the equilibrium state in the least possible time.

At first glance it might seem that we wish merely to find a shortest trajectory through phase space leading from the initial state i to the desired terminal state N. Actually, this is not so. Frequently, we shall have no prior knowledge of what the particular initial state i will be, so that we must be prepared to transform the system from any initial state i into the desired terminal state, and do so in minimum time. Now we may go one step further and note that if the system is in state i, it is not necessary to specify the entire trajectory from i to N. Indeed, all we need do is specify that if the system is in state i, then the next state into which to transform the system is a particular state j.

These considerations enable us to distinguish between open-loop control and feedback control. In an *open-loop* process, the entire sequence of transformations is specified ahead of time. This type of control is useful when one has great confidence that the system will perform as specified and that random external influences are negligible. In practice this means that all control influences will be exerted in a prescribed fashion as a function of time, that is, according to the reading of a clock, no notice being taken of the actual state of the system. A simple example of this is the control of the temperature of a room by turning a furnace on and off at certain fixed times. Open-loop control is usually economical to provide, but it may fail to yield satisfactory control because it neglects external factors.

A more sophisticated type of control is achieved by carrying out the following cycle of operation: (1) determine the state of the system (measure the temperature of the room); (2) decide on a control action (turn the furnace on or off); (3) return to (1). A control system that operates on this principle is termed a *feedback-control system*. There is an extensive theory of such processes, going back to Maxwell [6, 19, 20]. For the most part, this is a theory describing how a system would act if controlled in a certain way, the notion of stability playing a major role [18]. It is clear, however, that we wish to control in optimal fashion, so that emphasis has turned from stability to optimality considerations—although, of course, there is an intimate relation between the two.

8.3 FORMULATION

We introduce the following nomenclature [3]:

t_{ij} = the time required to transform the system
from state i to state j over a direct link.

u_i = the time to transform the system from state
i to state N (the desired state) in optimal
fashion, $i = 1, 2, \ldots, N$.

To derive relations among these quantities, we note that if the system is in state i and the decision to transform it into state j is made, $j \neq i$, then the process will have to continue optimally from state j to state N if the entire process is to be optimal. Furthermore, the choice of the next state, state j, is made on the basis of minimizing the sum

$$t_{ij} + u_j,$$

for t_{ij} is the time to pass directly from state i to state j, and u_j is the minimal time required to pass from state j to the terminal state N.

These observations, manifestations of Bellman's principle of optimality (see Chapter 7 or [4]), lead to the nonlinear system of equations

$$(8.2) \qquad u_i = \min_{j \neq i} (t_{ij} + u_j), \qquad i = 1, 2, \ldots, N - 1,$$

$$u_N = 0.$$

8.4 UNIQUENESS

Since we contemplate solving the equations (8.2) by way of various successive-approximation techniques, it is important to establish the uniqueness [3] of the solution. Suppose that $\{u_i\}$ and $\{U_i\}$ represent two solutions of the system of equations (8.2), and that k is an index for which the difference

$$u_i - U_i$$

achieves its maximum. Let

$$(8.3) \qquad u_k = \min_{j \neq k} \{t_{kj} + u_j\} = t_{kr} + u_r$$

and

$$(8.4) \qquad U_k = \min_{j \neq k} \{t_{kj} + U_j\} = t_{ks} + U_s.$$

Since we are assuming that

$$t_{ij} > 0,$$

it is clear that

$$r \neq k, \qquad s \neq k.$$

We have the relations

$$(8.5) \qquad u_k = t_{kr} + u_r \leq t_{ks} + u_s,$$

$$(8.6) \qquad U_k = t_{ks} + U_s,$$

which lead to the inequality

$$(8.7) \qquad u_k - U_k \leq u_s - U_s.$$

Since k is an index for which the difference $u_j - U_j$ is maximized, equality must hold in (8.7),

$$(8.8) \qquad u_k - U_k = u_s - U_s,$$

which implies that equality also holds in (8.5),

$$(8.9) \qquad u_k = t_{ks} + u_s.$$

Now, however, we may repeat the same reasoning for the state s and establish that there is another state, state m, for which

$$(8.10) \qquad u_m - U_m = u_s - U_s = u_k - U_k.$$

Furthermore, $m \neq s$ and $m \neq k$ since

$$(8.11) \qquad u_k = t_{ks} + t_{sm} + u_m.$$

Continuing in this way, we must eventually come on the state N, for which

$$(8.12) \qquad u_N - U_N = 0,$$

and this completes the proof.

8.5 SUCCESSIVE APPROXIMATIONS

We can use Picard's method of successive approximations to establish the existence of a solution of the system of equations (8.2) and to provide a practical computational scheme. As our initial approximation $u_i^{(0)}$, we make the decision to transform the system directly from state i to state N,

$$(8.13) \qquad u_i^{(0)} = t_{iN}, \qquad i = 1, 2, \ldots, N.$$

Of course, if no such direct link exists, we assume that the elapsed time is M, a suitably large number.

The higher-order approximations are obtained in the usual way,

$$(8.14) \qquad \begin{cases} u_i^{(k+1)} = \min_{j \neq i} \{t_{ij} + u_j^{(k)}\}, & i = 1, 2, \ldots, N - 1, \\ u_N^{(k+1)} = 0, \end{cases}$$

for $k = 0, 1, 2, \ldots$. The physics of the problem enables us to see some of the properties of the successive approximations. For example, since

$$u_i^{(1)} = \min_{j \neq i} \{t_{ij} + t_{jN}\}, \qquad i = 1, 2, \ldots, N - 1,$$

$$u_N^{(1)} = 0,$$

we see that

$u_i^{(1)} = $ the minimal time to transform the system from state i to state N by way of at most one intermediate state, $i = 1, 2, \ldots, N$.

In general, we have

$u_i^{(k)} = $ the minimal time to transform the system from state i to state N by way of at most k intermediate states.

From this it follows that the approximations are monotone decreasing,

$$(8.15) \qquad u_i^{(k+1)} \leq u_i^{(k)}, \qquad i = 1, 2, \ldots, N,$$

which is easy to establish by induction. Since $u_i^{(k)}$ is bounded from below,

$$(8.16) \qquad u_i^{(k)} \geq 0, \qquad i = 1, 2, \ldots, N,$$

the convergence of the approximating sequence is established. As a matter of fact, however, an optimal trajectory from any state i to state N has at most $N - 2$ intermediate states, since an optimal trajectory does not cross itself to form a loop. Thus the convergence of the process is assumed after at most $N - 2$ stages. That the limiting values satisfy equations (8.2) is clear.

8.6 OBSERVATIONS ON THE APPROXIMATION SCHEME

Use of the equations (8.14) involves only addition and comparison of numbers, two operations for which a digital computer is well suited. Furthermore, in calculating the value of $u_i^{(k+1)}$, only the ith row of the matrix (t_{ij}) and the vector $(u_1^{(k)}, u_2^{(k)}, \ldots, u_N^{(k)})$ need be in high-speed storage. In this way, problems for which N is of the order of several thousand can be solved by an IBM-7090 computer in a few minutes. Efficient programs will exploit special features of a given problem. This is certainly true if each state is directly connected to only a small number of its neighbors. In some instances the values of the successive approximations are more important than the solution of the original problem. If the optimal trajectory from state i to state N has many intermediate states, a complex instrumentation might be required to achieve it, so that a knowledge of both the successive approximations and the limiting values may be of importance in designing a control system.

Finally, let us note that our solution consists not so much in the production of the values $u_1, u_2, \ldots, u_N$, as in the knowledge of the value of j that minimizes the expression $t_{ij} + u_j$, for each value of i. This is precisely the knowledge that is required for determining optimal feedback control.

8.7 OTHER APPROACHES

Many other approaches to this problem have been devised, in particular by Dantzig [10] and Ford and Fulkerson [11]; see Chapter 12. In particular, it is possible to fan out from the destination and determine, one after the other, the nearest, second nearest, and so on, states to the terminal state.

A variety of analogue devices can be used. Some of these are described in [13], where many references are provided.

8.8 ARBITRARY TERMINAL STATES

If we wish to determine optimal trajectories from any initial state to any terminal state, we might apply the foregoing procedures N times, in each case letting a different state be the desired terminal state. Alternatively, we might let

$u_{ij}^{(k)} = $ the time to transform a system in state i into state j using a trajectory with at most k intermediate states.

Using the principle of optimality, we see that

$$(8.17) \qquad u_{ij}^{(2k+1)} = \min_{m \neq i} (u_{im}^{(k)} + u_{mj}^{(k)}), \qquad i \neq j.$$

Since k will be at most $N - 1$, and since we can easily determine the matrices $[u_{ij}^{(0)}], [u_{ij}^{(1)}], [u_{ij}^{(3)}], [u_{ij}^{(7)}], [u_{ij}^{(15)}], \ldots$, the problem is readily handled, at least computationally.

8.9 PREFERRED SUBOPTIMAL TRAJECTORIES

One of the key difficulties in applying these ideas to a concrete physical situation lies in deciding on the number and nature of the physical states of the system to be considered [16]. In particular, if the number of states chosen is too small, that is, if the grid in phase space is too coarse, then the time to traverse a next-to-shortest path may be considerably larger than the time to traverse an optimal path. On the other hand, if these times are not too different then one's confidence in the reasonableness of the mathematical model of the physical process may be increased.

In addition, we must recognize that even if we determine an exact solution to the mathematical problem, since we have neglected a variety of physical factors we have only an approximate solution of the physical problem. If a mathematically optimal trajectory that we have found is unsatisfactory from the physical viewpoint, we can either find a near-optimal trajectory for the mathematical problem, in hopes that it will be better from the physical viewpoint, or reformulate the mathematical problem. Let us show how we can determine second-best trajectories, once having determined optimal trajectories to the terminal state N.

Under the assumption that there is at least one trajectory from state i to state N that is not optimal, $i = 1, 2, \ldots, N - 1$, let us define the variables v_i, $i = 1, 2, \ldots, N - 1$, by

$v_i = $ the time that it takes to transform a system from state i to state N using a second-best trajectory, $i = 1, 2, \ldots, N - 1$,

$v_N = 0$.

Of course, we are assuming that there are no loops in the trajectory. Next notice that if we make the decision to transform the system from state i directly to state j, then the continuation must be along a trajectory from state j to state N that is either optimal or second best. This leads to the relations

$$(8.18) \qquad v_i = \min_2 \limits_{j \neq i} (t_{ij} + u_j, t_{ij} + v_j), \qquad i = 1, 2, \ldots, N - 1,$$

$$v_N = 0,$$

where we have used the notation

$\min_2 (a_1, a_2, \ldots, a_R) = $ the second smallest of $a_1, a_2, \ldots, a_R$ (under the assumption that they are not all equal).

Generalizations are given in [5], and are discussed at length in [22].

Pollack [23] has also observed that we may find second-best paths by first finding the optimal trajectory (assumed unique) from state i to state N. Then we eliminate the first link in the optimal path from the network and determine an optimal trajectory from state i to state N using only the remaining links in the network. This is done for each of the remaining links in the optimal trajectory being considered. Since there are at most $N - 1$ such links in the optimal trajectory, at most $N - 1$ such problems need be solved. A trajectory that yields the smallest of the numbers so found is a second-shortest trajectory. The proof is given by noting that a second-shortest trajectory must differ from an optimal trajectory in at least one link. With this method, there is no difficulty concerning the possible formation of trajectories with loops.

8.10 A STOCHASTIC TIME-OPTIMAL CONTROL PROCESS

Let us assume that the physical situation is such that the time involved in transforming the system from state i to state j, t_{ij}, is not known precisely [13]. Suppose, however, that we may consider it to be a random variable with a known probability density function $p_{ij}(t)$. This is a strong assumption that may or may not be justified in a given situation. Further-more, we shall assume that the time to traverse any link in a trajectory is independent of the time to traverse any other link in the trajectory, another severe restriction.

Under the assumptions just stated, our aim is to find the optimal feedback-control decision to make when the system is in state i. Let us now explain carefully what we mean by this. We shall assume that our objective is to maximize the probability of transforming the system into the desired state N in a time t or less, where

$$t \geq 0.$$

The sequence of operations to be carried out is this: First the current state of the system and the time are measured (by sensing equipment). Second, a decision is made as to the next state that the system is to occupy. Then a random time elapses until the system reaches this state. Next a measurement is made of the new time and state. On the basis of this knowledge of the new state and the new time, a decision is made as to what the next state is to be, and so on. Notice particularly that we do not attempt initially to lay out the entire sequence of decisions leading from state i to state N (as would be the case for open-loop control); rather we observe, decide, and act over and over again. We aim to determine the optimal decision to make under given circumstances, that is, for a given state and given time remaining in the process. Although for deterministic processes open loop and closed loop (feedback) control lead to identical results, for stochastic decision processes they are conceptually and actually quite different.

Let us now define the functions $u_i(t)$, $i = 1, 2, \ldots, N$, by the equations

> $u_i(t) =$ the probability of transforming a system from the initial state i to the desired terminal state N in time t or less, using an optimal feedback control policy.

If we use the principle of optimality once again, we can write the equations

$$(8.19) \quad u_i(t) = \max_{j \neq i} \int_0^t p_{ij}(t - s)u_j(s)\, ds, \qquad i = 1, 2, \ldots, N - 1,$$

$$u_N(t) = 1.$$

Equations (8.19) appear to be quite difficult to handle, both analytically and computationally. The appearance of the convolution integrals suggests the use of Laplace transforms [7], and the maximum transform discussed by Bellman and Karush [8] and others might also be useful.

Additional closely related stochastic control processes are discussed in [13] and [14] and in Chapter 9 of this book.

8.11 MINIMAX CONTROL PROCESSES

In some circumstances, such as we have discussed in Sec. 8.1, it is desirable to transform a system from an initial state to a terminal state in such a way as to minimize the maximum stress to which the system is exposed during the course of the process. We shall refer to such control processes as "minimax control processes." Once again we shall consider a system that may be in any of a finite number N of states, the states being numbered from 1 to N. We consider state N to be the desired terminal state. When the system is transformed directly from state i to state j, a maximum stress s_{ij} is encountered. Thus the stress s_{ij} is associated with the link (i, j). Our basic problem consists in finding a trajectory from state i to state N such that the maximum stress along this trajectory is as small as possible.

8.12 USE OF FUNCTIONAL EQUATIONS

Let us introduce the variables u_i, $i = 1, 2, \ldots, N - 1$, by the relations

$u_i =$ the maximum stress along an optimal trajectory
from state i to the terminal state N,
$i = 1, 2, \ldots, N - 1$,

$u_N = 0.$

Then use of the principle of optimality immediately leads to the relations

$$(8.20) \qquad u_i = \min_{j \neq i} \left[\max \left(s_{ij}, u_j \right) \right], \qquad i = 1, 2, \ldots, N - 1,$$

$$u_N = 0.$$

These are, of course, the analogues of equations (8.2). The results of the following sections could be obtained directly from these equations. Rather than pursue this method, we shall keep the network itself in the foreground.

8.13 A SPECIAL CASE

A very important and interesting special case arises when we stipulate the reversibility equality

$$s_{ij} = s_{ji}.$$

Thus one stress s_{ij} is connected with the arcs (i, j) and (j, i). We can carry through the analysis in some detail and establish a relationship with a seemingly unrelated problem in graph theory.

We first observe that the maximum stress encountered along any trajectory is a number of the set $\{s_{ij}\}$. Let us then arrange the (positive) numbers $\{s_{ij}\}$ in ascending order of magnitude and denote this sequence by $s_1, s_2, \ldots, s_R$, where $R \leq (N/2)(N + 1)$. Call the corresponding arcs $S_1, S_2, \ldots, S_R$. For convenience, we assume that the stresses associated with the arcs are all different from one another, a condition that can be attained by adding suitably small quantities, if necessary, to the given stresses. Then we observe that the states joined by the arc having the stress s_1 cannot be joined by any trajectory with a smaller maximum stress. This arc, S_1, constitutes an optimal trajectory for those states. Next we observe that the states joined by the arc of stress s_2 are also joined optimally by this arc S_2.

The situation becomes a little more complicated insofar as the states joined by arc S_3 are concerned. If the arcs S_1, S_2, and S_3 do not form a loop, then the states joined by arc S_3 are joined optimally by it. If, however, arcs S_1, S_2, and S_3 do form a loop, then the arcs S_1 and S_2, but not S_3, connect these states optimally.

Furthermore, we see that if we continue this process of selecting arcs from the sequence of arcs $S_1, S_2, \ldots, S_R$, making sure that no arc selected forms a loop with any of the arcs already selected, we shall eventually select $N - 1$ arcs containing no loops. We shall thus have formed a particular *spanning tree* in the network, and the unique trajectory in this tree that connects any two states is the minimax trajectory between those states.

8.14 MINIMAL SPANNING TREES

The construction that we have just indicated is well known to provide the solution to another seemingly unrelated problem: Under the conditions stated in Sec. 8.13, find the tree for which the sum of the stresses in its branches (arcs) is as small as possible. This tree is called the *minimal spanning tree*. In 1956 Kruskal [16] showed that the construction given solves this problem. Various other algorithms are given in [16] and [13].

The algorithm described is not satisfactory from the computational viewpoint, for testing to see whether or not an arc completes a loop when added to another set of arcs can be quite time-consuming. Prim [24] has given some very effective computational procedures.

8.15 COMMENTS AND INTERCONNECTIONS

Our result on minimax control processes, subject to the restrictions in Sec. 8.13, may be formulated thus: *Optimal minimax trajectories lie in the minimal spanning tree.* Let us now point up this result in several other ways.

First we give another proof. Consider an optimal minimax trajectory from state i to state N. Suppose that it is not the trajectory from state i to state N that lies in the minimal spanning tree. Then there is at least one arc in this trajectory that does not lie in the minimal spanning tree. Denote this as arc A. If we add this arc to the set of arcs in the minimal spanning tree, then exactly one loop is formed, say the loop with arcs $(A, A_1, \ldots, A_J)$. Note, however, that the following inequality must hold:

$$(8.21) \quad \text{stress}\,(A) \geq \max\,[\text{stress}\,(A_1),\, \text{stress}\,(A_2),\, \ldots,\, \text{stress}\,(A_J)].$$

If (8.21) did not hold, then it would be possible to lower the sum of the stresses in the branches of the minimal spanning tree by adding arc A to the minimal spanning tree and deleting an arc A_I for which

$$\text{stress}\,(A) < \text{stress}\,(A_I).$$

Inequality (8.21) shows that the arc A in the supposed minimax trajectory can be replaced by arcs in the minimal spanning tree without increasing the maximal stress encountered. Since arc A could be any arc in the minimax trajectory, the proof is complete.

Prim observed in [24] that the minimal spanning tree not only minimizes the sum of the stresses in its branches but also minimizes, among all tress, any monotone increasing and symmetric function of the stresses in the branches. In particular, we note, it minimizes the function

$$S_p = [s_1^p + s_2^p + \cdots + s_{N-1}^p]^{1/p},$$

for $p = 1, 2, \ldots$, and the limit function

$$(8.22) \quad \max\,(s_1, s_2, \ldots, s_{N-1}) = \lim_{p \to \infty} S_p,$$

as is easily proved.

This suggests that we consider the problem of determining a trajectory from state i to state N that minimizes the sum of the pth powers of the stresses in its arcs. Clearly for $p = 1$ this is equivalent to the problem of determining a time-optimal trajectory, which was considered earlier. On the other hand, in view of (8.22), for p sufficiently large this is the problem of determining a minimax trajectory. More precisely, we can show that for p sufficiently large a trajectory that minimizes the sum of the pth powers of the stresses in its arcs lies in the minimal spanning tree. To see this, consider an arc S that is in the optimal trajectory but that does not lie in the minimal spanning tree. Let its stress be s, and let the stresses of the arcs in the minimal spanning tree with which it forms a loop be $s_1, s_2, \ldots, s_r$. Then, as we observed earlier,

$$(8.23) \qquad s > \max(s_1, s_2, \ldots, s_r),$$

and for p sufficiently large,

$$(8.24) \qquad 1 > \left(\frac{s_1}{s}\right)^p + \left(\frac{s_2}{s}\right)^p + \cdots + \left(\frac{s_r}{s}\right)^p,$$

or

$$(8.25) \qquad s^p > s_1^p + s_2^p + \cdots + s_r^p.$$

This inequality shows that the arc S may be replaced by arcs in the minimal spanning tree and establishes the result.

In [21], Pollack poses the problem of determining a maximum-capacity path between two stations in a communications network and provides several solutions. This problem is equivalent to ours. Furthermore, D. R. Fulkerson has pointed out to the author that Hu [12], commenting on Pollack's paper, has obtained our result on minimax trajectories.

8.16 MULTIPLE STRESSES

It frequently happens that during a process, stresses arise from several different causes, for example, mechanical and thermal. Although, in general, a minimax trajectory for one is not a minimax trajectory for the other, we can attempt to determine trajectories that are optimal in the sense that no change in the trajectory can lower both maximal stresses.

On link (i, j) let the maximal thermal stress be t_{ij} and the maximal mechanical stress be m_{ij}. Then, if we introduce the Lagrange multipliers n_1 and n_2, we can associate the generalized stress s_{ij},

$$s_{ij} = n_1 t_{ij} + n_2 m_{ij},$$

with each arc. For example, by letting

$$n_1 = 1, \qquad n_2 = a,$$

and determining a minimax trajectory between two particular states, for which the maximum mechanical stress is, let us say, s, we can guarantee that, among all trajectories for which the maximal mechanical stress between the states is s, we will have found a trajectory for which the thermal stress is minimal. Then a parameter study, involving the determination of many minimal spanning trees, might yield useful design information concerning the possible trade-offs.

8.17 DISCUSSION

Our primary aim has been to show the close connection among several important classes of automatic-control problems and graph theory. These considerations have raised many additional questions. Let us conclude by stating some of these.

Formulation

How are we to decide how many states of a system need be considered, what the stresses or times are, and what the criterion is?

Analytical and computational treatment

What are the connections between the solution to the discrete and continuous problems? Myriad other problems, too varied to catalogue, arise.

Implementation

After the optimal feedback-control decisions have been determined, how can controllers be realized to carry out the programs?

It is felt that the answers to these questions would provide both automatic control and mathematics with additional interesting chapters.

MULTIPLE-CHOICE REVIEW PROBLEMS

1. Consider the rectangular-weighted network shown in the figure on page 251. Interpret the number on each arc to be the time required to pass over the arc. The minimal time to pass from node A to node B is

(a) 89, (b) 84, (c) 85, (d) 87.

2. Interpret the numbers on the arcs in the network of the figure shown on page 251 to be the maximal stresses encountered in passing over the arcs. The smallest possible maximal stress that can be encountered along a trajectory from node A to node B is

(a) 22, (b) 23, (c) 24, (b) 25.

3. The sum of the branches in the minimal spanning tree for the network of the figure shown below is

(a) 212,   (b) 211,   (c) 207,   (d) 208.

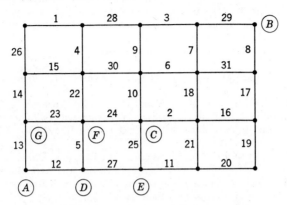

4. If, with the numerical interpretation of Problem 1, the system of the figure shown above is in state C, and if it is desired to transform the system into the state B as rapidly as possible the correct next state is obtained by moving

(a) up,   (b) down,   (c) right,   (d) left.

5. In the network of the figure shown above find the spanning tree for which the product of the numbers in its branches is a minimum. Consider the unique path in this tree from node A to node C. This path is

(a) $ADFC$,   (b) $AGFC$,   (c) $ADEC$,   (d) $AGFDEC$.

REFERENCES

1. Ash, M., R. Bellman, and R. Kalaba, "On Control of Reactor Shutdown Involving Minimal Xenon Poisoning," *Nuclear Science and Engineering* **6** (1959), 152–156.
2. Bellman, R., *Dynamic Programming*, Princeton University Press, Princeton, New Jersey, 1957.
3. Bellman, R., "On a Routing Problem," *Q. Appl. Math.* **16** (1958), 87–90.
4. Bellman, R., *Adaptive Control Processes: A Guided Tour*, Princeton University Press, Princeton, New Jersey, 1961.
5. Bellman, R., and R. Kalaba, "On kth Best Policies," *J. Soc. Ind. Appl. Math.* **8** (1960), 582–588.
6. Bellman, R., and R. Kalaba (editors), *Mathematical Trends in Control Theory*, Dover Publications, New York, 1964.
7. Bellman, R., R. Kalaba, and J. Lockett, *Numerical Solution of Functional Equations by Means of Laplace Transforms—VI: Stochastic Time-Optimal ,nirol*, RM-4119, the RAND Corporation, Santa Monica, Calif., 1964.

8. Bellman, R., and W. Karush, "On a New Functional Transform in Analysis: The Maximum Transform," *Bull. Amer. Math. Soc.* **67** (1961), 501–503.

9. Berge, C., *Théorie des graphes et ses applications*, Collection Universitaire de Mathematiques **2**, Dunod, Paris, 1958. *The Theory of Graphs and Its Applications*. Translated by Alison Doig, John Wiley and Sons, New York, 1962.

10. Dantzig, G., *On the Shortest Route through a Network*, P-1345, The RAND Corporation, Santa Monica, Calif., 1959.

11. Ford, L. R., Jr., and D. R. Fulkerson, *Flows in Networks*, Princeton University Press, Princeton, N.J., 1962.

12. Hu, T. C., "The Maximum Capacity Route Problem," *Operations Res.* **10** (1962), 898–900.

13. Kalaba, R., "On Some Communication Network Problems," Chapter 21 in R. Bellman and M. Hall, Jr. (editors), *Combinatorial Analysis, Proc. Sympos. Appl. Math.* **10**, American Mathematical Society, Providence, Rhode Island, 1960.

14. Kalaba, R., "Dynamic Programming and the Variational Principles of Classical and Statistisal Mechanics," Chapter 1 in J. E. Lay and L. E. Malvern (editors), *Developments in Mechanics* **1**, Proceedings of the Seventh Midwestern Mechanics Conference, Plenum Press, New York, 1962.

15. König, D., *Theorie der endlichen und unendlichen Graphen. Kombinatorische Topologie der Streckenkomplexe*, Mathematik in Monographien **16**, Akademische Verlagsgesellschaft, Leipsig, 1936; reprinted by Chelsea Publishing Co., New York, 1950.

16. Kruskal, J., "On the Shortest Spanning Subtree of a Graph and the Traveling Salesman Problem," *Proc. Amer. Math. Soc.* **7** (1956), 48–50.

17. LaSalle, J. P., "Time Optimal Control Systems," *Proc. Nat. Acad. Sci. U.S.A.* **45** (1959), 573–577.

18. LaSalle, J. P., and S. Lefschetz, *Stability by Liapunov's Direct Method with Applications*, Academic Press, New York, 1961.

19. Maxwell, J. C., "On Governors," *Proc. Royal Soc. London*, **16** (1868), 270–283.

20. Newton, G., L. Gould, and J. Kaiser, *Analytical Design of Linear Feedback Controls*, John Wiley and Sons, New York, 1957.

21. Pollack, M., "The Maximum Capacity through a Network," *Operations Res.* **8** (1960), 733–736.

22. Pollack, M., "Solutions of the kth Best Route Through a Network—A Review," *J. Math. Anal. Appl.* **3** (1961), 547–559.

23. Pollack, M., "The kth Best Route through a Network," *Operations Res.* **9** (1961), 578–580.

24. Prim, R., "Shortest Connection Networks and Some Generalizations," *Bell System Tech. J.* **36** (1957), 1389–1401.

CHAPTER 9

Optimum Multivariable Control

EDWIN L. PETERSON

9.1 PROBLEM STRUCTURE

Man has always sought ‛and will always continue to seek optimal solutions to his problems, as indicated in Chapters 7 and 8. If optimality were some unique state of affairs in the universe of possible courses of action, its determination would seem to present little difficulty, at least conceptually. We could simply examine all the alternatives.

The rapid growth in both the number and the complexity of situations over which man wishes to exercise control makes direct examination of alternatives increasingly difficult and often impossible. There is, in addition, the more fundamental problem of what is optimal with respect to a variety of interests. No amount of mathematical manipulation can provide answers to the subjective matter of balancing the interests.

Before mathematics can be of help in resolving the optimality problem, the problem must be rendered objective—a situation that is met provided the following three conditions prevail:

1. There exists a description of the state of affairs and the manner in which it is affected by any admissible course of action.

2. There exists a domain of admissible courses of action from which we are free to select a particular course in preference to all others.

3. There exists a criterion for optimality that reflects the desired state of affairs and forms the basis for selection of a course of action.

Even after the problem has been rendered objective in nature, its solution is far from trivial. Furthermore, there are various levels of solution.

To illustrate clearly the idea of various levels of solution, let us consider a concrete example. Suppose we desire the solution to a set of n simultaneous linear algebraic equations in n variables. One level of solution is knowledge that a solution exists. In this case, a unique solution exists provided the determinant constructed from the coefficients of the set of linear equations is nonvanishing. A second level of solution is its form. We know, for example, that if the coefficient determinant of our set of simultaneous linear algebraic equations is nonvanishing, then the form of the solution for any one of the variables is given by the ratio of two determinants, the denominator of which is the nonvanishing coefficient determinant. A third level of solution is the numerical solution value for each variable in the set of simultaneous equations. Since we know precisely the form of the solution for each variable, this would seem to present little difficulty. Actually, however, if we have a large number n of simultaneous linear equations, the time required to multiply out all the terms in even one determinant would preclude the possibility of obtaining numerical results in this manner.

In addition to the levels of solution, there is the further matter of the type of process represented by our description of the state of affairs. The first type of process, and by far the one that has received the greatest amount of attention, is the deterministic process. In this situation, we assume that we know precisely what state of affairs will result from any course of action. In other words, given a state and any decision, the outcome is known exactly. A second and somewhat more realistic situation occurs when there exists some degree of uncertainty regarding the outcome of the process describing the state of affairs. This second situation is characteristic of a stochastic process. Uncertainty can arise in any of several ways. It may be that even when given precise observations, we know only with some probability the state to be expected from a decision, or it may be that we know precisely the state resulting from a decision but have only observations with statistical uncertainty on which to base our decisions. There occurs yet a third kind of situation, known as an *adaptive process* [7, 8], in which a great deal of interest has manifested itself of late. In the adaptive process we start with an *a priori* estimate of the state of affairs resulting from a course of action, observe what happens as a result of such observations, and attempt to revise our *a priori* estimate of the process description. In effect, we attempt to *learn* something more about the process as we proceed and to incorporate the results of such learning in our future decisions about courses of action.

Before discussing the various methods of solution, we shall need a mathematical statement of our objective optimization problem. It is clear that there exists an almost infinite variety of objective types of optimization problems. In order to be fairly specific and at the same time sufficiently general to cover a rather broad range of interests, we shall take the following problem for purposes of discussion. Suppose that

(9.1) $$\frac{d\mathbf{x}}{dt} = \mathbf{G}(\mathbf{x}, \mathbf{y}), \qquad \mathbf{x}(0) = \mathbf{a},$$

where $\mathbf{x}$ is a p-dimensional state vector,

$$\mathbf{x} = (x_1, x_2, \ldots, x_p),$$

and $\mathbf{y}$ is a q-dimensional choice vector,

$$\mathbf{y} = (y_1, y_2, \ldots, y_q).$$

Let it be required to minimize the quantity

(9.2) $$J(\mathbf{y}) = \int_0^T F(\mathbf{x}, \mathbf{y}) \, dt,$$

subject to the restrictions

(9.3) $$c_i \le y_i(t) \le d_i, \qquad 0 \le t \le T, \qquad i = 1, 2, \ldots, l \le q,$$

with the terminal boundary conditions

(9.4) $$x_i(T) = b_i(T), \qquad i = 1, 2, \ldots, k \le p.$$

We suppose that the functions $b_i(t)$ are specified everywhere in the interval $(0, T)$.

The foregoing problem covers, for example, a variety of space interception and rendezvous problems. The vector $\mathbf{x}$ describes the state of affairs at any time, for example, position, velocity, and mass of the vehicle. The vector $\mathbf{y}$ represents the course of action that we are free to choose at any time, such as the magnitude and direction of thrust. The restrictions (9.3) represent certain real and practical physical limitations on our freedom to choose $\mathbf{y}$; for example, the magnitude of the thrust and/or angular rotation of the thrust vector may be restricted. The criterion (9.2) specifies what is to be minimized, perhaps the fuel expenditure. The equation $\mathbf{x}(0) = \mathbf{a}$ represents the initial state of affairs, and we require that certain of the state variables take on prescribed values at the end of the flight (9.4). For interception, we care only that the positions of the two objects be brought into coincidence at time T. For rendezvous, both the position and the velocity of the objects must be brought into correspondence at time T. If we were concerned, for example, with an interplanetary rendezvous with Mars, specification of $b_i(t)$ supposes that we know the position and velocity of the target planet as a function of time. The time duration of the encounter might be specified, or we might effect the encounter at an arbitrary time.

A word of caution is in order here. Solutions to problems of the foregoing type do not necessarily exist. A well-known case in point is the so-called minimum-time problem, with $F(\mathbf{x}, \mathbf{y}) = 1$. It may happen that from the admissible classes of $\mathbf{y}$ there simply are none that are capable of achieving the desired final state. This difficulty is bypassed if we regard the problem just stated as a special case of a more general problem that initially is *not* boundary-valued. Suppose that all elements of the problem remain the same except that the terminal boundary conditions (9.4) are removed, and instead of minimizing the quantity $J(\mathbf{y})$ given by (9.2), we take, for some $\beta \geq 0$,

$$(9.5) \qquad J_1(\mathbf{y}) = \int_0^T F(\mathbf{x}, \mathbf{y})\, dt + \beta \sum_{i=1}^k [x_i(T) - b_i(T)]^2$$

as the quantity to be minimized. In light of the earlier discussion, (9.5) implies that we are seeking to minimize a weighted combination of fuel expenditure and terminal error. For small values of β, we do not weight terminal error very heavily. As the value of β increases, the optimal choice for $\mathbf{y}$ becomes more strongly dependent on the terminal error, and as β becomes very large, the minimization of (9.5) results in a choice for $\mathbf{y}$ that tends to drive the terminal error to zero. In other words, if the solution exists, then

$$\lim_{\beta \to \infty} \min_{\mathbf{y}} [J_1(\mathbf{y})] \to \min [J(\mathbf{y})]$$

and

$$x_i(T) \to b_i(T), \qquad i = 1, 2, \ldots, k,$$

where we assume, of course, that T is sufficiently large. We call this process of inserting a special problem within the framework of a more general class one of *invariant imbedding*. Consequently, the solution of the previous boundary problem, if it exists, is contained within this problem as a special case. By computing the solution for a number of increasing values of β or T, not only is the boundary solution approached in a systematic way, but also we get a clear picture of the penalty in fuel expenditure that we must pay for the sake of reducing terminal error. Each value of β permits us to plot a point on the curve of minimum fuel expenditure as a function of allowable terminal error for a specified time duration of the encounter. We could also reduce the problem to one of minimizing the final value of an arbitrary function of the state. Suppose, for example, that we introduce the new state variable

$$\frac{dx_0}{dt} = F(\mathbf{x}, \mathbf{y}), \qquad x_0(0) = 0,$$

with

(9.6)
$$x_0(T) = \int_0^T F(\mathbf{x}, \mathbf{y}) \, dt.$$

If

$$\phi[\mathbf{x}(T)] = x_0(T) + \sum_{i=1}^{k} [x_i(T) - b_i(T)]^2,$$

then the function to be minimized is

$$J_1(\mathbf{y}) = \phi[\mathbf{x}(T)],$$

the final value of an arbitrary function of state.

We have observed how a rather arbitrary minimization problem can be reduced to minimization of the final value of an arbitrary function of state. We have also observed, through the concept of invariant imbedding, how a problem that is originally boundary-valued in nature is viewed as imbedded in a class of more general problems of which the boundary problem is simply a special limiting case. The imbedding procedure leads to additional applications as well. Consider, for example, a problem that is simply boundary-valued. Such a problem is one for which the constraints (9.1), (9.3), and (9.4) apply, but we have no variational minimization criterion such as (9.2). Under the imbedding procedure, the problem is viewed as a variational minimization problem in terms of (9.5), with F taken to be identically 0.

The reason for elimination of boundary-valuedness, of course, is the desire to achieve a reduction in complexity associated with the numerical solution of two-point boundary problems. From a computational point of view, however, the transformation of a boundary-value problem into a final-valued minimization is retained only when the numerical solution is undertaken by means of dynamic programming. Whether or not the original problem is boundary-valued, the computational solution by classical methods is always boundary-valued.

9.2 ALTERNATIVE APPROACHES

There are essentially only two approaches to the general optimization problem, the classical methods of variational calculus, and the functional-equation technique of dynamic programming. Up to the point where numerical results are sought by means of some computational algorithm, the methods are virtually indistinguishable. This fact has been well demonstrated in this country by Dreyfus [19], who has derived most of the conditional equations of classical variational calculus by means of the functional equations of dynamic programming. Similarly, the Soviet author Rozonoer [48] has demonstrated the correspondence between Pontriagin's classical maximum principle [47] and Bellman's dynamic programming [5].

Despite the similarity of the methods up to the point where numerical results are sought, the difficulties thereafter with either method stem from widely differing sources. Classical methods are always engulfed in the plague of boundary-valuedness, whereas dynamic programming is invariably beset by the curse of dimensionality. It is against either or both of these twin evils that our efforts must be concentrated if we are successfully to resolve the complex optimization problem presented by multidimensional nonlinear processes. Dimensionality and boundary-valuedness are precisely the problems with which we shall be concerned in this chapter.

To view the source and nature of these difficulties in somewhat greater perspective, let us consider the alternative approaches in some detail.

The general problem with which we are concerned has the form

$$(9.7) \quad \frac{dx_i}{dt} = G_i(x_1, x_2, \ldots, x_p, y_1, y_2, \ldots, y_q);$$

$$x_i(0) = a_i, \qquad i = 0, 1, 2, \ldots, p,$$

with the requirement to minimize

$$(9.8) \qquad J_1(\mathbf{y}) = \phi[\mathbf{x}(T)]$$

under the restrictions

$$(9.9) \quad c_i \leq y_i(t) \leq d_i, \qquad 0 \leq t \leq T, \qquad i = 1, 2, \ldots, l \leq q.$$

If the components of the vector $\mathbf{y}$ were unrestricted, that is, if we permitted $-\infty \leq y_i \leq \infty$, $i = 1, 2, \ldots, q$, then solution by classical methods would follow the well-known problem of Mayer in variational calculus [11, 12]. When restricted variations such as those specified by (9.9) must be included, the solution can still be approached along the lines of the problem of Mayer by means of the method of adjoined inequality constraints first proposed by Valentine [50].

A recent development using classical procedures, which permits arbitrary restrictions on the choice vector $\mathbf{y}$, is provided by the so-called *maximum principle* of the Soviet mathematician Pontriagin [47]. In any event, the problem of unrestricted variations is imbedded within Pontriagin's formulation as a special case. It seems likely therefore that formulation in terms of the maximum principle will find increasingly widespread acceptance among those pursuing the problem along classical lines.

In terms of the maximum principle, solution of the problem posed by (9.7), (9.8), and (9.9) requires the existence of an absolutely continuous and nonzero-valued vector function $\boldsymbol{\lambda}(t)$ such that the Hamiltonian function

$$H(\mathbf{x}, \mathbf{y}, \boldsymbol{\lambda}) = \boldsymbol{\lambda} \cdot \mathbf{G} = \sum_{j=0}^{p} \lambda_j G_j$$

has a constant maximum value in the choice variable $\mathbf{y}$ everywhere in the interval $0 \le t \le T$, which vanishes when T is free. Maximizing the function H generally leads to the requirement that

(9.10) $$\frac{\partial H}{\partial y_i} = \sum_{j=0}^{p} \lambda_j \frac{\partial G_j}{\partial y_i} = 0, \qquad i = 1, 2, \ldots, q.$$

The function H is also required to satisfy the Hamiltonian system of equations

(9.11) $$\frac{dx_i}{dt} = \frac{\partial H}{\partial \lambda_i} = G_i(\mathbf{x}, \mathbf{y}), \qquad x_i(0) = a_i, \qquad i = 0, 1, 2, \ldots, p,$$

and

(9.12) $$\frac{d\lambda_i}{dt} = \frac{-\partial H}{\partial x_i} = -\sum_{j=0}^{p} \lambda_j \frac{\partial G_j}{\partial x_i}, \qquad \lambda_i(T) = \frac{-\partial \phi}{\partial x_i}\bigg|_{t=T},$$
$$i = 0, 1, 2, \ldots, p.$$

The set (9.10) represents q simultaneous nonlinear algebraic equations in the q components of the choice vector $\mathbf{y}$ and implies the functional relationship

(9.13) $$y_i = Q_i(\mathbf{x}, \boldsymbol{\lambda}), \qquad i = 1, 2, \ldots, q.$$

If the q equations of (9.10) could be solved explicitly for the q variables $y_1, y_2, \ldots, y_q$, functions of the form given by (9.13) would be obtained. The equations (9.13) when substituted into (9.11) and (9.12) would give a $2(p + 1)$-dimensional boundary-valued problem in the pair of $(p + 1)$-dimensional vectors $\mathbf{x}$ and $\boldsymbol{\lambda}$.

There are difficulties in separating the elements y_i satisfying (9.10) into the explicit functional form (9.13) except in the simplest of cases, but even when it is possible to do so the problem of integrating the coupled sets of equations (9.11) and (9.12) remains. These latter equations are just ordinary first-order nonlinear differential equations which by themselves present no special difficulties. Rarely, of course, is it possible to integrate such sets of equations in closed form, but machine integration would be quite straightforward were it not for the fact that $\mathbf{x}(t)$ and $\boldsymbol{\lambda}(t)$ are specified at opposite ends of the interval $0 \le t \le T$.

Those readers who are unfamiliar with the maximum principle will recognize at once the similarities with ordinary classical procedures. The function H, for example, corresponds to the arbitrary vanishing function in the classical problem of Mayer. The components of the vector $\boldsymbol{\lambda}$

correspond to the time-variable Lagrangian multipliers. The equations (9.10) and (9.12) are quite similar to the classical Euler-Lagrange equations. On the other hand, the discontinuities in state derivatives $\dot{x}_i$, which may occur with discontinuities in the choice variable $\mathbf{y}$, and which are somewhat troublesome in the Euler-Lagrange equations, present no difficulties in the Hamiltonian system of equations. The prescribed values of the vector $\boldsymbol{\lambda}$ at time T are determined, as in all classical solutions, by the necessity for satisfying the so-called general transversality condition at the beginning and at the end of the problem, that is, at $t = 0$ and $t = T$. It should also be observed that whether we are using ordinary classical procedures or the maximum principle, solutions so obtained meet only the necessary conditions for a minimum. Further delicate sufficiency tests are required to establish that solutions meeting the necessary conditions are, in fact, optimal.

Despite arguments about necessary and sufficient conditions for optimality, or the relative virtues of one classical formulation over those of another, the singularly most important difficulty in all classical approaches is the inevitable boundary-valuedness of the solution. It is also significant to observe that this troublesome situation always prevails whether or not the original problem is initially boundary-valued. That is, it makes no difference whether we attempt to minimize $J_1(y)$ given by (9.8) with all elements of $\mathbf{x}(T)$ free or whether we try to minimize $J(\mathbf{y})$ given by (9.2) subject to the end conditions (9.4). The only difference is that instead of all elements of $\mathbf{x}$ being specified at $t = 0$ and all components of $\boldsymbol{\lambda}$ being specified at $t = T$, we use (9.4) and free the corresponding k components of the vector $\boldsymbol{\lambda}$ at $t = T$. The boundary-valuedness difficulty ultimately stems from the necessity of formulating an auxiliary set of equations, whether they be the Euler-Lagrange equations in ordinary classical procedures or the Hamiltonian system of equations arising in the maximum principle. In every classical approach to the solution, these auxiliary equations together with the original constraint equations lead to the computational requirement of solving the notoriously difficult two-point boundary-valued problem.

Virtually all efforts along classical lines have been concentrated on the deterministic problem. For that matter, it is not clear whether a meaningful formulation along classical lines is possible in nonlinear stochastic or adaptive processes. If or when such formulations appear, we can be assured that the formidable two-point boundary-valued problem will be present. Successful numerical solution of the general optimization problem requires satisfactory computational procedures to overcome boundary-valuedness if results are to be obtained from the classical formulation. We shall consider some interesting recent developments by Bryson et al.

[14] along the lines of linearization and successive approximation as a method of overcoming the difficulties presented by the two-point boundary-valued problem.

Dynamic programming, although presenting certain computational difficulties of its own, is not, however, plagued by boundary-valuedness. If the problem is not initially boundary-valued or if it can be reduced effectively to one that is not boundary-valued, as discussed earlier, no boundary-valuedness is encountered in the computational solution by dynamic programming. The reason for this situation is that dynamic programming attacks the minimization problem directly, without the introduction of an auxiliary set of equations such as the Euler-Lagrange or Hamiltonian system of equations in classical variational calculus. Furthermore, the minimum is sought in a manner both necessary and sufficient to ensure optimality since the minimum is sought by direct search over the admissible range of values for the choice variable. Moreover, it provides a standard formalism, within the framework of which it is possible to treat problems in which the constraints are algebraic or differential, linear or nonlinear, discrete or continuous, and deterministic or stochastic. It also holds considerable promise as a method of approaching the adaptive problem. In assessing the characteristics of dynamic programming, it should also be pointed out that the level of solution is numerical. That is, implementation is directly in terms of machine computation, so that any solutions obtained are automatically in terms of numerical results.

Suppose that we first inspect dynamic programming as a formalism before considering the nontrivial computational difficulties in some detail. The dynamic-programming view of the problem stated by (9.7), (9.8), and (9.9) is as follows. Instead of a fixed time duration T, consider the collection of all possible time durations from 0 up to T. Similarly, instead of a particular set of initial conditions $x(0) = a$, consider any set of initial conditions. The problem with a specific time duration and set of initial conditions is viewed as being imbedded within the problem of arbitrary time duration and initial conditions.

The functional equation of dynamic programming for this particular problem reads

$$f(\mathbf{a}; T) = \min_{\mathbf{y}} \{\phi[\mathbf{x}(T)]\}.$$

What is implied by the functional equation is the fact that if somehow we could search over the collection of all admissible values of $\mathbf{y}(t)$ in the interval $0 \leq t \leq T$ until we found the particular value minimizing $\varphi[\mathbf{x}(T)]$, the value assumed by the minimum return would ultimately depend on the initial conditions and the time duration of the process. This is clear from (9.7) since if $\mathbf{y}(t)$ were specified in the interval $0 \leq t \leq T$, we would simply integrate the equations (9.7), obtaining final values for all the state variables that depend only on the initial conditions and on the time duration.

Now all problems, whether they are continuous or discrete, must be rendered discrete for purposes of digital computation. We could, for example, divide time into increments of width Δ and write the constraints (9.7) in terms of first-order difference equations,

$$(9.14) \quad x_i^{k+1} = x_i^k + G_i(\mathbf{x}^k, \mathbf{y}^k)\Delta, \qquad \mathbf{x}^0(0) = \mathbf{a}, \qquad i = 0, 1, 2, \ldots, p.$$

Suppose now that $T = \Delta$. Then we have

$$\mathbf{x}(\Delta) = \mathbf{a} + \mathbf{G}(\mathbf{a}, \mathbf{y}^0)\Delta,$$

and thus

$$(9.15) \qquad f(\mathbf{a}; \Delta) = f_1(\mathbf{a}) = \min_{\mathbf{y}^0} \{\phi[\mathbf{a} + \mathbf{G}(\mathbf{a}, \mathbf{y}^0)\Delta]\}.$$

The solution is $\mathbf{y}^{01}(\mathbf{a})$, the optimum initial choice of $\mathbf{y}$ for a one-decision process starting from the initial state $\mathbf{x}(0) = \mathbf{a}$. Given $\mathbf{G}$ and $\mathbf{a}$, we must search (9.15) over the admissible values of $\mathbf{y}^0$ under the restrictions (9.9), obtaining, for each set $\mathbf{a}$, the minimum return $f_1(\mathbf{a})$ and the optimal choice $\mathbf{y}^{01}(\mathbf{a})$.

Suppose now that $T = 2\Delta$. The principle of optimality gives the functional recurrence formula of dynamic programming,

$$(9.16) \qquad f(\mathbf{a}; 2\Delta) = f_2(\mathbf{a}) = \min_{\mathbf{y}^0} \{f_1[\mathbf{a} + \mathbf{G}(\mathbf{a}, \mathbf{y}^0)\Delta]\}.$$

The solution is $\mathbf{y}^{02}(\mathbf{a})$, the optimum initial choice of $\mathbf{y}$ for a two-decision process starting from the initial state $\mathbf{x}(0) = \mathbf{a}$. The reasoning behind (9.16) can be stated rather clearly: Whatever the initial choice $\mathbf{y}^0$ may be, the state resulting from that decision is

$$(9.17) \qquad\qquad \mathbf{x}(\Delta) = \mathbf{a} + \mathbf{G}(\mathbf{a}, \mathbf{y}^0)\Delta.$$

At this point, we must make a decision $\mathbf{y}^1$ that minimizes $\phi[\mathbf{x}(2\Delta)]$. Whatever the state (9.17) may be, the best return from a single decision for any initial state is given by (9.15), which in this case is $f_1[\mathbf{x}(\Delta)]$. Since we are free to adjust $\mathbf{x}(\Delta)$ by the choice $\mathbf{y}^0$, it follows that $\mathbf{y}^0$ should be selected to minimize $f_1[\mathbf{x}(\Delta)]$, which is precisely the statement in (9.16). For the general case, we have $T = N\Delta$, and the functional recurrence formula becomes

$$(9.18) \qquad f(\mathbf{a}; N\Delta) = f_N(\mathbf{a}) = \min_{\mathbf{y}^0} \{f_{N-1}[\mathbf{a} + \mathbf{G}(\mathbf{a}, \mathbf{y}^0)\Delta]\}.$$

The solution is $\mathbf{y}^{0N}(\mathbf{a})$, the optimum initial choice of $\mathbf{y}$ for an N-stage decision process starting from the initial state $\mathbf{x}(0) = \mathbf{a}$.

Effectively, the minimum return and the optimal policy are generated from the sequence of functions $f_i(\mathbf{a})$ and $\mathbf{y}^{0i}(\mathbf{a})$. If $\mathbf{x}^*(t)$ and $\mathbf{y}^*(t)$ denote the optimal state and choice, respectively, then for any set of initial conditions and time duration $T = N\Delta$, the minimum return is

$$f_N(\mathbf{a}) = \min_{\mathbf{y}} \{\phi[\mathbf{x}(N\Delta)]\},$$

with

(9.19)
$$\mathbf{x}^*(0) = \mathbf{a},$$
$$\mathbf{y}^*(0) = \mathbf{y}^{0N}(\mathbf{a}).$$

Using the values in (9.19) and the state recurrence formula (9.14), we have

$$\mathbf{x}^*(\Delta) = \mathbf{a} + \mathbf{G}[\mathbf{a}, \mathbf{y}^*(0)]\Delta,$$
$$\mathbf{y}^*(\Delta) = \mathbf{y}^{0(N-1)}[\mathbf{x}^*(\Delta)].$$

Similarly, we find

$$\mathbf{x}^*(2\Delta) = \mathbf{x}^*(\Delta) + \mathbf{G}[\mathbf{x}^*(\Delta), \mathbf{y}^*(\Delta)]\Delta,$$
$$\mathbf{y}^*(2\Delta) = \mathbf{y}^{0(N-2)}[\mathbf{x}^*(2\Delta)],$$

and continue in a like manner, finding values of $\mathbf{x}^*(t)$ and $\mathbf{y}^*(t)$ at time intervals of length Δ.

The foregoing discussion has been carried out in terms of first-order difference equations. Questions can arise with regard to the increment size Δ for satisfactory convergence. Second, third, or higher-ordered difference equations could also be used, however, if needed. On the other hand, it is not necessary to use difference equations at all. Although it is always necessary to place the problem in discrete form for digital computation, difference equations represent only one such form. Various methods of rendering problems discrete are discussed by Greenstadt [26]. Questions about convergence and choice of a particular discrete form are certainly proper and significant. They are not the only questions to be asked, however, or the most important.

The most significant difficulty in the computational solution by dynamic programming stems from the multidimensional nature of the functional recurrence formulas (9.15) and (9.18). To compute f_1, we must search the space of $\mathbf{y}$ relative to each value selected for the vector $\mathbf{a}$. Then $f_1(\mathbf{a})$ is stored in the fast memory of the computer, where it is used to compute f_2. After f_2 is computed, $f_1(\mathbf{a})$ is printed out and $f_2(\mathbf{a})$ is placed in the fast memory, where it is used to compute f_3, etc. The corresponding optimal choices $\mathbf{y}^{01}(\mathbf{a})$, $\mathbf{y}^{02}(\mathbf{a})$, . . . , $\mathbf{y}^{0N}(\mathbf{a})$ are printed out as they are found, since we do not require $\mathbf{y}^{01}$ to compute f_2, or $\mathbf{y}^{02}$ to compute f_3, etc. It is not practical, in general, to store functions of more than about two dimensions because of the limited memory capacity even with currently available large-scale digital computers. Consequently, any approach to a workable computational solution must embody reduction in the dimensionality of the return functions $f_i(\mathbf{a})$. All known methods of reducing dimensionality of the return functions are based on the idea of exchanging computing time for memory. This is no doubt reasonable since in addition to logical functions we can perform only two things with a digital computer—compute and store. If storage capacity is limiting, it seems logical to look for ways of substituting computation for storage.

Storage problems with the return function are not the only reasons for desiring reduction in dimensionality. The search problem itself requires that we examine the admissible values of $\mathbf{y}$ for each vector $\mathbf{a}$ in the functional recurrence formula (9.18). If the dimensionality of the return function can be reduced, the time required to search the space of $\mathbf{y}$ is also reduced. If the dimensionality of $\mathbf{y}$ is large, considerable time is required to search for the optimal choice. It should be noted, however, that the principle of optimality permits a stage-by-stage solution. That is, at any stage in the solution of the functional equation, we are searching only over the admissible values of $\mathbf{y}$ at one instant in time rather than searching simultaneously over N sets of vectors $\mathbf{y}^0$, $\mathbf{y}^1$, . . . , $\mathbf{y}^{(N-1)}$, representing the discrete approximation to the continuous vector $\mathbf{y}(t)$ over the interval $0 \le t \le T$.

Although it is also possible to construct the functional recurrence relations of dynamic programming for stochastic and adaptive processes, successful numerical solution of the functional recurrence equations is still dependent on resolution of the dimensionality problem. We shall consider recent significant developments by Aoki [2] in terms of polynomial approximation as a means of overcoming the difficulties presented by dimensionality of the functional recurrence equations.

9.3 POLYNOMIAL APPROXIMATION

One of the more attractive methods of representing a function is in terms of a set of orthogonal functions. Many such representations are in common use, such as the Taylor series and the Fourier series. In approximating the minimum-return function of dynamic programming, several important factors are to be considered in choosing the representation. The accuracy of the approximation should have about the same validity at any point within the range of the variables on which the function is dependent.

Consequently, a Taylor series, expanded about some point in the interval, is not desirable.

We should also like to obtain approximations to the function from a knowledge of its values at only a limited number of discrete isolated points in order to diminish the fast-memory capacity requirements associated with the solution of the functional recurrence equation of dynamic programming. Consequently, expansion of the function in terms of an orthonormal set such as a Fourier series is objectionable because the orthogonality of the Fourier set is with respect to a weighted integration over an interval rather than with respect to a weighted summation over a discrete set of points within the interval.

If we choose a system of complete orthonormal polynomials, orthogonal with respect to a weighted summation over a discrete set of points in the interval, the function to be approximated can be represented with uniform accuracy throughout the normalized interval by storing the coefficients of the polynomial expansion and computing the function from a knowledge of the coefficients and the orthonormal set. The Chebyshev polynomials are particularly convenient because they satisfy a very simple recursive formula, they form a set orthogonal with respect to a weighted summation over a unique and easily determined set of discrete points in the normalized interval, and the formulas for coefficients of the polynomial representation are particularly simple.

The Chebyshev polynomials are defined in the interval $|x| \leq 1$ by

$$(9.20) \qquad p_n(x) = \cos{(n \ \mathrm{arc \ cos} \ x)} \qquad n = 0, 1, 2, \ldots .$$

Under the transformation

$$\alpha = \mathrm{arc \ cos} \ x,$$

it is clear that

$$(9.21) \qquad \qquad p_n(x) = \cos n\alpha.$$

It is readily proved that these polynomials satisfy the simple recursive formula

$$p_0(x) = 1 \qquad p_1(x) = x,$$

$$p_{n+1}(x) = 2xp_n(x) - p_{n-1}(x), \qquad n \geq 1.$$

Using (9.21), we can write

$$(9.22) \qquad p_{n+1}(x) = \cos{(n + 1)}\alpha = \cos n\alpha \cos \alpha - \sin n\alpha \sin \alpha$$

and

$$(9.23) \qquad p_{n-1}(x) = \cos{(n - 1)}\alpha = \cos n\alpha \cos \alpha + \sin n\alpha \sin \alpha.$$

Adding (9.22) and (9.23), we obtain

(9.24) $$p_{n+1}(x) + p_{n-1}(x) = 2 \cos n\alpha \cos \alpha.$$

Since $x = \cos \alpha$ and $p_n(x) = \cos n\alpha$, however, it follows that (9.24) can be written equivalently as

$$p_{n+1}(x) + p_{n-1}(x) = 2xp_n(x),$$

which is the Chebyshev polynomial recursive formula.

Using the definition (9.20), we observed that the zeros of the Chebyshev polynomials are uniquely determined by

$$n \text{ arc } \cos x_i = (2i + 1)\frac{\pi}{2}, \qquad i = 0, 1, 2, \dots, n - 1, \qquad n \geq 1;$$

thus the n zeros of the nth Chebyshev polynomial are

$$x_i = \cos \theta_i, \qquad \theta_i = \frac{(2i + 1)\pi}{2n}, \qquad i = 0, 1, 2, \dots, n - 1, \qquad n \geq 1.$$

Suppose we expand a function $f(x)$ into a finite series of Chebyshev polynomials,

(9.25) $$f(x) = \sum_{n=0}^{R} a_n p_n(x).$$

Let us now determine the $R + 1$ coefficients $a_0, a_1, a_2, \dots, a_R$. This is conveniently done in terms of the $R + 1$ zeros of the polynomial $p_{R+1}(x)$, namely

$$x_{i+1} = \cos \theta_i, \qquad \theta_i = \frac{(2i + 1)\pi}{2(R + 1)}, \qquad i = 0, 1, 2, \dots, R.$$

Since (9.25) is valid for any x, it follows that

(9.26) $$f(x_i) = \sum_{n=0}^{R} a_n p_n(x_i).$$

If we multiply both sides of (9.26) by $p_m(x_i)$ and then sum both sides of the equation over all $i = 1, 2, \dots, R + 1$, there results

$$\sum_{i=1}^{R+1} p_m(x_i) f(x_i) = \sum_{i=1}^{R+1} \sum_{n=0}^{R} a_n p_n(x_i) p_m(x_i).$$

Because of the orthogonality of the functions $p_k(x_i)$, we have

$$\sum_{i=1}^{R+1} p_n(x_i) p_m(x_i) = 0, \qquad n \neq m,$$

so that

(9.27) $$\sum_{i=1}^{R+1} p_m(x_i) f(x_i) = a_m \sum_{i=1}^{R+1} p_m^2(x_i).$$

Now it is readily shown by mathematical induction that

(9.28)
$$\sum_{i=1}^{R+1} p_n^2(x_i) = \frac{R+1}{2}, \qquad n \neq 0,$$

and it is quite obvious that

(9.29)
$$\sum_{i=1}^{R+1} p_0^2(x_i) = R + 1.$$

Therefore, with (9.28) and (9.29) in (9.27), the coefficients of the polynomial representation (9.26) satisfy the very simple relations

$$a_0 = \frac{1}{R+1} \sum_{i=1}^{R+1} f(x_i)$$

and

$$a_n = \frac{2}{R+1} \sum_{i=1}^{R+1} f(x_i) p_n(x_i), \qquad n \neq 0.$$

Although one-dimensional dynamic-programming problems can be handled in a straightforward manner without resort to schemes such as polynomial approximation to resolve dimensionality problems, it will be instructive to observe exactly how the procedure of polynomial approximation is applied to such a case.

Suppose the problem is

$$\frac{dx}{dt} = G(x, y), \qquad x(0) = a,$$

with the requirement to minimize

$$J(y) = \phi[x(T)], \qquad c \leq y \leq d,$$

some arbitrary function of the outcome. The following step-by-step procedure would be employed to generate the solution to polynomial approximation:

1. Normalize the range of the state variable so that $|x| \leq 1$.
2. Select an integer R that fixes the degree of the polynomial approximation to the minimum return function.
3. Compute the angles

$$\theta_i = \frac{(2i+1)\pi}{2(R+1)}, \qquad i = 0, 1, 2, \ldots, R.$$

4. Compute the $R + 1$ zeros of $p_{R+1}(x)$ from

$$x_{i+1} = \cos \theta_i, \qquad i = 0, 1, 2, \ldots, R.$$

5. Determine

$$f_1(x_i) = \min_{y_0} \{\phi[x_i + G(x_i, y_0)\Delta]\}$$

for $i = 1, 2, \ldots, R + 1$ by searching the function ϕ over the admissible range of values y_0. For each value of state x_i, the value of y_0 minimizing is $y_{01}(x_i)$, the optimal initial choice starting in state x_i with a one-decision process (that is, $T = \Delta$).

6. Using the $R + 1$ values $f_1(x_i)$ obtained in step 5, compute the coefficient

$$a_{01} = \frac{1}{R + 1} \sum_{i=1}^{R+1} f_1(x_i),$$

and with the aid of the recursive formula for the Chebyshev polynomials given by

$$p_0(x) = 1, \qquad p_1(x) = x,$$
$$p_{n+1}(x) = 2xp_n(x) - p_{n-1}(x), \qquad n \geq 1,$$

compute the remaining coefficients

$$a_{n1} = \frac{2}{R + 1} \sum_{i=1}^{R+1} f_1(x_i)p_n(x_i)$$

for $n = 1, 2, 3, \ldots, R$.

7. Store the $R + 1$ coefficients $a_{01}, a_{11}, \ldots, a_{R1}$ computed in step 6 in the fast memory of the digital computer.

8. Determine

$$f_2(x_i) = \min_{y_0} \sum_{n=0}^{R} a_{n1}p_n[x_i + G(x_i, y_0)\Delta],$$

using the coefficients stored in step 7 and the recursive formula for the Chebyshev polynomials by searching over the admissible range of values for y_0, for each $i = 1, 2, \ldots, R + 1$. For each value of state x_i, the value of y_0 minimizing the above function is $y_{02}(x_i)$, the optimum initial choice starting in state x_i with a two-decision process (that is $T = 2\Delta$).

9. Using the $R + 1$ values $f_2(x_i)$ obtained in step 8, compute the new set of coefficients

$$a_{02} = \frac{1}{R + 1} \sum_{i=1}^{R+1} f_2(x_i),$$

$$a_{n2} = \frac{2}{R + 1} \sum_{i=1}^{R+1} f_2(x_i)p_n(x_i), \qquad n \geq 1.$$

10. The coefficients a_{n1} can now be transferred from fast memory and printed out on tape, and the new set of coefficients a_{n2} are then deposited in fast memory.

We now continue the process as described, computing $f_k(x_i)$ and the associated set of coefficients a_{nk}, which are used to determine $f_{k+1}(x_i)$ in terms of the recurrence formula of dynamic programming,

$$f_{k+1}(x_i) = \min_{y_0} \{f_k[x_i + G(x_i, y_0)\Delta]\}$$

$$= \min_{y_0} \sum_{n=0}^{R} a_{nk} p_n[x_i + G(x_i, y_0)\Delta],$$

terminating, of course, at $k + 1 = N$ (that is, $T = N\Delta$).

It should be observed that the net effect of polynomial approximation is effectively to exchange computation for memory. Instead of $f(x)$ being stored over a large number of grid points, $f(x)$ is generated over the whole range of the state variable by storing a set of $R + 1$ coefficients of the polynomial representation for $f(x)$.

The vast reduction in fast-memory capacity offered through polynomial approximation will become clear now as we consider a multidimensional problem. Suppose that $\mathbf{x} = (x_1, x_2, \ldots, x_p)$ is a p-dimensional state vector and $\mathbf{y} = (y_1, y_2, \ldots, y_q)$ is a q-dimensional choice vector with the defining equations of state given by

$$\frac{d\mathbf{x}}{dt} = \mathbf{G}(\mathbf{x}, \mathbf{y}), \qquad \mathbf{x}(0) = \mathbf{a},$$

and that the requirement is to minimize

$$J(\mathbf{y}) = \phi[\mathbf{x}(T)], \qquad \mathbf{c} \le \mathbf{y} \le \mathbf{d}.$$

The functional equation of dynamic programming gives

$$f(a_1, a_2, \ldots, a_p; T) = \min_{\mathbf{y}} [J(\mathbf{y})],$$

a p-dimensional minimum-return function. Direct application of the principle of optimality gives the recurrence formula

$$f_1(a_1, a_2, \ldots, a_p) = \min_{\mathbf{y}} \{\phi[\mathbf{x}(\Delta)]\},$$

$$f_{k+1}(a_1, a_2, \ldots, a_p) = \min_{\mathbf{y}} \{f_k[x_1(\Delta), x_2(\Delta), \ldots, x_p(\Delta)]\},$$

where

$$x_i(\Delta) = a_i + G_i(\mathbf{x}, \mathbf{y})\Delta, \qquad i = 1, 2, \ldots, p.$$

If straightforward solution of this functional recurrence formula is undertaken by storing the function at grid points, the storage requirements become enormous. If $p = 5$ and we desire the function at 100 values for each of its arguments a_1, a_2, a_3, a_4, a_5, then the memory capacity required is

$$M = (100)^5 = 10^{10},$$

a physical impossibility for fast-memory units on any current or contemplated computer.

Consider now the requirements if we use multidimensional polynomial approximation. We approximate the multidimensional function, as before, by a series of Chebyshev polynomials and write

(9.30) $$f(a_1, a_2, \ldots, a_p) = \sum_{i_1=0}^{R} \sum_{i_2=0}^{R} \cdots \sum_{i_p=0}^{R} P_{i_1 i_2 \ldots i_p},$$

where

$$P_{i_1 i_2 \ldots i_p} = a_{i_1 i_2 \ldots i_p} p_{i_1}(a_1) p_{i_2}(a_2) \cdots p_{i_p}(a_p).$$

If $R + 1$ zeros of $p_{R+1}(z)$ be denoted by

(9.31) $$z_{i+1} = \cos \theta_i, \qquad \theta_i = \frac{(2i + 1)\pi}{2(R + 1)}, \qquad i = 0, 1, 2, \ldots, R,$$

then the coefficients in (9.30) are given by

(9.32) $$a_{i_1 i_2 \cdots i_p} = \frac{2^{g(i_1, i_2, \ldots, i_p)}}{(R + 1)^p} \sum_{j_1=1}^{R+1} \sum_{j_2=1}^{R+1} \cdots \sum_{j_p=1}^{R+1} Q_{j_1 j_2 \ldots j_p},$$

where

$$Q_{j_1 j_2 \ldots j_p} = f(z_{j_1}, z_{j_2}, \ldots, z_{j_p}) p_{i_1}(z_{j_1}) p_{i_2}(z_{j_2}) \cdots p_{i_p}(z_{j_p}),$$

and we have

(9.33)
$$0 \leq g(i_1, i_2, \ldots, i_p) = u(i_1 - 1) + u(i_2 - 1) + \cdots + u(i_p - 1) \leq p,$$

where $u(v - w)$ is the translated unit step function with value 0 if $v - w < 0$, and value 1 if $v - w \geq 0$.

Let us illustrate the proof of these relations for the case $p = 2$, from which it will be clear that the proof of (9.32) can likewise be made for any value of p. If $p = 2$, we have a two-dimensional return function

(9.34) $$f(a_1, a_2) = \sum_{m=0}^{R} \sum_{n=0}^{R} a_{mn} p_m(a_1) p_n(a_2)$$

If the $R + 1$ zeros of $p_{R+1}(z)$ are denoted by (9.31), it is clear that the relation

$$f(z_i, z_j) = \sum_{m=0}^{R} \sum_{n=0}^{R} a_{mn} p_m(z_i) p_n(z_j)$$

is valid since (9.34) holds for any values of a_1 and a_2 within their range.

If both sides of (9.34) are multiplied by $p_k(z_i)p_l(z_j)$ and the result summed over all the values of i and j, there results

$$(9.35) \quad \sum_{i=1}^{R+1}\sum_{j=1}^{R+1} f(z_i, z_j)p_k(z_i)p_l(z_j)$$
$$= \sum_{m=0}^{R}\sum_{n=0}^{R} a_{mn} \sum_{i=1}^{R+1} p_m(z_i)p_k(z_i) \sum_{j=1}^{R+1} p_n(z_j)p_l(z_j).$$

Now, because of the orthogonality relations

$$(9.36) \qquad \sum_{i=1}^{R+1} p_m(z_i)p_k(z_i) = 0, \qquad m \neq k,$$

and

$$(9.37) \qquad \sum_{j=1}^{R+1} p_n(z_j)p_l(z_j) = 0, \qquad n \neq l,$$

substitution of (9.36) and (9.37) into (9.35) gives

$$a_{kl} = \frac{\displaystyle\sum_{i=1}^{R+1}\sum_{j=1}^{R+1} f(z_i, z_j)p_k(z_i)p_l(z_j)}{\left[\displaystyle\sum_{i=1}^{R+1} p_k^2(z_i)\right]\left[\displaystyle\sum_{j=1}^{R+1} p_l^2(z_j)\right]}.$$

In view of (9.28) and (9.29), it is clear that

$$(9.38) \qquad a_{00} = \frac{1}{(R+1)^2} \sum_{i=1}^{R+1}\sum_{j=1}^{R+1} f(z_i, z_j),$$

$$(9.39) \qquad a_{01} = \frac{2}{(R+1)^2} \sum_{i=1}^{R+1}\sum_{j=1}^{R+1} f(z_i, z_j)p_1(z_j),$$

$$(9.40) \qquad a_{10} = \frac{2}{(R+1)^2} \sum_{i=1}^{R+1}\sum_{j=1}^{R+1} f(z_i, z_j)p_1(z_i),$$

and

$$(9.41) \quad a_{mn} = \frac{2^2}{(R+1)^2} \sum_{i=1}^{R+1}\sum_{j=1}^{R+1} f(z_i, z_j)p_m(z_i)p_n(z_j), \qquad m \geq 1, \qquad n \geq 1.$$

The results (9.38) through (9.41) can be obtained directly from (9.32) and (9.33) if $p = 2$.

The reduction in dimensionality given by multidimensional polynomial approximation is clear when we observe that the multidimensional function (9.30) is obtained by storing only the coefficients, and the required memory capacity becomes

$$M = (R+1)^p$$

for a p-dimensional return function approximated in terms of Rth degree polynomials.

9.4 LINEARIZATION AND SUCCESSIVE APPROXIMATION

Suppose that we seek to determine the minimum-return function for a case in which the criterion function ϕ satisfies the relation $\phi[\mathbf{x}(T)] = x_0(T)$, the final value of a single element of the state vector. Then we have

$$(9.42) \qquad \min_{\mathbf{y}} [x_0(T)] = f(\mathbf{a}; T) = h(u; T),$$

where

$$(9.43) \qquad u = \mathbf{k} \cdot \mathbf{a},$$

the dot product of a constant vector $\mathbf{k}$ and the initial state $\mathbf{x}(0) = \mathbf{a}$. As a matter of fact, if the problem is linear, the solution exists precisely in the form given by (9.42) and (9.43). Suppose that $\mathbf{c} \leq \mathbf{y} \leq \mathbf{d}$, and that the state vector $\mathbf{x} = (x_0, x_1, \ldots, x_p)$ satisfies the equation

$$(9.44) \qquad \frac{d\mathbf{x}}{dt} = A(t)\mathbf{x} + B(t)\mathbf{y}, \qquad \mathbf{x}(0) = \mathbf{a},$$

where A and B are matrices. Let $\boldsymbol{\lambda}(t)$ be a vector such that

$$(9.45) \qquad \frac{d\boldsymbol{\lambda}}{dt} = -A'(t)\boldsymbol{\lambda},$$

where $A'(t)$ is the transpose of A. If we multiply (9.44) by the transpose of $\boldsymbol{\lambda}$ and (9.45) by the transpose of $\mathbf{x}$ and add, then there results

$$\boldsymbol{\lambda}' \frac{d\mathbf{x}}{dt} + \mathbf{x}' \frac{d\boldsymbol{\lambda}}{dt} = \boldsymbol{\lambda}'A\mathbf{x} - \mathbf{x}'A'\boldsymbol{\lambda} + \boldsymbol{\lambda}'B\mathbf{y}.$$

Since

$$\frac{d(\boldsymbol{\lambda}'\mathbf{x})}{dt} = \boldsymbol{\lambda}' \frac{d\mathbf{x}}{dt} + \mathbf{x}' \frac{d\boldsymbol{\lambda}}{dt},$$

and

$$\boldsymbol{\lambda}'A\mathbf{x} - \mathbf{x}'A'\boldsymbol{\lambda} = 0,$$

we have

$$\frac{d(\boldsymbol{\lambda}'\mathbf{x})}{dt} = \boldsymbol{\lambda}'B\mathbf{y}.$$

It follows that

$$\boldsymbol{\lambda}'x \Big|_0^T = \int_0^T \boldsymbol{\lambda}'B\mathbf{y} \, dt,$$

and therefore we obtain

$$(9.46) \qquad \boldsymbol{\lambda}'(T)\mathbf{x}(T) = \boldsymbol{\lambda}'(0)\mathbf{a} + \int_0^T \boldsymbol{\lambda}'(t)B(t)\mathbf{y}(t) \, dt.$$

If we choose the elements of $\boldsymbol{\lambda}$ at time $t = T$ such that

(9.47) $\qquad \lambda_i(T) = 0, \quad i \neq 0, \quad \text{and} \quad \lambda_0(T) = 1,$

then (9.46) becomes

(9.48) $\qquad x_0(T) = \boldsymbol{\lambda}'(0)\mathbf{a} + \int_0^T \boldsymbol{\lambda}'(t)B(t)\mathbf{y}(t)\, dt.$

If

$$\boldsymbol{\lambda}(0) = \mathbf{k},$$

then

(9.49) $\qquad \boldsymbol{\lambda}'(0)\mathbf{a} = \boldsymbol{\lambda}(0) \cdot \mathbf{a} = \mathbf{k} \cdot \mathbf{a} = u,$

and it is clear from (9.48) and (9.49) that

(9.50) $\quad \min_{\mathbf{y}} [x_0(T)] = \min_{\mathbf{y}} \left[u + \int_0^T \boldsymbol{\lambda}'(t)B(t)\mathbf{y}(t)\, dt \right] = h(u; T).$

The elements of $\boldsymbol{\lambda}$ are completely determined for any time $0 \leq t \leq T$ by integrating the so-called adjoint equation (9.45) backward from time $t = T$, using the specified conditions (9.47).

For this simple problem, the optimal choice is bang-bang [6] and the switching instants for $\mathbf{y}$ are readily found by examination of the integrand in (9.50). For example, the element in the ith row of the column vector $B(t)\mathbf{y}(t)$ is

$$[B(t)\mathbf{y}(t)]_i = \sum_{j=1}^q b_{ij}(t)y_j(t)$$

if $\mathbf{y}(t)$ is a q-dimensional choice vector. Hence the scalar $\boldsymbol{\lambda}'(t)B(t)\mathbf{y}(t)$ is

$$\boldsymbol{\lambda}'(t)B(t)\mathbf{y}(t) = \sum_{i=0}^p \sum_{j=1}^q \lambda_i(t)b_{ij}(t)y_j(t),$$

so that

$$\int_0^T \boldsymbol{\lambda}'(t)B(t)\mathbf{y}(t)\, dt = \int_0^T \sum_{j=1}^q \left[\sum_{i=0}^p \lambda_i(t)b_{ij}(t) \right] y_j(t)\, dt,$$

and the optimal choices are governed by the sign of the bracketed term. For example, if $c_j < 0$ and $d_j > 0$, then the optimal choices are

$$y_j(t) = c_j \qquad \text{for} \quad \sum_{i=1}^p \lambda_i(t)b_{ij}(t) > 0,$$

$$= d_j \qquad \text{for} \quad \sum_{i=1}^p \lambda_i(t)b_{ij}(t) < 0,$$

and $y_j(t)$ is arbitrary for

$$\sum_{i=1}^p \lambda_i(t)b_{ij}(t) = 0.$$

Suppose that we now consider the constrained nonlinear variational problem with $\mathbf{x} = (x_1, x_2, \ldots, x_p)$ and $\mathbf{y} = (y_1, y_2, \ldots, y_q)$, in which

$$(9.51) \qquad \frac{d\mathbf{x}}{dt} = \mathbf{G}(\mathbf{x}, \mathbf{y}), \qquad \mathbf{x}(0) = \mathbf{a},$$

with the requirement to minimize, for restricted choices $\mathbf{c} \leq \mathbf{y} \leq \mathbf{d}$, the quantity

$$J(\mathbf{y}) = \phi[\mathbf{x}(T)],$$

where ϕ is any arbitrary scalar function of the final state vector. In terms of Pontriagin's maximum principle, there must exist an absolutely continuous and nonzero-valued vector function $\boldsymbol{\lambda}(t)$ such that the Hamiltonian function

$$(9.52) \qquad H(\mathbf{x}, \mathbf{y}, \boldsymbol{\lambda}) = \boldsymbol{\lambda}'(t)\mathbf{G}(\mathbf{x}, \mathbf{y}) = \sum_{i=1}^{p} \lambda_i(t)G_i(\mathbf{x}, \mathbf{y})$$

is maximized under optimal conditions, and the Hamiltonian system of equations

$$(9.53) \qquad \dot{x}_i = \frac{\partial H}{\partial \lambda_i}, \qquad \dot{\lambda}_i = -\frac{\partial H}{\partial x_i} = -\sum_{j=1}^{p} \frac{\partial G_j}{\partial x_i} \lambda_j$$

must be satisfied everywhere in $0 \leq t \leq T$. Furthermore, under optimal conditions, the equation

$$(9.54) \qquad \frac{\partial H}{\partial y_i} = \sum_{j=1}^{q} \lambda_j(t) \frac{\partial G_j(\mathbf{x}, \mathbf{y})}{\partial y_i} = \psi_i(t)$$

is required to vanish. It is also necessary that the general transversality condition

$$(9.55) \qquad \left[\delta\phi + H\,\delta t + \sum_{i=1}^{p} \frac{\partial H}{\partial \dot{x}_i} (\delta x_i - \dot{x}_i\,\delta t) \right]_0^T = 0$$

be satisfied for any admissible endpoint variations. From the Hamiltonian (9.52), however, we can see that

$$\frac{\partial H}{\partial \dot{x}_i} = \lambda_i, \qquad H = \sum_{i=1}^{p} \frac{\partial H}{\partial \dot{x}_i} \dot{x}_i,$$

and since

$$\delta\phi[\mathbf{x}(t)] = \sum_{i=1}^{p} \frac{\partial \phi}{\partial x_i} \delta x_i(t),$$

it is clear that (9.55) is satisfied if

$$(9.56) \qquad \left[\sum_{i=1}^{p} \left[\lambda_i(t) + \frac{\partial \phi}{\partial x_i} \right] \delta x_i \right]_0^T = 0.$$

With $x(0) = a$ and $t = 0$ given, (9.56) is satisfied for any independent variations $\delta x(T)$ whenever

$$(9.57) \qquad \lambda_i(T) = -\left.\frac{\partial \phi}{\partial x_i}\right|_{t=T}.$$

Now it is possible to satisfy all conditions for optimality except the vanishing of the elements $\psi_i(t)$ [see (9.54)] for any choice $y_0(t)$ even if $y_0(t)$ is not optimal. Let us therefore select any arbitrary policy $y_0(t)$ and by successive approximation seek the minimum value of ϕ.

For any arbitrary choice $y_0(t)$, subject of course to the restriction $c \leq y_0(t) \leq d$, the associated state is determined by

$$\frac{dx_0}{dt} = G(x_0, y_0), \qquad x_0(0) = a,$$

which satisfies the original constraint (9.51) and is the same as the first Hamiltonian equation in (9.53). Using $x_0(t)$ and $y_0(t)$, we solve the second Hamiltonian equation in (9.53) subject to the conditions (9.57), so that

$$(9.58) \qquad \dot{\lambda}_i = -\left(\frac{\partial H}{\partial x_i}\right)_0 = -\sum_{j=1}^{p}\left(\frac{\partial G_j}{\partial x_i}\right)_0 \lambda_j,$$

$$\lambda_i(T) = -\left(\frac{\partial \phi}{\partial x_i}\right)_0\bigg|_{t=T},$$

where the subscript 0 denotes that all quantities are evaluated at the arbitrary nonoptimal solution $[x_0(t), y_0(t)]$. We also have

$$(9.59) \qquad \psi_i(t) = \sum_{j=1}^{q} \lambda_j(t)\left(\frac{\partial G_j}{\partial y_i}\right)_0,$$

where the functions $\psi_i(t)$ are, of course, nonvanishing under nonoptimal conditions.

If we now consider variations in the choice vector, where

$$y_1(t) = y_0(t) + \delta y(t), \qquad x_1(t) = x_0(t) + \delta x(t),$$

then the perturbed linearized form of the constraint equations (9.51) becomes

$$(9.60) \qquad \frac{d(\delta x)}{dt} = A_0(t)\,\delta x + B_0(t)\,\delta y, \qquad \delta x(0) = \delta a,$$

where the elements of the ith row and jth column of matrices $A_0(t)$ and $B_0(t)$ are

$$(9.61) \qquad [A_0(t)]_{ij} = \left(\frac{\partial G_i}{\partial x_j}\right)_0, \qquad [B_0(t)]_{ij} = \left(\frac{\partial G_i}{\partial y_j}\right)_0.$$

With the vector $\boldsymbol{\lambda} = (\lambda_1, \lambda_2, \ldots, \lambda_p)$, it is clear from (9.61) that (9.58) can be written as

$$(9.62) \qquad \frac{d\boldsymbol{\lambda}}{dt} = -A_0'(t)\boldsymbol{\lambda},$$

so that (9.60) and (9.62) are similar to the linear problem we previously considered. Consequently, it follows that

$$(9.63) \qquad \boldsymbol{\lambda}' \, \delta\mathbf{x} \bigg|_0^T = \int_0^T \boldsymbol{\lambda}' B_0 \, \delta\mathbf{y} \, dt.$$

We note, however, that we have

$$(9.64) \quad \boldsymbol{\lambda}' \, \delta\mathbf{x}\big|_{t=T} = [\lambda_1 \, \delta x_1 + \lambda_2 \, \delta x_2 + \cdots + \lambda_p \, \delta x_p]_{t=T}$$

$$= -\left[\left(\frac{\partial \phi}{\partial x_1}\right)_0 \delta x_1 + \left(\frac{\partial \phi}{\partial x_2}\right)_0 \delta x_2 + \cdots + \left(\frac{\partial \phi}{\partial x_p}\right)_0 \delta x_p\right]_{t=T}$$

$$= -\delta\phi[\mathbf{x}_0(T)],$$

and from (9.60) we obtain

$$(9.65) \qquad \boldsymbol{\lambda}' \, \delta\mathbf{x}\big|_{t=0} = \boldsymbol{\lambda}'(0) \, \delta\mathbf{a}.$$

If we expand the integrand in (9.63) and substitute (9.59), there results

$$(9.66) \qquad \boldsymbol{\lambda}' B_0 \, \delta\mathbf{y} = \sum_{i=1}^{p} \lambda_i(t) \sum_{j=1}^{q} \left(\frac{\partial G_i}{\partial y_j}\right)_0 \delta y_j = \sum_{j=1}^{q} \psi_j(t) \, \delta y_j(t).$$

Using (9.66), (9.65), and (9.64) in (9.63) gives

$$(9.67) \qquad -\delta\phi[\mathbf{x}_0(T)] = \boldsymbol{\lambda}'(0) \, \delta\mathbf{a} + \int_0^T \sum_{j=1}^{q} \psi_j(t) \, \delta y_j(t)$$

for any variations in the initial conditions or the choice vector.

Now, we have

$$\phi[\mathbf{x}_1(T)] = \phi[\mathbf{x}_0(T) + \delta\mathbf{x}(T)] \cong \phi[\mathbf{x}_0(T)] + \delta\phi[\mathbf{x}_0(T)],$$

from which, for sufficiently small $\delta\mathbf{x}(T)$, we obtain

$$(9.68) \qquad \phi[\mathbf{x}_1(T)] \le \phi[\mathbf{x}_0(T)] \quad \text{if} \quad \delta\phi[\mathbf{x}_0(T)] \le 0.$$

If in the perturbed solution we take the same initial conditions $\mathbf{x}_1(0) = \mathbf{a}$, so that $\delta\mathbf{x}(0) = \delta\mathbf{a} = 0$, then we have

$$(9.69) \qquad -\delta\phi[\mathbf{x}_0(T)] = \sum_{j=1}^{q} \int_0^T \psi_j(t) \, \delta y_j(t) \, dt, \qquad \delta\mathbf{a} = 0.$$

In view of (9.69), the requirement in (9.68) can be met under a variety of circumstances. A sufficient condition to ensure that $\delta\phi[\mathbf{x}_0(T)] \le 0$ is

$$(9.70) \qquad \text{sign } [\delta y_j(t)] = \text{sign } [\psi_j(t)],$$

which is satisfied, for example, by

(9.71) $$\delta y_j(t) = \epsilon \psi_j(t),$$

where ϵ is any arbitrary positive number. For the choice in (9.71), we obtain

$$\delta \phi[\mathbf{x}_0(T)] = -\epsilon \sum_{j=1}^{q} \int_0^T \psi_j^2(t)\, dt \le 0, \qquad \delta \mathbf{a} = 0.$$

The choice in (9.71) is a logical one, too, since $\psi_j(t)$ is, according to (9.54), a measure of how much the jth element of $\mathbf{y}_0(t)$ differs from the optimal choice in that $\psi_j(t)$ measures the sensitivity of the Hamiltonian to the choice element $y_j(t)$. For a fixed ϵ, the change in $y_j(t)$ is proportionately large if $\psi_j(t)$ is large and proportionately small if $\psi_j(t)$ is small.

If we attempt to make $\delta \phi[\mathbf{x}_0(T)]$ negative and as large as possible in absolute value, then we can choose the sign of $\delta y_j(t)$ according to (9.70) and make $\delta y_j(t)$ as large as possible (the bang-bang problem). This, however, would probably violate the linearization. In addition, we may not be able to choose $\delta y_j(t)$ according to (9.71) everywhere, since we require

(9.72) $$\mathbf{y}_1(t) = \mathbf{y}_0(t) + \delta \mathbf{y}(t),$$

and yet we must not violate the restriction

$$\mathbf{c} \le \mathbf{y}_1(t) \le \mathbf{d},$$

or equivalently,

$$\mathbf{c} - \mathbf{y}_0(t) \le \delta \mathbf{y}(t) \le \mathbf{d} - \mathbf{y}_0(t).$$

A logical choice then is

(9.73)
$$\begin{aligned}
\delta \mathbf{y}(t) &= \epsilon \boldsymbol{\psi}(t), & \mathbf{c} - \mathbf{y}_0(t) &\le \epsilon \boldsymbol{\psi}(t) \le \mathbf{d} - \mathbf{y}_0(t), \\
\delta \mathbf{y}(t) &= \mathbf{c} - \mathbf{y}_0(t), & \mathbf{c} - \mathbf{y}_0(t) &> \epsilon \boldsymbol{\psi}(t), \\
\delta \mathbf{y}(t) &= \mathbf{d} - \mathbf{y}_0(t), & \mathbf{d} - \mathbf{y}_0(t) &< \epsilon \boldsymbol{\psi}(t),
\end{aligned}$$

for which the right-hand member of (9.69) is nonnegative, ensuring that $\delta \phi(\mathbf{x}_0(T)) \le 0$.

Equations (9.72) and (9.73) yield a new choice vector, which when substituted into the original nonlinear state equation gives us a new state equation,

$$\frac{d\mathbf{x}_1}{dt} = \mathbf{G}(\mathbf{x}_1, \mathbf{y}_1), \qquad \mathbf{x}_1(0) = \mathbf{a}.$$

We must then evaluate $\phi[\mathbf{x}_1(T)]$ to see that the condition

(9.74) $$\phi[\mathbf{x}_1(T)] \le \phi[\mathbf{x}_0(T)]$$

is satisfied. Now (9.74) will generally be true in view of (9.68) and the choice (9.73) unless there has been a gross violation of linearity. If this happens, it simply means that $\delta \mathbf{y}(t)$ is too large and we must take a smaller value of ϵ. Assuming that (9.74) is satisfied, or that we choose a smaller value of ϵ so that it is, we repeat the process using linearization about the set $[\mathbf{x}_1(t), \mathbf{y}_1(t)]$. The procedure continues until the minimum of ϕ is reached, a condition in which the elements $\psi_i(t) \rightarrow 0$ on the nth approximation, so that, with $\delta \mathbf{a} = 0$,

$$\delta \phi[\mathbf{x}_n(T)] \rightarrow 0 \qquad \text{for any } \delta \mathbf{y}(t),$$
and

(9.75) $$\min_{\mathbf{y}} \{\phi[\mathbf{x}(T)]\} = \phi[\mathbf{x}_n(T)] = f(\mathbf{a}; T).$$

If we desire an estimate of the minimum return in some neighborhood of the initial conditions $\mathbf{a}$, then, according to (9.67) and (9.75), we have

$$f(\mathbf{a} + \delta \mathbf{a}; T) \cong \phi[\mathbf{x}_n(T)] - \boldsymbol{\lambda}'(0)\delta \mathbf{a},$$

where $\boldsymbol{\lambda}(t)$ is the vector determined on the nth approximation.

The solution for $\mathbf{y}(t)$ obtained on the nth approximation with $\mathbf{x}(0) = \mathbf{a}$ is the optimal choice of $\mathbf{y}(t)$ only for the initial conditions $\mathbf{x}(0) = \mathbf{a}$. It is a good zero-order choice, however, for the problem in which $\mathbf{x}(0) = \mathbf{a} + \delta \mathbf{a}$. In order to find the optimal choice for the initial condition $\mathbf{x}(0) = \mathbf{a} + \delta \mathbf{a}$, we would therefore take $\mathbf{y}_0(t)$ to be the optimal $\mathbf{y}(t)$ for the problem with initial conditions $\mathbf{x}(0) = \mathbf{a}$ and compute the corresponding zero-order state from

$$\frac{d\mathbf{x}_0}{dt} = \mathbf{G}(\mathbf{x}_0, \mathbf{y}_0), \qquad \mathbf{x}_0(0) = \mathbf{a} + \delta \mathbf{a}.$$

We would then compute the optimal choice with initial state $\mathbf{x}(0) = \mathbf{a} + \delta \mathbf{a}$ by linearization about $[\mathbf{x}_0(t), \mathbf{y}_0(t)]$ and successive approximation, just as before.

The approach outlined here is similar to that used by Bryson [14] based on the method of steepest descent. Kelley [32] has also done work along the same lines as Bryson. In contrast to Bryson, however, Kelley breaks the choice variable into two components, one of which is sacrificed to meet the boundary conditions, whereas the other is varied to give optimality.

The method we suggest here for the successive approximation nevertheless differs in certain essential respects from those that have been suggested by Bryson. In his work, Bryson deals only with unrestricted variations in the choice vector $\delta y(t)$, and he computes the changes $\delta y(t)$ to be made for a specified change in the criterion $\delta \phi$. This amounts to fixing the value of ϵ in the problem discussed here by computing

$$\epsilon = - \frac{\delta \phi}{\displaystyle\int_0^T \sum_{j=1}^q \psi_j^2(t) \, dt},$$

assuming unrestricted choices, so that $\delta y_j(t) = \epsilon \psi_j(t)$. As is clear from the earlier discussion, however, restrictions on the choice vector $y(t)$ render the computation of $\delta y(t)$ necessary to produce a given $\delta \phi$ much more complicated. In any event, the actual change $\delta \phi$ will not be the same as that computed from perturbation theory, and it is always necessary to check that we have

$$\phi[x_1(T)] \le \phi[x_0(T)].$$

Regardless of how the steps may be taken, the method of linearization and successive approximation appears to be a powerful way of overcoming the inherent obstacles presented by multidimensionality and boundary-valuedness. In view of the present state of the art, it is one of the most promising methods of obtaining computationally feasible and practical solutions to the problem of optimization of multidimensional nonlinear processes.

9.5 PHYSICAL IMPLEMENTATION

Regardless of the means by which numerical solutions of the multi-variable control problem are determined, the value of the solutions will necessarily be measured in terms of reduction to practice. It is in this area that we shall find the ultimate benefit to mankind.

In the general nonlinear problem, optimal choices are functions of the state existing at the time. In principle, the optimal choice is therefore a nonlinear function state. Before man can instrument truly optimal non-linear control, it will be necessary to resolve the problem of how the multi-dimensional nonlinear feedback function is to be represented. Once the numerical solution is known, however, much can be done in the way of approximating the solution physically.

Merriam's method of parametric expansion [41] can provide linearized least-squared error control about the particular nonlinear solution. Such an approach seeks a control that holds the system to a region in close proximity to the optimum nonlinear solution. Interest in a linearized version of control stems, of course, from the ease with which such feedback functions can be achieved. They are obtained simply from a matrix operation on the state vector. There are, however, delicate matters of controllability that cannot be lightly dismissed. Linear control in the sense just described may not always be possible unless the control matrix is well behaved.

There is need for imagination and creative ingenuity in the matter of reduction to practice. Progress in implementation is likely to come slowly, little by little, rather than in large abrupt strides.

MULTIPLE-CHOICE REVIEW PROBLEMS

1. The foremost problem in the dynamic-programming solution of multivariable control problems in optimization results from

    (a) restrictions on the allowable choice or control variables,
    (b) boundary-valuedness in the solutions,
    (c) dimensionality in the functional equations,
    (d) nonlinearity in the state equations.

2. The foremost problem in the classical solution of multivariable control problems in optimization results from

    (a) restrictions on the allowable choice or control variables,
    (b) boundary-valuedness in the solutions,
    (c) dimensionality in the functional equations,
    (d) nonlinearity in the state equations.

3. Aoki's method of polynomial approximation is one of the most useful methods of overcoming

    (a) restrictions on the allowable choice or control variables,
    (b) boundary-valuedness in the solutions,
    (c) dimensionality in the functional equations,
    (d) nonlinearity in the state equations.

4. Bryson's method of linearization and successive approximation is one of the most useful methods of overcoming

    (a) restrictions on the allowable choice or control variables,
    (b) boundary-valuedness in the solutions,
    (c) dimensionality in the functional equations,
    (d) nonlinearity in the state equations.

5. The maximum principle of the Soviet mathematician, Pontriagin, is

    (a) a computational algorithm for carrying out the formalism of dynamic programming,
    (b) a compact formulation in classical variational theory that states necessary conditions for optimality in a wide variety of problems including those with restricted variations,
    (c) a method of eliminating the traditional two-point boundary problem,
    (d) none of the above.

REFERENCES

1. Andreev, N. I., "A Method of Determining the Optimum Dynamic System from the Criterion of the Extreme of a Functional Which Is a Given Function of Several Other Functions," *Proceedings of The First International Congress on Automatic Control*, Moscow, 1960.

2. Aoki, M., "Dynamic Programming and Numerical Experimentation as Applied to Adaptive Control," Doctoral Dissertation, University of California, Los Angeles, 1959.

3. Aseltine, J. A., A. R. Mancini, and C. W. Sarture, "A Survey of Adaptive Control Systems," PGAC6, *IRE Trans., Automatic Control* (1958), 102–108.

4. Beckwith, R. E., *Analytic and Computational Aspects of Dynamic Programming Processes of High Dimension*, Purdue University, Lafayette, Indiana, 1959.

5. Bellman, R., *Dynamic Programming*, Princeton University Press, Princeton, New Jersey, 1957.

6. Bellman, R., "Some New Techniques in the Dynamic Programming Solution of Variational Problems," *Quart. Appl. Math.* **16** (1958), 295–305.

7. Bellman, R., *Adaptive Control Processes: A Guided Tour*, Princeton University Press, Princeton, New Jersey, 1961.

8. Bellman, R., and R. Kalaba, *Dynamic Programming and Adaptive Processes, I*, P-1416, The RAND Corporation, Santa Monica, Calif., 1959.

9. Bellman, R., and R. Kalaba, *On Adaptive Control Processes*, P-1610, The RAND Corporation, Santa Monica, Calif., 1959.

10. Bellman, R., and R. Kalaba, *Reduction of Dimensionality, Dynamic Programming, and Control Processes*, P-1964, The RAND Corporation, Santa Monica, Calif., 1960.

11. Bliss, G. A., *Calculus of Variations*, The Open Court Publishing Company, Chicago, 1925.

12. Bliss, G. A., *Lectures on the Calculus of Variations*, University of Chicago Press, Chicago, 1946.

13. Breakwell, J. V., *The Optimization of Trajectories*, AL-2706, North American Aviation, Los Angeles, 1957.

14. Bryson, A. E., W. F. Denham, F. J. Carroll, and K. Mikami, "Determination of the Lift or Drag Program That Minimizes Re-entry Heating with Acceleration or Range Constraints Using a Steepest Descent Computation Procedure," IAS 29th Annual Meeting, New York, 1961.

15. Buscher, R. G., and J. Hamill, *Possible Mechanizations for Memory-Multiplier Functions in Self-adaptive Control*, LMED R59APS100, General Electric Company, Schenectady, New York, 1959.

16. Cavotti, C., and A. Miele, *Optimum Thrust Programming Along Arbitrarily Inclined Rectilinear Paths*, ASTIA Document AD 148088, Armed Services Technical Information Agency, Washington, D.C., 1957.

17. Chestnut, H., *Adaptive Control—An Appraisal*, General Engineering Laboratory Report 59GL133, General Electric Company, Schenectady, New York, 1959.

18. Chestnut, H., R. R. Duersch, and R. Rustay, *Automatic Optimizing of a Poorly Defined Process, Part I*, General Engineering Laboratory Report 61GL167, General Electric Company, Schenectady, New York, 1961.

19. Dreyfus, S., *Dynamic Programming and the Calculus of Variations*, P-1464, The RAND Corporation, Santa Monica, Calif., 1960.

20. Duersch, R. R., *Gradient Methods for Optimization*, General Engineering Laboratory Report 60GL211, General Electric Company, Schenectady, New York, 1960.

21. Ellert, F. J., and C. W. Merriam, III., *Design of a Linear Time-Varying Feedback Control System Using Optimization Theory*, General Engineering Laboratory Report 61GL67, General Electric Company, Schenectady, New York, 1961.

22. Fend, F. A., and C. B. Chandler, *Numerical Optimization For Multidimensional Problems*, General Engineering Laboratory Report 61GL78, General Electric Company, Schenectady, New York, 1961.

23. Freimer, M., *Topics in Dynamic Programming*, Lincoln Laboratory Reports 54–16 and 54 G-0020, Massachusetts Institute of Technology, Cambridge, Mass., 1960.

24. Fried, B. D., "General Formulation of Powered Flight Trajectory Optimization Problems," *J. Appl. Phys.* **29** (1958), 1203–1209.

25. Gamkrelidze, R. V., "On the General Theory of the Optimal Process," *Dokl. Akad. Nauk* (2) **123** (1958), 223–226. *Automation Express* **1** (1959), 37–39.

26. Greenstadt, J., "On the Reduction of Continuous Problems to Discrete Form," *IBM J. Research and Development* **3** (1959), 355–363.

27. Ho, Yu-Chi, "A Successive Approximation Technique for Optimal Control Systems Subject to Input Saturation," ASME 61-JAC-10, Joint Automatic Control Conference, Boulder, Colorado, 1961. *ASME Trans.* (D) **84** (1962), 33–40.

28. Ivakhnenko, A. G., *Engineering Cybernetics*, JPRS: 6650, Washington, D.C., Translation of *Teknicheskaya Kibernetika. Sistemy Avtomaticheskogo Upravleniya s Presposobleniyem Khzrakteristik*, State Publishing House of Technical Literature, Kiev, U.S.S.R., 1959.

29. Jakowatz, C. V., R. L. Shuey, and G. M. White, *Adaptive Waveform Recognition*, General Electric Research Laboratory Report 60-RL-2353E, General Electric Company, Schenectady, New York, 1960.

30. Kalaba, R., "On Nonlinear Differential Equations, the Maximum Operation and Monitone Convergence," *J. Math. Mech.* **8** (1959), 519–574.

31. Kashmar, C. M., E. L. Peterson, and F. X. Remond, *A General Approach to the Numerical Solution of Multidimensional, Nonlinear Boundary-Valued Variational Problems*, GE-TEMPO Report R 60TMP-27, General Electric Company, Santa Barbara, Calif., 1960.

32. Kelley, H. J., "Gradient Theory of Optimal Flight Paths," *ARS J.* **30** (1960), 947–954.

33. Landyshev, A. N., and E. L. Peterson, *Recent Soviet Progress in Adaptive and Optimal Control*, GE-TEMPO Report SP-121, General Electric Company, Santa Barbara, Calif., 1961.

34. Lawden, D., "Inter-Orbital Transfer of a Rocket," *J. British Interplanetary Soc.* **11** (1952), 321–333.

35. Lawden, D., "Optimal Transfer between Circular Orbits about Two Planets," *Astronautica Acta* **1** (1955), 89–99.

36. Leitmann, G., *On a Class of Variational Problems in Rocket Flight*, Report 5067, Lockheed Missile and Space Systems Division, Lockheed Aircraft Corporation, Los Angeles, Calif., 1958.

37. Marx, H. F., *General Electric Self-adaptive Control System Description*, LMED, General Electric Company, Schenectady, New York, 1960.

38. Marx, H. F., *Recent Adaptive Control Work at the General Electric Company*, LMED, General Electric Company, Schenectady, New York, undated.

39. Marx, H. F., *Navy Self-adaptive Control Flight Test Evaluation*, Final Report BuWeaps, as 59-6078-C, LMED, General Electric Company, Schenectady, New York, 1960.

40. McCausland, I., "Adaptation in Feedback Control Systems," *J. Franklin Inst.* **268** (1959), 143–147.
41. Merriam, C. W., "A Class of Optimum Control Systems," *J. Franklin Inst.* **267** (1959), 267–281.
42. Merriam, C. W., "Use of a Mathematical Error Criterion in the Design of Adaptive Control Systems," *Trans. AIEE, Applications and Industry,* **46** (1960), 506–512.
43. Miele, A., *General Variational Theory of the Flight Paths of Rocket-Powered Aircraft, Missiles, and Satellite Carriers,* A-58-2, Purdue Research Foundation, Lafayette, Indiana, 1958.
44. Mishkin, E., and L. Braun, *Adaptive Control Systems,* McGraw-Hill Book Company, New York, 1961.
45. Perry, H. R., *A Self-adaptive Missile Guidance System for Statistical Inputs,* NASA TN D-343, Ames Research Center, Ames, Iowa, 1960.
46. Peterson, E. L., *Optimization Theory—Methods, Implications, and Difficulties,* GE-TEMPO Report SP-137, General Electric Company, Santa Barbara, Calif., 1961.
47. Peterson, E. L., *Statistical Analysis and Optimization of Systems,* John Wiley and Sons, New York, 1961.
48. Pontryagin, L. S., "Optimal Control Processes," *Uspehi Mat. Nauk* (1) **14** (85) (1959), 3-20. *Automation Express,* Part I, **1** (1959), 15–18; Part II, **2** (1959), 26–30.
49. Rozonoer L. I., "The Maximum Principle of L. S. Pontryagin in Optimal-System Theory, III," *Avtomatika Telemehanika* **20** (1959), 1561–1578. *Automation and Remote Control* **20** (1960), 1517–1532.
50. Smith, K. C., "Adaptive Control Through Sinusoidal Response," 60AC-14, IRE Conf. Automatic Control, Massachusetts Institute of Technology, Cambridge, Mass., 1960.
51. Valentine, F. A., "The Problem of Lagrange with Differential Inequalities as Added Side Conditions," *Contributions to the Calculus of Variations,* 1933–1937, University of Chicago Press, Chicago, 1937.

CHAPTER *10*

Stopping-Rule Problems

LEO BREIMAN

10.1 HOW TO RECOGNIZE A STOPPING-RULE PROBLEM

Why should we study stopping-rule problems? One reason is simply that, aside from their own intrinsic interest, they form the simplest subclass of problems in the rapidly growing field of multiple-decision sequential problems, which include such interesting aspects as dynamic programming [1] and sequential tests of hypotheses and estimation.

As a very serious and warlike example, consider a radar antenna periodically sweeping the horizon. The operator sees a bleep reoccurring and after every sweep has a choice of two decisions, either to report an object sighted or to continue observing and gather more information.

For a more frivolous and amusing example, consider the device shown in Fig. 10.1. Simple as it appears, it does illustrate all the essential features of a stopping-rule problem. To summarize its well-known operation, we observe the present state of the device, that is, the symbols appearing in the

Fig. 10.1 The one-armed bandit!

window, and we have our choice of at most two decisions: to take our payoff up to the present time and quit or to pay the entrance fee—nickle, dime, quarter, or more—and be allowed a transition to the next state. This example is frivolous only in that no one appears to have determined the transition probabilities from state to state.

284

The essential features of a stopping-rule problem, then, can roughly be split into two parts:

1. A probabilistic mechanism, that is, a random device, that moves from state to state under a known, partially known, or unknown probability law.

2. A payoff and decision structure such that, after observing the current state, we have our choice of *at most two* decisions:

(*a*) Take our accumulated payoff to date and quit.

(*b*) Pay an entrance fee for the privilege of watching one more transition.

10.2 EXAMPLES

Now let us consider some examples that are more serious than one-armed bandits—more serious in the sense that they can be formulated mathematically. Many such examples have been investigated and solved in the literature; a very incomplete list of references is given at the end of the chapter.

Example 10.1   The Unrestricted Coin-Tossing Problem [7]. A fair coin is tossed repeatedly; heads we win one chip, tails we lose one. That is, if X_k is defined by $X_k = +1$ provided the kth toss results in heads and -1 provided it results in tails, then

$$(10.1) \qquad S_n = \sum_{k=1}^{n} X_k$$

is our accumulated winnings after n tosses. This describes the probabilistic structure of the problem. To specify the decision structure, we assume that our opponent is infinitely obliging. He is willing to continue or stop as we wish, and in addition he is also infinitely wealthy. To dwell further in fantasy, let us assume that our initial fortune is infinite. The implications of these assumptions are that after each toss we have available *exactly* two decisions: stop and collect our winnings to date or continue for at least one more toss. (Note that there is no entrance fee here; we do not have to ante up to play another game.) A famous strategy for playing this game is to stop when you are ahead. Although to sophisticated mathematicians this is a discredited axiom, it will later lead to an interesting remark, and we file this strategy for future reference.

In the foregoing description of a stopping-rule problem, notice that it was said that *at most two* decisions were available at each point. In Example 10.1, because of the vast amounts of assumed wealth and time, *exactly two* decisions were always available.

Example 10.2   The Restricted Coin-Tossing Problem. This is the same problem as the previous one except that we assume a finite initial fortune S. If the fortunes of gambling are such that at some stage we have $S_n = -S$, then we no longer have two decisions available. We have lost all our money and we must quit.

The fascinating and informative book [7] contains a great deal of material on the coin-tossing game. An advanced treatment of theoretical aspects of stopping-rule problems, under the heading of *Martingales*, will be found in [6].

Example 10.3   The House-Hunting Problem [2, 13]. Here we sample independently from the same population with known distribution.

More specifically, the kth draw results in the integer i with known probability p_i, so that if X_k is the result of the kth draw, then

$$P(X_k = i) = p_i.$$

The payoff structure is specified as follows: If we stop after n samples, we receive the quantity max $(X_1, \ldots, X_n)$, the best number we have seen to date. If, however, we wish to continue, an amount C must be paid for one sample. The labeling of this problem as the *house-hunting problem* is amusing but tenuous. It has given rise to the interesting comment that only a resident of Los Angeles could consider sampling from an infinite population (since no one draw affects the distribution of the remainder) as a realistic model of house hunting.

Example 10.4   The E.S.P. Problem. In this problem, it is desired to show that a fair coin is actually biased toward heads by stopping at an appropriate point. Thus, if $X_k = +1$ provided the kth toss results in heads, and $X_k = 0$ otherwise, then the X_k are independent,

$$P(X_k = 1) = \tfrac{1}{2}, \ P(X_k = 0) = \tfrac{1}{2}.$$

If we let

$$S_n = \sum_{k=1}^{n} X_k,$$

our payoff, if we quit after the nth toss, is S_n/n; there is no entrance fee. Is it really possible to induce a bias by selective stopping? Consider the following rule: Stop if $X_1 = 1$; otherwise, stop after exactly two tosses. Then

$$(10.2) \qquad \frac{S_n}{n} = 1 \qquad \text{for } X_1 = 1,$$

$$= 0 \qquad \text{for } X_1 = 0, \ X_2 = 0,$$

$$= \tfrac{1}{2} \qquad \text{for } X_1 = 0, \ X_2 = 1.$$

Hence the expected payoff is $1 \cdot \frac{1}{2} + 0 \cdot \frac{1}{4} + \frac{1}{2} \cdot \frac{1}{4} = \frac{5}{8}$. The solution of this problem, that is, the best stopping rule and the value of the maximum expected payoff, appears to be unknown.

Example 10.5   The Parking-Place Problem. Each of a sequence of independent random variables, $X_{-N}, X_{-N+1}, \ldots, X_0, X_1, \ldots$, takes on the values 0 or 1 with the same probability, that is,

$$P(X_k = 0) = p, \qquad P(X_k = 1) = 1 - p.$$

We can stop at X_k only if $X_k = 0$, and if $X_k = 0$ and we do stop, we pay the penalty $|k|$. More picturesquely, we start at a distance N parking places from our destination. As we cruise along, we can see only one parking place at a time. Naturally, if a place is occupied, we cannot stop there, and if a place is empty and we do park there, then our penalty or loss is the distance we walk to our destination. If we add that empty places occur independently with probability p, then we have the parking-place problem. This example was communicated to the author by Dr. Stanley Frankel.

10.3 · FORMULATION

Now that we have in hand an adequate supply of examples, the next project is a precise but general formulation of what is meant by a stopping-rule problem. We have pointed out that two distinct structures must be specified, a probabilistic device that moves from state to state and the payoff and decision structure.

The probabilistic device

We shall assume that the device is a Markov chain with a countable number of states and specified stationary transition probabilities. The theory of such devices has been extensively studied [7]. Here is a brief recapitulation of the pertinent definitions. The device has states that can be labeled by a subset of the integers (positive, zero, and negative). The *Markov property* [4, 6, 7, 10] is defined by the assumption that if S_n is the state of the device at time n, then

$$P(S_{n+1} = j, \text{ given } S_n = i \text{ and all past history up to time } n) = P(j \mid i),$$

where the $P(j \mid i)$, the specified transition probabilities of moving to state j from state i, satisfy

$$P(j \mid i) \geq 0, \sum_j P(j \mid i) = 1,$$

the latter since the sum of the probabilities of moving to all possible states must be unity. This definition will be a cornerstone of much that follows. The implication is that if at the present time we are in state i, then no matter how many transitions we have made to get to this state, no matter what our exact route was, whether circuitous or direct, the probability of going to state j on the next transition is given by the number $P(j \mid i)$. Heuristically, the present state contains *all* the pertinent information regarding the past. It may seem that the Markov property is very restrictive and that the random device occurring in a given problem does not satisfy the given definition. We assert, however, and shall illustrate in examples to follow, that by appropriate choice of the set of states, every device involving sampling from a countable population can be embedded in a Markov device of the type defined. Finally, by specifying the starting state i_0 of the device, we complete the probabalistic description.

The payoff and decision structure

This is specified by two functions defined on the states of the system, the terminal payoff function $F(i)$ and the entrance fee $f(i)$. If at time n we are in state i, then we have available at most two decisions:

1. Stop and receive the amount $F(i)$.
2. Pay $f(i)$ and continue for at least one more step.

There may, however, be states at which at most one decision is available. We formulate this by saying that there may be specified a set T_s of *forced stopping states* such that if $i \in T_s$, then we must stop and collect $F(i)$, and a disjoint set T_c of *forced continuation states* such that if $i \in T_c$, then we must pay $f(i)$ and continue. Notice here that $f(i)$ need not be defined on T_s nor $F(i)$ on T_c.

Let us return now to the examples and place them into the present framework.

For Example 10.1, let the state of our system be the total accumulated winning to date. If we are in state i, that is, if our winnings to date are i, then, regardless of how many plays it has taken us to accumulate the amount i or the previous visissitudes of our fortune, the probability of a transition to $i + 1$ is $1/2$, and the transition to $i - 1$ is similarly $1/2$. Thus this device has the Markov property, with

$$
\begin{aligned}
P(j \mid i) &= \tfrac{1}{2} \quad \text{for } j = i + 1, \\
&= \tfrac{1}{2} \quad \text{for } j = i - 1, \\
&= 0 \quad \text{for other values of } j.
\end{aligned}
$$

(10.3)

The starting state is $i_0 = 0$. The payoff if we stop in state i is simply our winnings to date; that is, we have $F(i) = i$. There is no entrance fee, so $f(i) \equiv 0$, and there are no forced stopping or continuation states.

In Example 10.2, we have the same situation as in Example 10.1, except that forced stopping states are present, and T_s consists of all states $i \leq -S$.

Again, in Example 10.3 we must carefully define the state of the system in order to get the Markov property. The right choice is to take the state S_n as

$$S_n = \max (X_1, \ldots, X_n).$$

By direct computation, we then have

$$P(S_{n+1} = j, \text{ given } S_n = i \text{ and past history}) = 0 \qquad \text{for } j < i,$$
$$= P(X_{n+1} \leq i) \quad \text{for } j = i,$$
$$= P(X_{n+1} = j) \quad \text{for } j > i,$$

so that we have the Markov property with $P(j \mid i)$ defined in terms of p_i by

$$\begin{aligned} P(j \mid i) &= 0 \qquad \text{for } j < i, \\ &= \sum_{k \leq i} p_k \text{ for } j = i, \\ &= p_j \qquad \text{for } j > i. \end{aligned}$$

(10.4)

It should be clear that $F(i) = i$, that there is a constant entrance fee $f(i) = C$, and that the sets T_s and T_c are empty.

For Example 10.4, encouraged by Example 10.1, we could try to define the state S_n by

$$S_n = \sum_{k=1}^{n} X_k.$$

The device so defined certainly has the Markov property, with

$$P(S_{n+1} = j, \text{ given } S_n = i) = \tfrac{1}{2} \qquad \text{for } j = i + 1,$$
$$= \tfrac{1}{2} \qquad \text{for } j = i,$$

but the payoff if we stop at state i after n steps is i/n. This is not of the desired form, because we wish the payoff to depend only on the state in which we stop, not on how many steps we have previously taken. To avoid this difficulty, we use the device of putting a little clock on each state, so enlarging our set of states that, by looking at a state, we know not only the magnitude of S_n but also the number of transitions. Briefly, we take as states all ordered pairs of integers (i, n), $i = 0, 1, \ldots$; $n = 0, 1, \ldots$, with starting state $(0, 0)$. The first integer i keeps track of S_n, and n keeps account of time. The transition rule is that (i, n) goes to

$(i + 1, n + 1)$ with probability $1/2$, and goes to $(i, n + 1)$ with probability $1/2$. The payoff $F(i, n)$ is given by i/n, the entrance fee is given by $f(i, n) = 0$, and T_s and T_c are empty.

Again, Example 10.5 is an exercise in choosing states correctly. We take as states the pairs of integers (i, k), where i ranges forward from $-N$, and k takes only the values 0 and 1. Intuitively, i measures the distance of the space from the destination, and k registers whether the space is vacant or occupied. The transition law is that (i, k) goes into $(i + 1, 0)$ with probability p, and into $(i + 1, 1)$ with probability $1 - p$. This example differs from the previous ones in that there is a set of forced continuation states. That is, T_c consists of all states $(i, 1)$, $i = -N$, $-N + 1, \ldots$. The payoff is given by

$$F(i, 0) = -|i|,$$

where the minus sign is used to indicate that payoff is the negative of the penalty. Once more, the entrance fee is 0 and we take the starting state as $(-N - 1, 0)$.

10.4 WHAT IS A STOPPING RULE?

For any stopping-rule problem, we ask: What does it mean to know when to stop? What is the mathematical formulation of a stopping rule? First, let us observe that we do not consider prophesy; the decision of whether to stop or continue at the nth step must be based only on knowledge of history up to the nth step and on the specified distributions and payoffs.

Since we are dealing with a probabilistic device rather than a deterministic one, there are a great many different possible sequences of states following i_0. A well-defined stopping rule must tell us when to stop along each possible sequence of states, since otherwise it would be possible for the device to produce a sequence of states along which our rule would not hold. Now, supposing we do have a rule that covers every possible sequence, we shall call a sequence of states $i_0, i_1, \ldots, i_n$ a *stopping sequence* if along this sequence we decided to stop at i_n, but not before. Then *we can fully describe our rule by producing a list of all stopping sequences* since, if by any stage the device has produced a certain sequence of states, to decide whether or not to stop we simply check to see whether our sequence is on the list.

There are a few requirements with which any such list must be compatible. First, if $i_0, i_1, \ldots, i_n$ is a stopping sequence, then no continuation of $i_0, i_1, \ldots, i_n$ can appear on the list. Also, no state in the sequence except possibly i_n, can be in T_s, and i_n cannot be in T_c. Finally, there is the condition that a reasonable stopping rule should eventually stop us. One way to give this meaning is to say that the probability of stopping is 1. More precisely, the probability $P(i_0, i_1, \ldots, i_n)$ of a sequence of states $i_0, i_1, \ldots, i_n$ can be computed from the transition probabilities by the formula

$$(10.5) \qquad P(i_0, i_1, \ldots, i_n) = P(i_1 \mid i_0)P(i_2 \mid i_1) \cdots P(i_n \mid i_{n-1});$$

that is, the probability of the transitions $i_0 \to i_1 \to i_2 \to \cdots \to i_n$ is the probability of the transition $i_0 \to i_1$ times the probability of the transition $i_1 \to i_2$, etc. We insist now that our list, to be reasonable, have the property that

$$(10.6) \qquad \sum P(i_0, i_1, \ldots, i_n) = 1,$$

where the summation extends over all stopping sequences.

To summarize, any list of stopping sequences satisfying the foregoing conditions will be said to define a *stopping rule*.

10.5 WHAT IS A SOLUTION?

With all the foregoing explanations and definitions—knowing what the problem is and what the rules are that may be applied to it—we finally have arrived at the point at which we can begin to discuss what it is we wish to do, that is, what it is we wish to solve and what a solution is.

Along any stopping sequence $i_0, i_1, \ldots, i_n$, the *total payoff*, denoted by

$$Z(i_0, i_1, \ldots, i_n),$$

is equal to $F(i_n) - f(i_0) - \cdots - f(i_{n-1})$. In other words, the profit realized is the terminal payoff $F(i_n)$ decreased by the entrance fees we paid at states $i_0, i_1, \ldots, i_{n-1}$. Roughly, we want a rule that makes the total profit as large as possible. A rule, however, must be judged by its over-all performance, so that if it does very well along some sequences having small probability and poorly along the more probable sequences, it is not a very good rule. That is, a rule must be judged by the *expected total payoff EZ* it produces; under a given rule, this is defined as

$$EZ = \sum Z(i_0, i_1, \ldots, i_n)P(i_0, i_1, \ldots, i_n),$$

where the sum is taken over all stopping sequences on the list, and E is the classical symbol for expectation. Thus EZ is obtained by averaging the total payoff along a stopping sequence as weighted by the probability of the sequence.

The *best stopping rule*, the solution we are seeking, makes EZ as large as possible. Now the problem is finally completely formulated, and there remains only the task of finding solutions. A stopping rule, however, as we have defined it, is a bulky and often unmanageable object. It would be a huge chore to give a complete list of all stopping sequences and furthermore to choose the best among such lists. It is at this point that the Markov property that we so painstakingly introduced enters to accomplish an enormous reduction of the work.

We assert that a best stopping rule will be found among those rules in which we make our decision only on the basis of the state in which we currently find ourselves. Now this can be proved rigorously, if a best stopping rule exists, but the proof is difficult and technical. Yet the essential idea is clear: If, after any number of transitions, we are in state i, then we are affected by past history only to the extent of the amount of entrance fees we have paid to date; but these fees are water under the bridge, for we have paid them and they will never be returned to us. Whether or not we should continue from i depends on a comparison of what we would get if we quit, namely $F(i)$, with our prospects if we continue. Because of the Markov property, however, the entire future probabilistic development depends only on the present state i, not on past history. Hence our decision should rest only on our prospects in starting out, as it were, afresh from state i.

What does it mean to have a stopping rule in which our decision is always based only on the current state? Each state can be put in just one of two categories: states at which we stop and states from which we continue. Mathematically, therefore, such a rule is nothing more than a partition of all possible states into two sets, the *stopping-state set T* and the *continuation-state set $\bar{T}$*, which is the complement of T. The rule is then given as follows: Stop the first time that T is entered.

We recapitulate with the following definitions.

A *stopping rule* is any set T of states satisfying

1. $T_s \subset T$.
2. $T_c \subset \bar{T}$.
3. $P(\text{entering } T \text{ from } i_0) = 1$.

A *solution* to a stopping-rule problem is a stopping rule T^*, such that, if $E_T Z$ denotes the expected total payoff using the stopping rule T, then for each T we have

(10.7)
$$E_{T^*} Z \geq E_T Z.$$

10.6 STOP WHEN YOU ARE AHEAD; THE STABILITY PROBLEM

As pointed out in various places in this chapter, there are some serious and deep problems concerning the nature and existence of solutions and the efficacy of methods of solution. For the most part, we shall here ignore these, referring the more mathematically concerned reader to [14], and relying instead on that famous observation, well known to engineers and physicists, that every problem drawn from the real world, *when properly formulated*, has a unique and reasonable solution.

There is one aspect of the existence and nature of solutions that, because of its interest, and even more because it is illustrative of the difficulties that beset a thorough mathematical treatment, we shall briefly discuss. This is the so-called question of stability. For any stopping rule T, $T^{(N)}$ will denote the truncation of this rule at step N. That is, we use rule T for $N - 1$ steps, and if it has not yet stopped us, we stop automatically after the Nth step. Another way of saying this is to state that the list of stopping sequences for $T^{(N)}$ includes all stopping sequences on the T list that are less than N in length plus all other sequences terminated at length N. We shall call the rule T *stable* if

$$(10.8) \qquad \lim_{N \to \infty} E_{T^{(N)}} Z = E_T Z.$$

Stability of a stopping rule means, therefore, that it can be approximated, in terms of payoff, by rules in which we decide to quit, if we have not already done so, after a large but fixed number of plays. For example, in the house-hunting problem, a truncation at $N = 1{,}000{,}000$ would mean that if we have not settled on a house after sampling 1,000,000 times, then we automatically quit and take the best we have seen to date.

In addition, we now define a stopping-rule problem to be *stable* if it has a stable solution. For a stable solution, we can do almost as well as possible in the given problem, even if we decide beforehand that we can play at most N steps.

Of the various examples in Sec. 10.2, one is not stable, namely Example 10.1, the unrestricted coin-tossing problem. Now we refer back to our remark concerning a proposed rule for this game, to stop when ahead. This is a good strategy! To formulate it, we set a quitting level $\alpha > 0$ and decide to stop the first time our fortune exceeds or equals the value α. Thus our stopping set T consists of all states i such that $i \geq \alpha$. There is one question as to whether T is an admissible stopping set. Does it eventually stop us with probability 1? Are we certain that eventually there will come a time at which $S_n \geq \alpha$? The answer is an affirmative one (see [7] for a proof). Heuristically, we can argue that if we plot our fortune as a function of the number of plays, as indicated in Fig. 10.2, then the graph oscillates up and down, but that the oscillations are centered about 0, and eventually an oscillation will occur that carries above α. Thus T is a valid stopping rule. Clearly, we have

$$(10.9) \qquad\qquad E_T Z \geq \alpha,$$

since with probability 1 we always win at least the amount α with this rule. It can nevertheless be proved, for this example, that with any rule T we have

$$(10.10) \qquad E_{T(N)}Z = 0.$$

In other words, for any way of playing this game such that we decide beforehand that we can play at most a week, a month, a year, etc., then the game is fair—the expected payoff is zero. This can be explained somewhat by noticing that if α is reasonable in size and N large, then the probability is nearly 1 that by the time N we would have exceeded the

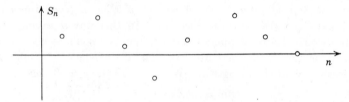

Fig. 10.2 Fortune S_n as function of number n of plays.

amount α and quit; but there is still that small remaining probability that we are still playing at time N and are forced to quit losers. It can be shown that, in this latter alternative, the chances are that we will be very heavy losers, and this large loss with small probability exactly cancels out the win α with large probability.

At any rate, Example 10.1 furnishes us with an example of an unstable stopping-rule problem. Without further discussion, then, we shall henceforth assume that our problems are stable and that solutions exist and are unique.

10.7 THE FUNCTIONAL EQUATION

There are various ways of trying to solve stopping-rule problems. The first method we shall explore is similar to the methods used by Richard Bellman in dynamic programming [1]. It involves setting up a functional equation and using an optimization argument.

Let $H(i)$ be defined as the payoff when we use the best possible stopping rule with the starting state $i_0 = i$. Then it can be shown that $H(i)$ satisfies the equations

$$H(i) = F(i) \qquad\qquad\qquad \text{for } i \in T_s,$$
$$(10.11) \qquad = \sum_j H(j)P(j \mid i) - f(i) \qquad\qquad \text{for } i \in T_c,$$
$$= \max_j \left[F(i), \sum_j H(j)P(j \mid i) - f(i) \right] \qquad \text{for } i \in \overline{(T_s \cup T_c)}.$$

Again, we dodge a rigorous proof, but here give only an outline around
which a proof can be constructed. If we are at state i, and if $i \in T_s$, then,
of course, we must stop; thus $H(i) = F(i)$. If we are to continue, however,
let us compute the expected payoff. We have expended the entrance fee
$f(i)$ and go to j with probability $P(j \mid i)$. If we continue from j using the
best available strategy, then our payoff is $H(j)$. Putting things together,
we see that our expected payoff, starting from i, if we continue one more
step and then do the smartest thing, is

$$(10.12) \qquad H_1(i) = \sum_j H(j)P(j \mid i) - f(i).$$

In the states of T_c, from which we must continue one more step, $H(i)$
must be exactly equal to this expression. In $\overline{T_s \cup T_c}$, the states in which
two decisions are available, if we quit we get $F(i)$, and if we continue we
get $H_1(i)$. Since $H(i)$ is the payoff when we do the best thing, it follows
that

$$(10.13) \qquad H(i) = \max\,[F(i), H_1(i)] \qquad \text{on } \overline{T_s \cup T_c},$$

and this is the functional equation to which we previously alluded.

Still, of what use is a functional equation? What is desired is the best
stopping set. The point is that if we can solve this functional equation,
then we can get the best stopping set. In fact, let

$$T^* = \{\text{all states } i \text{ not in } T_c \text{ such that } H(i) = F(i)\}.$$

Now, we claim that if T^* is a stopping set at all, that is, if

$$P \text{ (entering } T^* \text{ from } i_0) = 1,$$

then T^* is a *best* stopping set, because if the maximum payoff we can
get with initial state i is exactly what we can get by quitting at that
state, then we should quit.

In a word, solving the functional equation will bring us into our kingdom. But this gateway is far from satisfactory. The ratio of functional
equations of the preceding type with known solutions to those with
unknown solutions is very small. It is almost axiomatic that to solve a
functional equation exactly, either the functional equation itself must be
quite simple or one must be next to a genius. We therefore ask about
approximation methods and computational methods. One usual iteration
technique for approaching functional equations is quickly applicable, and
we illustrate it in the case in which T_c is empty. Let $H^{(0)}(i) = F(i)$, and
define the successive approximations by

$$(10.14) \quad H^{(n+1)}(i) = F(i) \qquad\qquad\qquad\qquad\qquad \text{on } T_s,$$
$$= \max\,[F(i), \sum_j H^{(n)}(j)P(j \mid i) - f(i)] \quad \text{on } \overline{T}_s.$$

It is not difficult to prove that

$$H^{(n+1)}(i) \geq H^{(n)}(i),$$

and that the process converges to $H(i)$. There is a grave difficulty here, namely, that it is not $H(i)$ that is desired but T^*. We could try to approximate T^* by

$$T_n^* = \{\text{all } i \text{ such that } H^{(n)}(i) = F(i)\}.$$

This generally is useless, however, because if $H^{(n)}(i) < H(i)$, then both equalities

$$H^{(n)}(i) = F(i) \quad \text{and} \quad H(i) = F(i)$$

cannot simultaneously be satisfied, and thus i cannot be in both T^* and T_n^*. Evidently, the thing to do is to say that T_n^* consists of all states i such that $H^{(n)}(i)$ is sufficiently close to $F(i)$, and now we face the question: How close is sufficiently close? No doubt this procedure, intelligently implemented, can give good approximations; but it is not to our liking, and we leave it now for other attractive considerations.

I0.8 ELIMINATION OF FORCED CONTINUATION

Eventually, we shall be able to formulate stopping-rule problems in linear-programming form. The first steps in this project are to reduce the problems to a standard form. To begin, let us eliminate the forced continuation states T_c. Assuming that i_0 is not in T_c, we replace the original problem by an altered problem that has states $\overline{T_c}$ and that has no forced continuation states, but that is equivalent to the original problem. Note that for any state $i \in \overline{T_c}$, such that

$$P \text{ (ever entering } i \text{ from } i_0) > 0,$$

we must have

$$P \text{ (ever returning to } \overline{T_c} \text{ from } i) = 1,$$

since otherwise there would be a positive probability of remaining forever in T_c starting from i_0, and therefore never being able to stop. On the states of $\overline{T_c}$, we define altered transition probabilities $\hat{P}(j \mid i)$ by

$$\hat{P}(j \mid i) = P \text{ (1st state entered in } \overline{T_c} \text{ from } i \text{ is } j).$$

These may be computed from the equations

$$(10.15) \qquad \hat{P}(j \mid i) = \sum_{n=0}^{\infty} q^{(n)}(j \mid i),$$

where $q^{(n)}(j \mid i)$ is the probability, starting from i, of going into T_c and remaining there for n steps and then making a transition to j; that is,

$$(10.16) \qquad q^{(n)}(j \mid i) = \sum P(i, i_1, \ldots, i_n, j),$$

where the sum is taken over all sequences $i, i_1, \ldots, i_n, j$ such that $i_1 \in T_c, \ldots, i_n \in T_c$.

Along with these changes in the transition probabilities, there is one alteration in the payoff structure. The $F(i)$ and T_s remain, but a substitute is needed for the entrance-fee function $f(i)$. The idea is that actually we do not pay for one more transition; rather, we pay the expected total of entrance fees until we return to a state in which stopping is possible. Another way of saying this is the following: With the altered transition probabilities, we are really considering the successive visits of the device to the states not in T_c. Thus if we are in a state $i \in \overline{T_c}$ and we wish to continue, our entrance fee $\hat{f}(i)$ is the expected total of entrance fees until we next return to $\overline{T_c}$. Analytically, therefore, we replace $f(i)$ by

$$\hat{f}(i) = \sum [f(i) + f(i_1) + \cdots + f(i_n)] P(i, i_1, \ldots, i_n, k),$$

where the sum is taken over all sequences $i, i_1, \ldots, i_n, k$ such that $i_1 \in T_c, \ldots, i_n \in T_c$, and $k \in \overline{T_c}$.

It should be intuitively clear that the foregoing two problems are equivalent in the sense that a stopping set T will produce the same expected payoff in each of them. A solution T^* for one is a solution also for the other.

Finally, by way of illustration let us apply this reduction to Example 10.5. The states of T_c are $(i, 1)$, $i = -N, -N + 1, \ldots,$ and $\overline{T_c}$ consists of $(i, 0)$, $i = -N, -N + 1, \ldots$. We obtain the altered transition probabilities on $\overline{T_c}$ by asking: If we are in state $(i, 0)$, what is the probability that the next state we encounter in $\overline{T_c}$ is $(j, 0)$? More concretely, if the ith parking place is empty, what is the probability that the jth place is the next empty place to which we come? Clearly, we have

$$\hat{P}[(j, 0) \mid (i, 0)] = 0 \qquad \text{for } j \leq i,$$
$$= p(1 - p)^{j-i-1} \qquad \text{for } j > i.$$

As for the altered entrance fee, fortune is with us; since $f(i) \equiv 0$, also $\hat{f}(i) \equiv 0$. We drop 0 from $(i, 0)$, and the altered problem now reads: Given states $-N, -N + 1, \ldots,$ with transition probabilities

$$(10.17) \qquad P(j \mid i) = 0 \qquad \text{for } j \leq i,$$
$$= p(1 - p)^{j-i-1} \qquad \text{for } j > i,$$

and with $F(i) = |i|, f(i) \equiv 0$, maximize the expected payoff.

Henceforth, we shall suppose this reduction to have been carried out, and take T_c to be empty.

10.9 ENTRANCE-FEE PROBLEMS AND REDUCTION TO THEM

The next, and very important step is the introduction of a special class of stopping-rule problems.

A stopping-rule problem is said to be an *entrance-fee problem* if the terminal payoff $F(i)$ vanishes identically: $F(i) \equiv 0$.

Therefore, in an entrance-fee problem one gets nothing for stopping. The immediate response is that if one gets nothing for stopping but must pay $f(i)$ to continue, then why should one ever continue? The answer is that nothing has been said about the sign of $f(i)$; it may be either positive, zero, or negative, and a negative $f(i)$ means that we are *being*

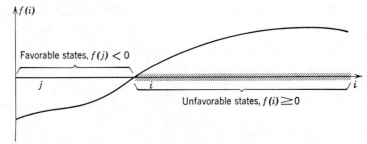

Fig. 10.3   Entrance-fee problem, with fee $f(i)$ for continuing from state i.

paid to continue. Denote the favorable states as those states such that $f(i) < 0$, and the unfavorable as those such that $f(i) \geq 0$. One might think now that the way to treat an entrance-fee problem would be to continue as long as we are in the favorable states (are being paid to continue) and stop as soon as we come to an unfavorable state. This is not necessarily true. Suppose that $f(i)$ has the appearance indicated in Fig. 10.3, where we have artificially made the graph continuous. If we are at state i, even though it is unfavorable, we may wish to continue, paying a small fee, because there is a large probability of a transition to state j, in which case we shall be paid a large amount of money to continue. One point is clear—*we should certainly continue as long as i is favorable*, because as long as we are being paid to continue and can quit after the transition with no penalty, it would be foolish not to continue.

There is one obvious special case in which the stopping rule discussed above is best. If we can never get back from the unfavorable states to the favorable states, then there is no motive for not quitting as soon as we hit an unfavorable state. We formulate this simple remark as follows:

Theorem 10.1.   In an entrance-fee problem, let $T^* = \{$all unfavorable states $i\}$. If $P(j \mid i) = 0$ for each i, j such that i is unfavorable and j is favorable, and T^* satisfies all requirements for stopping rules, then T^* is a solution.

The case that this theorem covers we shall call the *absolutely monotonic* case. Although it seems trivial, it will eventually provide solutions to many of our examples.

Besides their own importance, there is a very strong reason for introducing the concept of entrance-fee problems—*every stopping-rule problem can be reduced to an equivalent entrance-fee problem.* The exact statement is this: Given a stopping-rule problem with $P(j \mid i)$, T_s, $F(i)$, $f(i)$, construct the entrance-fee problem (the so-called prime problem) with the same transition probabilities and the same T_s, but with entrance fee $f'(i)$ given by

$$f'(i) = f(i) - \left[\sum_j F(j)P(j \mid i) - F(i) \right],$$

and, of course, $F'(i) = 0$. Then these two problems are equivalent in the sense of this theorem:

Theorem 10.2. Let T be any stopping rule for a stopping-rule problem and the corresponding entrance-fee problem. If T is stable for both problems, then we have

$$E_T Z = E_T Z' + F(i_0),$$

where $E_T Z'$ is the payoff in the entrance-fee problem using T. Therefore, if both problems are stable and T^* is a solution for one problem, then it is also a solution for the other.

Before discussing the proof of this theorem, let us note its application to some of the foregoing examples.

Examples 10.1 and 10.2 are equivalent to the entrance-fee problem with

$$f(i) = 0 - \left[\sum_j F(j)P(j \mid i) - F(i) \right]$$
$$= 0 - [\tfrac{1}{2}(i+1) + \tfrac{1}{2}(i-1) - i] = 0.$$

Surprisingly, then, these examples are both equivalent to a problem in which $F(i) \equiv 0, f(i) = 0$. The theorem states that for any stable rule T, $E_T Z = E_T Z' + F(i_0)$; since $E_T Z = 0$, and $F(i_0) = 0$, the result is that, for any stable rule, $E_T Z = 0$.

Example 10.3 is equivalent to the entrance-fee problem with

$$f(i) = C - \left[\sum_{j \geq i} jP(j \mid i) - i \right]$$
$$= C - \left(\sum_{j > i} jp_j - \sum_{k > i} ip_k \right)$$
$$= C - \sum (j - i)p_j,$$

since $\sum p_k = 1$. We note further, for future reference, that

$$f(i + 1) - f(i) = -\sum_{j>i+1}(j - i - 1)p_j + \sum_{j>i}(j - i)p_j$$
$$= \sum_{j>i} p_j \geq 0,$$

so that

$$f(i + 1) \geq f(i).$$

The already-reduced version of the parking-place problem, Example 10.5, with T_c empty, is equivalent to the entrance-fee problem with

$$f(i) = \sum_{j>i} |j|\, p(1 - p)^{j-i-1} - |i|.$$

$i \geq 0$, then

$$f(i) = \sum_{j>i}(j - i)p(1 - p)^{j-i-1} = p\sum_{k=1}^{\infty} k(1 - p)^{k-1}$$

$$= -p\frac{d}{dp}\left[\sum_{k=1}^{\infty}(1 - p)^k\right]$$

$$= -p\frac{d}{dp}\left(\frac{1}{p} - 1\right) = \frac{1}{p}.$$

A similar computation for $i < 0$ gives

$$f(i) = \frac{1}{p}[2(1 - p)^{|i|} - 1].$$

Again we have $f(i + 1) \geq f(i)$.

Now, as a pleasant surprise, we notice that in Examples 10.3 and 10.5 we are in the absolutely monotonic case, since in both problems we have

$$f(i + 1) \geq f(i),$$
$$P(j\,|\,i) = 0, \qquad j < i.$$

Thus the simple principle embodied in Theorem 10.1 enables us to write (see also [6, 14]) the following solutions to Examples 10.3 and 10.5, respectively:

$$T^* = \left\{\text{all } i \text{ such that } C > \sum_{j=i+1}^{\infty}(j - i)p_j\right\},$$

$$T^* = \{\text{all } i \geq 0 \text{ together with those negative } i \text{ for which } (1 - p)^{|i|} > \tfrac{1}{2}\}.$$

Rather than give in detail the proof of Theorem 10.2, we shall exhibit its mechanics on an example. The idea is to prove it for truncated stopping rules and then pass to the limit; that is, we prove first that

$$E_{T(N)}Z = E_{T(N)}Z' + F(i_0)$$

for all N, and then, if the problem is stable, let $N \to \infty$. To illustrate, let us prove the result for $N = 2$. Now $T^{(2)}$ is the rule with the following list of stopping sequences:

$$(i_0, i_1) \quad \text{for } i_1 \in T, \quad \text{and} \quad (i_0, i_1, i_2) \quad \text{for } i_1 \in \bar{T}.$$

The expected *terminal payoff* using $T^{(2)}$ is

$$(10.18) \qquad \sum_{i_1 \in T} F(i_1)P(i_0, i_1) + \sum_{i_1 \in \bar{T}} F(i_2)P(i_0, i_1, i_2),$$

where the second sum is taken over all sequences (i_0, i_1, i_2) such that $i_1 \in \bar{T}$. We rewrite this second sum by noticing that

$$P(i_0, i_1, i_2) = P(i_2 \mid i_1)P(i_0, i_1),$$

and therefore

$$\sum_{i_1 \in \bar{T}} F(i_2)P(i_0, i_1, i_2) = \sum_{i_1 \in \bar{T}}\left[\sum_{\text{all } i_2} F(i_2)P(i_2 \mid i_1)\right]P(i_0, i_1).$$

Let $G(i) = \sum_j F(j)P(j \mid i) - F(i)$. Then we have

$$\sum_{i_1 \in \bar{T}} F(i_2)P(i_0, i_1, i_2) = \sum_{i_1 \in \bar{T}} [G(i_1) + F(i_1)]P(i_0, i_1),$$

and (10.18) becomes

$$\sum_{i_1 \in T} F(i_1)P(i_0, i_1) + \sum_{i_1 \in \bar{T}} [G(i_1) + F(i_1)]P(i_0, i_1)$$

$$= \sum_{\text{all } i_1} F(i_1)P(i_1 \mid i_0) + \sum_{i_1 \in \bar{T}} G(i_1)P(i_0, i_1)$$

$$= G(i_0) + F(i_0) + \sum_{i_1 \in \bar{T}} G(i_1)P(i_0, i_1).$$

The expected entrance fees paid total

$$f(i_0) + \sum_{i_1 \in \bar{T}} f(i_1)P(i_0, i_1).$$

Thus we have

$$E_{T^{(2)}} = [G(i_0) - f(i_0)] + \sum_{i_1 \in \bar{T}} [G(i_1) - f(i_1)]P(i_0, i_1) + F(i_0).$$

If we use the entrance-fee problem with

$$f'(i) = f(i) - G(i),$$

it is quite clear that the preceding expression is exactly $E_{T^{(2)}}Z' + F(i_0)$. The proof for general N follows exactly the same idea; we omit it only because of the complexity of notation.

There is one further result that we shall present in this section for later reference. Heuristically, the idea is that if the entrance fee $f(i)$, in an entrance-fee problem, is increasing with i, and if there is a consistent trend toward larger i, that is, if it is more probable to go to larger values of i

than to smaller values, then the solution T^* should be of the form: There is a i^* such that we stop at the first $i \geq i^*$. A formal statement is the following:

Theorem 10.3. If $f(i + 1) \geq f(i)$, $F(i) \equiv 0$, and the transition probabilities are such that, for any nondecreasing function $h(i)$, the function $\sum_j h(j)P(j \mid i)$ is also nondecreasing, then there is an i^* such that a solution T^* is given by the set $\{$all i such that $i \geq i^*\}$.

One method of proof proceeds by using the functional equation. In this case, we write

$$H(i) = \max \left[0, \sum_j H(j)P(j \mid i) - f(i) \right]$$

and

$$T^* = \{\text{all } i \text{ such that } H(i) = 0\}.$$

Notice from the assumptions of the theorem, by changing sign, that for a nonincreasing function $h(i)$, the function $\sum_j h(j)P(j \mid i)$ is again nonincreasing. Since $H(i) \geq 0$, the theorem will be proved if we show that $H(i)$ is nonincreasing, since then $H(i^*) = 0$ will imply $H(i) = 0$ for all $i \geq i^*$. We argue from the successive approximation method, with

$$H^{(0)}(i) \equiv 0, \quad H^{(n+1)}(i) = \max \left[0, \sum_j H^{(n)}(j)P(j \mid i) - f(i) \right], \quad n \geq 0,$$

that if $H^{(n)}(i)$ is nonincreasing, then, by the hypothesis of the theorem, so is $\sum_j H^{(h)}(j)P(j \mid i)$ nonincreasing. Hence by the nondecreasing nature of $f(i)$, so is $\sum_j H^{(n)}(j)P(j \mid i) - f(i)$ nonincreasing, and thus $H^{(n+1)}(i)$, being the maximum of 0 and a nonincreasing function, is also nonincreasing; that is, $H^{(n+1)}(i + 1) \leq H^{(n+1)}(i)$.

Since $H^{(0)}(i) = 0$, $H^{(n)}(i)$ is nonincreasing for all n, and so is the limit function $H(i)$.

We call the case covered by this theorem the *monotonic case*. It is similar to the absolutely monotonic situation, but it leaves unresolved the sometimes very difficult problem of computing the desired i^*.

10.10 THE LINEAR-PROGRAMMING SOLUTION

There are a number of good and sufficient reasons for the interesting observation that a stopping-rule problem can be formulated as a linear-programming problem. Not the least of these is the practical consideration that many machine programs exist and that for such problems modern machinery can handle variables in numbers that appear to a mathematician as almost incredible.

To leave our disbelief, however, we place two restrictions on the stopping-rule problems, which will be formulated in this way.

1. They are entrance-fee problems with T_c empty; that is, they have previously been reduced.

2. There are only a finite number of states, say $1, \ldots, M$, in $\bar{T}_s$.

Consider the linear-programming problem with a variable x_i associated with each state in $\bar{T}_s$, so that we have altogether the M variables $x_1, \ldots,$ x_M. These are subject to the inequalities

$$(10.19) \qquad x_i \geq 0, \qquad i = 1, \ldots, M \qquad \text{(nonnegativity)},$$

$$x_i \geq -f(i) + \sum_{j \in T_s} x_j P(j \mid i), \qquad i = 1, \ldots, M.$$

We add now: Under the inequalities (10.19), minimize x_{i_0}, where i_0 is the starting state. Since x_{i_0} is certainly a linear function of $x_1, \ldots, x_M$ ($x_{i_0} = a_1 x_1 + \cdots + a_M x_M$, with $a_i = 0, i \neq i_0, a_{i_0} = 1$), this problem is a standard linear-programming problem. The connection with the original stopping-rule problem is in the following result.

Theorem 10.4. Let $(x_1^*, \ldots, x_M^*)$ be a solution of the linear-programming problem, and let $T = \{\text{all } i \text{ such that } x_i^* = 0\}$. Then $T^* = T \cup T_s$ is a solution of the entrance-fee problem, and

$$x_{i_0}^* = E_{T^*} Z.$$

To sketch a proof of this theorem, assume that all solutions are unique. Let G be the set of all vectors $(x_1, \ldots, x_M)$ satisfying the inequalities (10.19), and let $(x_1, \ldots, x_M)$ and $(y_1, \ldots, y_M)$ both be in G; then the vector $(z_1, \ldots, z_M)$, defined by

$$z_i = \min (x_i, y_i),$$

is also in G since, certainly, $z_i \geq 0$, and

$$(10.20)$$
$$z_i = \min (x_i, y_i) \geq \min \left[-f(i) + \sum_j x_j P(j \mid i), -f(i) + \sum_j y_j P(j \mid i) \right].$$

The expression on the right-hand side of (10.20) equals

$$-f(i) + \min \left[\sum_j x_j P(j \mid i), \sum_j y_j P(j \mid i) \right]$$
$$\geq -f(i) + \sum_j \min (x_j, y_j) P(j \mid i),$$

where all sums are understood to be taken over the states $1, \ldots, M$ of $\bar{T}_s$.

Let $(x_1^*, \ldots, x_M^*)$ be the solution, and define $(y_1, \ldots, y_M)$ by

$$y_i = \max\left[0, -f(i) + \sum_j x_j^* P(j \mid i)\right].$$

Now $(y_1, \ldots, y_M)$ is in G, since $y_i \geq 0$; hence, noticing that $y_i \leq x_i^*$, we have

$$y_i \geq -f(i) + \sum_j x_j^* P(j \mid i) \geq -f(i) + \sum_j y_j P(j \mid i).$$

Therefore $(z_1, \ldots, z_M)$, defined by $z_i = \min(x_i^*, y_i)$, is also a solution. If the solution is unique, then $z_i = x_i^*$, which implies, since $y_i \leq x_i^*$, that $y_i = x_i^*$. Thus x_i^* satisfies

$$x_i^* = \max\left[0, \sum_{j \in \bar{T}_s} x_j^* P(j \mid i) - f(i)\right].$$

Defining $L(i)$ by

$$L(i) = x_i^* \qquad \text{for} \quad i \in \bar{T}_s,$$
$$= 0 \qquad \text{for} \quad i \in T_s,$$

we see that $L(i)$ satisfies the functional equation

$$L(i) = 0 \qquad\qquad\qquad\qquad \text{for} \quad i \in T_s,$$
$$= \max\left[0, \sum_j L(j) P(j \mid i) - f(i)\right] \qquad \text{for} \quad i \in \bar{T}_s.$$

This is the equation satisfied by $H(i)$, so that with uniqueness we have

$$L(i) = H(i),$$
$$T^* = \{\text{all } i \text{ such that } L(i) = 0\}.$$
$$x_{i_0}^* = H(i_0) = E_{T^*} Z.$$

Let us make a few last remarks on the preceding theorem and proof. A careful reading of the proof indicates that for these inequalities we would get the same solution minimizing any linear function $a_1 x_1 + \cdots + a_M x_M$ with nonnegative coefficients a_i. This is true if the solutions are unique, and still essentially true as regards T^* even in the nonunique case. The choice of x_{i_0} was dictated by a connection to be unearthed in Sec. 10.17. Secondly, in general $\bar{T}_s$ does not have a finite number of states—then what? One possibility is that a crude initial assessment of the problem will allow the identification of a set of states T_0 on which $f(i)$ is so large that it is fairly evident that we should stop on all states in T_0, if we have not already stopped. Now take an enlarged $T_s' = T_s \cup T_0$, and if $\overline{T_s'}$ has only a finite number of states then we are in the previously treated case.

BINARY DECISION RENEWAL PROBLEMS

10.11 INTRODUCTION AND EXAMPLES

We turn at this point to what is seemingly an entirely new class of problems. The reasons are mixed; first of all, although stopping-rule problems are often amusing and attractive, they seldom appear intrinsically in industrial and engineering applications. There is, however, a wide class of applied problems that are closely related—that generate mathematical situations suggestive of stopping-rule problems. Our second motive will be to show that problems of this new class have not only a family resemblance but an even more familiar relation.

The type of problem we call a *binary decision renewal problem* occurs in many contexts [11]: replacement problems, inspection and repair problems, quality-control problems, etc. We first present a few familiar examples.

Example 10.6   The Failure and Replacement Problem [5]. An item, say a light bulb, has probability p_k of failing between age k and age $k + 1$. There are periodic replacement opportunities at times $0, 1, 2, \ldots$, and the bulb is new at time 0. At any time n, it may be replaced at cost C_1, if it is good, and the replacement bulb is considered new at time $n + 1$. If, however, the bulb fails between time n and $n + 1$, then the replacement cost is $C_1 + C_2$ and a new bulb is in operation at time $n + 2$. Thus the replacement time is here taken to be one period. The objective is to find a replacement policy that will minimize the cost per unit time (see Sec. 10.13).

Example 10.7   The Inspection and Repair Problem [3]. A machine has M internal states $1, 2, \ldots, M$. It is periodically inspected at times $0, 1, 2, \ldots$, and the inspection reveals the present state i of the machine. There are known transition probabilities $P(j \mid i)$ that a machine in state i at time n will go to state j by time $n + 1$. The machine starts in state 1, the state of perfect repair, and if at time n it is in state $j \neq M$, then it may be repaired at cost C_1 and will be in state 1 at time $n + l_1$; but if at time n it is in state M, the breakdown state, then it must be repaired, the cost is $C_1 + C_2$, and the machine is put into state 1 at time $n + l_2$. Here the repair time is variable, and the objective again is to minimize the cost per unit time.

Example 10.8   The Replacement Problem [8]. If an item, say a truck, is i years old, then its maintenance costs over the next year of operation is a random variable M_i with known distribution

$$P(M_i = r) = p_r(i), \qquad r = 0, 1, \ldots .$$

If a truck is k years old at time n, then we have the option of replacing it and starting with a new truck at time n, at cost C, or of keeping the old truck for one more year. The problem is to find a replacement policy that will minimize the cost per unit time.

10.12 FORMULATION

What are the essential features of Examples 10.6 to 10.8? As with stopping-rule problems, there are two elements, a probabilistic device moving from state to state and a payoff and decision structure. To specify a binary decision renewal problem, we must specify the following items.

1. An underlying random device, which we shall assume is of the same type as in the stopping-rule problems. That is, it is a Markov device with its states some subset of the integers, known transition probabilities $P(j \mid i)$, and a specified starting state i_0, which we shall call the *origin*.
2. Two functions $G(i)$ and $g(i)$ taking arbitrary values, and a function $l(i)$ taking only the values $0, 1, 2, \ldots$, such that if we are in state i at time n then we have available at most two decisions.

   (a) Pay the amount $G(i)$ and be returned to state i_0 at time $n + l(i)$, and the system is considered inoperative between times n and $n + l(i)$. Then $G(i)$ will be called the *return to origin cost*, and $l(i)$ the *time of return to origin*.
   (b) Receive the incentive fee $g(i)$ for proceeding one more step.

Just as with stopping-rule problems, there may be states at which at most one decision is available. Accordingly, we postulate that there may be present a set of forced renewal states U_r such that if we are in state $i \in U_r$, then we must pay $G(i)$ and start at the origin $l(i)$ periods later. Similarly, we postulate a set of *forced continuation* states U_c in which we must continue for one more step, receiving the appropriate incentive fee.

Let us now fit the examples into this framework. Again care must be taken in the selection of states.

For Example 10.6, we take the states to be the age of the current bulb, together with a $+1$ if it is burning, and a -1 if it has failed. Thus the states are $(k, 1)$, $(k, -1)$, $k = 0, 1, \ldots$, with $i_0 = (0, 1)$. The transition probabilities are given by

$$P[(k+1, 1) \mid (k, 1)] = 1 - p_k,$$
$$P[(k+1, -1) \mid (k, 1)] = p_k,$$

and naturally,

$$P[(k+1, -1) \mid (k, -1)] = 1.$$

There are no continuation states, but the states $(k, -1)$ are all forced renewal states. The time of return is $l(i) \equiv 1$. The costs are given by $g \equiv 0$, and $G[(k, 1)] = C_1$, $G[(k, -1)] = C_1 + C_2$.

Example 10.7 already is almost entirely formulated. There are no continuation states; U_r consists of the single state M, $G(i) = C_1$ for the states $1, \ldots, M - 1$, and $G(M) = C_1 + C_2$. The time of return is given by $l(i) = l_1$ for $i \neq M$, and $l(M) = l_2$.

For Example 10.8, we take the states to be the age of the present truck. Hence i ranges over the values $0, 1, 2, \ldots$, and we take $i_0 = 0$. The cost of return is $G(k) = C$, and the time of return is 0. If we decide to continue, then $k \to k + 1$. The chain here is purely deterministic; that is, we have

$$P(k + 1 \mid k) = 1, \quad P(j \mid k) = 0 \qquad \text{for all } j \neq k + 1.$$

The incentive fee we receive for continuing is the negative of the expected cost of repair over the kth period, namely,

$$g(k) = -C_k = -\sum_r r p_r(k).$$

This last remark is not obvious and needs some clarification. The idea is that, although the cost of going from i to $i + 1$ varies from truck to truck, since we are interested in long-term average cost, we may replace this varying cost by its expected value.

10.13 WHAT IS A SOLUTION?

A solution here is a *renewal rule*, that is, a rule that tells us when to return to the origin so as to minimize the cost per unit time. By an argument similar to those given earlier, we shall say that sensible renewal rules are just a set of states U such that $U_r \subset U$, $U_c \subset \overline{U}$, and P (entering U from i_0) $= 1$ (we eventually must renew).

There is here, however, an essential difficulty not encountered in stopping-rule problems. This is in the nature of the thing we wish to minimize. In stopping-rule problems, it is simple to define the total expected payoff using a given stopping rule. In this class of problems, however, the process goes on forever, now and then being returned to the origin and restarting. The expression we wish to minimize is a long-term average, and we must do work to formulate the concept of an expression we may call the *cost per unit time*. For a given renewal rule U, let $C(i)$ be defined as

$$
\begin{aligned}
C(i) &= \quad G(i) \qquad \text{for} \quad i \in U, \\
&= -g(i) \qquad \text{for} \quad i \in \overline{U},
\end{aligned}
$$

and let $t(i)$ be unity if $i \in \overline{U}$, but $t(i) = l(i)$, the time of return to the origin, if $i \in U$. Then for a sequence of states $i_0, i_1, \ldots, i_n$, the total cost is $C(i_0) + \cdots + C(i_n)$, and the total time taken to move through this sequence is $t(i_0) + \cdots + t(i_n)$. The average cost per unit time in moving through this sequence is

$$
\frac{C(i_0) + \cdots + C(i_n)}{t(i_0) + \cdots + t(i_n)}.
$$

We appeal now to a result that is pure magic in many fields, the ergodic theorem. Although the random device and the renewal rule combine to give many different possible sequences of states, the ergodic theorem states that with probability 1 we have

$$
\lim_{n \to \infty} \frac{C(i_0) + \cdots + C(i_n)}{t(i_0) + \cdots + t(i_n)} = W_U,
$$

where W_U is a constant depending on the renewal rule U. Put otherwise, the long-term average cost per unit time does exist in a very strong sense and has the same value, with probability 1, along the various sequences that are ground out.

A solution, then, is a renewal rule U^* such that for any other renewal rule U, we have

$$
W_{U^*} \leq W_U.
$$

10.14 REDUCTION TO A STOPPING-RULE PROBLEM BY CYCLE ANALYSIS

We hinted, at the introduction of this class of problems, that there is a closer relation to stopping-rule problems than immediately appears. This we now uncover. The connection has been established and used, for example, in [12] and [9].

Suppose that, by some renewal rule, our device generates a sequence of states $i_0, i_1, i_2, \ldots$. With probability 1, i_0 will appear an infinite number of times in this sequence. By the *first cycle* will be meant the sequence of states commencing with the initial i_0 and ending at the first state that is in U, so that the next state to appear is i_0. The *second cycle* is the sequence of states starting with the i_0 following the first cycle and continuing until the next entry into U, and so on. Thus if $i_0 = 0$, and we have the sequence

$0, 3, 2, 7, 0, 1, 6, 0, 2, \ldots$, where 7 and 6 are in U, then the first cycle is
0 3 2 7, the second is 0 1 6, etc. Along any sequence, let D_k denote the
cost of the kth cycle; that is, if the kth cycle is $i_0, \ldots, i_n$, let $D_k = C(i_0) + \cdots + C(i_n)$. Furthermore, let L_k denote the *length* of the kth
cycle; that is, let $L_k = t(i_0) + \cdots + t(i_n)$. For n cycles, the average cost
per unit time is

(10.21)
$$\frac{D_1 + \cdots + D_n}{T_1 + \cdots + T_n}.$$

The different cycles, their costs, and their lengths are random variables,
but they all have the same probabilistic distribution. Furthermore, successive cycles are independent in a probabilistic sense, since once we return
to i_0 and begin a new cycle, its development is independent of the history
of past cycles. Using the law of large numbers gives the conclusion that,
with probability 1,

(10.22)
$$\frac{D_1 + \cdots + D_n}{T_1 + \cdots + T_n} \to \frac{E_U D_1}{E_U T_1},$$

where E_U denotes expectation under the renewal rule U. This gives the
equation

$$W_U = \frac{E_U D_1}{E_U T_1}.$$

A significant reduction has been achieved in that we have reduced the
problem to an expression involving only the first cycle. In a sense, it is
now reduced to something like a stopping-rule problem in that we might
as well consider the game to be finished when the first cycle is completed.
If the expression to be minimized were only $E_U D_1$ instead of a quotient,
then it would be exactly a stopping-rule problem, with $F(i) = -G(i)$ and
$f(i) = -g(i)$, so that $Z = -D_1$ and we want to maximize $E_U Z$. However,
the quotient prevents such a direct identification. Instead, we go through
the following computations: Let $\alpha^* =$ the minimum of W_U, so that

$$\alpha^* \le \frac{E_U D_1}{E_U T_1} \qquad \text{for all } U, \text{ with equality if } U = U^*.$$

Therefore, as before, we have

$$E_U D_1 \ge \alpha^* E_U T_1 \qquad \text{or} \qquad E_U D_1 - \alpha^* E_U T_1 \ge 0,$$

with equality if $U = U^*$.
If the cycle is $i_0, i_1, \ldots, i_n$, then $i_n \in U$, and we have

$$D_1 = G(i_n) - g(i_{n-1}) - \cdots - g(i_0),$$
$$T_1 = t(i_n) + t(i_{n-1}) + \cdots + t(i_0) = t(i_n) + n,$$
$$D_1 - \alpha^* T_1 = [G(i_n) - \alpha^* t(i_n)] - [\alpha^* + g(i_{n-1})] - \cdots - [\alpha^* + g(i_0)].$$

Rewriting further, we see that

$$E_U(-D_1 + \alpha^* T_1) \leq 0,$$

and

$$\max E_U(-D_1 + \alpha^* T_1) = 0.$$

Now, look at the stopping-rule problem defined by

$$T_s = U_r, \quad T_c = U_c, \quad F(i) = \alpha^* l(i) - G(i), \quad f(i) = -g(i) - \alpha^*$$

If we use U as a stopping set in this problem, then we have

$$E_U Z = E_U(-D_1 + \alpha^* T_1).$$

The conclusion is that *for any stopping set T, we have $E_T Z \leq 0$; in partic- ular, for $T^* = U^*$, we have $E_{T_*} Z = 0$. Thus U^* is a solution to this stopping-rule problem.* Accordingly, the following result holds.

Theorem 10.5. For a binary decision renewal problem with cost or return $G(i)$, incentive fee $g(i)$, and time of return to the origin $l(i)$, consider a one-parameter family of stopping-rule problems with the same transition probabilities, $T_s = U_r$, $T_c = U_c$, and

$$F(i) = \alpha l(i) - G(i), \quad f(i) = -g(i) - \alpha.$$

Letting $\beta(\alpha) = \max E_T Z$, choose α^* such that $\beta(\alpha^*) = 0$. Then for this value of α, the solution to the stopping-rule problem is also the solution to the renewal problem, and $W_{U*} = \alpha^*$.

For Example 10.6, the corresponding family of stopping-rule problems is described as follows:

> States $(k, \pm 1)$; $T_s = (k, -1)$;
> $F(k, +1) = -C_1 - \alpha$; $F(k, -1) = -C_1 - C_2 - \alpha$;
> $f(k, +1) = -\alpha.$

For Example 10.7, the family is characterized thus:

> States $1, \ldots, M$; $F(i) = -C_1 - \alpha$, $i = 1, \ldots, M - 1$;
> $F(M) = -C_1 - C_2 - \alpha$; $f(i) = -\alpha$, $i = 1, 2, \ldots, M - 1.$

Similarly, for Example 10.8, the conditions are these:

> States $0, 1, \ldots$; $F(i) = -C$, $f(i) = C_i - \alpha.$

Under reasonable assumptions about p_k and C_i, Examples 10.6 and 10.8

can be solved. In Example 10.6, we reduce the problem to an entrance-fee problem. The entrance fee is given by

$$\begin{aligned} f'(k, 1) &= -\alpha - [(-C_1 - C_2 - \alpha)p_k \\ &\quad + (-C_1 - \alpha)(1 - p_k) - (-C_1 - \alpha)] \\ &= -\alpha + (C_2 + \alpha)p_k. \end{aligned}$$

If the p_k are increasing, $p_{k+1} > p_k$, that is to say, if that the light bulb is more likely to fail after $k + 1$ periods of life than after k periods, then we are in the absolutely monotonic case, and the solution is

$$T^* = \left\{ \text{all } (k, +1) \quad \text{such that} \quad p_k > \frac{\alpha}{C_2 + \alpha} \right\} \cup T_s.$$

Let P_k be the probability that the light bulb is still burning at time k, $P_k = (1 - p_{k-1}) \cdots (1 - p_0)$. If by x^+ is meant the function of x defined as $x^+ = x$ for $x \geq 0$, $x^+ = 0$ for $x < 0$, then the expected payoff in the entrance-fee problem using T^* is

$$\sum_{k=0}^{\infty} [\alpha - (C_2 + \alpha)p_k]^+ P_k.$$

The solution to the renewal problem is: Replace the light bulb as soon as the probability of a failure in the next period is greater than $\alpha^*/C_2 + \alpha^*$.

The entrance-fee problem for Example 10.8 is given by

$$f'(i) = C_i - \alpha;$$

and if $C_i > C_{i-1}$, then Theorem 10.1 holds and T^* becomes {all i such that $C_i > \alpha$}. The payoff is

$$\sum_{i=0}^{\infty} (\alpha - C_i)^+.$$

As $F(0) = -C$, the equation for α^* is

$$C = \sum_{0}^{\infty} (\alpha - C_i)^+.$$

The best replacement rule now reads: Buy a new truck as soon as $C_i > \alpha^*$.

Example 10.7 is generally not in the absolutely monotonic class. Reducing it to an entrance-fee problem, we obtain

$$\begin{aligned} f'(i) &= -\alpha - \left[\sum_{j \leq M-1} (-C_1 - \alpha)P(j \mid i) \right. \\ &\quad \left. + (-C_1 - C_2 - \alpha)P(M \mid i) - (-C_1 - \alpha) \right] \\ &= -\alpha + (C_2 + \alpha)P(M \mid i). \end{aligned}$$

If it is assumed that the $P(j \mid i)$ take nondecreasing functions into non-decreasing functions, that is, that the second hypothesis of Theorem 10.3 applies, then, automatically, $P(M \mid i)$ is nondecreasing in i, and the conclusion of the theorem is valid; there is an i^*, depending on α, such that the best policy is to repair the machine whenever it is in state $i \geq i^*$. This latter treatment was suggested by C. Derman [3].

10.15 A DIRECT APPROACH

There is another approach that will carry us back to stopping-rule problems, and although it may be more difficult, it is amusing and of independent interest. It is based on a well-known result [7] regarding Markov chains: Let a chain have transition probabilities $\hat{P}(j \mid i)$ and be such that from any states i, j there is a sequence of states connecting i and j, say $(i, j_1, \ldots, i_{n-1}, j)$, with $P(i, i_1, \ldots, i_{n-1}, j) > 0$. For any starting state i_0, let $\pi_i(n)$ be the expected proportion of times we are in state i during the first n transitions. Then, under very weak assumptions, which will always be satisfied in the cases we study, the limits

$$\lim_n \pi_i(n) = \pi_i$$

exist, and are the unique solutions of the equations

(10.23)
$$\sum_i \pi_i = 1,$$

$$\pi_j = \sum_i \hat{P}(j \mid i)\pi_i.$$

Returning to W_U, we have

$$W_U = \lim_{n \to \infty} \frac{\dfrac{1}{n} \displaystyle\sum_{k=0}^{n-1} C(i_k)}{\dfrac{1}{n} \displaystyle\sum_{k=0}^{n-1} t(i_k)}.$$

The succession of states is governed by the transition probabilities

(10.24)
$$\hat{P}(j \mid i) = P(j \mid i) \qquad \text{for } i \in \bar{U},$$
$$= \delta(j, i_0) \qquad \text{for } i \in U,$$

where $\delta(j, i_0) = 1$ for $j = i_0$, and 0 otherwise. That is, if i is in U, then the next state is i_0. We may write

$$\frac{1}{n} \sum_{k=0}^{n-1} C(i_k) = \sum_i C(i)\pi_i(n),$$

$$\frac{1}{n} \sum_{k=0}^{n-1} t(i_k) = \sum_i t(i)\pi_i(n).$$

Thus we convert the problem to the minimization of

(10.25)
$$\frac{\sum_i C(i)\pi_i}{\sum_i t(i)\pi_i},$$

where the π_i are the solution of equation (10.23) with the $\hat{P}(j \mid i)$ given by (10.24).

This presents a straightforward but computationally awkward method of solution: For all possible U, compute π_i, substitute into (10.25), and search for the minimizing U^*. Hardly a feasible method! All is not lost, however. Once more we inquire into the possibility of replacing (10.23) and (10.24) by a system of linear inequalities, with a view to the eventual conversion of the problem into a linear-programming problem. Of course, we already have one route to a linear-programming solution—convert to the family of stopping-rule problems, and then convert these to inequalities. The difficulty is that we obtain not one, but a family of linear-programming problems. Actually this is only a first-glance difficulty (see Sec. 10.16), but we nevertheless wish to carry on with the direct approach at this point.

To implement the direct approach, it is convenient to reduce the problem. By the exact reduction used before, we may turn to a problem such that U_c is empty. Now we define an *incentive-fee* problem as a binary-decision renewal problem for which $G(i)$ is constant. There is not quite such a universal result as the reduction to entrance-fee problems, but sufficient for our purposes is the following:

Theorem 10.6. Given a binary decision renewal problem with U_c empty and $l(i)$ constant, $l(i) = l$, construct the incentive-fee problem with the same U_s and $l(i)$, but with an incentive fee $\tilde{g}(i)$, given by

$$\tilde{g}(i) = g(i) - \left[\sum_j G(j)p(j \mid i) - G(i) \right]$$

and
$$\tilde{G}(i) = G(i_0).$$

Then the two problems are equivalent in the sense that any solution U^* for one is a solution for the other, and $W_{U^*} = \tilde{W}_{U^*}$.

PROOF. To the first problem is associated the family of stopping-rule problems with
$$F(i) = \alpha l - G(i), \qquad f(i) = -g(i) - \alpha.$$

If we convert these to entrance-fee problems, we obtain

$$f'(i) = -g(i) - \alpha + \left[\sum_j G(j)P(j \mid i) - G(i) \right],$$

and the payoffs are given by

$$E_T Z = E_T Z' + \alpha l - G(i_0).$$

If we look at the family of stopping-rule problems associated with the incentive-fee problem, we have

$$\tilde{F}(i) = \alpha l - G(i_0), \qquad \tilde{f}(i) = -g(i) - \alpha + \left[\sum_j G(j) P(j \mid i) - G(i) \right],$$

and this reduces to exactly the same family of entrance-fee problems, with the same payoffs.

For an incentive-fee problem, with $l(i) = l$ and $G(i) = C$, (10.25) reduces to

(10.26)
$$\frac{- \sum_{i \in \overline{U}} g(i) \pi_i + C \sum_{i \in \overline{U}} \pi_i}{\sum_{i \in \overline{U}} \pi_i + l \sum_{i \in \overline{U}} \pi_i},$$

and this quotient may be minimized by the simple linear-programming approach of Sec. 10.16.

10.16 THE LINEAR-PROGRAMMING SOLUTION

The work in this section applies only to incentive-fee problems with a finite number of states, say $i, \ldots, M$, in $\overline{U}_r$.

Letting

$$\lambda = \sum_{i \in U} \pi_i,$$

rewrite (10.24) and (10.25) as follows: Given the constraints

$$\pi_j = \delta(j, i_0) \lambda + \sum_{i \in \overline{U}} P(j \mid i) \pi_i, \qquad \lambda + \sum_{i \in \overline{U}} \pi_i = 1,$$

minimize

$$\frac{- \sum_{i \in \overline{U}} g(i) \pi_i + C\lambda}{\sum_{i \in \overline{U}} \pi_i + l\lambda} = \frac{- \sum_{i \in \overline{U}} g(i) \pi_i + C\lambda}{(l - 1)\lambda + 1}.$$

To linearize this problem, we introduce new variables r_i, σ related to π_i, λ by

$$r_i = 0 \qquad\qquad \text{for } i \in U,$$
$$= [1 - (l - 1)\sigma] \pi_i \qquad \text{for } i \in \overline{U},$$
$$\sigma = \frac{\lambda}{1 + (l - 1)\lambda}.$$

Since $0 \le \lambda \le 1$ and $l \ge 0$, it follows that σ is nonnegative; and since

$$\sigma = \lambda[1 - (l - 1)\sigma] \qquad \text{and} \qquad \pi_i \ge 0,$$

the r_i also are nonnegative. We adopt now the notation of writing $\sum_i$ instead of $\sum_{i \in \bar{U}_r}$. With this, the equations satisfied by r_i, σ are

(10.27) $r_j = 0$ for $j \in U$,

$\qquad\qquad = \delta(j, i_0)\sigma + \sum_i P(j \mid i)r_i$ for $j \in \bar{U}$,

$\qquad l\sigma + \sum_i r_i = 1$.

The expression to be minimized becomes, on multiplying numerator and denominator by $[1 - (l - 1)\sigma]$,

(10.28) $$\frac{-\sum_i g(i)r_i + C\sigma}{(l - 1)\sigma + [1 - (l - 1)\sigma]} = -\sum_i g(i)r_i + C\sigma.$$

Thus far, we have a problem that, in principle, can be solved as follows: For each renewal set U, solve (10.27) and substitute in (10.28) to get the cost. Do this for all possible U and find a solution U^* by direct comparison. No small task!

Now consider a linear-programming problem such that with each state $i = 1, \ldots, M$ is associated a variable $y_1, \ldots, y_M$ together with one additional variable σ. The restrictions are

(10.29)   (a) $y_i \geq 0$ for all i, $\sigma \geq 0$,

$\qquad$ (b) $y_j \leq \delta(j, i_0)\sigma + \sum_i P(j \mid i)y_i$ for all j,

$\qquad$ (c) $l\sigma + \sum_i y_i = 1$,

and we are to minimize

$$-\sum_i g(i)y_i + C\sigma.$$

How is this problem related to (10.27)? One way is clear: For any U, the corresponding solution (r_i, σ) to (10.27) satisfies the restriction (10.29); in fact, it has the property that there is equality in (10.29a) except on U, where $r_i = 0$ and the right-hand side is nonnegative. Hence the set of feasible vectors for (10.29) includes everything that satisfies (10.27) for any U, so that the minimum in problem (10.29) is less than or equal to the minimum under (10.27). It is not so clear that a solution of (10.29) will lead back to (10.27); this we prove below.

We should comment here, before proceeding, on the vast improvement disguised in the transition from (10.27) to (10.29). The former presents formidable difficulties; the latter, being a standard linear-programming problem, would be considered almost trivial in this transistor age.

The identification of (10.29) with (10.27) results as follows: We cannot argue that every feasible vector of (10.29) corresponds to a solution of

(10.27) with some U, but we do know that in solving (10.29) we may restrict attention to the extreme points of the set of feasible vectors of (10.29). Hence it suffices to prove the following result.

Theorem 10.7. Let (y_i^0, σ^0) be an extreme point of the set of feasible vectors of (10.29). Then (y_i^0, σ^0) is a solution of (10.27) corresponding to the renewal set U, with $\bar{U}$ given by {states i such that $y_i^0 > 0$}.

PROOF. Let (y_i, σ) be any feasible vector, and define

$$z_j = y_j - \delta(j, i_0)\sigma - \sum_i P(j \mid i)y_i,$$
$$A = \{i \text{ such that } y_i = 0\},$$
$$B = \{i \text{ such that } z_i = 0\}.$$

If all states are in either A or B, then we have finished, since, defining $\bar{U} = \{i \text{ such that } y_i > 0\}$, we have

$$y_i = 0 \qquad\qquad \text{for } i \in U,$$
$$y_j = \delta(j, i_0)\sigma + \sum_i P(j \mid i)y_i \qquad \text{for } j \in \bar{U},$$

so that (y_i, σ) is a solution of (10.27). Assume, therefore, that there is at least one state not in $A \cup B$; then the set of equations for the M variables $v_1, \ldots, v_M$, where

$$v_i = 0 \qquad\qquad \text{for } i \in A,$$
$$v_j = \sum_i P(j \mid i)v_i \qquad \text{for } j \in B, j \notin A,$$

have a solution not identically 0. Take $\epsilon_j = \epsilon v_j$, with ϵ so small that for all j we have

$$|\epsilon_j| \leq \tfrac{1}{2}y_j, \qquad \left| \sum_i P(j \mid i)\epsilon_i - \epsilon_j \right| \leq \tfrac{1}{2}z_j.$$

This is possible since we have defined v_j to be zero when $y_j = 0$, and

$$\sum_i P(j \mid i)v_i - v_j$$

is 0 for $y_j > 0$ and $z_j = 0$. If $y_j = 0$ and $z_j = 0$, then

$$0 = \sum_{i \in A} P(j \mid i)y_i,$$

so that $P(j \mid i) = 0$ for all $j \in A$ and $i \in \bar{A}$. Hence if $j \in A \cap B$, then

$$\sum_i P(j \mid i)v_i - v_j$$

is automatically 0, and therefore

$$\sum_i P(j \mid i)v_i - v_j$$

is 0 whenever $z_j = 0$. Define the number a as $1 - \sum_i \epsilon_i$, so that

$a \geq \frac{1}{2}$ and $(2-a) \geq \frac{1}{2}$. Now we assert (see the following paragraph) that the two vectors

$$(ay_i + \epsilon_i, a\sigma), \qquad [(2-a)y_i - \epsilon_i, (2-a)\sigma]$$

are feasible. From this it follows, since their average is exactly (y_i, σ), that (y_i, σ) is not extreme, and thus for an extreme point it is necessary that every state be in $A \cup B$.

To demonstrate the foregoing assertion, we work first with $(ay_i + \epsilon_i, a\sigma)$. Condition (10.29c) is

$$l(a\sigma) + \sum_i (ay_i + \epsilon_i) = 1,$$

or

$$a\left(l\sigma + \sum_i y_i\right) + \sum_i \epsilon_i = 1,$$

which is automatically satisfied by the selection of a. The nonnegativity of $ay_i + \epsilon_i$ follows from $|\epsilon_i| \leq \frac{1}{2}y_i$. The restriction (10.29b) becomes

$$ay_j + \epsilon_j \leq a\sigma\delta(j, i_0) + a\sum_i P(j \mid i)y_i + \sum_i P(j \mid i)\epsilon_i,$$

or

$$\epsilon_j - \sum_i P(j \mid i)\epsilon_i \leq az_j,$$

which is satisfied through the inequality $\left|\epsilon_j - \sum_i P(j \mid i)\epsilon_i\right| \leq \frac{1}{2}z_j$. Therefore feasibility follows.

The same argument yields the feasibility of $[(2-a)y_i - \epsilon_i, (2-a)\sigma]$. The only thing left to check is that $(y_i, \sigma) \neq (ay_i + \epsilon_i, a\sigma)$. Equality holds only if $a = 1$, but then the ϵ_i must be identically 0, contradicting our construction.

10.17 A DUALITY RELATION

For a concluding observation, suppose we consider the linear-programming problem dual to (10.29). Thus we introduce $M + 1$ variables $x_1, \ldots, x_M, -\alpha$, and the dual is

$$(a) \quad x_i \geq 0, i = 1, \ldots, M, \quad \alpha \text{ unrestricted,}$$

$$(10.30) \quad (b) \quad x_i \geq g(i) + \alpha + \sum_{j=1}^M x_j P(j \mid i), \qquad i = 1, \ldots, M,$$

$$(c) \quad -x_{i_0} - l\alpha \geq -C;$$

we are to maximize α.

Slightly rephrased, this is the following: For fixed α, minimize x_{i_0} under the restrictions (10.30a) and (10.30b). Let the value of this minimum be $\phi(\alpha)$. Then find the largest α such that $C - l\alpha \geq \phi(\alpha)$; in other words, if α^* is the largest solution of $\beta(\alpha) = 0$, where $\beta(\alpha) = \phi(\alpha) + l\alpha - C$, then the solution of the dual problem is given by minimizing x_{i_0} under the restrictions (10.30a) and (10.30b) is *exactly the linear-programming problem* of solving the family of entrance-fee problems with $f(i) = -g(i) - \alpha$; here $\phi(\alpha) + l\alpha - C$ is exactly $E_{T_*}Z' + l\alpha - G(i_0)$, the payoff in the original nonreduced class of stopping-rule problems that were associated with the renewal problem by the method of cycle analysis.

Is the conclusion of the duality theorem recognizable? The answer is that we have

$$\alpha^* = \max \alpha = \min \left[-\sum g(i)y_i + \sigma C\right],$$

where the right-hand side is the minimum cost; comparing this with Theorem 10.5, we see that $\alpha^* = W_{U^*}$.

Our path has now gone through a full circle. Two seemingly different methods have led from a renewal problem to an associated class of stopping-rule problems. It is at once surprising and not surprising; surprising in the sense that reduction-using cycles, and the much longer route of the direct approach plus linear programming, plus duality, reach the same destination; not surprising in that, unless one has made a mistake somewhere, in mathematics usually all roads lead to Rome.

MULTIPLE-CHOICE REVIEW PROBLEMS

1. A stopping-rule problem may best be described as a multistage decision problem in which

    (a) after a specified number n of stages, one must stop,
    (b) after each stage, one is allowed to choose between continuing or stopping,
    (c) after some stages, one is allowed to choose between continuing or stopping,
    (d) all of the above are false.

2. The term "Markov property" refers most specifically to

    (a) the property that decision making depends only on the present state,
    (b) the fact that the transitions from state to state are not deterministic,
    (c) random devices of such a sort that their past history does not affect their future,
    (d) none of the above.

3. A stopping-rule problem may be solved by

    (a) reducing the problem to a problem involving a functional equation and then solving the functional equation by linear programming,
    (b) reducing it to an equivalent problem with $F(i) = 0$ and then formulating this as a linear-programming problem,
    (c) finding a related problem with $f(i) = 0$ and solving this latter problem by linear programming,
    (d) sheer inspiration, that is, none of the above.

4. One way in which a binary-decision renewal problem differs from a stopping rule problem is that

(a) the underlying probabilistic device is different,
(b) one wishes to stop so as to maximize expected payoff per unit time,
(c) it uses a different notation for essentially the same situation,
(d) the process continues *ad infinitum* instead of stopping.

5. A binary-decision renewal problem is similar to a stopping-rule problem in

(a) that the solution to both is a set of states,
(b) that one maximizes total payoff and the other minimizes total cost,
(c) that in both of them one has the option of stopping or continuing,
(d) none of the above ways.

REFERENCES

1. Bellman, R., "The Theory of Dynamic Programming," Chapter 11 in E. F. Beckenbach (editor), *Modern Mathematics for the Engineer, First Series*, McGraw-Hill Book Company, New York, 1956.
2. Chow, Y. S., and H. E. Robbins, "A Martingale System Theorem and Applications," pp. 93–104 in J. Neyman (editor), *Proceedings of the Fourth Berkeley Symposium on Mathematical Statistics and Probability* **1**, University of California Press, Berkeley and Los Angeles, Calif., 1961.
3. Derman, C., "Optimal Replacement Rules When Changes of State Are Markovian," Chapter X in R. Bellman (editor), *Mathematical Optimization Techniques*, University of California Press, Berkeley and Los Angeles, Calif., 1963.
4. Derman, C., "On Sequential Decisions and Markov Chains," to be published.
5. Derman, C., and J. Sacks, "Replacement of Periodically Inspected Equipment," *Naval Research Logist. Quart.* **7** (1960), 597–608.
6. Doob, J. L., *Stochastic Processes*, John Wiley and Sons, New York, 1953.
7. Feller, W., *An Introduction to Probability Theory and Its Applications*, John Wiley and Sons, New York, 1950.
8. Gumbel, H., and R. J. Frisbee, *Repair Versus Replacement*, Tech. Memorandum PMR-TM-GIR, Naval Air Missile Test Center, Point Mugu, Calif., 1961.
9. Jorgenson, D. W., and R. Radner, *Optimal Replacement and Inspection of Stochastically Failing Equipment*, P-2074, The RAND Corporation, Santa Monica, Calif., 1960.
10. Kemeny, J. G., and J. L. Snell, "Semi-Martingales on Markov Chains," *Amer. Math. Stat.* **29** (1958), 143–154.
11. Klein, M. *Inspection-Maintenance-Replacement Schedule under Markovian Deterioration*, Tech. Report 14, Statistical Engineering Group, Columbia University, New York.
12. MacQueen, J. B., *Sequences of Independent Time Variable Games*, Working Paper 4, Western Management Science Institute, The University of California, Los Angeles, 1962.
13. MacQueen, J. B., and R. G. Miller, "Optimal Persistence Policies," *J. Op. Res.* **8** (1960), 312–380.
14. Snell, J. L., "Applications of Martingale System Theorems," *Trans. Amer. Math. Soc.* **73** (1952), 293–312.

CHAPTER 11

Combinatorial Algebra of Matrix Games and Linear Programs

ALBERT W. TUCKER

11.1 INTRODUCTION

It was observed by John von Neumann, and substantiated by others (see [2,7]), that the problem of solving a matrix game and that of solving a feasible pair of dual linear programs are essentially the same. The common structure of the two problems is set forth here in terms of a "combinatorial" linear algebra, mainly by examples.

11.2 GAME EXAMPLE

Players I and II choose independently a row and a column, respectively, of the matrix

(11.1)
$$\begin{bmatrix} 2 & 3 & 7 \\ 6 & 5 & 4 \end{bmatrix},$$

and then player II pays player I an amount equal to the entry in the chosen row and chosen column. This is a small-scale example of a *matrix game* (or, in the technical terminology of the theory of games, a two-person zero-sum game in normal form). It is known by the celebrated minimax theorem, first proved in 1928 by John von Neumann, that there must exist a (unique) number w, nonnegative numbers x_1, x_2, and nonnegative y_3, y_4, y_5 such that

(11.2)
$$2x_1 + 6x_2 \geq w,$$
$$3x_1 + 5x_2 \geq w,$$
$$7x_1 + 4x_2 \geq w,$$
$$x_1 + x_2 = 1,$$

and

$$2y_3 + 3y_4 + 7y_5 \leq w,$$

(11.3)
$$6y_3 + 5y_4 + 4y_5 \leq w,$$

$$y_3 + y_4 + y_5 = 1.$$

Note that the left-hand side of each inequality in (11.2) is the inner product of a column of (11.1) with x_1, x_2, and that the left-hand side of each inequality in (11.3) is the inner product of a row of (11.1) with y_3, y_4, y_5. The two x's (nonnegative probabilities with sum equal 1) constitute a *mixed strategy* by which player I achieves an expected gain of at least w, and the three y's (nonnegative probabilities with sum equal 1) constitute a *mixed strategy* by which player II ensures himself against an expected loss of any more than w. Such mixed strategies are termed *optimal*; the number is the (equilibrium) *value* of the matrix game.

Our aim is to "solve" the matrix game determined by (11.1), that is, to find $x_1 \geq 0$, $x_2 \geq 0$, $y_3 \geq 0$, $y_4 \geq 0$, $y_5 \geq 0$, and w satisfying (11.2) and (11.3). To this end, we form the system of linear equations

(11.4)
$$x_3 = 2x_1 + 6x_2 - u,$$
$$x_4 = 3x_1 + 5x_2 - u,$$
$$x_5 = 7x_1 + 4x_2 - u,$$
$$1 = x_1 + x_2,$$

so that $x_3 \geq 0$, $x_4 \geq 0$, $x_5 \geq 0$ correspond to the inequalities in (11.2) with u substituted for w, and we form the system of linear equations

(11.5)
$$-y_1 = 2y_3 + 3y_4 + 7y_5 - v,$$
$$-y_2 = 6y_3 + 5y_4 + 4y_5 - v,$$
$$-1 = -y_3 - y_4 - y_5,$$

so that $-y_1 \leq 0$, $-y_2 \leq 0$ (that is, $y_1 \geq 0$, $y_2 \geq 0$) correspond to the inequalities in (11.3) with v substituted for w. We seek a solution x_1, x_2, x_3, x_4, x_5, u of (11.4) with all five x's ≥ 0, termed a *feasible* solution of (11.4), and a solution y_1, y_2, y_3, y_4, y_5, v of (11.5) with all five y's ≥ 0, termed a *feasible* solution of (11.5), which are such that $u = v$ ($= w$).

In the next sections it will develop that

$$x_1 = 1/5, \quad x_2 = 4/5, \quad x_3 = 3/5, \quad x_4 = 0, \quad x_5 = 0, \quad u = 23/5$$

and

$$y_1 = 0, \quad y_2 = 0, \quad y_3 = 0, \quad y_4 = 3/5, \quad y_5 = 2/5, \quad v = 23/5$$

are feasible solutions of (11.4) and (11.5), respectively, with $u = v$. Then

$$x_1 = 1/5, \quad x_2 = 4/5, \quad y_3 = 0, \quad y_4 = 3/5, \quad y_5 = 2/5, \quad w = 23/5$$

satisfy (11.2) and (11.3). Hence 1/5, 4/5 and 0, 3/5, 2/5 are optimal mixed strategies for players I and II, respectively, and 23/5 is the value of the game.

11.3 KEY EQUATION AND BASIC SOLUTIONS

For any solutions of (11.4) and (11.5), we have

$$
\begin{aligned}
x_1(-y_1) &+ x_2(-y_2) + u(-1) \\
&= x_1(2y_3 + 3y_4 + 7y_5 - v) \\
&\quad + x_2(6y_3 + 5y_4 + 4y_5 - v) + u(-y_3 - y_4 - y_5) \\
&= (2x_1 + 6x_2 - u)y_3 + (3x_1 + 5x_2 - u)y_4 \\
&\quad + (7x_1 + 4x_2 - u)y_5 + (x_1 + x_2)(-v) \\
&= x_3y_3 + x_4y_4 + x_5y_5 + 1(-v).
\end{aligned}
$$

That is, the "key equation"

$$(11.6) \qquad v - u = x_1y_1 + x_2y_2 + x_3y_3 + x_4y_4 + x_5y_5$$

holds for any solutions of (11.4) and (11.5). For feasible solutions with $u = v$, the right-hand side of (11.6) consists of nonnegative terms with sum 0. This requires each term x_iy_i to be 0, which implies $x_i = 0$ or $y_i = 0$ (or both). Since the system (11.4) has two "degrees of freedom" (four equations in six unknowns), and the system (11.5) has three "degrees of freedom" (three equations in six unknowns), we look for a feasible solution of the system (11.4) that has two of its x's = 0 and a feasible solution of the system (11.5) that has the complementary three of its y's = 0.

A *basic* solution of the system (11.4) is defined to be a solution (if existent and unique) of a system of six equations in six unknowns obtained by adjoining to the system (11.4) a pair of equations $x_g = 0$, $x_h = 0$, where $\{g, h\}$ is any two-element subset of $\{1, 2, 3, 4, 5\}$. Similarly, a *basic* solution of the system (11.5) is defined to be a solution (if existent and unique) of a system of six equations in six unknowns obtained by adjoining to the system (11.5) a triple of equations $y_i = 0$, $y_j = 0$, $y_k = 0$, where $\{i, j, k\}$ is any three-element subset of $\{1, 2, 3, 4, 5\}$. Basic solutions of the two systems are *complementary* if $\{g, h\}$ and $\{i, j, k\}$ are complementary subsets of $\{1, 2, 3, 4, 5\}$; clearly there are at most 5!/(2! 3!), or 10, pairs of complementary basic solutions. To solve our game, we seek complementary basic *feasible* solutions.

For example, the complementary basic solutions for the case $g = 3$, $h = 5$; $i = 1$, $j = 2$, $k = 4$ can be determined as follows. Solve the system (11.4) for x_1, x_2, x_4, and u in terms of x_3, x_5 to obtain (by some elementary algebra)

(11.7)
$$x_1 = -\tfrac{1}{7}x_3 + \tfrac{1}{7}x_5 + \tfrac{2}{7},$$
$$x_2 = \tfrac{1}{7}x_3 - \tfrac{1}{7}x_5 + \tfrac{5}{7},$$
$$x_4 = \tfrac{5}{7}x_3 + \tfrac{2}{7}x_5 - \tfrac{3}{7},$$
$$u = -\tfrac{3}{7}x_3 - \tfrac{4}{7}x_5 + \tfrac{34}{7}.$$

Now, adjoining $x_3 = 0$, $x_5 = 0$, read off the (nonfeasible) basic solution

(11.8) $x_1 = 2/7$, $x_2 = 5/7$, $x_3 = 0$, $x_4 = -3/7$, $x_5 = 0$, $u = 34/7$.

Similarly, solve the system (11.5) for $-y_3$, $-y_5$, and v in terms of y_1, y_2, y_4, to obtain

(11.9)
$$-y_3 = -\tfrac{1}{7}y_1 + \tfrac{1}{7}y_2 + \tfrac{5}{7}y_4 - \tfrac{3}{7},$$
$$-y_5 = \tfrac{1}{7}y_1 - \tfrac{1}{7}y_2 + \tfrac{2}{7}y_4 - \tfrac{4}{7},$$
$$v = \tfrac{2}{7}y_1 + \tfrac{5}{7}y_2 - \tfrac{3}{7}y_4 + \tfrac{34}{7}.$$

Now, adjoining $y_1 = 0$, $y_2 = 0$, $y_4 = 0$, read off the (feasible) basic solution

(11.10) $y_1 = 0$, $y_2 = 0$, $y_3 = 3/7$, $y_4 = 0$, $y_5 = 4/7$, $v = 34/7$.

For our game example, there are nine pairs of complementary basic solutions, as listed in Table 11.1. The case $g = 1$, $h = 2$, $i = 3$, $j = 4$, $k = 5$ does not occur because $x_1 = 0$, $x_2 = 0$ is inconsistent with $x_1 + x_2 = 1$ in (11.4) and because $y_3 = 0$, $y_4 = 0$, $y_5 = 0$ is inconsistent with $y_3 + y_4 + y_5 = 1$ in (11.5).

TABLE 11.1
Complementary Basic Solutions for Game Example

gh	u	$x_1,$	$x_2,$	$x_3,$	$x_4,$	x_5	Feasible?	ijk	v	$y_1,$	$y_2,$	$y_3,$	$y_4,$	y_5	Feasible?
13	6	0,	1,	0,	-1,	-2	No	245	6	4,	0,	1,	0,	0	Yes
14	5	0,	1,	1,	0,	-1	No	235	5	2,	0,	0,	1,	0	Yes
15	4	0,	1,	2,	1,	0	Yes	234	4	-3,	0,	0,	0,	1	No
23	2	1,	0,	0,	1,	5	Yes	145	2	0,	-4,	1,	0,	0	No
24	3	1,	0,	-1,	0,	4	No	135	3	0,	-2,	0,	1,	0	No
25	7	1,	0,	-5,	-4,	0	No	134	7	0,	3,	0,	0,	1	Yes
34	4	1/2,	1/2,	0,	0,	3/2	Yes	125	4	0,	0,	-1,	2,	0	No
35	34/7	2/7,	5/7,	0,	$-3/7$,	0	No	124	34/7	0,	0,	3/7,	0,	4/7	Yes
45	23/5	1/5,	4/5,	3/5,	0,	0	Yes	123	23/5	0,	0,	0,	3/5,	2/5	Yes

In the last line of Table 11.1 (that is, the case $g = 4$, $h = 5$; $i = 1$, $j = 2$, $k = 3$) the complementary basic solutions are both feasible. Hence the value w of our matrix game is $23/5$, and optimal mixed strategies for players I and II are, respectively,

$$x_1 = 1/5, \quad x_2 = 4/5, \quad \text{and} \quad y_3 = 0, \quad y_4 = 3/5, \quad y_5 = 2/5.$$

II.4 SCHEMATIC REPRESENTATION

The "schema"

$$(11.11)$$

	y_3	y_4	y_5	$-v$	
x_1	2	3	7	1	$= -y_1$
x_2	6	5	4	1	$= -y_2$
u	-1	-1	-1	0	$= -1$
	$= x_3$	$= x_4$	$= x_5$	$= 1$	

exhibits the linear-equation systems (11.4) and (11.5) simultaneously and compactly. The system (11.4) is the *column system* of (11.11): each equation of (11.4) is obtained by forming the inner product of a coefficient column of (11.11) with the symbol column at the left margin and then equating this inner product to the marginal symbol below that coefficient column. The system (11.5) is the *row system* of (11.11): each equation of (11.5) is obtained by forming the inner product of a coefficient row of (11.11) with the symbol row at the top margin and then equating this inner product to the marginal symbol at the right of that coefficient row. Note that the coefficient matrix of schema (11.11) is the game matrix (11.1) bordered on the right by a column of 1's and bordered on the bottom by a row of -1's, with a 0 entry in the lower right-hand corner. Furthermore, note that the key equation (11.6) is obtained by equating the inner product of the symbol column at the left-hand margin of (11.11) with the symbol column at the right-hand margin to the inner product of the symbol row at the bottom margin with the symbol row at the top margin, that is,

$$x_1(-y_1) + x_2(-y_2) + u(-1) = x_3y_3 + x_4y_4 + x_5y_5 + 1(-v).$$

Similarly, the schema

$$(11.12)$$

	y_1	y_2	y_4	1	
x_3	$-1/7$	$1/7$	$5/7$	$-3/7$	$= -y_3$
x_5	$1/7$	$-1/7$	$2/7$	$-4/7$	$= -y_5$
1	$2/7$	$5/7$	$-3/7$	$34/7$	$= v$
	$= x_1$	$= x_2$	$= x_4$	$= u$	

exhibits the linear-equation systems (11.7) and (11.9). The schema (11.12) is *equivalent* to the schema (11.11) in the sense that any solution of the column or row system of one schema is a solution of the column or row system, respectively, of the other schema. From the schema (11.12), the complementary basic solutions (11.8) and (11.10) can be read off: take $x_3 = 0$, $x_5 = 0$ and thus get $x_1 = 2/7$, $x_2 = 5/7$, $x_4 = -3/7$, $u = 34/7$; take $y_1 = 0$, $y_2 = 0$, $y_4 = 0$ and thus get $y_3 = 3/7$, $y_5 = 4/7$, $v = 34/7$.

Table 11.2 exhibits all schemata of the form

$$
\begin{array}{c|ccc|c}
 & y_i & y_j & y_k & 1 \\
\hline
x_g & & & & = -y_g \\
x_h & & & & = -y_h \qquad (g < h, i < j < k) \\
1 & & & & = v \\
\hline
 & = x_i & = x_j & = x_k & = u
\end{array}
$$

(11.13)

that are equivalent to the schema (11.11); each schema in Table 11.2 corresponds to a pair of complementary basic solutions in Table 11.1. The two digits at the upper left-hand corner of each schema in Table 11.2 are the values of g and h that pertain to the schema. These digits are underlined if the schema is "column-system basic-feasible," that is, if the basic solution of the column system given by $x_g = 0$, $x_h = 0$ is feasible; these digits have a line above them if the schema is "row-system basic-feasible," that is, if the basic solution of the row system given by $y_i = 0$, $y_j = 0$, $y_k = 0$ is feasible. The schema $\overline{45}$ at the lower right in Table 11.2 is basic-feasible both ways. It yields complementary basic feasible solutions

$$x_1 = 1/5, \ x_2 = 4/5, \ x_3 = 3/5, \ x_4 = 0, \quad x_5 = 0, \quad u = 23/5,$$
$$y_1 = 0, \quad y_2 = 0, \quad y_3 = 0, \quad y_4 = 3/5, \ y_5 = 2/5, \ v = 23/5,$$

which solve our game problem. To check this, we substitute these numerical results back in schema (11.11), as follows:

$$
\begin{array}{c|cccc|c}
 & 0 & 3/5 & 2/5 & -23/5 & \\
\hline
1/5 & 2 & 3 & 7 & 1 & = 0 \\
4/5 & 6 & 5 & 4 & 1 & = 0 \\
23/5 & -1 & -1 & -1 & 0 & = -1 \\
\hline
 & = 3/5 & = 0 & = 0 & = 1 &
\end{array}
$$

That is, the value of the game is 23/5, and optimal mixed strategies for players I and II are 1/5, 4/5 and 0, 3/5, 2/5, respectively.

TABLE 11.2
Equivalent Schemata

$\overline{13}$

	y_2	y_4	y_5	1	
x_1	-1	2	7	-4	$= -y_1$
x_3	0	1	1	-1	$= -y_3$
1	1	-1	-2	6	$= v$
	$= x_2$	$= x_4$	$= x_5$	$= u$	

$\overline{14}$

	y_2	y_3	y_5	1	
x_1	-1	-2	5	-2	$= -y_1$
x_4	0	1	1	-1	$= -y_4$
1	1	1	-1	5	$= v$
	$= x_2$	$= x_3$	$= x_5$	$= u$	

$\overline{15}$

	y_2	y_3	y_4	1	
x_1	-1	-7	-5	3	$= -y_1$
x_5	0	1	1	-1	$= -y_5$
1	1	2	1	4	$= v$
	$= x_2$	$= x_3$	$= x_4$	$= u$	

$\overline{23}$

	y_1	y_4	y_5	1	
x_2	-1	-2	-7	4	$= -y_2$
x_3	0	1	1	-1	$= -y_3$
1	1	1	5	2	$= v$
	$= x_1$	$= x_4$	$= x_5$	$= u$	

$\overline{24}$

	y_1	y_3	y_5	1	
x_2	-1	2	-5	2	$= -y_2$
x_4	0	1	1	-1	$= -y_4$
1	1	-1	4	3	$= v$
	$= x_1$	$= x_3$	$= x_5$	$= u$	

$\overline{25}$

	y_1	y_3	y_4	1	
x_2	-1	7	5	-3	$= -y_2$
x_5	0	1	1	-1	$= -y_5$
1	1	-5	-4	7	$= v$
	$= x_1$	$= x_3$	$= x_4$	$= u$	

$\overline{34}$

	y_1	y_2	y_5	1	
x_3	$-1/2$	$1/2$	$-5/2$	$2/2$	$= -y_3$
x_4	$1/2$	$-1/2$	$7/2$	$-4/2$	$= -y_4$
1	$1/2$	$1/2$	$3/2$	$8/2$	$= v$
	$= x_1$	$= x_2$	$= x_5$	$= u$	

$\overline{35}$

	y_1	y_2	y_4	1	
x_3	$-1/7$	$1/7$	$5/7$	$-3/7$	$= -y_3$
x_5	$1/7$	$-1/7$	$2/7$	$-4/7$	$= -y_5$
1	$2/7$	$5/7$	$-3/7$	$34/7$	$= v$
	$= x_1$	$= x_2$	$= x_4$	$= u$	

$\overline{45}$

	y_1	y_2	y_3	1	
x_4	$-1/5$	$1/5$	$7/5$	$-3/5$	$= -y_4$
x_5	$1/5$	$-1/5$	$-2/5$	$-2/5$	$= -y_5$
1	$1/5$	$4/5$	$3/5$	$23/5$	$= v$
	$= x_1$	$= x_2$	$= x_3$	$= u$	

11.5 EQUIVALENCE BY PIVOT STEPS; SIMPLEX METHOD

If $a \neq 0$, then the schemata

$$
\begin{array}{cc}
 & \eta^* \quad \eta \\
\begin{array}{c} x^* \\ x \end{array} &
\boxed{\begin{array}{cc} a & b \\ c & d \end{array}}
\begin{array}{c} = -y^* \\ = -y \end{array} \\
 & = \xi^* = \xi
\end{array}
\qquad \text{and} \qquad
\begin{array}{cc}
 & y^* \quad\quad \eta \\
\begin{array}{c} \xi^* \\ x \end{array} &
\boxed{\begin{array}{cc} a^{-1} & a^{-1}b \\ -ca^{-1} & d - ca^{-1}b \end{array}}
\begin{array}{c} = -\eta^* \\ = -y \end{array} \\
 & \;= x^* \quad\; = \xi
\end{array}
$$

are equivalent. This can be shown as follows. Solve the first column equation in the left-hand schema, $x^*a + xc = \xi^*$, for x^* to get the first column equation in the right-hand schema, $x^* = \xi^*a^{-1} - xca^{-1}$, and then substitute from this for x^* in the second column equation in the left-hand schema, $x^*b + xd = \xi$, to get the second column equation in the right-hand schema, $\xi^*a^{-1}b + x(d - ca^{-1}b) = \xi$. In addition, solve the first row equation in the left-hand schema, $a\eta^* + b\eta = -y^*$, for η^* to get the first row equation in the right-hand schema, $a^{-1}y^* + a^{-1}b\eta = -\eta^*$, and then substitute from this for η^* in the second row equation in the left-hand schema, $c\eta^* + d\eta = -y$, to get the second row equation in the left-hand schema, $-ca^{-1}y^* + (d - ca^{-1}b)\eta = -y$.

The transformation from the schema at the left above to the equivalent schema at the right is a *pivot step*, with the nonzero entry a as *pivot*. More generally, if $a \neq 0$, then the schemata

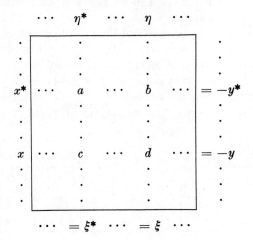

and

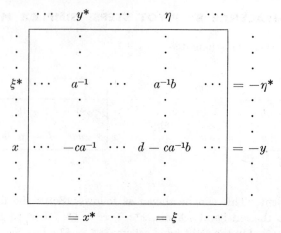

are equivalent. Here the marginal labels x^*, ξ^* and $-y^*$, η^* are replaced by ξ^*, x^* and $-\eta^*$, y^*, and all other labels, such as x, ξ and $-y$, η, are unchanged. This interchange of starred marginal labels (and no others) signalizes the fact that the transformation from the general schema on the left to that on the right is accomplished by solving the ξ^*-equation (column) and the y^*-equation (row) on the left for x^* and η^*, which requires $a \neq 0$, and then substituting for x^* and η^* in all the remaining column and row equations to eliminate x^* and η^* from them. The formal rules for the transformation of coefficients in the schema by the pivot step may be summarized as follows: Replace the *pivot* $a (\neq 0)$ by $1/a$; multiply each remaining entry b in a's row by $1/a$ and each remaining entry c in a's column by $-1/a$; add to every other entry d the product $(-c/a)b$ of the new entry $-c/a$ in d's row and a's column and the old entry b in d's column and a's row.

Now, returning to the schemata of the preceding section, we point out that each of the top six schemata in Table 11.2 can be obtained (except for possible rearrangement in order of rows and columns) from the initial schema (11.11) by two pivot steps, using as pivots one of the right-hand border 1's and one of the bottom border -1's. For example, we can pass from the schema (11.11) to the schema $\overline{13}$ of Table 11.2, pivoting on starred entries (and rearranging) as follows:

	y_3	y_4	y_5	$-v$	
x_1	2	3	7	1	$= -y_1$
x_2	6	5	4	1^*	$= -y_2$
u	-1	-1	-1	0	$= -1$
	$= x_3$	$= x_4$	$= x_5$	$= 1$	

$\rightarrow$

	y_3	y_4	y_5	y_2	
x_1	-4	-2	3	-1	$= -y_1$
1	6	5	4	1	$= v$
u	-1^*	-1	-1	0	$= -1$
	$= x_3$	$= x_4$	$= x_5$	$= x_2$	

$$
\begin{array}{c|cccc|l}
 & 1 & y_4 & y_5 & y_2 & \\
\hline
x_1 & -4 & 2 & 7 & -1 & = -y_1 \\
\rightarrow 1 & 6 & -1 & -2 & 1 & = v \\
x_3 & -1 & 1 & 1 & 0 & = -y_3 \\
\hline
 & = u & = x_4 & = x_5 & = x_2 &
\end{array}
\qquad \text{or} \qquad
\begin{array}{c|cccc|l}
\overline{13} & y_2 & y_4 & y_5 & 1 & \\
\hline
x_1 & -1 & 2 & 7 & -4 & = -y_1 \\
x_3 & 0 & 1 & 1 & -1 & = -y_3 \\
1 & 1 & -1 & -2 & 6 & = v \\
\hline
 & = x_2 & = x_4 & = x_5 & = u &
\end{array}
$$

Note that the entry 6 in this initial schema, unchanged by pivoting on the entry 1 in its row and on the entry -1 in its column, becomes the entry 6 at the lower right-hand corner of the final schema $\overline{13}$ above. One of the remaining two entries in the last column of schema $\overline{13}$ is the entry -1 used as pivot, and the other is the difference $-4 = 2 - 6$ between the entries 2 and 6 in the first column of the game matrix (11.1). Thus, since the entry 6 is maximal in its column of the game matrix (11.1), the schema $\overline{13}$ is row-system basic-feasible (as indicated by the line above). Similarly, since the entries 5 and 7 in the game matrix (11.1) are maximal in their columns, the schemata $\overline{14}$ and $\overline{25}$ (with entries 5 and 7, respectively, in the lower right-hand corner) are row-system basic-feasible. In the same way, since the entries 2 and 4 in the game matrix (11.1) are minimal in their rows, the schemata $\underline{23}$ and $\underline{15}$ (with entries 2 and 4, respectively, in the lower right-hand corner) are column-system basic-feasible. The schema 24 is neither row- nor column-system basic-feasible, because the entry 3 is neither minimal in its row nor maximal in its column of the game matrix (11.1).

All possible pivot steps from one schema of Table 11.2 to another are diagrammed in Fig. 11.1. The nine nodes or vertices of the network in Fig. 11.1 correspond to the nine schemata of Table 11.2, as indicated by their labels; the branches, or lines, of the network represent pivot steps from one schema to another (with change of just one digit in the label). Double-line branches with single arrows represent pivot steps from one schema of type $\overline{gh}$ to another of the same type, the direction of the arrows being toward the schema having the lesser entry in the lower right-hand corner. Double-line branches with double arrows represent pivot steps from one schema of type $\underline{gh}$ to another of the same type, the direction of the arrows being toward the schema having the greater entry in the lower right-hand corner.

We observe in Fig. 11.1 that there is at least one double-line path from any node $\overline{gh}$ or $\underline{gh}$ to the solution node $\overline{45}$. The construction of one such path will of itself solve our problem; the remainder of the network is not needed (except for better over-all understanding). Fortunately, especially for large-scale problems (in which the full network may be hopelessly intricate), there is a highly effective algorithm for computing such a path without knowing the remainder of the network. This algorithm is the essence of the ingenious simplex method devised in 1947 by George B. Dantzig, the major originator of linear programming. We illustrate the

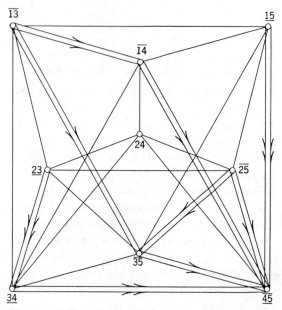

Fig. 11.1   Pivot steps between schemata of Table 11.2.

algorithm by means of the path $\overline{13} \to \overline{35} \to \overline{45}$, pivoting on starred entries as follows:

$\overline{13}$	y_2	y_4	y_5	1	
x_1	-1	2	7^*	-4	$= -y_1$
x_3	0	1	1	-1	$= -y_3 \to x_3$
1	1	-1	-2	6	$= v$
	$= x_2$	$= x_4$	$= x_5$	$= u$	

	y_2	y_4	y_1	1	
x_5	$-1/7$	$2/7$	$1/7$	$-4/7$	$= -y_5$
x_3	$1/7$	$5/7$	$-1/7$	$-3/7$	$= -y_3$ or
1	$5/7$	$-3/7$	$2/7$	$34/7$	$= v$
	$= x_2$	$= x_4$	$= x_1$	$= u$	

$\overline{35}$	y_1	y_2	y_4	1	
x_3	$-1/7$	$1/7$	$5/7^*$	$-3/7$	$= -y_3$
x_5	$1/7$	$-1/7$	$2/7$	$-4/7$	$= -y_5 \to x_5$
1	$2/7$	$5/7$	$-3/7$	$34/7$	$= v$
	$= x_1$	$= x_2$	$= x_4$	$= u$	

$\overline{45}$	y_1	y_2	y_3	1	
x_4	$-1/5$	$1/5$	$7/5$	$-3/5$	$= -y_4$
x_5	$1/5$	$-1/5$	$-2/5$	$-2/5$	$= -y_5$
1	$1/5$	$4/5$	$3/5$	$23/5$	$= v$
	$= x_1$	$= x_2$	$= x_3$	$= u$	

In each case, the pivot is a positive entry taken in a column (other than the last) with a negative entry at the bottom. Having chosen the third coefficient column in $\overline{13}$, we select the entry 7 as pivot, rather than the entry 1, because $-4/7 > -1/1$. In $\overline{35}$ there is just one coefficient column with bottom entry negative; we select 5/7 as pivot, rather than 2/7, because $(-3/7)/(5/7) > (-4/7)/(2/7)$. This selection rule for the pivot among the positive entries in a column (other than the last) with bottom entry negative makes sure that we move to a schema that remains row-system basic-feasible and has a lesser (or equal) entry in the lower right-hand corner. When we reach a schema with no bottom entries negative (except perhaps the last), the procedure terminates in a schema that is both row- and column-system basic-feasible, and so is optimal.

The procedure just illustrated is the "primal" form of the simplex algorithm, which generates a path of the single-arrow type. There is a "dual" (or negative-transposed) form of the simplex algorithm that generates a path of the double-arrow type.

11.6 LINEAR-PROGRAMMING EXAMPLES

A *linear program* is a problem of "optimizing" (that is, maximizing or minimizing) a linear "objective" function of many (real) variables subject to a system of linear "constraints," each of which is a linear inequality or linear equation. The following is a small-scale example: Maximize the objective function

(11.14) $$3\lambda + 2\mu,$$

subject to the constraints

(11.15) $$-\lambda - \mu + 2 \geq 0,$$
(11.16) $$\lambda + \mu - 1 \geq 0,$$
(11.17) $$\lambda + 2\mu - 2 \geq 0,$$
(11.18) $$-\lambda + 1 \geq 0,$$
(11.19) $$-6\lambda - 5\mu + 10 \geq 0.$$

This miniature linear program can readily be analyzed graphically. In a (λ, μ)-coordinate plane (see Fig. 11.2) we plot the "halfplanes" (1) to (5) corresponding to equations (11.15) to (11.19). In each case the halfplane (which includes its boundary line) is labeled by its number along its boundary line on the side of the line in which the halfplane lies; for example, the label (1) appears along the line $-\lambda - \mu + 2 = 0$ within the halfplane $-\lambda - \mu + 2 \geq 0$. The five constraints together determine the shaded region in Fig. 11.2 [extending to infinity between the parallel lines (1) and (2)]. The objective function (11.14) has contour (or level) lines parallel to the broken line (0) shown passing through the origin in Fig. 11.2, with values increasing in the direction of the arrow. The desired maximum appears to occur at the point of intersection of lines (4) and (5), namely at the point 45, with coordinates $\lambda = 1$, $\mu = 4/5$, where the objective function $3\lambda + 2\mu$ takes the value $23/5$.

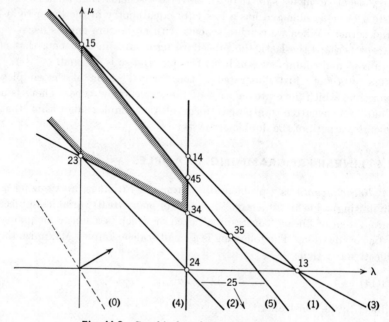

Fig. 11.2   Graphical analysis of linear program.

Another small-scale example, of a different sort, is the following: Minimize the objective function

(11.20) $$2y_1 - y_2 - 2y_3 + y_4 + 10y_5,$$

subject to the constraints

(11.21) $$-y_1 + y_2 + y_3 - y_4 - 6y_5 + 3 = 0,$$

(11.22) $$-y_1 + y_2 + 2y_3 \cdot -5y_5 + 2 = 0,$$

(11.23) $y_1 \geq 0,$ $y_2 \geq 0,$ $y_3 \geq 0,$ $y_4 \geq 0,$ $y_5 \geq 0.$

This linear program cannot readily be analyzed graphically because it involves five variables. It is related, however, to the previous example by the following joint schema (with each $y \geq 0$):

		y_1	y_2	y_3	y_4	y_5	1	
	λ	-1	1	1	-1	-6	3	$= 0$
(11.24)	μ	-1	1	2	0	-5	2	$= 0$
	1	2	-1	-2	1	10	0	$= \min$
		≥ 0	≥ 0	≥ 0	≥ 0	≥ 0	$= \max$	

The six coefficient columns of (11.24), along with the marginal material at the left and bottom, exhibit the constraints (11.15)–(11.19) and objective (11.14) of the previous (λ, μ)-program; the three coefficient rows of (11.24), along with the marginal material at the top and right, exhibit the constraints (11.21) and (11.22), and the objective (11.20), of the above y-program.

In the terminology of G. B. Dantzig, the row program of (11.24) is a *primal* program "in standard form," and the column program of (11.24) is its *dual* program. We shall see in the next section that these two companion programs can be solved simultaneously by finding "complementary basic feasible solutions" exactly parallel to those in our game example.

11.7 KEY EQUATION AND BASIC SOLUTIONS FOR LINEAR PROGRAMS

Let x_1, x_2, x_3, x_4, x_5 denote the linear functions of λ, μ constrained to be nonnegative by (11.15)–(11.19), and let u and v denote the objective functions (11.14) and (11.20). Then the two programs of (11.24) can be restated in terms of the schema

		y_1	y_2	y_3	y_4	y_5	1	
	λ	-1	1	1	-1	-6	3	$= 0$
(11.25)	μ	-1	1	2	0	-5	2	$= 0$
	1	2	-1	-2	1	10	0	$= v$
		$= x_1$	$= x_2$	$= x_3$	$= x_4$	$= x_5$	$= u$	

The row (or primal) program is to find a solution y_1, y_2, y_3, y_4, y_5, v of the row system of (11.25) such that each y is nonnegative and v is minimal. The column (or dual) program is to find a solution λ, μ, x_1, x_2, x_3, x_4, x_5, u

of the column system of (11.25) such that each x is nonnegative and u is maximal. A solution of the row system for which each y is nonnegative is said to be *feasible*; if also v is minimal, then the solution is *optimal*. Similarly, a solution of the column system for which each x is nonnegative is said to be *feasible*; if also u is maximal, the solution is *optimal*.

We point out that

$$x_1y_1 + x_2y_2 + x_3y_3 + x_4y_4 + x_5y_5 + u$$

becomes

$$\lambda(-y_1 + y_2 + y_3 - y_4 - 6y_5 + 3) + \mu(-y_1 + y_2 + 2y_3 - 5y_5 + 2)$$
$$+ (2y_1 - y_2 - 2y_3 + y_4 + 10y_5)$$

by substituting for x_1, x_2, x_3, x_4, x_5, u from the column system of (11.25). By the row system of (11.25), however, this reduces simply to v. Note that this "key equation"

(11.26) $\qquad x_1y_1 + x_2y_2 + x_3y_3 + x_4y_4 + x_5y_5 + u = v$

(in which λ, μ do *not* appear explicitly) is obtained by setting the inner product of the bottom marginal symbols with the top marginal symbols of the schema (11.25) equal to the inner product of the left-hand marginal symbols with the right-hand marginal symbols of the schema.

The "key equation" (11.26) holds for any solutions (feasible or not) of the column and row systems of the schema (11.25). If these solutions are both feasible, then

$$x_1y_1 + x_2y_2 + x_3y_3 + x_4y_4 + x_5y_5 \geq 0, \qquad \text{and so} \qquad u \leq v.$$

Therefore the value of u in any feasible solution of the column program is a lower bound for the value of v in any feasible solution of the row program, and hence a lower bound for the optimal value of v. Likewise, the value of v in any feasible solution of the row program is an upper bound for the value of u in any feasible solution of the column program, and hence an upper bound for the optimal value of u. The fundamental existence theorem of linear programming (not to be discussed here) asserts that, provided feasible solutions exist at all for *both* the row system and the column system of schema (11.25), there must exist optimal solutions for the row (or primal) program and the column (or dual) program such that $u = v$.

We examine, therefore, the problem of finding feasible solutions of the column and row systems of (11.25) such that

$$x_1y_1 + x_2y_2 + x_3y_3 + x_4y_4 + x_5y_5 = v - u = 0.$$

The sum of the nonnegative terms x_iy_i is zero if and only if each term $x_iy_i = 0$. However, $x_iy_i = 0$ if and only if $x_i = 0$ or $y_i = 0$ (or both).

The column system of (6.1) has *two* "degrees of freedom" since it consists of six equations in eight unknowns, and the row system has *three* "degrees of freedom" since it consists of three equations in six unknowns. Thus we look for a feasible solution of the column system that has two of its x's equal 0, and a feasible solution of the row system that has the complementary three of its y's equal 0.

A *basic* solution of the column system of (11.25) is defined to be a solution (if existent and unique) of a system of eight equations in eight unknowns obtained by adjoining to the column system a pair of equations $x_g = 0$, $x_h = 0$, where $\{g, h\}$ is a two-element subset of $\{1, 2, 3, 4, 5\}$. Similarly, a *basic* solution of the row system of (11.25) is defined to be a solution (if existent and unique) of a system of six equations in six unknowns obtained by adjoining to the row system a triple of equations $y_i = 0$, $y_j = 0$, $y_k = 0$, where $\{i, j, k\}$ is any three-element subset of $\{1, 2, 3, 4, 5\}$. Basic solutions of the two systems are *complementary* if $\{g, h\}$ and $\{i, j, k\}$ are complementary subsets of $\{1, 2, 3, 4, 5\}$. To solve simultaneously our two programs, we seek complementary basic *feasible* solutions.

For example, the complementary basic solutions for the case $g = 1$, $h = 3$; $i = 2$, $j = 4$, $k = 5$ can be determined as follows. Solve the column system of (11.25) for λ, μ, x_2, x_4, x_5, and u in terms of x_1, x_3, to obtain (by some elementary algebra)

$$\lambda = -2x_1 - x_3 + 2,$$
$$\mu = x_1 + x_3,$$
$$x_2 = -x_1 + 1,$$
(11.27)
$$x_4 = 2x_1 + x_3 - 1,$$
$$x_5 = 7x_1 + x_3 - 2,$$
$$u = -4x_1 - x_3 + 6.$$

Now, adjoining $x_1 = 0$, $x_3 = 0$, read off the (nonfeasible) basic solution

$$\lambda = 2, \mu = 0; \qquad x_1 = 0, x_2 = 1, x_3 = 0, x_4 = -1, x_5 = -2, u = 6.$$

Similarly, solve the row system of (11.25) for $-y_1$, $-y_3$, and v in terms of y_2, y_4, y_5 to obtain

$$-y_1 = -y_2 + 2y_4 + 7y_5 - 4,$$
(11.28)
$$-y_3 = y_4 + y_5 - 1,$$
$$v = y_2 - y_4 - 2y_5 + 6.$$

Now, adjoining $y_2 = 0$, $y_4 = 0$, $y_5 = 0$, read off the (feasible) basic solution

$$y_1 = 4, y_2 = 0, y_3 = 1, y_4 = 0, y_5 = 0, v = 6.$$

Altogether there are nine pairs of complementary basic solutions, as listed in Table 11.3. The case $g = 1$, $h = 2$ and $i = 3$, $j = 4$, $k = 5$ does not occur because the equations $-\lambda - \mu + 2 = x_1$, $\lambda + \mu - 1 = x_2$ are inconsistent with $x_1 = 0$, $x_2 = 0$.

In the last line of Table 11.3 (that is, the case $g = 4$, $h = 5$ and $i = 1$, $j = 2$, $k = 3$), the complementary basic solutions are both feasible. Hence the objective function v of the row (or primal) program attains its minimum of $23/5$ at $y_1 = 0$, $y_2 = 0$, $y_3 = 0$, $y_4 = 3/5$, $y_5 = 2/5$, and the

TABLE 11.3
Complementary Basic Solutions for Linear Programs

gh	$\lambda,$	μ	u	$x_1,$	$x_2,$	$x_3,$	$x_4,$	x_5	Feasible?	ijk	v	$y_1,$	$y_2,$	$y_3,$	$y_4,$	y_5	Feasible?
13	2,	0	6	0,	1,	0,	$-1,$	-2	No	245	6	4,	0,	1,	0,	0	Yes
14	1,	1	5	0,	1,	1,	0,	-1	No	235	5	2,	0,	0,	1,	0	Yes
15	0,	2	4	0,	1,	2,	1,	0	Yes	234	4	$-3,$	0,	0,	0,	1	No
23	0,	1	2	1,	0,	0,	1,	5	Yes	145	2	0,	$-4,$	1,	0,	0	No
24	1,	0	3	1,	0,	$-1,$	0,	4	No	135	3	0,	$-2,$	0,	1,	0	No
25	5,	-4	7	1,	0,	$-5,$	$-4,$	0	No	134	7	0,	3,	0,	0,	1	Yes
34	1,	1/2	4	1/2,	1/2,	0,	0,	3/2	Yes	125	4	0,	0,	$-1,$	2,	0	No
35	10/7,	2/7	34/7	2/7,	5/7,	0,	$-3/7,$	0	No	124	34/7	0,	0,	3/7,	0,	4/7	Yes
45	1,	4/5	23/5	1/5,	4/5,	3/5,	0,	0	Yes	123	23/5	0,	0,	0,	3/5,	2/5	Yes

objective function u of the column (or dual) program attains its maximum of $23/5$ at $\lambda = 1$, $\mu = 4/5$. The nine basic solutions of the column (or dual) program correspond to the nine points of intersection in Fig. 11.2, as indicated by the labels 13, 14, etc., in that figure.

Compare Table 11.3 with Table 11.1! They are identical except for the additional λ, μ column in Table 11.3. This is because our linear-programming examples have been "rigged" to demonstrate as vividly as possible that the problem of solving a matrix game and the problem of solving a feasible pair of linear programs are essentially the same.

11.8 CANONICAL FORMS

The systems (11.27) and (11.28), equivalent to the column and row systems of schema (11.25), can be exhibited simultaneously and compactly as the column and row systems of the schema

$$
(11.29) \quad
\begin{array}{c}
\begin{array}{cccccc}
0 & 0 & \dot{y}_2 & y_4 & y_5 & 1
\end{array} \\
\begin{array}{c} x_1 \\ x_3 \\ 1 \end{array}
\left|
\begin{array}{cccccc}
-2 & 1 & -1 & 2 & 7 & -4 \\
-1 & 1 & 0 & 1 & 1 & -1 \\
2 & 0 & 1 & -1 & -2 & 6
\end{array}
\right|
\begin{array}{c} = -y_1 \\ = -y_3 \\ = v \end{array} \\
\begin{array}{cccccc}
= \lambda & = \mu & = x_2 & = x_4 & = x_5 & = u
\end{array}
\end{array}
$$

This schema (except for rearrangement) can be obtained from schema (11.25) by two pivot steps, pivoting on starred entries as follows:

	y_1	y_2	y_3	y_4	y_5	1	
λ	-1^*	1	1	-1	-6	3	$= 0$
μ	-1	1	2	0	-5	2	$= 0$
1	2	-1	-2	1	10	0	$= v$

$$= x_1 \quad = x_2 \quad = x_3 \quad = x_4 \quad = x_5 \quad = u$$

$$\downarrow$$

	0	y_2	y_3	y_4	y_5	1	
x_1	-1	-1	-1	1	6	-3	$= -y_1$
μ	-1	0	1^*	1	1	-1	$= 0$
1	2	1	0	-1	-2	6	$= v$

$$= \lambda \quad = x_2 \quad = x_3 \quad = x_4 \quad = x_5 \quad = u$$

$$\downarrow$$

	0	y_2	0	y_4	y_5	1	
x_1	-2	-1	1	2	7	-4	$= -y_1$
x_3	-1	0	1	1	1	-1	$= -y_3$
1	2	1	0	-1	-2	6	$= v$

$$= \lambda \quad = x_2 \quad = \mu \quad = x_4 \quad = x_5 \quad = u$$

or

	0	0	y_2	y_4	y_5	1	
x_1	-2	1	-1	2	7	-4	$= -y_1$
x_3	-1	1	0	1	1	-1	$= -y_3$
1	2	0	1	-1	-2	6	$= v$

$$= \lambda \quad = \mu \quad = x_2 \quad = x_4 \quad = x_5 \quad = u$$

We observe that schema (11.29) can be partitioned into two schemata:

(11.30)

x_1	-2	1
x_3	-1	1
1	2	0

$$= \lambda \quad = \mu$$

and

(11.31)

	y_2	y_4	y_5	1	
x_1	-1	2	7	-4	$= -y_1$
x_3	0	1	1	-1	$= -y_3$
1	1	1	-2	6	$= v$

$$= x_2 \quad = x_4 \quad = x_5 \quad = u.$$

Schema (11.30) serves to express λ, μ in terms of x_1, x_3; schema (11.31) exhibits the result of eliminating λ, μ by way of x_1, x_3 from the column system of (11.29) and also the result of solving the row system of (11.29) for $-y_1$, $-y_3$, v in terms of y_2, y_4, y_5. Schema (11.31) is precisely the schema $\overline{13}$ at the beginning of Table 11.2.

Table 11.4 exhibits the nine schemata that accompany the nine schemata of Table 11.2 in exactly the way in which (11.30) accompanies (11.31). Each schema (ij) in Table 11.4 can be combined with the corresponding schema ij of Table 11.2 to make a schema of the form

	0	0	y_i	y_j	y_k	1	
x_g							$= -y_g$
x_h							$= -y_h$
1							$= v$

$$= \lambda \quad = \mu \quad = x_i \quad = x_j \quad = x_k \quad = u,$$

which expresses the result of reducing the column and row systems of schema (11.25) to a "canonical form" by a suitable pair of pivot steps (reflecting the classical "elimination" of Gauss and Jordan).

From schema (45) of Table 11.4 we obtain the basic values

$$\lambda = 1, \qquad \mu = 4/5,$$

which accompany the complementary basic *feasible* solutions

$$x_1 = 1/5, \quad x_2 = 4/5, \quad x_3 = 3/5, \quad x_4 = 0, \quad x_5 = 0, \quad u = 23/5$$

and

$$y_1 = 0, \quad y_2 = 0, \quad y_3 = 0, \quad y_4 = 3/5, \quad y_5 = 2/5, \quad v = 23/5,$$

read from schema $\overline{45}$ of Table 11.2. These constitute optimal solutions

TABLE 11.4
Schemata to Accompany Those of Table 11.2

(13)

x_1	-2	1
x_3	-1	1
1	2	0
	$= \lambda$	$= \mu$

(14)

x_1	0	-1
x_4	-1	1
1	1	1
	$= \lambda$	$= \mu$

(15)

x_1	5	-6
x_5	-1	1
1	0	2
	$= \lambda$	$= \mu$

(23)

x_2	2	-1
x_3	-1	1
1	0	1
	$= \lambda$	$= \mu$

(24)

x_2	0	1
x_4	-1	1
1	1	0
	$= \lambda$	$= \mu$

(25)

x_2	-5	6
x_5	-1	1
1	5	-4
	$= \lambda$	$= \mu$

(34)

x_3	$0/2$	$1/2$
x_4	$-2/2$	$1/2$
1	$2/2$	$1/2$
	$= \lambda$	$= \mu$

(35)

x_3	$-5/7$	$6/7$
x_5	$-2/7$	$1/7$
1	$10/7$	$2/7$
	$= \lambda$	$= \mu$

(45)

x_4	$-5/5$	$6/5$
x_5	$0/5$	$-1/5$
1	$5/5$	$4/5$
	$= \lambda$	$= \mu$

for the two linear programs with which we have been dealing. As explained in Sec. 11.7, a feasible value of u cannot exceed a feasible value of v. Therefore, if a feasible u equals a feasible v, optimality is necessarily achieved in both programs.

II.9 INVERSE BASIS PROCEDURE

Let schema (11.25) be rewritten as

$$
\begin{array}{c|ccccccc}
 & y_1 & y_2 & y_3 & y_4 & y_5 & 1 & -v \\
\hline
\lambda & -1 & 1 & 1 & -1 & -6 & 3 & 0 \\
\mu & -1 & 1 & 2 & 0 & -5 & 2 & 0 \\
1 & 2 & -1 & -2 & 1 & 10 & 0 & 1 \\
\end{array}
\begin{array}{c}
= 0 \\
= 0 \\
= 0 \\
\end{array}
$$
$$
= x_1 \quad = x_2 \quad = x_3 \quad = x_4 \quad = x_5 \quad = u \quad = 1
$$

(11.32)

The row system of (11.32) is the same as the row system of (11.25), with v on the right-hand side of the last equation of (11.25) transferred to $-v$ on the left-hand side of the last equation of (11.32); the column system of (11.32) is just the column system of (11.25), with the trivial equation $1 = 1$ adjoined. Now extract from (11.32) the schema

(11.33)

$$
\begin{array}{c|ccc}
\lambda & -1 & 1 & 0 \\
\mu & -1 & 2 & 0 \\
1 & 2 & -2 & 1 \\
\end{array}
$$
$$
= x_1 \quad = x_3 \quad = 1
$$

and compute its inverse schema

(11.34)

$$
\begin{array}{c|ccc}
x_1 & -2 & 1 & 0 \\
x_3 & -1 & 1 & 0 \\
1 & 2 & 0 & 1 \\
\end{array}
$$
$$
= \lambda \quad = \mu \quad = 1 .
$$

This computation can be made by pivot steps (with starred entries as pivots):

$$
\begin{array}{c|ccc}
\lambda & -1^* & 1 & 0 \\
\mu & -1 & 2 & 0 \\
1 & 2 & -2 & 1 \\
\end{array}
\to
\begin{array}{c|ccc}
x_1 & -1 & -1 & 0 \\
\mu & -1 & 1^* & 0 \\
1 & 2 & 0 & 1 \\
\end{array}
\to
\begin{array}{c|ccc}
x_1 & -2 & 1 & 0 \\
x_3 & -1 & 1 & 0 \\
1 & 2 & 0 & 1 \\
\end{array}
$$
$$
= x_1 \; = x_3 \; = 1 \qquad = \lambda \; = x_3 \; = 1 \qquad = \lambda \; = \mu \; = 1 .
$$

The three-by-three matrix of schema (11.34) is the inverse matrix of the three-by-three matrix of schema (11.33), that is,

$$
\begin{bmatrix} -1 & 1 & 0 \\ -1 & 2 & 0 \\ 2 & -2 & 1 \end{bmatrix}
\begin{bmatrix} -2 & 1 & 0 \\ -1 & 1 & 0 \\ 2 & 0 & 1 \end{bmatrix}
= \begin{bmatrix} 1 & 0 & 0 \\ 0 & 1 & 0 \\ 0 & 0 & 1 \end{bmatrix}.
$$

The schema (11.34) is seen to be schema (11.30), rewritten with the trivial equation $1 = 1$ adjoined.

We now observe that the product of the matrix of schema (11.34) and the matrix of schema (11.32), that is,

$$
\begin{bmatrix} -2 & 1 & 0 \\ -1 & 1 & 0 \\ 2 & 0 & 1 \end{bmatrix}
\begin{bmatrix} -1 & 1 & 1 & -1 & -6 & 3 & 0 \\ -1 & 1 & 2 & 0 & -5 & 2 & 0 \\ 2 & -1 & -2 & 1 & 10 & 0 & 1 \end{bmatrix}
$$

$$
= \begin{bmatrix} 1 & -1 & 0 & 2 & 7 & -4 & 0 \\ 0 & 0 & 1 & 1 & 1 & -1 & 0 \\ 0 & 1 & 0 & -1 & -2 & 6 & 1 \end{bmatrix},
$$

is the matrix of the schema

	y_1	y_2	y_3	y_4	y_5	1	$-v$	
x_1	1	-1	0	2	7	-4	0	$= 0$
(11.35) x_3	0	0	1	1	1	-1	0	$= 0$
1	0	1	0	-1	-2	6	1	$= 0$
	$= x_1$	$= x_2$	$= x_3$	$= x_4$	$= x_5$	$= u$	$= 1$	

This schema can be condensed to

	y_2	y_4	y_5	1	
x_1	-1	2	7	-4	$= -y_1$
(11.36) x_3	0	1	1	-1	$= -y_3$
1	1	-1	-2	6	$= v$
	$= x_2$	$= x_4$	$= x_5$	$= u$	

by eliminating the three trivial column equations $x_1 = x_1$, $x_3 = x_3$, $1 = 1$ from (11.35) and by transferring y_1, y_2, and $-v$ to the right-hand sides of the row equations of (11.35). Schema (11.36) is precisely the schema (11.31) previously obtained otherwise.

More generally, we may extract from schema (11.25) a schema

(11.37) λ 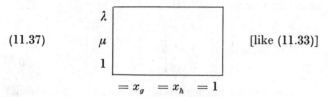   [like (11.33)]

(11.38)

[The case $g = 1$, $h = 2$ fails because then (11.37) is singular and possesses no inverse.] Now premultiply the matrix of schema (11.32) by the matrix of schema (11.38) to obtain the matrix of a schema

(11.39)

which may be condensed to

Schema (11.38) is schema (ij) of Table 11.4, with the trivial equation $1 = 1$ adjoined, and schema (11.40) is schema ij of Table 11.2. Through this "inverse basis" procedure, the data of Tables 11.4 and 11.2 can be computed from schema (11.32) by a combination of matrix inversion, as from (11.33) to (11.34), and matrix multiplication, as from (11.34) and

(11.32) to (11.35). [The uncondensed schema (11.39), corresponds to the "tableau" (in canonical form) ordinarily employed in linear-programming textbooks.]

11.10 GENERAL FORMULATIONS

The schema

(11.41)

$$
\begin{array}{c|ccccc|l}
 & y_1 & y_2 & \cdots & y_N & 1 & \\
\hline
\lambda_1 & a_{11} & a_{12} & \cdots & a_{1N} & b_1 & = 0 \\
\lambda_2 & a_{21} & a_{22} & \cdots & a_{2N} & b_2 & = 0 \\
\cdot & \cdot & \cdot & & \cdot & \cdot & \cdot \\
\cdot & \cdot & \cdot & & \cdot & \cdot & \cdot \\
\cdot & \cdot & \cdot & & \cdot & \cdot & \cdot \\
\lambda_M & a_{M1} & a_{M2} & \cdots & a_{MN} & b_M & = 0 \\
1 & c_1 & c_2 & \cdots & c_N & d & = v \\
\hline
 & = x_1 & = x_2 & \cdots & = x_N & = u &
\end{array}
$$

pertains to a general pair of linear programs "in standard form," just as schema (11.25) pertains to the particular pair of linear programs used as small-scale examples in Secs. 11.6 to 11.9. The primal program is to minimize v for nonnegative y's subject to the system of row equations in schema (11.41), and the dual program is to maximize u for nonnegative x's subject to the system of column equations in schema (11.41). Any solution of the column system and any solution of the row system are related by the equation

$$x_1 y_1 + x_2 y_2 + \cdots + x_N y_N + u = v,$$

in which the λ's (parameters, unrestricted in sign) do not appear.

Motivated by this "key equation" (just as in Sec. 11.7), we set ourselves the goal of finding "complementary basic feasible solutions." It has been proven that such solutions must exist *provided* both programs are feasible [that is, the row and column systems of schema (11.41) have solutions in which all the y's and x's are nonnegative]. Let m and n be the numbers of "degrees of freedom" in the column and row systems of schema (11.41): m equals the rank of the M by N matrix of the a's, and n equals $N - m$. Then complementary basic solutions (with λ's eliminated) have the form

$$x_{\bar{1}} = 0,\, x_{\bar{2}} = 0,\, \ldots,\, x_{\bar{m}} = 0,\, x_{\overline{m+1}} = \bar{c}_1,\, x_{\overline{m+2}} = \bar{c}_2,\, \ldots,\, x_{\overline{m+n}} = \bar{c}_n,\, u = \bar{d}$$

and

$$y_{\bar{1}} = -\bar{b}_1,\, y_{\bar{2}} = -\bar{b}_2,\, \ldots,\, y_{\bar{m}} = -\bar{b}_m,\, y_{\overline{m+1}} = 0,$$
$$y_{\overline{m+2}} = 0,\, \ldots,\, y_{\overline{m+n}} = 0,\, v = \bar{d},$$

where $\bar{1}, \bar{2}, \ldots, \overline{m+n}$ denote some permutation of the $N = m + n$ subscripts $1, 2, \ldots, m + n$. Corresponding to each such pair of complementary basic solutions there is a (reduced) schema "in canonical form" [like (11.31)],

(11.42)

	$y_{\overline{m+1}}$	$y_{\overline{m+2}}$	$\cdots$	$y_{\overline{m+n}}$	1	
$x_{\bar{1}}$	$\bar{a}_{11}$	$\bar{a}_{12}$	$\cdots$	$\bar{a}_{1n}$	$\bar{b}_1$	$= -y_{\bar{1}}$
$x_{\bar{2}}$	$\bar{a}_{21}$	$\bar{a}_{22}$	$\cdots$	$\bar{a}_{2n}$	$\bar{b}_2$	$= -y_{\bar{2}}$
$\cdot$	$\cdot$	$\cdot$		$\cdot$	$\cdot$	$\cdot$
$\cdot$	$\cdot$	$\cdot$		$\cdot$	$\cdot$	$\cdot$
$\cdot$	$\cdot$	$\cdot$		$\cdot$	$\cdot$	$\cdot$
$x_{\bar{m}}$	$\bar{a}_{m1}$	$\bar{a}_{m2}$	$\cdots$	$\bar{a}_{mn}$	$\bar{b}_m$	$= -y_{\bar{m}}$
1	$\bar{c}_1$	$\bar{c}_2$	$\cdots$	$\bar{c}_n$	d	$= v$
	$= x_{\overline{m+1}}$	$= x_{\overline{m+2}}$	$\cdots$	$= x_{\overline{m+n}}$	$= u$	

in which the row system is equivalent to the row system of schema (11.41), and the column system is equivalent to the column system of schema (11.41) with the λ's eliminated. If the $\bar{b}$'s are nonpositive, schema (11.42) is row-system basic-feasible; if the $\bar{c}$'s are nonnegative, schema (11.42) is column-system basic-feasible.

The Dantzig simplex method (in primal form) moves by a finite succession of pivot steps, as exemplified in Sec. 11.5, from one column-system basic-feasible schema (11.42) to another such schema in which d is less (in any case not greater), and so on, until a schema that is both row- and column-system basic-feasible is attained (usually by less than $m + n$ pivot steps). In this way, provided both programs are feasible, complementary basic feasible solutions are obtained that constitute optimal solutions for the primal and dual programs (just as with the examples treated in Secs. 11.6 to 11.9).

The schema

(11.43)

	y_{m+1}	y_{m+2}	$\cdots$	y_{m+n}	$-v$	
x_1	g_{11}	g_{12}	$\cdots$	g_{1n}	1	$= -y_1$
x_2	g_{21}	g_{22}	$\cdots$	g_{2n}	1	$= -y_2$
$\cdot$	$\cdot$	$\cdot$		$\cdot$	$\cdot$	$\cdot$
$\cdot$	$\cdot$	$\cdot$		$\cdot$	$\cdot$	$\cdot$
$\cdot$	$\cdot$	$\cdot$		$\cdot$	$\cdot$	$\cdot$
x_m	g_{m1}	g_{m2}	$\cdots$	g_{mn}	1	$= -y_m$
u	-1	-1	$\cdots$	-1	0	$= -1$
	$= x_{m+1}$	$= x_{m+2}$	$\cdots$	$= x_{m+n}$	1	

pertains to the problem of solving the general game specified by the m by n "payoff matrix" $\|g_{ij}\|$, just as schema (11.11) pertains to the particular game specified by (11.1). By using one of the border 1's and one of the border -1's as pivots (and then transferring the u-column to the last and the v-row to the last), we pass to a schema like (11.42). This schema can be made row-system basic-feasible by selecting the pivots 1 and -1 in the same row and column as a payoff entry g_{ij} that is greater than (or equal to) every other payoff entry in the same column. Then, by the Dantzig simplex method (in primal form), it is possible to move by pivot steps through a finite succession of row-system basic-feasible schemata with monotonically decreasing corner entries d until a schema is attained that is both row- and column-system basic-feasible. In this way, for any matrix game, complementary basic feasible solutions can be obtained that provide optimal strategies for the two players and the (equilibrium) value of the game.

The linear methods presented here are clearly combinatorial: Each pair of complementary basic solutions (and accompanying schema) corresponds to a partition of the key inner product

$$x_1y_1 + x_2y_2 + \cdots + x_{m+n}y_{m+n}$$

into a sum of m terms, in which the x's are equated to 0, and a complementary sum of n terms, in which the y's are equated to 0. The Dantzig simplex method is a combinatorial algorithm that employs simple pivot steps to construct a small part of a large finite network so as to get a path that is nonrepeating (with suitable precautions), leading from any one-sided feasible state to a two-sided feasible (therefore optimal) state—if such exists.

MULTIPLE-CHOICE REVIEW PROBLEMS

1. The two schemata

$$
\begin{array}{c}
\begin{array}{cc} & y_3 \quad\; y_4 \end{array} \\
\begin{array}{c} x_1 \\ x_2 \end{array}
\begin{array}{|cc|} \hline 2 & -1 \\ -4 & 3 \\ \hline \end{array}
\begin{array}{l} = -y_1 \\ = -y_2 \end{array} \\
\quad = x_3 \; = x_4
\end{array}
\quad \text{and} \quad
\begin{array}{c}
\begin{array}{cc} & y_1 \quad\;\; y_4 \end{array} \\
\begin{array}{c} x_3 \\ x_2 \end{array}
\begin{array}{|cc|} \hline 1/2 & -1/2 \\ 2 & k \\ \hline \end{array}
\begin{array}{l} = -y_3 \\ = -y_2 \end{array} \\
\quad = x_1 \; = x_4
\end{array}
$$

are equivalent if

(a) $k = 3/2$, (b) $k = 1$, (c) $k = 0$, (d) $k = -1$.

2. The game with payoff matrix

$$\begin{array}{cc} 2 & -1 \\ -4 & 3 \end{array}$$

has (minimax) *value* equal to

(a) 0.2,   (b) 0,   (c) −0.2,
(d) none of the above.

3. The linear program to maximize u for $x_1 \geq 0$, $x_2 \geq 0$, $x_3 \geq 0$, $x_4 \geq 0$ subject to the column equations in the schema

$$
\begin{array}{c c}
 & \begin{array}{ccc} y_3 & y_4 & 1 \end{array} \\
\begin{array}{c} x_1 \\ x_2 \\ 1 \end{array} &
\begin{array}{|ccc|}
\hline
1 & 0 & -2 \\
1 & -1 & -1 \\
-4 & 3 & 9 \\
\hline
\end{array}
\begin{array}{l} = -y_1 \\ = -y_2 \\ = v \end{array}
\end{array}
$$

$$= x_3 \;\; = x_4 \;\; = u$$

has as optimum value

(a) $u = 1$,   (b) $u = 4$,   (c) $u = 5$,   (d) $u = 6$.

4. The linear progam to minimize v for $y_1 \geq 0$, $y_2 \geq 0$, $y_3 \geq 0$, $y_4 \geq 0$ subject to the row equations in Problem 3 has its optimum at the basic solution given by

(a) $y_2 = 0$, $y_3 = 0$,
(b) $y_2 = 0$, $y_4 = 0$,
(c) $y_1 = 0$, $y_2 = 0$,
(d) $y_1 = 0$, $y_4 = 0$,

5. Let the (real) coefficients in the schema

$$
\begin{array}{c c}
 & \begin{array}{ccc} y_{m+1} & \cdots\cdots & y_{m+n} \end{array} \\
\begin{array}{c} x_1 \\ \cdot \\ \cdot \\ \cdot \\ \cdot \\ x_m \end{array} &
\begin{array}{|ccc|}
\hline
a_{11} & \cdots\cdots & a_{1n} \\
\cdot & \cdot & \cdot \\
\cdot & \cdot & \cdot \\
\cdot & \cdot & \cdot \\
\cdot & \cdot & \cdot \\
a_{m1} & \cdots\cdots & a_{mn} \\
\hline
\end{array}
\begin{array}{l} = -y_1 \\ \cdot \\ \cdot \\ \cdot \\ \cdot \\ = -y_m \end{array}
\end{array}
$$

$$= x_{m+1} \;\; \cdots\cdots \;\; = x_{m+n}$$

be such that the x-system possesses a solution $(x_1, \ldots, x_{m+n})$ in which each $x > 0$. Then the y-system possesses a solution $(y_1, \ldots, y_{m+n})$ in which

(a) each $y > 0$,
(b) each $y \geq 0$ and some $y > 0$,
(c) no $y \geq 0$,
(d) none of the above hold.

REFERENCES

Items 5, 8, 10, and 17 in the following selective list are quite elementary. Items 1, 3, 6, and 16 are not too technical. Items 4 and 9 are comprehensive textbooks, with extensive bibliographies. Items 2, 7, 11, 12, 13, and 14 are technical papers.

1. Bohnenblust, H. F., "The Theory of Games," Chapter 9 in *Modern Mathematics for the Engineer*, First Series, E. F. Beckenbach (editor), McGraw-Hill Book Company, New York, 1956, pp. 191–210.

2. Dantzig, G. B., "Constructive Proof of the Min-Max Theorem," *Pacific J. Math.* **6** (1956), 25–33.

3. Dantzig, G. B., "Formulating and Solving Linear Programs," Chapter 9 in *Modern Mathematics for the Engineer*, Second Series, E. F. Beckenbach (editor), McGraw-Hill Book Company, New York, 1961, pp. 213–227.

4. Dantzig, G. B., *Linear Programming and Extensions*, Princeton University Press, Princeton, New Jersey, 1963.

5. Glicksman, A. M., *An Introduction to Linear Programming and the Theory of Games*, John Wiley and Sons, New York, 1963.

6. Kemeny, J. G., A. Schleifer, J. L. Snell, and G. L. Thompson, *Finite Mathematics with Business Applications*, Prentice-Hall, Englewood Cliffs, New Jersey, 1962, pp. 364–455.

7. Koopmans, T. C. (editor), *Activity Analysis of Production and Allocation*, Cowles Commission Monograph **13**, John Wiley and Sons, New York, 1951. See Chapter XIX, D. Gale, H. W. Kuhn, and A. W. Tucker, "Linear Programming and the Theory of Games," p. 326; Chapter XX, G. B. Dantzig, "Equivalence of Programming and Game Problems," p. 331; Chapter XXII, R. Dorfman, "Application of the Simplex Method to a Game Theory Problem," p. 349.

8. Kuhn, H. W., and A. W. Tucker, "Games, Theory of" *Encyclopaedia Britannica*, 1956 (or later).

9. Luce, R. D., and H. Raiffa, *Games and Decisions*, John Wiley and Sons, New York, 1957.

10. Richardson, M., *Fundamentals of Mathematics*, The Macmillan Co., New York, revised edition, 1958, pp. 259–265 and 389–401.

11. Tucker, A. W., "A Combinatorial Equivalence of Matrices," *Combinatorial Analysis*, R. Bellman and M. Hall (editors), *Proceedings of Symposia in Applied Mathematics* **10**, American Mathematical Society, Providence, Rhode Island, 1960, pp. 129–140.

12. Tucker, A. W., "Solving a Matrix Game by Linear Programming," *IBM J. Research and Development* **4** (1960), 507–517.

13. Tucker, A. W., "Simplex Method and Theory," *Mathematical Optimization Techniques*, R. Bellman (editor), University of California Press, Berkeley, Calif., 1963, pp. 213–231.

14. Tucker, A. W., "Combinatorial Theory Underlying Linear Programs," *Recent Advances in Mathematical Programming*, R. L. Graves and P. Wolfe (editors), McGraw-Hill Book Co., New York, 1963, pp. 1–16.

15. Vajda, S., *The Theory of Games and Linear Programming*, John Wiley and Sons, New York, 1956.

16. Vajda, S., *Readings in Mathematical Programming*, John Wiley and Sons, New York, 1962.

17. Williams, J. D., *The Compleat Strategyst*, McGraw-Hill Book Company, New York, 1954.

CHAPTER 12

Network Flow Problems

EDWIN F. BECKENBACH

12.1 INTRODUCTION

The basic combinatorial problem considered in the first part of this chapter can be viewed as that of maximizing the rate of flow of material from one depot to another over a given railroad complex [8, 5]. Thus it is a problem of extremizing a linear function subject to linear inequality constraints [2], and as such it can be effectively treated by the general linear-programming methods developed in Chapter 11.

Alternatively, as we shall see in this chapter, a constructive proof [6] in the form of a geometric algorithm can be given for this and other related military and industrial problems in the broad field of operational analysis.

12.2 NETWORKS

Let there be given a finite set N of elements $p, q, \ldots, r$, and a subset A of the ordered pairs (p, q), $p \neq q$, of elements of N. The set $\{N; A\}$ is called a *directed network*. The elements of N are called the *nodes* of $\{N; A\}$, the elements of A are called the *arcs* of $\{N; A\}$, and the arc (p, q) is said to *extend from p to q*.

The nodes might, for example, be thought of as cities, and the arcs as railroad lines. In any case, we shall represent a node p by a circle in the plane, and an arc (p, q) by an arrow from p to q, as illustrated in the simple network of Fig. 12.1.

If, for a given sequence $p_1, p_2, \ldots, p_n$ of distinct nodes $\in N$, the arcs

(p_i, p_{i+1}), $i = 1, 2, \ldots, n - 1$, are all $\in N$, then the sequence

$$S: p_1, (p_1, p_2), p_2, \ldots, (p_{n-1}, p_n), p_n$$

of nodes and arcs is said to be a *directed path from p_1 to p_n* in $\{N; A\}$.
For example, in Fig. 12.1,

$$s, (s, p), p, (p, q), q, (q, r), r$$

is a directed path. Ordinarily, however, we would denote this directed path simply by $s\, p\, q\, r$.

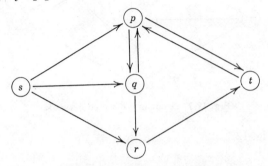

Fig. 12.1   A directed network.

If the sequence $p_1, p_2, \ldots, p_n$ of distinct nodes $\in N$ is such that for each i, $i = 1, 2, \ldots, n - 1$, either $(p_i, p_{i+1}) \in A$ or $(p_{i+1}, p_i) \in A$, and in each instance for which these arcs are both $\in A$ we specify a particular one of the two, then the resulting sequence of nodes and arcs is called a *path from p_1 to p_n* in $\{N; A\}$.
For example, in Fig. 12.1, the sequence

$$s, (s, r), r, (q, r), q$$

is a path. Here (s, r) is called a *forward arc*, and (q, r) is called a *backward arc*, for obvious reasons.

For a given $p \in N$, the set $A(p)$ (read "after p") is the set of nodes $q \in N$ such that $(p, q) \in A$. Similarly, $B(p)$ (read "before p") is the set of nodes $q \in N$ such that $(q, p) \in A$.

In Fig. 12.1, for example, we have

$$A(s) = \{p, q, r\}, \qquad A(p) = \{q, t\}, \qquad A(q) = \{p, r\}$$

and

$$B(s) = \phi \quad \text{(the null set)}, \qquad B(p) = \{s, q, t\}, \qquad B(q) = \{s, p\}.$$

12.3 CUTS

In the network $\{N; A\}$, let $P \subset N$; that is, let P be a subset of nodes $p \in N$. Furthermore, for $P \subset N$ and $Q \subset N$, let $(P, Q) \subset A$ denote the set of arcs $(p, q) \in A$ for which $p \in P$ and $q \in Q$.

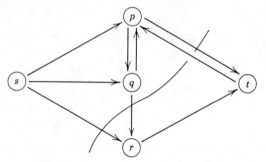

Fig. 12.2   A cut in a directed network.

Thus, in Fig. 12.1, if

$$P = \{s, p, q\} \qquad \text{and} \quad Q = \{s, r, t\},$$

then

$$(P, Q) = \{(s, r), (q, r), (p, t)\}.$$

By $\bar{P}$ we shall denote the complement of P in N, $\bar{P} = N - P$, so that $\bar{P}$ is the set of nodes $\in N$ but $\notin P$.

For example, in Fig. 12.1, if

$$P = \{s, p, q\}, \qquad \text{then} \quad \bar{P} = \{r, t\}$$

and

$$(P, \bar{P}) = \{(s, r), (q, r), (p, t)\}.$$

Since $(P, \bar{P})$ contains every arc extending from any node $\in P$ to any node $\in \bar{P}$, it follows in particular that every directed path extending from a node $\in P$ to a node $\in \bar{P}$ contains an arc $\in (P, \bar{P})$. For this reason, we say that $(P, \bar{P})$ is a *cut separating P from $\bar{P}$ in* $\{N; A\}$. Similarly, for any $p \in P$ and $q \in \bar{P}$, we say that $(P, \bar{P})$ is a *cut separating p from q in* $\{N; A\}$.

This is illustrated, for

$$P = \{s, p, q\}, \qquad \bar{P} = \{r, t\},$$

in Fig. 12.2. Note, however, in this figure, that $(t, p) \notin (P, \bar{P})$; we "cut" only forward arcs from nodes $\in P$ to nodes $\in \bar{P}$. Thus the cut $(P, \bar{P})$ "separates" P from $\bar{P}$, not $\bar{P}$ from P.

12.4 FLOWS

To each arc $(p, q) \in A$, let us now associate a nonnegative real number $c(p, q)$, and let us call this number the *capacity* of (p, q).

Adjoining the set C of capacities to the set N of nodes and the set A of directed arcs, we now have a *capacitated* directed network $\{N; A; C\}$, as illustrated in Fig. 12.3.

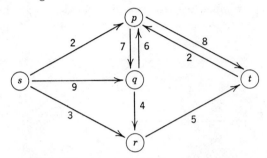

Fig. 12.3 A capacitated directed network.

Relative to our railroad illustration, the capacity function $c(p, q)$ might be thought of as the maximum amount of some commodity that can be conveyed from p to q over (p, q) in a unit of time.

The railroad example might also help in understanding the significance of the following definition of a *static flow in* $\{N; A; C\}$, *of value v, from a given node* $s \in N$ (called the *source*) *to a different given node* $t \in N$ (called the *sink*). Such a flow is defined to be a nonnegative real-valued function $f(p, q)$, with domain A, that has the following properties:

(12.1) $f(p, q) \leq c(p, q)$ for all $(p, q) \in A$,

$$\sum_{q \in A(p)} f(p, q) - \sum_{q \in B(p)} f(q, p) = v \qquad \text{for } p = s,$$

(12.2) $= 0 \qquad \text{for } p \neq s \text{ or } t,$

                                              $= -v \quad \text{for } p = t.$

The inequality (12.1) expresses the condition that the flow f in (p, q) does not exceed the capacity c of (p, q).

The sums

$$\sum_{q \in B(p)} f(q, p) \quad \text{and} \quad \sum_{q \in A(p)} f(p, q)$$

are called the *flow into* p and the *flow from* p, respectively. The difference

$$\sum_{q \in A(p)} f(p, q) - \sum_{q \in B(p)} f(q, p) \qquad .$$

is then called the *net flow from* p.

The conditions (12.2) indicate that the net flow from the source s is the value v of the flow, the net flow from the sink t is $-v$ (that is, the net flow *into t* is v), and the net flow from any *intermediate p* (that is, any $p \in N$ other than s or t) is 0. Thus, for any intermediate p, the flow into p and the flow out of p satisfy the *conservation equation*

$$\sum_{q \in B(p)} f(q, p) = \sum_{q \in A(p)} f(p, q).$$

Consider, for example, the situation illustrated in Fig. 12.4. Here the first numbers beside the arcs give the arc capacities in the capacitated

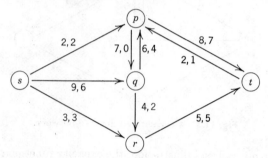

Fig. 12.4   A static flow in a capacitated directed network.

directed network $\{N; A; C\}$ of Fig. 12.3, and—as we might verify by checking the conditions (12.1) and (12.2)—the second numbers give a static flow of value 11 from s to t in $\{N; A; C\}$.

12.5   FLOWS AND CUTS

In a capacitated directed network $\{N; A; C\}$, the *capacity* $c(P, \bar{P})$ of a cut $(P, \bar{P})$ is defined by

$$c(P, \bar{P}) = \sum_{p \in P, q \in \bar{P}} c(p, q).$$

Thus when the cut illustrated in Fig. 12.2 is applied to the capacitated directed network of Fig. 12.3, the result shown in Fig. 12.5 is obtained. The capacity of the cut is

$$c(P, \bar{P}) = c(s, r) + c(q, r) + c(p, t) = 3 + 4 + 8 = 15.$$

Since, as shown in Sec. 12.3, every directed path extending from s to t must contain one of the arcs (s, r), (q, r), (p, t), it should be intuitively obvious that no static flow from s to t could be of value greater than 15. For the same reason, in fact, it would appear, and will be proved in Sec. 12.6, that in any capacitated directed network $\{N; A; C\}$ no static **flow** from one node s to another node t in $\{N; A; C\}$ could be of value greater than the least capacity of any cut separating s from t in $\{N; A; C\}$.

Now let us do a bit of experimenting with cuts $(P, \bar{P})$ separating s

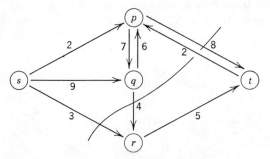

Fig. 12.5 A cut in a capacitated directed network.

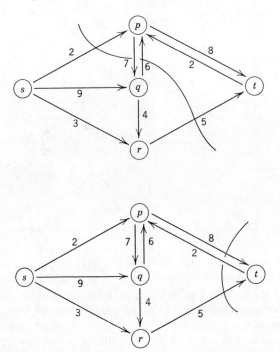

Fig. 12.6 Minimal cuts in a capacitated directed network.

from t in the network illustrated in Fig. 12.5. Which, if any, of the nodes p, q, r should be included with the source s in the set P to obtain a cut of least capacity separating s from t? Trial-and-error methods lead, in this instance, to two alternative *minimal* cuts (cuts of least capacity),

as indicated in Fig. 12.6. The capacity of the first of these cuts is
$2 + 6 + 5 = 13$, and of the second is $5 + 8 = 13$.

It might be noted that if, for example, the capacity of the arc (s, p)
were changed from 2 to 3, then there would be a unique minimal cut,
namely, the one with $\bar{P} = \{t\}$. Its capacity would still be 13.

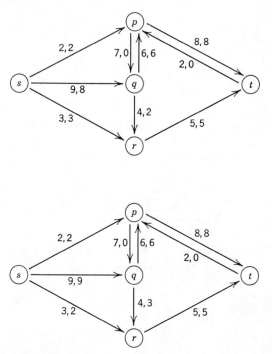

Fig. 12.7 Maximal flows in a capacitated directed network.

The value of the static flow illustrated in Fig. 12.4 is 11, which is less
than the capacity 13 of the minimal cut. It is natural to ask whether or
not there is a static flow of value equal to the capacity of the minimal cut.
A bit of experimentation leads to an affirmative answer, with two alternative *integer-valued maximal static flows* (static flows of greatest values,
with integer-valued flows in all arcs), as illustrated in Fig. 12.7. Intermediate between these two integer-valued maximal flows is a continuum
of real-valued maximal flows.

It might be noted that if, for example, the capacity of the arc (s, p)
were changed from 2 to 3, then there would be additional integer-valued
maximal static flows. The value of each maximal static flow would still
be 13.

12.6 BOUNDS ON CAPACITIES OF CUTS AND ON VALUES OF FLOWS

Let $\{N; A; C\}$ be a capacitated directed network, and let $s \in N$, $t \in N$, $s \neq t$.

We note first that there *is* a cut $(P, \bar{P})$ separating s from t in $\{N; A; C\}$, for the choice $P = \{s\}$ achieves this, yielding a cut of capacity

$$\sum_{q \in A(s)} c(s, q).$$

Similarly, the cut determined by $\bar{P} = \{t\}$ also separates s from t, and the capacity of this cut is

$$\sum_{q \in B(t)} c(q, t).$$

Since there can be only a finite number of cuts in $\{N; A; C\}$, it follows that there must be a minimizing cut in this network. Accordingly, if $(P_0, \bar{P}_0)$ is a minimizing cut, then we have

(12.3) $$0 \leq c(P_0, \bar{P}_0) \leq \min \left[\sum_{q \in A(s)} c(s, q), \sum_{q \in B(t)} c(q, t) \right],$$

in which the first inequality follows from the fact that $c(P_0, \bar{P}_0)$ is a sum (perhaps with no addends) of nonnegative addends.

Again, there *is* a static flow from s to t in $\{N; A; C\}$, for the choice $f(p, q) = 0$ for all $(p, q) \in A$ satisfies the defining conditions (12.1) and (12.2) for such a flow. The value of this flow is 0.

By (12.1) and the first equation in (12.2), the value v of any static flow f from s to t in $\{N; A; C\}$ satisfies

(12.4)
$$v = \sum_{q \in A(s)} f(s, q) - \sum_{q \in B(s)} f(q, s)$$
$$\leq \sum_{q \in A(s)} f(s, q) \leq \sum_{q \in A(s)} c(s, q).$$

Similarly, by (12.1) and the third equation in (12.2), we have

(12.5)
$$v = \sum_{q \in B(t)} f(q, t) - \sum_{q \in A(t)} f(t, q)$$
$$\leq \sum_{q \in B(t)} f(q, t) \leq \sum_{q \in B(t)} c(q, t).$$

Since, in accordance with (12.1) and (12.2), the flow $f(p, q)$ in the arcs (p, q) varies on closed and bounded real-number intervals, by continuity there exists a maximal static flow f_0 in $\{N; A; C\}$. The same conclusion holds if, as assumed in Sec. 12.8, the flows are restricted to integral values on bounded real-number intervals. By (12.4) and (12.5) and the fact that there is a static flow of value 0, the value v_0 of f_0 satisfies the inequalities

$$(12.6) \qquad 0 \le v_0 \le \min\left[\sum_{q \in A(s)} c(s, q), \sum_{q \in B(t)} c(q, t)\right].$$

Thus in (12.3) and (12.6) we have the same lower bound and the same upper bound, for the capacity of a minimal cut separating s from t in $\{N; A; C\}$, as we have for the value of a maximal static flow from s to t in $\{N; A; C\}$.

The following result further relates the capacity of any cut separating s from t, and the value of any static flow from s to t, in $\{N; A; C\}$.

Theorem 12.1. If v is the value of any static flow from s to t in $\{N; A; C\}$, and $c(P, \bar{P})$ is the capacity of any cut separating s from t in $\{N; A; C\}$, then

$$(12.7) \qquad v \le c(P, \bar{P}).$$

PROOF. By (12.2), we have

$$(12.8) \qquad \begin{aligned} \sum_{q \in A(s)} f(s, q) - \sum_{q \in B(s)} f(q, s) &= v, \\ \sum_{q \in A(p)} f(p, q) - \sum_{q \in B(p)} f(q, p) &= 0, \qquad p \ne s \quad \text{or} \quad t. \end{aligned}$$

If for $p \in N$, $q \in N$, $(p, q) \notin A$, *we define* $f(p, q)$ by

$$f(p, q) = 0,$$

then (12.8) can be written equivalently as

$$(12.9) \qquad \begin{aligned} \sum_{q \in N} f(s, q) - \sum_{q \in N} f(q, s) &= v, \\ \sum_{q \in N} f(p, q) - \sum_{q \in N} f(q, p) &= 0, \qquad p \ne s \quad \text{or} \quad t. \end{aligned}$$

Summing (12.9) over P with respect to p, and noting that $t \notin P$, we obtain

$$\sum_{p \in P, q \in N} f(p, q) - \sum_{p \in P, q \in N} f(q, p) = v,$$

which can be written as

$$\sum_{p \in P, q \in P} f(p, q) - \sum_{p \in P, q \in P} f(q, p) + \sum_{p \in P, q \in \bar{P}} f(p, q) - \sum_{p \in P, q \in \bar{P}} f(q, p) = v.$$

By symmetry, however, we have

$$\sum_{p \in P, q \in P} f(p, q) - \sum_{p \in P, q \in P} f(q, p) = 0,$$

and therefore

$$(12.10) \qquad \sum_{p \in P, q \in \bar{P}} f(p, q) - \sum_{p \in P, q \in \bar{P}} f(q, p) = v.$$

Accordingly, we obtain

$$v = \sum_{p \in P, q \in \bar{P}} f(p, q) - \sum_{p \in P, q \in \bar{P}} f(q, p)$$
$$\leq \sum_{p \in P, q \in \bar{P}} f(p, q) \leq \sum_{p \in P, q \in \bar{P}} c(p, q) = c(P, \bar{P}),$$

as desired.

12.7 THE MAX-FLOW—MIN-CUT THEOREM

Since the inequality (12.7) holds in general, it holds in particular for the value v_0 of a maximal static flow f_0 from s to t in $\{N; A; C\}$ and the capacity $c(P_0, \bar{P}_0)$ of a minimal cut separating s from t in $\{N; A; C\}$:

(12.11) $v_0 \leq c(P_0, \bar{P}_0).$

Actually, however, the sign of equality holds in (12.11), as expressed in the following max-flow–min-cut theorem.

Theorem 12.2. If v_0 is the value of a maximal static flow from s to t in $\{N; A; C\}$, and $c(P_0, \bar{P}_0)$ is the capacity of a minimal cut separating s from t in $\{N; A; C\}$, then

$$v_0 = c(P_0, \bar{P}_0).$$

PROOF. By (12.7), we need only to establish the existence of a flow from s to t in $\{N; A; C\}$, and a cut separating s from t in $\{N; A; C\}$, such that the value of the flow is equal to the capacity of the cut.

For this, let f_0 be a maximal flow from s to t in $\{N; A; C\}$, and define P recursively as follows [5]:

(a) $s \in P$;

(b) if $p \in P$ and either $f_0(p, q) < c(p, q)$ or $f_0(q, p) > 0$, then $q \in P$.

The node t cannot be in the set P thus constructed. If it were, then by conditions (a) and (b), there would be a path W in P from s to t. Let

$$\delta = \min_{(p, q) \in W} [c(p, q) - f_0(p, q), f_0(q, p)] > 0.$$

Then on W the flow in each forward arc could be increased, and the flow in each backward arc decreased, by an amount δ. The conditions (12.1) and (12.2) would still be satisfied, and accordingly we would have a flow from s to t in $\{N; A; C\}$ of value $v_0 + \delta > v_0$, a contradiction of the assumption that f_0 is a maximal flow.

Thus the cut $(P, \bar{P})$ separates s from t in $\{N; A; C\}$. Furthermore, by condition (b), for each $p \in P$ and $q \in \bar{P}$ we have

$$f(p, q) = c(p, q), \qquad f(q, p) = 0.$$

Hence, from (12.10), we obtain

$$v = \sum_{p \in P, q \in \bar{P}} f(p, q) - \sum_{p \in P, q \in \bar{P}} f(q, p)$$

$$= \sum_{p \in P, q \in \bar{P}} c(p, q) = c(P, \bar{P}),$$

as desired.

12.8 THE FORD-FULKERSON ALGORITHM

The following constructive algorithm [6] for determining a maximal flow from the source s to the sink t in $\{N ; A ; C\}$ is based on the foregoing proof of the max-flow—min-cut theorem.

To assure termination of the process, we assume that the capacity function $c(p, q)$ is integer-valued. This assumption is a natural one for applications to combinatorial problems.

We might start with the 0 flow, since this flow exists for any $\{N; A; C\}$. Any initial integer-valued flow from s to t in $\{N; A; C\}$ whatsoever might be adopted, however, and a more speedy termination can often be achieved if we start with a flow that seems to be a reasonably close approximation to the maximal flow.

The algorithm consists of the alternate application of the following labeling and flow-augmenting processes until a maximal flow and a minimal cut have been achieved.

Labeling process

At the start of this process, no nodes are labeled. During the process, a node might or might not become *labeled*, and if labeled it might or might not become *scanned*. First the source s is labeled $(-, \infty)$. Next, all nodes $q \neq s$ for which $f(s, q) < c(s, q)$ are labeled $[s^+, \delta(q)]$, where

$$\delta(q) = c(s, q) - f(s, q),$$

and then all presently unlabeled nodes q for which $f(q, s) > 0$ are labeled $[s^-, \delta(q)]$, where

$$\delta(q) = f(q, s).$$

The source s is now said to be scanned. In general, as soon as one labeled node has been scanned, we fix our attention on another node p that has been labeled, but not yet scanned. To each node q that has not yet been labeled and for which $f(p, q) < c(p, q)$, we apply the label $[p^+, \delta(q)]$, where

$$\delta(q) = \min [\delta(p), c(p, q) - f(p, q)],$$

and then to each node q that has still not been labeled and for which

$f(q, p) > 0$ we apply the label $[p^-, \delta(q)]$, where

$$\delta(q) = \min [\delta(p), f(q, p)].$$

When all this has been done, we say that p has been scanned. The process is now repeated, with another labeled but not yet scanned node in place of p, until either (1) the sink t is labeled or (2) the sink t is not labeled, but no more labels can be assigned.

In case (1), we turn to the following flow-augmenting process, whereas in case (2) we terminate the algorithm.

Flow-augmenting process

If, in the labeling process, the sink t has been labeled $[q^+, \delta(t)]$, then $f(q, t)$ is replaced by $f(q, t) + \delta(t)$; but if t has been labeled $[q^-, \delta(t)]$, then $f(t, q)$ is replaced by $f(t, q) - \delta(t)$. Next, if the previously mentioned node q is labeled $[p^+, \delta(q)]$ or $[p^-, \delta(q)]$, then $f(p, q)$ is replaced by $f(p, q) + \delta(t)$ or $f(q, p)$ is replaced by $f(q, p) - \delta(t)$, respectively. The process is continued back to the source s, and then the present labels are discarded and the labeling process repeated.

Since an application of the flow-augmenting process increases the flow by the positive integral amount $\delta(t)$, and since by Sec. 12.6 the values of all flows in $\{N; A; C\}$ from s to t are uniformly bounded, the algorithm terminates, yielding a maximal flow. The last application of the labeling process yields a minimal cut $(P, \bar{P})$, in which the set P consists of the labeled nodes, and the set $\bar{P}$ consists of the unlabeled ones.

It will be recalled that we started the algorithm with an integer-valued flow. In the process of applying the algorithm, each flow augmentation also is integer-valued. Accordingly, we have the following important combinatorial consequence.

Theorem 12.3. If in $\{N; A; C\}$ the capacity function $c(p, q)$ is integer-valued, then for any $s \in N$, $t \in N$, $s \neq t$, there is an integer-valued maximal flow from s to t in $\{N; A; C\}$.

12.9 EXAMPLE

Consider the capacitated directed network shown in Fig. 12.8. In this, there is indicated a flow of value 5.

An application of the labeling process yields the labels that also are shown in Fig. 12.8. This shows a *breakthrough* to t of value 6.

Now the flow-augmenting process results in the flow of value 11 that is shown in Fig. 12.9.

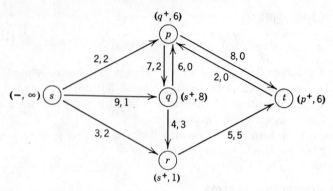

Fig. 12.8 An application of the labeling process to a directed network with a given flow.

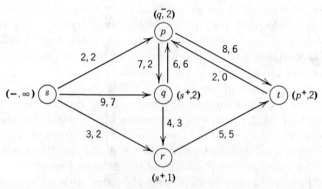

Fig. 12.9 An application of the flow-augmenting process to the labeled network of Fig. 12.8, followed by a second application of the labeling process.

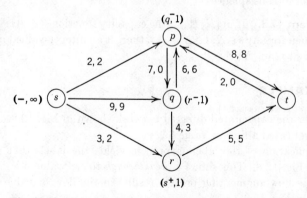

Fig. 12.10 A terminating application of the labeling process.

Next, a second application of the labeling process also is depicted in Fig. 12.9.

A second application of the flow-augmenting process yields the flow of value 13 that is shown in Fig. 12.10.

Finally, a third application of the labeling process terminates the algorithm, as shown in Fig. 12.10, thus formally corroborating our earlier

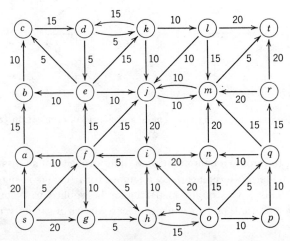

Fig. 12.11   A more complex capacitated directed network.

observations (see Figs. 12.6 and 12.7) that $\bar{P} = \{t\}$ determines a minimal cut in this network and that the maximal flow has value 13.

As an exercise, one might apply the algorithm to determine a maximal flow and a minimal cut for the more involved network shown in Fig. 12.11.

12.10 EXTENSIONS AND APPLICATIONS

The methods and results that have been presented in this chapter admit many extensions and combinatorial applications. Some are quite immediate, others less so. We shall mention just a few. Excellent extensive treatments and bibliographies can be found in [1] and [7].

Suppose, for example, that in $\{N; A; C\}$ we want to determine a maximal flow into the sink $t \in N$ from a *set* $S = \{s_1, s_2, \ldots, s_k\}$ *of sources* $\in N$. To define such a flow, we replace the first equation in (12.2) by

$$\sum_{s \in S, q \in A(s)} f(s, q) - \sum_{s \in S, q \in B(s)} f(q, s) = v.$$

Although this problem might appear at first glance to be basically more difficult than the one we have been discussing, it is easily reducible to the earlier problem by adjoining one fictitious node s to the network, together with the arcs and unlimited capacities indicated in Fig. 12.12. A similar device for treating multisink flow problems is also shown in the figure.

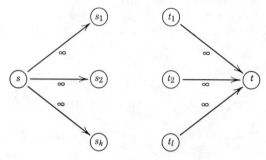

Fig. 12.12 Devices for treating multisource and multisink problems in capacitated directed networks.

Similarly, to impose a commodity-handling capacity of c units per time period on a node p, we replace p by a subnetwork such as is suggested in Fig. 12.13. This device might in particular be applied to a source that can produce material, or to a sink that can absorb it, at only a limited rate.

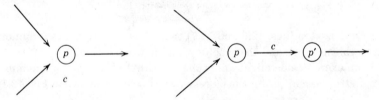

Fig. 12.13 A device for treating capacitated nodes.

For further extensions, and applications to assignment problems, warehousing problems, catering problems, supply-demand problems, minimal-cost transportation problems, etc., see the excellent treatment in [7].

12.11 GRAPH THEORY AND COMBINATORIAL PROBLEMS

Many combinatorial problems involving linear graphs are closely related to the max-flow—min-cut theorem. We shall conclude this chapter by mentioning a few of them.

A *graph* can be considered as a network $\{N; A\}$ in which the arcs $(p, q) \in A$ are not necessarily considered as being directed; that is, the pairs (p, q) are not necessarily considered as being ordered.

A *bipartite graph* is a graph $\{N; A\}$ in which the set N of nodes consists of two disjoint subsets, $N = S \cup T$, $S \cap T = \phi$, and all the arcs of A extend from nodes $p \in S$ to nodes $q \in T$.

A set of nodes *disconnects* S from T if each path from S to T contains one of these nodes.

The König-Egerváry theorem [9, 4] is the following.

Theorem 12.4. If $\{S, T; A\}$ is a bipartite graph, then the maximal number of arcs $\in A$ that are node-disjoint by pairs is equal to the minimal number of nodes in a set of nodes disconnecting S from T.

PROOF. Consider the capacitated directed network obtained from $S, T; A\}$ by adjoining a source s and a sink t to $N = S \cup T$, adjoining arcs (s, p) and (q, t) to A for each $p \in S$ and each $q \in T$, and assigning capacities as follows:

$$c(s, p) = 1 \qquad \text{for all } p \in S,$$

$$c(q, t) = 1 \qquad \text{for all } q \in T,$$

$$c(p, q) = \infty \qquad \text{for all } (p, q) \in A.$$

The theorem now follows directly from Theorems 12.2 and 12.3.

The following theorem of Menger [9] is a generalization of the foregoing König-Egerváry theorem.

Theorem 12.5. If S and T are disjoint subsets of nodes in the graph $\{N; A\}$, then the maximal number of paths from S to T in $\{N; A\}$ that are node-disjoint by pairs is equal to the minimal number of nodes in a set of nodes disconnecting S from T.

PROOF. Construct a capacitated directed network as in the proof of Theorem 12.4 by adjoining a source s and a sink t to N, and arcs (s, p) and (q, t) to A for each $p \in S$ and $q \in T$, but this time assign infinite capacity to all arcs and unit capacity to all nodes except s and t. An application of Theorems 12.2 and 12.3 completes the proof.

A *finite partially ordered set* $\{N; \succ\}$ is a finite set of elements $p, q, \ldots, r$, with transitive order relation $p \succ q$ (read "p precedes q"), in which both $p \succ q$ and $q \succ p$ never occur. Such a set can be considered as an ordered network $\{N; A\}$, in which (p, q) replaces $p \succ q$. A *chain* in $\{N; \succ\}$ is a subset $\{p_1, p_2, \ldots, p_n\}$ of N, with

$$p_1 \succ p_2 \succ \cdots \succ p_n, \qquad n \geq 1.$$

A *decomposition* of $\{N; \succ\}$ is a partition of $\{N; \succ\}$ into disjoint chains. Elements $p \in N$ and $q \in N$ are *unrelated* if neither $p \succ q$ nor $q \succ p$. As an immediate consequence of these definitions, the number of mutually unrelated elements of N is less than or equal to the number of chains in a minimal decomposition of $\{N; \succ\}$; according to Dilworth's *chain-decomposition theorem* [3], which can be derived from Theorems 12.2 and 12.3, these two numbers actually are equal.

MULTIPLE-CHOICE REVIEW PROBLEMS

1. Let the capacitated directed network and the static flow shown in the

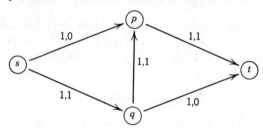

figure be given. In the labeling process, the label $(p^{-1}, 1)$ would be applied to the node

(a) s,   (b) p,   (c) q,   (d) t.

2. For the flow indicated in the figure of Problem 1, a flow-augmenting path is

(a) $s\,p\,t$,   (b) $s\,q\,t$,   (c) $s\,p\,q\,t$,   (d) $s\,q\,p\,t$.

3. For the capacitated directed network of Problem 1, $(P, \overline{P})$ is not a minimal cut if

(a) $P = \{s\}$,   (b) $P = \{s, p\}$,   (c) $P = \{s, q\}$,   (d) $P = \{s, p, q\}$.

4. For the capacitated directed network of Problem 1, the value of the maximal flow is

(a) 0,   (b) 1,   (c) 2,   (d) 3.

5. For the capacitated directed network of Problem 1, there is no flow of value

(a) 0,   (b) 1,   (c) 2,   (d) 3.

REFERENCES

1. Berge, C., *Théorie des graphes et ses applications*, Collection Universitaire de Mathematiques **2**, Dunod, Paris, 1958. Translated by Alison Doig, *The Theory of Graphs and Its Applications*, John Wiley and Sons, New York, 1962.
2. Dantzig, G. B., and D. R. Fulkerson, "On the Max-Flow Min-Cut Theorem of Networks," *Linear Inequalities and Related Systems*, Annals of Mathematics Studies, No. 38, Princeton University Press, Princeton, New Jersey, 1956.

3. Dilworth, R. P., "A Decomposition Theorem for Partially Ordered Sets," *Ann. Math.* **51** (1950), 161–166.
4. Egerváry, J., "Matrixok kombinatorikus tulajonsagairól," *Mat. és Fiz. Lapok* **38** (1931), 16–28. Translated by H. W. Kuhn, "On Combinatorial Properties of Matrices," *George Washington University Logistics Papers* **11** (1955).
5. Ford, L. R., Jr., and D. R. Fulkerson, "Maximal Flow through a Network," *Canadian J. Math.* **8** (1956), 399–404.
6. Ford, L. R., Jr., and D. R. Fulkerson, "A Simple Algorithm for Finding Maximal Network Flows and an Application to the Hitchcock Problem," *Canadian J. Math.* **9** (1957), 210–218.·
7. Ford, L. R., Jr., and D. R. Fulkerson, *Flows in Networks*, Princeton University Press, Princeton, New Jersey, 1962.
8. Harris, T. E., and F. S. Ross, *Fundamentals of a Method for Evaluating Rail Net Capacities*, (U) RM-1573 (Secret), The RAND Corporation, Santa Monica, Calif., 1956.
9. König, D., *Theorie der endlichen und unendlichen Graphen*, Chelsea Publishing Company, New York, 1950.

PART 3

CONSTRUCTION AND EXISTENCE

Whenever you can settle a question by explicit construction,
be not satisfied with purely existential arguments.

HERMANN WEYL

(*from an address delivered at the Princeton University Bicentennial
Conference on the Problems of Mathematics, December 17–19, 1946*)

CHAPTER 13

Block Designs

MARSHALL HALL, JR.

13.1 INTRODUCTION

Block designs are arrangements of objects into sets, called blocks, such that various conditions on the number of occurrences of objects, of pairs of objects, and sometimes of other things are satisfied. They have arisen in the study of algebraic geometry, which was the source of Steiner's [27] original problem. They occur also in the theory of the design of experiments. For this aspect of the theory the reader is referred to the excellent book by Henry B. Mann [20]. Finite geometrical systems are special kinds of block designs, as we see if we compare the fundamental papers by Bruck and Ryser [8] and by Chowla and Ryser [9]. There is a relationship between number theory and block designs, which is certainly deep but hardly clear cut. This appears in papers [6, 8, 9, 14, 26], although these give an incomplete indication of the relationship. More recently block designs have been of interest with respect to error-correcting codes (see Chapter 14).

Section 13.2 deals with examples and definitions, and Sec. 13.3 gives some explanation of the applications of block designs to problems in error-correcting codes and the design of statistical experiments. Section 13.4 presents the general theorems on existence and relations to matrices and quadratic forms. The next section consists of a brief survey of the extensive variety of known block designs and the methods by which they are constructed.

In some instances, it has not been possible to give full details. A fuller account of many of the proofs can be found in the author's forthcoming book [15].

13.2 BLOCK DESIGNS AND LATIN SQUARES; EXAMPLES AND DEFINITIONS

A block design is an arrangement of objects into blocks subject to certain rules concerning the occurrence of objects and pairs of objects. Three examples follow.

Example 13.1   *Seven Objects* $1, \ldots, 7$ *in Seven Blocks* $B_1, \ldots, B_7$.

B_1: $1, 2, 4$; B_3: $3, 4, 6$; B_5: $5, 6, 1$; B_7: $7, 1, 3$.
B_2: $2, 3, 5$; B_4: $4, 5, 7$; B_6: $6, 7, 2$;

Example 13.2   *Nine Objects* $1, \ldots, 9$ *in Twelve Blocks* $B_1, \ldots, B_{12}$.

B_1: $1, 2, 3$; B_4: $1, 4, 7$; B_7: $1, 5, 9$; B_{10}: $1, 6, 8$;
B_2: $4, 5, 6$; B_5: $2, 5, 8$; B_8: $2, 6, 7$; B_{11}: $2, 4, 9$;
B_3: $7, 8, 9$; B_6: $3, 6, 9$; B_9: $3, 4, 8$; B_{12}: $3, 5, 7$.

Example 13.3   *Eleven Objects* $1, \ldots, 11$ *in Eleven Blocks* $B_1, \ldots, B_{11}$.

B_1: $1, 3, 4, 5, 9$; B_5: $5, 7, 8, 9, 2$; B_9: $9, 11, 1, 2, 6$;
B_2: $2, 4, 5, 6, 10$; B_6: $6, 8, 9, 10, 3$; B_{10}: $10, 1, 2, 3, 7$;
B_3: $3, 5, 6, 7, 11$; B_7: $7, 9, 10, 11, 4$; B_{11}: $11, 2, 3, 4, 8$.
B_4: $4, 6, 7, 8, 1$; B_8: $8, 10, 11, 1, 5$;

In the first two examples, each pair of distinct objects i, j occurs together in exactly one block; and in the third example, each pair occurs in exactly two blocks. In each of the three examples, the blocks are of uniform size; and each object occurs the same number of times, respectively three times, four times, and five times.

These block designs are instances of a general system called an *incidence system*.

Definition. An *incidence system* S consists of blocks $B_1, \ldots, B_b$ and objects $a_1, \ldots, a_v$ and an incidence relation $a_i \in B_j$ between certain pairs of objects a_i and blocks B_j. If $a_i \in B_j$ we say that a_i *belongs to* B_j or that B_j *contains* a_i.

In this very general definition of an incidence system, as extreme cases we might have every object belonging to every block or no object belonging to any block. Even without these extremes, however, it is permitted that two different blocks, say B_1 and B_2, contain exactly the same objects. In this respect the blocks are not strictly sets of objects. If we wish to incorporate the incidence systems into general set theory, we must say that a block B_j is a function $f(j)$ of its index j, having as its value a subset of the set of objects $a_1, \ldots, a_v$; but there seems to be no particular virtue in using such terminology.

From the objects $a_1, a_2, \ldots, a_v$, we may form a subset by either including or excluding each object $a_1, a_2, \ldots, a_v$ in turn; thus, with two choices for each a_i, we have a total of 2^v choices and so 2^v subsets including the empty set and the set of all a_i. Since each of $B_1, B_2, \ldots, B_b$ may contain any one of these subsets, there are $(2^v)^b = 2^{vb}$ different incidence systems of objects $a_1, \ldots, a_v$ and blocks $B_1, \ldots, B_b$. Apart from the fact, however, that the total number of these systems is very large, there seem to be no particular properties of interest that unrestricted incidence systems have.

A *block design* is an incidence system in which the objects are distributed among the blocks with a certain uniformity.

Definition. *A balanced incomplete block design* is an incidence system of v objects $a_1, \ldots, a_v$ and b blocks $B_1, \ldots, B_b$, such that

1. Each block B_j contains the same number k of objects.
2. Each object a_i is in the same number r of blocks.
3. For each unordered pair a_i, a_j of distinct objects, the number of blocks containing both of them is the same number λ.

In Example 13.1, we have $v = b = 7$, $k = r = 3$, $\lambda = 1$. In Example 13.2, we have $v = 9$, $b = 12$, $r = 4$, $k = 3$, $\lambda = 1$, and in Example 13.3, $v = b = 11$, $k = r = 5$, $\lambda = 2$. Unless otherwise specified, the term *block design* will be used here to mean balanced incomplete block design.

There are two relations on the five parameters which can easily be established:

$$(13.1) \qquad\qquad bk = vr, \qquad r(k - 1) = \lambda(v - 1).$$

The first of these counts the total number of incidences $a_i \in B_j$ in two ways, there being b blocks, each of which containing k objects, and v objects, each occurring r times. For the second relation, we consider the pairs containing a particular object a_1. Here a_1 appears in r blocks and in each of these is paired with $k - 1$ further objects. But a_1 must be paired with each of the remaining $v - 1$ objects exactly λ times. Comparing these counts, we have $r(k - 1) = \lambda(v - 1)$.

The conditions (13.1) on the parameters b, v, r, k, λ are necessary for the existence of a block design with these parameters; but, as we shall see later, they are not sufficient. Specifically, $v = b = 22$, $r = k = 7$, $\lambda = 2$ satisfy (13.1), but there is no design with these parameters. The question as to the existence of designs with parameters satisfying (13.1) remains a mysterious one.

Let us turn to Latin squares. An n by n square in which objects $a_1, \ldots,$ a_n are placed in such a way that each object appears exactly once in each row and exactly once in each column is called a *Latin square*. In the following example, we give three 4 by 4 squares.

Example 13.4   Latin Squares.

$$\begin{bmatrix} 1 & 2 & 3 & 4 \\ 2 & 1 & 4 & 3 \\ 3 & 4 & 1 & 2 \\ 4 & 3 & 2 & 1 \end{bmatrix} \quad \begin{bmatrix} 1 & 2 & 3 & 4 \\ 3 & 4 & 1 & 2 \\ 4 & 3 & 2 & 1 \\ 2 & 1 & 4 & 3 \end{bmatrix} \quad \begin{bmatrix} 1 & 2 & 3 & 4 \\ 4 & 3 & 2 & 1 \\ 2 & 1 & 4 & 3 \\ 3 & 4 & 1 & 2 \end{bmatrix}$$

Two Latin squares are said to be *orthogonal* if we can superpose the squares, placing the objects of both squares in the same n^2 cells, in such a way that each object of the first square occurs exactly once with each object of the second square. A set of r Latin squares is said to be *mutually orthogonal* if each pair of them are orthogonal. We note that the squares of Example 13.4 are mutually orthogonal. The orthogonality of the first two can be seen in the superposed form:

$$\begin{bmatrix} 11 & 22 & 33 & 44 \\ 23 & 14 & 41 & 32 \\ 34 & 43 & 12 & 21 \\ 42 & 31 & 24 & 13 \end{bmatrix}$$

Latin squares, and indeed orthogonal sets of Latin squares, can be regarded as partially balanced block designs. In a partially balanced block design each block has the same number of objects, and each object is in the same number of blocks; but certain pairs of objects occur with one frequency, whereas others occur with another frequency, or there may even be several prescribed frequencies for pairs.

For a set of three mutually orthogonal n by n squares, we take $5n$ objects, $r_1, \ldots, r_n$ corresponding to rows, $c_1, \ldots, c_n$ corresponding to columns, and $f_1, \ldots, f_n, s_1, \ldots, s_n, t_1, \ldots, t_n$ corresponding to objects of the first, second, and third squares. For each of n^2 cells we form a block of five objects containing r_i, c_j, f_u, s_v, t_w if the entry in the ith row and jth column is f_u in the first square, s_v in the second, and t_w in the third. Here there are n^2 blocks, each with five objects, and each object is in n blocks. To express the fact that two objects of the same kind ($r, c, f, s,$ or t) do not occur together at all, we write $\lambda_1 = 0$; and to indicate that two objects of different kinds occur together exactly once, we write $\lambda_2 = 1$. It is easy to see that a partially balanced block design of this kind, with $b = n^2$, $v = 5n$, $k = 5$, $r = n$, $\lambda_1 = 0$, $\lambda_2 = 1$, conversely yields a set of three mutually orthogonal squares. This approach can, of course, be extended to any number of mutually orthogonal Latin squares. The partially balanced block design corresponding to Example 13.4 contains sixteen blocks of five objects each, and is given in the following example.

Example 13.5   The Subscripts For a Partially Ordered Block Design.

	r	c	f	s	t			r	c	f	s	t			r	c	f	s	t
B_1:	1,	1,	1,	1,	1		B_7:	2,	3,	4,	1,	2		B_{12}:	3,	4,	2,	1,	3
B_2:	1,	2,	2,	2,	2		B_8:	2,	4,	3,	2,	1		B_{13}:	4,	1,	4,	2,	3
B_3:	1,	3,	3,	3,	3		B_9:	3,	1,	3,	4,	2		B_{14}:	4,	2,	3,	1,	4
B_4:	1,	4,	4,	4,	4		B_{10}:	3,	2,	4,	3,	1		B_{15}:	4,	3,	2,	4,	1
B_5:	2,	1,	2,	3,	4		B_{11}:	3,	3,	1,	2,	4		B_{16}:	4,	4,	1,	3,	2
B_6:	2,	2,	1,	4,	3														

In this representation of Latin squares, we can see that the rows and columns play a role exactly similar to that of the objects.

13.3   APPLICATIONS OF BLOCK DESIGNS

In communication with a satellite it is to be expected that, because of the enormous distances involved, reception will be imperfect. For this reason it is desirable to use a code of signals sufficiently different from each other so that the presence of a moderate number of errors can be detected and corrected. Example 13.6 is an error-correcting code of zeros and ones. The patterns have the property that no two of them agree in more than half of their positions.

Example 13.6   An Error-Correcting Code. This is shown in Table 13.1. We may use the rows of this array as the words of an error-correcting code. Since any two words differ in at least six places, it follows that if a word is received with at most two errors, we can determine without ambiguity which the correct word must have been. In other words, with two places changed we cannot change two others and obtain another row. With three errors it may not be clear as to which was the original row. For example,

$$0 \quad 0 \quad 0 \quad 0 \quad 0 \quad 0 \quad 0 \quad 0 \quad 1 \quad 0 \quad 1 \quad 1$$

differs in three places from

$$0 \quad 0 \quad 0 \quad 0 \quad 0 \quad 0 \quad 0 \quad 0 \quad 0 \quad 0 \quad 0 \quad 0,$$

and in three other places from

$$0 \quad 0 \quad 1 \quad 0 \quad 0 \quad 0 \quad 1 \quad 1 \quad 1 \quad 0 \quad 1 \quad 1.$$

Accordingly, although we can detect that the signal received is in error, we cannot correct it with certainty. Thus this code detects up to five errors and corrects up to two errors.

This code can be constructed from the block design of Example 13.3. The first row consists entirely of 1's. Rows 2 through 12 are constructed from the block design of Example 13.3 by putting a 1 in column 0 and 1's in further columns in which the numbers are the numbers in a block of the

TABLE 13.1

An Error-Correcting Code

Row number	Column number											
	0	1	2	3	4	5	6	7	8	9	10	11
0	1	1	1	1	1	1	1	1	1	1	1	1
1	1	1	0	1	1	1	0	0	0	1	0	0
2	1	0	1	0	1	1	1	0	0	0	1	0
3	1	0	0	1	0	1	1	1	0	0	0	1
4	1	1	0	0	1	0	1	1	1	0	0	0
5	1	0	1	0	0	1	0	1	1	1	0	0
6	1	0	0	1	0	0	1	0	1	1	1	0
7	1	0	0	0	1	0	0	1	0	1	1	1
8	1	1	0	0	0	1	0	0	1	0	1	1
9	1	1	1	0	0	0	1	0	0	1	0	1
10	1	1	1	1	0	0	0	1	0	0	1	0
11	1	0	1	1	1	0	0	0	1	0	0	1
12	0	0	0	0	0	0	0	0	0	0	0	0
13	0	0	1	0	0	0	1	1	1	0	1	1
14	0	1	0	1	0	0	0	1	1	1	0	1
15	0	1	1	0	1	0	0	0	1	1	1	0
16	0	0	1	1	0	1	0	0	0	1	1	1
17	0	1	0	1	1	0	1	0	0	0	1	1
18	0	1	1	0	1	1	0	1	0	0	0	1
19	0	1	1	1	0	1	1	0	1	0	0	0
20	0	0	1	1	1	0	1	1	0	1	0	0
21	0	0	0	1	1	1	0	1	1	0	1	0
22	0	0	0	0	1	1	1	0	1	1	0	1
23	0	1	0	0	0	1	1	1	0	1	1	0

design, and putting 0's in all remaining columnar positions. Rows 13 through 24 are complements of the first twelve rows, with 0's replaced by 1's and 1's by 0's.

Suppose that we have a block design with $v = b = 4t - 1$, $k = r = 2t - 1$, and $\lambda = t - 1$. Then any two different blocks have exactly $\lambda = t - 1$ objects in common. (This property is proved in Sec. 13.4.) From such a design we can construct an error-correcting code of $8t$ words of length $4t$. If the objects of the block design are the numbers $1, \ldots,$ $4t - 1$, we number our columns $0, 1, \ldots, 4t - 1$. The first row consists entirely of 1's. Rows $2, \ldots, 4t$ are constructed from the blocks $B_1, \ldots,$ B_{4t-1} of our design. In row i we place a 1 in column 0, 1's in columns of which the numbers are the objects of B_{i-1}, and 0's in the remaining columns. This we do for $i = 2, \ldots, 4t$. Rows $4t + 1, \ldots, 8t$ are the complements of rows $1, \ldots, 4t$, respectively; that is, we interchange the bits, replacing 0's by 1's and 1's by 0's. We note that each of the words given by rows $2, \ldots, 4t$ contains 1's in $1 + 2t - 1 = 2t$ places and also 0's in $2t$ places, and so each agrees with the first row in $2t$ places and disagrees in $2t$ places.

If i and $j \neq i$ are in the range $2, \ldots, 4t$, then the corresponding blocks have $t - 1$ objects in common, and so the rows W_i and W_j both have 1's in column 0 and in $t - 1$ additional positions. Hence W_i and W_j both contain 1's in exactly t places, W_i has t additional 1's in places where W_j has 0's, W_j has t additional 1's in places where W_i has 0's. This accounts for $3t$ of the $4t$ columns, and in the remaining t columns both W_i and W_j have 0's. Hence W_i and W_j agree in $2t$ places and disagree in $2t$ places. Since $W_{4t+1}, \ldots, W_{8t}$ are complements of the first $4t$, any two of these agree in half of their positions and disagree in the other half. Between a word W_i in the top half and a word W_j in the bottom half, if W_j is the complement of W_i, they disagree in all $4t$ positions. If not, W_j agrees with W_i in the $2t$ positions where its complement disagrees, and disagrees in the $2t$ positions where its complement agrees. The code constructed in this way will correct up to $t - 1$ errors. Thus if a word W_i is in error in at most $t - 1$ places, then up to $t - 1$ changes in the erroneous form will give the correct word and also words differing from it in at most $2t - 2$ places, and hence will give no other word in the list since any two words in the code differ in at least $2t$ places.

An important area of application for block designs is in the design of experiments for statistical studies. For a thorough treatment of this subject the reader is referred to Henry B. Mann [20]. The use of Latin squares and block designs eliminates unwanted effects in statistical studies and permits analysis of the quantities in which there is interest. For example, if it is desired to examine the yield of m varieties of wheat by planting in a plot, then it is desirable to eliminate variations in fertility within the plot. This may be done by subdividing the plot into m^2 sections and planting so that each variety is used once in each row and once in each column of the corresponding square division. Thus an m by m Latin square is the underlying pattern for such an experiment. We may also wish to test the effect of m fertilizers on the varieties of wheat. Here we wish to use a Latin square for the fertilizers, with the property that each fertilizer is applied exactly once to each variety. With five varieties of wheat $v_1, \ldots, v_5$ and five fertilizers $f_1, \ldots, f_5$, we need two orthogonal 5 by 5 Latin squares. The following orthogonal squares satisfy our requirements.

Example 13.7   Orthogonal Squares of Order 5.

$$
\begin{array}{ccccc}
v_1 f_1 & v_2 f_2 & v_3 f_3 & v_4 f_4 & v_5 f_5 \\
v_2 f_3 & v_3 f_4 & v_4 f_5 & v_5 f_1 & v_1 f_2 \\
v_3 f_5 & v_4 f_1 & v_5 f_2 & v_1 f_3 & v_2 f_4 \\
v_4 f_2 & v_5 f_3 & v_1 f_4 & v_2 f_5 & v_3 f_1 \\
v_5 f_4 & v_1 f_5 & v_2 f_1 & v_3 f_2 & v_4 f_3
\end{array}
$$

When it is necessary to test a relatively large number of varieties, we may conduct our experiment by planting in a number of plots and assuming that the fertility is constant in each individual plot. If we assign v varieties to b plots using a balanced incomplete block design for our assignment rule, then we can make statistical studies of the yields that are independent of the differences in fertility between the different plots. For this purpose any of Examples 13.1 to 13.5 will serve as an illustration.

Sometimes a block design with special properties meets the requirements of a particular test. Let us suppose that a testing center wishes to investigate various kinds of toothpaste, giving each one of the subjects several kinds to try. The tubes that each receives are distinguished by colors. It is desirable that (1) each subject shall receive the same number of brands of toothpaste, (2) each brand shall be used by the same number of subjects, (3) each pair of brands shall be compared by a subject the same number of times, and (4) each brand shall be given each color the same number of times, so that no color preference will influence the findings. The first three requirements can be met by using a block design to determine the assignment of brands to subjects, but the fourth requirement is an additional restriction.

With thirteen different brands of toothpaste, denoted by the numbers $0, 1, \ldots, 12$, and thirteen subjects represented by the letters $A, B, \ldots, M$, we can give each subject four brands, see that each brand is used by four subjects, and have every pair of brands compared by one subject, through use of a block design with $v = b = 13$, $r = k = 4$, $\lambda = 1$. In addition, we can arrange to have each brand used exactly once in a tube of each of the colors red, white, blue, and green. The brand number is under the color and is opposite a letter for the subject.

Example 13.8   The Distribution of Toothpaste in Colored Tubes among Subjects. This is described on page 376, and is shown in Table 13.2.

Here we see that the cyclic pattern of the blocks permits the color assignment. The designs of Example 13.1 and Example 13.3 also have this cyclic pattern.

TABLE 13.2

Distribution of Brands and Colors among Subjects

	Tube color			
Subject	Red	White	Blue	Green
A	0	1	3	9
B	1	2	4	10
C	2	3	5	11
D	3	4	6	12
E	4	5	7	0
F	5	6	8	1
G	6	7	9	2
H	7	8	10	3
I	8	9	11	4
J	9	10	12	5
K	10	11	0	6
L	11	12	1	7
M	12	0	2	8

13.4   GENERAL THEORY OF BLOCK DESIGNS

It is possible to describe the properties of block designs in terms of matrices and also in terms of quadratic forms. These have furnished fruitful methods for developing the theory of block designs, since both of these subjects have been extensively studied.

With any incidence system S (and so in particular any block design), we may associate an *incidence matrix*.

Definition. If S is an incidence system with blocks $B_1 \ldots, B_b$ and objects $a_1, \ldots, a_v$, the incidence matrix $M = (m_{ij})$, where $i = 1, \ldots, v$ and $j = 1, \ldots, b$, of S is defined by the rules

$$m_{ij} = 1 \qquad \text{if } a_i \in B_j,$$

$$m_{ij} = 0 \qquad \text{if } a_i \notin B_j.$$

Let us recall that if a block design D has parameters v, b, r, k, λ, then the two relations (13.1) hold:

$$bk = vr, \qquad r(k - 1) = \lambda(v - 1).$$

It may happen that $b = v$, and so $r = k$. Here we call the design a *symmetric design*, and the first relation of (13.1) becomes trivial, whereas the second reduces to

$$(13.2) \qquad k(k - 1) = \lambda(v - 1).$$

We shall exclude the trivial case in which $k = v$, that is, the one in which each block consists of all objects.

Let $A = [a_{ij}]$, where $i = 1, \ldots, v$ and $j = 1, \ldots, b$, be the incidence matrix of the block design D. Then we shall show that

$$(13.3) \qquad AA^T = (r - \lambda)I + \lambda J,$$

where A^T is the transpose of A, I is the v by v identity matrix, and J is the v by v matrix in which all entries are 1's. To see this, note that if $AA^T = B = [b_{st}]$ for $s, t = 1, \ldots, v$, then b_{st} is the inner product of the sth and tth rows of A. Since each entry of A is 0 or 1, an inner product of a row with itself counts the number of 1's in a row. Hence $b_{ss} = r$ for $s = 1, \ldots, v$, since each object of D occurs in exactly r blocks, and a row of A lists the occurrences of an object in the different blocks. Similarly, if $s \neq t$, then b_{st} is the inner product of different rows of A, and b_{st} counts the number of columns in which both rows have 1 as an entry. This, however, is counting the number of blocks in which both objects occur. By definition, this is the parameter λ, and so $b_{st} = \lambda$ if $s \neq t$. Thus in the matrix $B = [b_{st}]$, we have $b_{ss} = r$, $b_{st} = \lambda$, for $s \neq t$. This establishes the relation (13.3).

We have already used the fact that each row of A contains 1's in exactly r places. It is also true that each column of A contains 1's in exactly k places, since a column lists the objects in a particular block. We may express these facts in terms of matrices. Let J_{bb}, J_{vv}, and J_{vb} be the matrices consisting entirely of 1's, the subscripts indicating the number of rows and columns. Since A is a v by b matrix, the relations are

$$AJ_{bb} = rJ_{vb}, \qquad J_{vv}A = kJ_{vb}.$$

The first of these relations indicates that 1 occurs exactly r times in each row of A, and the second that 1 occurs exactly k times in each column of A.

The incidence matrix corresponding to the block design of Example 13.1, with seven objects $1, \ldots, 7$ and seven blocks $B_1, \ldots, B_7$, is the following:

$$A = \begin{bmatrix} 1 & 0 & 0 & 0 & 1 & 0 & 1 \\ 1 & 1 & 0 & 0 & 0 & 1 & 0 \\ 0 & 1 & 1 & 0 & 0 & 0 & 1 \\ 1 & 0 & 1 & 1 & 0 & 0 & 0 \\ 0 & 1 & 0 & 1 & 1 & 0 & 0 \\ 0 & 0 & 1 & 0 & 1 & 1 & 0 \\ 0 & 0 & 0 & 1 & 0 & 1 & 1 \end{bmatrix}$$

Here the third row shows that object 3 occurs in blocks B_2, B_3, B_7, and the fifth column shows that block B_5 contains objects $1, 5, 6$. For this matrix, we can check directly that

$$A A^T = 2I + J, \qquad AJ = 3J, \qquad JA = 3J.$$

Here in each case the J is a 7 by 7 matrix.

We can also describe a block design D in terms of a quadratic form. Let $x_1, \ldots, x_v$ be indeterminates corresponding to the objects. Then if $A = [a_{ij}]$ is the incidence matrix of D, we write

$$L_j = \sum_{i=1}^{v} a_{ij} x_i, \qquad j = 1, \ldots, b.$$

Here L_j corresponds to the block B_j and is simply the sum of the indeterminates corresponding to the objects contained in B_j. Then we have

(13.4)

$$L_1^2 + L_2^2 + \cdots + L_b^2 = (r - \lambda)(x_1^2 + \cdots + x_v^2) + \lambda(x_1 + \cdots + x_v)^2.$$

Thus we observe that in $L_1^2 + \cdots + L_b^2$ the coefficient of x_i^2 is the number of times that x_i occurs in the L's, namely r, the number of occurrences of the object a_i in the blocks. Similarly, a cross product $2x_i x_j$ arises whenever an L contains both x_i and x_j with coefficient 1, and this happens λ times for each pair x_i, x_j. Hence in $L_1^2 + \cdots + L_b^2$ the term $x_i x_j$ occurs with coefficient 2λ. Thus the coefficients of all terms agree, and (13.4) is an identity. Of course, the matrix relation (13.3) and the quadratic identity (13.4) are equivalent to each other.

We now turn to the derivation of properties of block designs from the matrix formulation (13.3) and the quadratic formulation (13.4). We calculate the determinant of the following matrix B:

$$AA^T = B = \begin{bmatrix} r & \lambda & \cdots & \lambda \\ \lambda & r & \cdots & \lambda \\ \cdot & \cdot & \cdots & \cdot \\ \lambda & \lambda & \cdots & r \end{bmatrix},$$

obtaining

(13.5) $\det B = [r + (v - 1)\lambda](r - \lambda)^{v-1}$.

This is easily seen by subtracting the first column of B from each of the others and then adding the second, third, and remaining rows to the first row. The matrix then becomes triangular, with $r + (v - 1)\lambda$ in the upper left-hand corner and $r - \lambda$ in the other positions on the main diagonal. This gives the value of $\det B$ in (13.5).

Theorem 13.1. In any block design we must have $b \geq v$. In a symmetric block design, if v is even, then $k - \lambda$ is a square.

PROOF. If $r = \lambda$, then an object occurs with each other object every time it occurs. This would mean that each block contains each object, a case that has been excluded. Hence we have $r > \lambda$ and $\det B \neq 0$. Since the rank of B cannot exceed the rank of A, we must have $b \geq v$, the first assertion of the theorem.

This may also be seen in the following way: If $b < v$, we may adjoin $v - b$ columns of zeros to A to form a square matrix A_1. Then $A_1 A_1^T = AA^T = B$, as is easily verified. But then $\det B = (\det A_1)^2$, and since A_1 has a column of 0's we have $\det A_1 = 0$, conflicting with $\det B \neq 0$. Hence the assumption $b < v$ is impossible, and so $b \geq v$.

For the second part of the theorem, we note that A is a square matrix and

$$\det B = (\det A)^2 = [k + (v - 1)\lambda](k - \lambda)^{v-1}.$$

For a symmetric design, the basic relation $k(k - 1) = \lambda(v - 1)$ gives $k + (v - 1)\lambda = k^2$, and therefore we obtain

$$(\det A)^2 = k^2(k - \lambda)^{v-1}.$$

Thus $(k - \lambda)^{v-1}$ is a square, and so if v is odd then $k - \lambda$ must be a square. This is the second part of the theorem.

In 1949 Bruck and Ryser [8] made the observation that if (13.4) holds with the L's linear forms having 0's and 1's as coefficients, then necessarily it must hold if we merely require that the L's be linear forms with rational coefficients. For symmetric block designs, this amounts to the rational equivalence of two quadratic forms, and this question may be answered by application of a general theory due to Hasse and Minkowski.

This is a deep theory that makes use of the Hilbert norm-residue symbol. Here we shall give a more elementary derivation of the result, following Chowla and Ryser [9].

We begin with the identity

(13.6) $(b_1^2 + b_2^2 + b_3^2 + b_4^2)(x_1^2 + x_2^2 + x_3^2 + x_4^2) = y_1^2 + y_2^2 + y_3^2 + y_4^2,$

where

(13.7)
$$y_1 = b_1 x_1 - b_2 x_2 - b_3 x_3 - b_4 x_4,$$
$$y_2 = b_2 x_1 + b_1 x_2 - b_4 x_3 + b_3 x_4,$$
$$y_3 = b_3 x_1 + b_4 x_2 + b_1 x_3 - b_2 x_4,$$
$$y_4 = b_4 x_1 - b_3 x_2 + b_2 x_3 + b_1 x_4.$$

This may be verified directly. If b_1, b_2, b_3, b_4 are rational, we can solve (13.7) for the x's in terms of the y's, the denominator being $(b_1^2 + b_2^2 + b_3^2 + b_4^2)^2$.

By means of this identity, it was shown by Lagrange that every positive integer n has a representation

(13.8) $n = b_1^2 + b_2^2 + b_3^2 + b_4^2,$

the b's being integers. For a proof of this, the reader is referred to [18, pp. 300, 301].

For a symmetric block design, (13.4) becomes

(13.9) $L_1^2 + \cdots + L_v^2 = (k - \lambda)(x_1^2 + \cdots + x_v^2) + \lambda(x_1 + \cdots + x_v)^2.$

We write $k - \lambda = n$ and consider cases in which v is odd. First we take $v \equiv 1 \pmod 4$. Using (13.8), we can apply the identity (13.6) to the right-hand side of (13.9), taking four terms at a time. This gives

(13.10) $L_1^2 + \cdots + L_v^2 = y_1^2 + \cdots + y_{v-1}^2 + n x_v^2 + \lambda(x_1 + \cdots + x_v)^2.$

When we write $y_v = x_v$ and express $L_1, \ldots, L_v$ and $x_1 + \cdots + x_v = w$ as rational linear forms in the y's, (13.10) becomes a rational identity in $y_1, \ldots, y_v$, namely,

(13.11) $L_2^2 + \cdots + L_v^2 = y_1^2 + \cdots + y_{v-1}^2 + n y_v^2 + \lambda w^2.$

Now we have $L_1 = C_{11} y_1 + C_{12} y_2 + \cdots + C_{1v} y_v$, with the C's rational numbers. If $C_{11} \neq 1$, we put $L_1 = y_1$, and if $C_{11} = 1$, we put $L_1 = -y_1$. In either case, we can solve rationally for y_1 in terms of $y_2, \ldots, y_v$ and have $L_1^2 = y_1^2$. Then (13.11) becomes

$$L_2^2 + \cdots + L_v^2 = y_2^2 + \cdots + y_{v-1}^2 + n y_v^2 + \lambda w^2,$$

an identity in the independent indeterminates $y_2, \ldots, y_v$. Similarly, we may set $L_2 = \pm y_2$, $L_3 = \pm y_3, \ldots, L_{v-1} = \pm y_{v-1}$ in turn, and ultimately find

$$L_v^2 = ny_v^2 + \lambda w^2,$$

where L_v and w are some rational multiples of the indeterminate y_v. Taking $y_v = x \neq 0$ as an integral multiple of the denominators in L_v and w, we have, in integers with $x \neq 0$,

(13.12) $$z^2 = nx^2 + \lambda y^2.$$

In the same way if $v \equiv 3 \pmod 4$, let us take a new indeterminate x_{v+1} and add nx_{v+1}^2 to both sides of (13.9), obtaining

$$L_1^2 + \cdots + L_v^2 + nx_{v+1}^2 = n(x_1^2 + \cdots + x_{v+1}^2) + \lambda(x_1 + 4x_v)^2.$$

Using (13.6), we have

$$L_1^2 + \cdots + L_v^2 + nx_{v+1}^2 = y_1^2 + \cdots + y_{v+1}^2 + \lambda w^2,$$

and proceeding as before we find a solution in integers, with $x \neq 0$, of

(13.13) $$nx^2 = z^2 + \lambda y^2.$$

We combine (13.12) and (13.13), for v odd, into

$$z^2 = nx^2 + (-1)^{(v-1)/2}\lambda y^2.$$

We can now combine these results into a single theorem:

Theorem 13.2. If there is a symmetric block design D with parameters v, k, λ, then necessarily

(a) for v even, $k - \lambda$ is a square,
(b) for v odd, the equation

$$z^2 = (k - 1)x^2 + (-1)^{(v-1)/2}\lambda y^2$$

has a solution in integers x, y, z, with $x \neq 0$.

By means of the deep Hasse-Minkowski theory it can be shown that the conditions of Theorem 13.2 are in addition sufficient for the existence of a rational matrix A satisfying (13.3) or of rational linear forms $L_1, \ldots, L_v$ satisfying (13.9). In short, Theorem 13.2 gives substantially all the information about block designs that can be obtained by considering only the rational field.

The following theorem due to Ryser [25] shows how certain properties of symmetric block designs are purely matrix properties.

Theorem 13.3. Let A be a real nonsingular v by v matrix that satisfies either

(13.14) $$AA^T = (k - \lambda)I + \lambda J$$

or

(13.15) $$.A^T A = (k - \lambda)I + \lambda J,$$

and either

(13.16) $$AJ = kJ$$

or

(13.17) $$JA = kJ.$$

Then A satisfies all four relations, and v, k, λ satisfy the equation

$$k^2 - k = \lambda(v - 1).$$

PROOF. Since $\det B = [k + \lambda(v - 1)](k - \lambda)^{v-1}$, the nonsingularity of A means that $k - \lambda$ and $k + \lambda(v - 1)$ are different from zero.

Let us first suppose that (13.14) and (13.16) hold. Multiplying (13.16) on the left by A^{-1}, we obtain $J = kA^{-1}J$, and so $k \neq 0$ and $A^{-1}J = k^{-1}J$. Now multiplying (13.14) on the left, we get

(13.18) $$A^T = (9k - \lambda)A^{-1} + \lambda A^{-1}J = (k - \lambda)A^{-1} + k^{-1}\lambda J.$$

Taking the transpose of (13.16) and noting that $J^T = J$, we have $JA^T = kJ$. Multiplying (13.18) on the left by J gives

$$kJ = JA^T = (k - \lambda)JA^{-1} + k^{-1}\lambda J^2$$
$$= (k - \lambda)JA^{-1} + k^{-1}v\lambda J,$$

since $J^2 = vJ$. This yields

(13.19) $$JA^{-1} = \frac{(k - k^{-1}v\lambda)}{k - \lambda} J = mJ,$$

where we have written m for the constant. Then we have

$$J = mJA,$$

and therefore

$$vJ = J^2 = (mJA)J = mJ(AJ)$$
$$= mJ(kJ) = mkJ^2 = mkvJ.$$

From this we get $mk = 1$, and so $m = k^{-1}$. Now, putting $m = k^{-1}$ in (13.19) and comparing the constants gives

$$k^{-1}(k - \lambda) = k - k^{-1}v\lambda,$$

which yields

$$k^2 - k = \lambda(v - 1),$$

the equation of the theorem. In addition, with $m = k^{-1}$ in (13.19), from $JA^{-1} = k^{-1}J$, we obtain

(13.20) $$JA = kJ,$$

which is (13.17). Now multiplying (13.18) on the right by A and using (13.20) gives

$$A^T A = (k - \lambda)I + \lambda J,$$

which is (13.15), as desired.

We have now shown that (13.14) and (13.16) imply the remaining relations (13.15), (13.17), and $k^2 - k = \lambda(v - 1)$.

Now let us suppose that (13.14) and (13.17) hold. Then we have

$$J(AA^T) = (k - \lambda)J + \lambda J^2,$$
$$kJA^T = (k - \lambda)J + \lambda vJ = mJ, \qquad m = k - \lambda + \lambda v,$$
$$kJ(A^TJ) = mJ^2,$$
$$kJ(JA)^T = mJ^2,$$
$$kJ(kJ)^T = mJ^2,$$
$$k^2J^2 = mJ^2,$$

and therefore $k^2 = m = k - \lambda + \lambda v$, which is the relation $k^2 - k = \lambda(v - 1)$. Also this shows that $kJA^T = k^2J$. From (13.17) we get $k \neq 0$ since A is nonsingular. Thus we have $JA^T = kJ$, and taking transposes we obtain $AJ = kJ$, which is (13.16). Then also we have

(13.21) $A^T A = A^{-1}(AA^T)A = (k - \lambda)I + A^{-1}\lambda JA.$

Since $AJ = kJ = JA$, however, it follows that $A^{-1}JA = J$, and so (13.21) becomes

$$A^T A = (k - \lambda)I + \lambda J,$$

which is (13.15) as we wished to prove. Thus (13.14) and (13.17) imply the remaining relations. With A^T taking the place of A, the same argument shows that (13.15) and either (13.16) or (13.17) imply the rest. This proves all parts of the theorem.

From our original definition of block designs, the incidence matrix, a symmetric block design with parameters v, k, λ, satisfies (13.14), (13.16), and (13.17). Theorem 13.3 shows that (13.15) is also satisfied. Hence A^T is also the incidence matrix of a symmetric v, k, λ design, which is dual to the original design in the sense that the roles of objects and blocks are interchanged. This proves the statement made in Sec. 13.2 that any two blocks of a symmetric design have exactly λ objects in common.

The foregoing property of a symmetric design enables us to construct two nonsymmetric designs from a symmetric one. If $B_1, \ldots, B_v$ are the blocks of a symmetric design D, we choose any block, say B_v, and delete from each of $B_1, \ldots, B_{v-1}$ the λ objects in common with B_v. The deleted blocks $B_1^*, \ldots, B_{v-1}^*$ form what is called the *residual design* D^* in which

TABLE 13.3
Construction of Two Nonsymmetric
Designs from a Symmetric Design

Residual design						Derived design		
1	2	5	6	11	12	17	20	23
1	2	9	10	15	16	17	21	25
1	2	7	8	13	14	17	22	24
3	4	7	8	9	10	17	20	23
3	4	11	12	13	14	17	21	25
3	4	5	6	15	16	17	22	24
1	4	5	8	10	11	19	22	25
1	4	9	12	14	15	19	20	24
1	4	6	7	13	16	19	21	23
2	3	6	7	9	12	19	22	25
2	3	10	11	13	16	19	21	24
2	3	5	8	14	15	19	21	23
1	3	5	7	10	12	18	21	24
1	3	9	11	14	16	18	22	23
1	3	6	8	13	15	18	20	25
2	4	6	8	9	11	18	21	24
2	4	10	12	13	15	18	22	23
2	4	5	7	14	16	18	20	25
5	6	9	10	13	14	17	18	19
5	7	9	11	13	15	20	21	22
5	8	9	12	13	16	23	24	25
7	8	11	12	15	16	17	18	19
6	8	10	12	14	16	20	21	22
6	7	10	11	14	15	23	24	25

the parameters b_1, v_1, k_1, r_1, and λ_1 are related to those of D with parameters $b = v, r = k$, and λ by the relations

(13.22) D^*: $b_1 = v - 1, v_1 = v - k, r_1 = k, k_1 = k - \lambda, \lambda_1 = \lambda.$

This is easily seen, since exactly λ objects are deleted from each of $B_1, \ldots, B_{v-1}$, and since the $v - k$ objects of D not in B_v remain unaltered in their occurrences in $B_1, \ldots, B_{v-1}$. We may also form the *derived design* D' by taking, as our blocks $B_1', \ldots, B_{v-1}'$, the λ objects that $B_1, \ldots, B_{v-1}$, respectively, have in common with B_v. This again is a design, with parameters b_2, v_2, r_2, k_2, λ_2 having the values

$$D': \quad b_2 = v - 1, v_2 = k, r_2 = k - 1, k_2 = \lambda, \lambda_2 = \lambda - 1.$$

Example 13.9   Construction of Nonsymmetric Designs from Symmetric Designs. We illustrate the foregoing construction with an example due to Bhattacharya [3] for $v = b = 25$, $k = r = 9$, $\lambda = 3$. Here B_{25} consists of the objects $17, \ldots, 25$, D^* is the design with 24 blocks in which the objects are the numbers $1, \ldots, 16$, and D' is the design with 24 blocks

TABLE 13.4
Nonresidual Design with Parameters of a Residual Design

Block No.	Block						Block No.	Block					
1	1	2	7	8	14	15	13	1	4	7	8	11	16
2	3	5	7	8	11	13	14	2	4	8	10	12	14
3	2	3	8	9	13	16	15	5	6	8	10	15	16
4	3	5	8	9	12	14	16	1	6	8	10	12	13
5	1	6	7	9	12	13	17	1	2	3	11	12	15
6	2	5	7	10	13	15	18	2	6	7	9	14	16
7	3	4	7	10	12	16	19	1	4	5	13	14	16
8	3	4	6	13	14	15	20	2	5	6	11	12	16
9	4	5	7	9	12	15	21	1	3	9	10	15	16
10	2	4	9	10	11	13	22	4	6	8	9	11	15
11	3	6	7	10	11	14	23	1	5	9	10	11	14
12	1	2	3	4	5	6	24	11	12	13	14	15	16

in which the objects are the numbers $17, \ldots, 25$, as shown in Table 13.3. The blocks of D^* are to the left of the vertical line, the blocks of D' to the right; D^* has parameters $b_1 = 24$, $v_1 = 16$, $r_1 = 9$, $k_1 = 6$, $\lambda_1 = 3$, and D' has parameters $b_2 = 24$, $v_2 = 9$, $r_2 = 8$, $\lambda_2 = 2$.

Example 13.10   Nonresidual Design with Parameters of a Residual Design. The nature of the derived and residual design obtained from a design D depends on which block is taken as B_v to be eliminated. The different residual and derived designs obtained by eliminating different blocks from D need not be isomorphic. Bhattacharya [2] has also found an example of a design with the parameters of a residual design that is not the residual design of any symmetric design. The design, with $b = 24$, $v = 16$, $r = 9$, $k = 6$, $\lambda = 3$, is shown in Table 13.4.

Here the two blocks 1, 6, 7, 9, 12, 13 and 1, 6, 8, 10, 12, 13 have four objects in common. Hence it is not possible for this design to be the derived design of a symmetric design with $v = b = 25$, $r = k = 9$, $\lambda = 3$, since in such a symmetric design any two distinct blocks have exactly three objects in common.

If, however, $\lambda = 1$ or $\lambda = 2$, then a design with parameters as given by (13.22) is necessarily the residual design of a unique symmetric design. For $\lambda = 1$, this is essentially the assertion that an affine Euclidean plane can be extended to a projective plane by adding a line at infinity, and the proof is fairly simple. But for $\lambda = 2$, the situation is considerably more complicated. This has been proved by Connor and Hall [11], who used conditions originally developed by Connor [10], these being necessary conditions on blocks $B_1, \ldots, B_t$ to be part of a design with blocks $B_1, \ldots, B_b$. The proof depends in part on the fact that for $\lambda = 1$ or 2 the derived design is trivial.

13.5 CONSTRUCTION OF BLOCK DESIGNS AND ORTHOGONAL LATIN SQUARES

The methods that have been used to construct block designs are so varied that it is tempting to lump them all together into a single category labeled "miscellaneous." We can single out three general categories, however, which we shall describe as (1) recursive methods, (2) arithmetical methods, and (3) group-theoretical methods.

Recursive methods

There are several methods of constructing block designs that can loosely be described as *recursive methods*. The first is a method of *composition*, in which two designs D_1 and D_2 are combined in a way to form a third design D_3.

A block design with $k = 3$ and $\lambda = 1$ is called a *Steiner triple* system. From the basic relations (13.1) we find

$$3b = vr, \qquad 2r = v - 1,$$

and therefore $\qquad v = 2r + 1, \qquad 3b = r(2r + 1)$

Thus v is odd, and either r or $2r + 1$ is a multiple of 3. These conditions combined show that either $v = 6t + 1$ [if $r = 3t$] or $v = 6t + 3$ [if $2r + 1 = 3(2t + 1)$]. Hence for a Steiner triple system we have either

(13.23) $\quad b = 6t^2 + t, \qquad v = 6t + 1, \qquad r = 3t, \qquad k = 3, \qquad \lambda = 1,$

or

(13.24)

$b = 6t^2 + 5t + 1, \qquad v = 6t + 3, \qquad r = 3t + 1, \qquad k = 3, \qquad \lambda = 1.$

It is true that for every number v of the form $6t + 1$ or $6t + 3$ there is a Steiner triple system with v objects and with parameters as given by (13.23) or (13.24). From a Steiner triple system S with v objects, we can construct a multiplicative system M on the v objects of S by putting $x^2 = x$ for every x, and putting $xy = z$ if $x \neq y$ and if x, y, z is the unique triple of S containing the pair x, y. The systems M can be characterized in the following ways.

1. We have $x^2 = x$ for every $x \in M$.

2. If $x \neq y$, then $xy = z$ is a unique element of M and $z \neq x$, $z \neq y$, and also $yx = z$, $xz = zx = y$, $yz = zy = x$.

We can easily verify that, given a system M with a binary multiplication satisfying (1) and (2), the triples x, y, z determined by $x \neq y$, $xy = z$ form a Steiner triple system with the same number of objects as the number of elements in M.

If M_1 is the multiplicative system for a Steiner triple system S_1 with v_1 objects, and M_2 is the multiplicative system for S_2 with v_2 objects, then we can construct a system $M = M_1 \times M_2$ in which the elements are the $v_1 v_2$ ordered pairs (x_1, x_2), $x_1 \in M_1$, $x_2 \in M_2$. If we define our product in M by the rule

$$(x_1, x_2)(y_1, y_2) = (x_1 y_1, x_2 y_2),$$

then it is immediate that the multiplication in $M = M_1 \times M_2$ has the properties (1) and (2), and so M yields a Steiner triple system with $v = v_1 v_2$ objects. Thus we conclude that if there is a Steiner triple system S_1 with v_1 objects, and another S_2 with v_2 objects, then there is a third, $S = S_1 \times S_2$, with $v_1 v_2$ objects. The reader may verify that the trivial Steiner triple system S_1, with three objects 1, 2, 3 and the single triple 1, 2, 3, is such that $S_1 \times S_1$ is the system S with 9 objects as given in Example 13.2. We note in (13.23) and (13.24) that the subsystems with x_1 or x_2 fixed correspond to subsystems of S isomorphic to S_2 and S_1, respectively.

A more complicated recursive construction for Steiner triple systems is due to E. H. Moore [21]. We describe this by a theorem.

Theorem 13.4. If S_2 is a Steiner triple system with v_2 objects, if S_2 contains a Steiner triple subsystem S_3 with v_3 objects, and if S_1 is also a Steiner triple system, with $v_1 > 1$ objects, then we can construct a Steiner triple system S with $v = v_3 + v_1(v_2 - v_3)$ objects, such that S has subsystems isomorphic to S_1, S_2, and S_3.

PROOF. Here we regard a system S_3 with one object x and no triples as satisfying our requirements. Thus $v_3 = 1$ is permitted for any S_2 with $v_2 > 1$. We construct an array with $v = v_3 + v_1(v_2 - v_3)$ objects listed in $v_1 + 1$ sets,

$$T_0: \quad a_1, \ldots, a_{v_3},$$
$$T_1: \quad b_{11}, \ldots, b_{1s},$$
(13.25)
$$T_2: \quad b_{21}, \ldots, b_{2s},$$
$$\cdots \cdots \cdots \cdots$$
$$T_{v_1}: \quad b_{v_1 1}, \ldots, b_{v_1 s}, \qquad s = v_2 - v_3.$$

The $v_3 + v_1(v_2 - v_3)$ objects in (13.25) are all distinct. We form triples of these objects by three rules: (1) We associate the objects of S_0 with the system having v_3 objects, and we take as a triple of a new system S a triple (a_i, a_j, a_k) if (i, j, k) is a triple of S_3. (2) We make T_0 and any T_i, $i = 1, \ldots, v_1$, correspond to the system of order v_2, T_0 corresponding to the subsystem of v_3 objects. Triples of a's are already determined by rule (1). Other triples contain at most one a, and we form triples (a_m, b_{ij}, b_{ik}) or (b_{ij}, b_{ik}, b_{ir}) according to the correspondence with the system S_2 having v_2 elements. (3) If (j, k, r) is a triple of the system having v_1 objects, we form all triples (b_{jx}, b_{ky}, b_{rz}) in which the second subscripts satisfy $x + y + z \equiv 0 \pmod{s}$.

These three rules taken together give the triples of a Steiner system S with $v = v_3 + v_1(v_2 - v_3)$ objects. We can prove the existence of systems S of all orders $v = 6t + 1$ or $6t = 3$ by using several specializations of this rule. Note that the system S thus constructed has subsystems isomorphic to S_1, S_2, S_3.

We consider five special cases of the general theorem, giving

$$\begin{aligned}
&(a)\ v_1 = t, &&v_2 = 3, &&v_3 = 1, &&v = 2t + 1, &&t \geq 3;\\
&(b)\ v_1 = 3, &&v_2 = t, &&v_3 = 1, &&v = 3t - 2, &&t \geq 3;\\
&(c)\ v_1 = 3, &&v_2 = t, &&v_3 = 3, &&v = 3t - 6, &&t \geq 7;\\
&(d)\ v_1 = t, &&v_2 = 9, &&v_3 = 3, &&v = 6t + 3, &&t \geq 3;\\
&(e)\ v_1 = 3, &&v_2 = t, &&v_3 = 7, &&v = 3t - 14, &&t \geq 15.
\end{aligned}$$

In case e, it is necessary that the system of order v_2 contain a subsystem of order 7. These rules, when applicable, enable us to construct further Steiner triple systems as shown in Table 13.5.

Examples 13.1 and 13.2 are the systems $S(7)$ and $S(9)$ of orders 7 and 9. We must also construct the system $S(13)$ (there are two nonisomorphic systems of order 13). If we form $S(21)$ as the direct product of $S(3)$ and $S(7)$, and construct systems $S(25)$, $S(33)$, and $S(37)$ containing subsystems $S(7)$, then the above rules suffice to construct Steiner triple systems $S(v)$ of all orders $v = 6t + 1$ or $v = 6t + 3$; furthermore, for $v \geq 15$, $S(v)$ always contains an $S(7)$, this last being necessary whenever we use rule e.

We can avoid the labor of constructing $S(25)$, $S(27)$, $S(33)$, and $S(37)$ with a subsystem $S(7)$ if we use composition or the rules of the theorem to construct $S(v)$'s so that, whenever possible, $S(v)$ contains an $S(7)$. With a little calculation, we find that $S(v)$'s exist with a subsystem $S(7)$ except for $v = 1, 3, 9, 13, 25, 27, 33, 37, 67, 69, 75, 81, 97, 109, 139, 201, 289, 321, 643$. We must use the theorem in a different way whenever t is in this list and

TABLE 13.5

Construction of Steiner Triple Systems

Form of v	Rule	Value of t
$36k + 1$	b	$12k + 1$
$36k + 3$	a	$18k + 1$
$36k + 7$	a	$18k + 3$
$36k + 9$	d	$6k + 1$
$36k + 13$	e	$12k + 9$
$36k + 15$	a	$18k + 7$
$36k + 19$	b	$12k + 7$
$36k + 21$	d	$6k + 3$
$36k + 25$	b	$12k + 9$
$36k + 27$	a	$18k + 13$
$36k + 31$	a	$18k + 15$
$36k + 33$	c	$12k + 3$

v is not. As an example, Table 13.5 gives $85 = 3 \cdot 33 - 14$ by application of e, which we may not use unless we have an $S(33)$ containing an $S(7)$. For instance, the values obtained from $t = 643$ by the rules are

$$2t + 1 = 1287 = 3 \cdot 429,$$
$$3t - 14 = 1915 = 1 + 319(7 - 1),$$
(13.26)
$$3t - 6 = 1923 = 1 + 2 \cdot 961,$$
$$3t - 2 = 1927 = 1 + 321(7 - 1),$$
$$6t + 3 = 3861 = 3 \cdot 1287.$$

The alternative expressions for v yield another way of finding an $S(v)$ containing an $S(7)$. This proves that Steiner triple systems of all orders $v = 6t + 1$, $6t + 3$ exist. The original proof of the existence of Steiner triple systems of all orders was due to M. Reisz [24]; his proof used a recursive method of going from a system $S(v)$ to $S(2v + 1)$ or to $S(2v - 5)$ by explicitly listing triples to be deleted from and added to $S(v)$ to give the desired system.

More recent use of recursive methods has been made by H. Hanani [17].

An *Hadamard matrix* H of order n is an n by n matrix of $+1$'s and -1's such that

(13.27) $$HH^T = nI.$$

Now (13.27) is equivalent to the assertion that any two rows of H are orthogonal. Clearly, permuting rows and columns of H or multiplying a row or column of H by -1 leaves the property (13.27) unchanged, and so such matrices are considered equivalent to H. If we wish to replace H by an equivalent matrix, we may assume H *normalized* so that the first row and first column consist entirely of $+1$'s. Let us suppose that $n \geq 3$ and that H is normalized. In the first three rows, consider the columns of types

$$\begin{bmatrix} 1 \\ 1 \\ 1 \end{bmatrix}, \quad \begin{bmatrix} 1 \\ 1 \\ -1 \end{bmatrix}, \quad \begin{bmatrix} 1 \\ -1 \\ 1 \end{bmatrix}, \quad \begin{bmatrix} 1 \\ -1 \\ -1 \end{bmatrix},$$

and suppose there are x, y, z, w of these types, respectively. Then we have

$$x + y + z + w = n,$$
$$x + y - z - w = 0,$$
$$x - y + z - w = 0,$$
$$x - y - z + w = 0.$$

The first equation indicates that there are n columns, the second and third express the orthogonality of the first row with the second row and with the third row, and the last expresses the orthogonality of the second and third rows. From these equations, we find that

(13.28) $$x = y = z = w = \frac{n}{4}.$$

Hence we conclude that if H is an Hadamard matrix of orders $n > 2$, then n is a multiple of 4. We readily find Hadamard matrices of orders 1 and 2, namely

(13.29) $$H_1 = [1], \qquad H_2 = \begin{bmatrix} 1 & 1 \\ 1 & -1 \end{bmatrix}.$$

In an Hadamard matrix H of order $n = 4t$, we suppose that H is normalized, and we number the rows and columns $0, 1, \ldots, 4t - 1$. From the ith row, $i = 1, \ldots, 4t - 1$, we construct an incidence system of blocks B_i, in which $j \in B_i$ if and only if $j > 0$ and $h_{ij} = +1$. We have $b = 4t - 1$ blocks B_i, and thus by the orthogonality of the ith row with the 0th row, each block contains $k = 2t - 1$ objects since the ith row must contain $+1$ in column 0 and $2t - 1$ additional $+1$'s. Furthermore, from

(13.28) we see that any two distinct blocks B_i and B_k have exactly $n/4 - 1 = t - 1$ objects in common. In terms of the incidence matrix A of this coincidence system, we thus see that (13.15) and (13.16) are satisfied with the parameters $v = 4t - 1$, $k = 2t - 1$, $\lambda = t - 1$. By Theorem 13.3, all four equations (13.14), (13.15), (13.16), and (13.17) are satisfied, and so we have a symmetric block design with parameters $v = 4t - 1$, $k = 2t - 1$, $\lambda = t - 1$. Conversely, it is easy to construct a normalized Hadamard matrix H of order $4t$ from such a block design, since the block design specifies the positions of the $+1$'s inside the border of $+1$'s in row 0 and column 0. The block-design properties correspond exactly to the orthogonality of H.

There is a very interesting composition for Hadamard matrices. This we state as a theorem.

Theorem 13.5. The Kronecker product $H_1 \times H_2$ of an Hadamard matrix H_1 of order n_1 and a Hadamard matrix H_2 of order n_2 is an Hadamard matrix $H = H_1 \times H_2$ of order $n_1 n_2$.

In general, if $A = (a_{ij})$, where $i, j = 1, \ldots, n$, and $B = (b_{rs})$, where $r, s = 1, \ldots, m$, are two matrices of order n and m, respectively, then the *Kronecker product* $A \times B$ is the mn by mn matrix

$$A \times B = \begin{bmatrix} a_{11}B & a_{12}B & \cdots & a_{1n}B \\ a_{21}B & a_{22}B & \cdots & a_{2n}B \\ \cdots\cdots\cdots\cdots\cdots\cdots\cdots \\ a_{n1}B & a_{n2}B & \cdots & a_{nn}B \end{bmatrix}.$$

It is not hard to show that $B \times A$ can be obtained from $A \times B$ by permuting rows and columns. If A and B are both Hadamard matrices, then trivially each element of $A \times B$ is a $+1$ or a -1. Now consider the inner product of two rows of $A \times B$. This is

$$\sum_{j=1}^{n} a_{ij} a_{kj} (x_r, x_s),$$

where x_r and x_s are rows of B. If $x_r \neq x_s$, then $(x_r, x_s) = 0$ since B is an Hadamard matrix. If $x_r = x_s$, then $(x_r, x_s) = m$ and $i \neq k$, and the sum becomes

$$m \sum_{j=1}^{n} a_{ij} a_{kj} = m \cdot 0 = 0$$

since A is an Hadamard matrix. This proves the theorem.

As an easy application of the theorem, we calculate $H_4 = H_2 \times H_2$. By (13.29), this is

$$H_4 = \begin{bmatrix} 1 & 1 & 1 & 1 \\ 1 & -1 & 1 & -1 \\ 1 & 1 & -1 & -1 \\ 1 & -1 & -1 & 1 \end{bmatrix}.$$

In particular, Hadamard matrices of order 2^m exist for all m.

It is plausible that Hadamard matrices exist for all $n = 4t$. At the present this is known for $n = 4, 8, \ldots, 112$, the first multiple of 4 in doubt being 116. The value 92 was in doubt for many years, but a solution was recently found by Baumert, Golomb, and Hall [1].

Arithmetical methods

The second general category of constructive methods has been labeled *arithmetical*. As an instance of this kind of method, let us consider a way of constructing certain Hadamard matrices given by R. E. A. C. Paley [22]. Let $p \equiv 3 \pmod 4$ be a prime number. For $x \equiv 1, \ldots, p - 1 \pmod p$, there are exactly $(p - 1)/2$ values $x^2 \equiv a \pmod p$; these are called the *quadratic residues* of p. Then it is true that, with $p = 4t - 1$, the $k = 2t - 1$ quadratic residues $a_1, \ldots, a_k \pmod p$ are such that B_{i+1}: $a_1 + i$, $a_2 + i, \ldots, a_k + i \pmod p$, for $i = 0, \ldots, 4t - 2 = p - 1$, form the blocks of a symmetric design with $v = 4t - 1$, $k = 2t - 1$, $\lambda = t - 1$. For a proof of this, the reader is referred to the author's book [15], which has already been mentioned. For example, with $p = 11$ we have $a_1 = 1$, $a_2 = 3$, $a_3 = 4$, $a_4 = 5$, $a_5 = 9$, and this yields the blocks

B_1: 1, 3, 4, 5, 9; B_5: 5, 7, 8, 9, 2; B_9: 9, 11, 1, 2, 6;
B_2: 2, 4, 5, 6, 10; B_6: 6, 8, 9, 10, 3; B_{10}: 10, 1, 2, 3, 7;
B_3: 3, 5, 6, 7, 11; B_7: 7, 9, 10, 11, 4; B_{11}: 11, 2, 3, 4, 8.
B_4: 4, 6, 7, 8, 1; B_8: 8, 10, 11, 1, 5;

These are the blocks that appeared in Example 13.3. This method yields an Hadamard matrix of order $4t$ whenever $4t - 1$ is a prime. More generally, if we use the finite fields with p^r elements rather than the residues modulo p, then we have an Hadamard matrix of order $4t$ whenever $4t - 1$ is a prime power. Thus since $28 - 1 = 3^3$, there is an Hadamard matrix of order 28.

The arithmetical methods are related to the finite fields $GF(p^r)$ and to the geometries over these fields. Since, whenever a number n is a prime power, $n = p^r$, there is a finite field $GF(p^r)$, with $n = p^r$ elements, unique to within isomorphism, it follows that we may use certain arithmetical, or geometrical, methods of construction. If $n = p^r$, let $0, 1, x_2, \ldots, x_{n-1}$ be the elements of the finite field $GF(p^r)$. Then $(uy_0, uy_1, \ldots, uy_m)$, for u, $y_i \in GF(p^r)$, $u \neq 0$, $(y_0, \ldots, y_m) \neq (0, \ldots, 0)$, is a point in the finite projective geometry, $PG(m, p^r)$, of the m-dimensional projective geometry over $GF(p^r)$. The $(m-1)$-dimensional subspaces of $PG(m, p^r)$ form a symmetric block design with

$$v = \frac{p^{(m+1)r} - 1}{p^r - 1}, \qquad k = \frac{p^{mr} - 1}{p^r - 1}, \qquad \lambda = \frac{p^{(m-1)r} - 1}{p^r - 1}.$$

For example, if $p^r = 2^2 = 4$, $m = 2$, we have a symmetric block design with $v = 21$, $k = 5$, $\lambda = 1$.

Under the general heading of arithmetical methods, we shall include the use of finite fields. If p is a prime, the residues $0, 1, \ldots, p - 1$ (mod p) form a finite field in that addition, subtraction, multiplication, and division, excluding division by zero, are all unique operations, provided we assume the familiar commutative and associative laws for addition and multiplication, as well as the distributive law $(a + b)c = ac + bc$.

Besides the residues modulo p, there are also fields with p^r elements, p being a prime and r any positive integer. The finite field $GF(p^r)$ exists and is unique to within isomorphism, for each prime power p^r. This field can be represented in the following way: There exists at least one polynomial

$$f(x) = x^r + a_1 x^{r-1} + \cdots + a_r,$$

the a_i being residues modulo p, which is irreducible modulo p. This means that it is impossible to have

$$f(x) \equiv g(x)h(x) \;(\text{mod } p),$$

where

$$g(x) = x^s + b_1 x^{s-1} \cdots + b_s, \qquad h(x) = x^t + c_1 x^{t-1} + \cdots + c_t,$$

of degrees $s \geq 1$ and $t \geq 1$. If by $A(x) \equiv B(x) \;[\text{modd } p, f(x)]$ we mean that

$$A(x) - B(x) = pR(x) + f(x)S(x),$$

where $A(x)$, $B(x)$, $R(x)$, $S(x)$ are polynomials with integral coefficients, then the polynomials $c_0 + c_1 x + \cdots + c_{r-1} x^{r-1} \;[\text{modd } p, f(x)]$, with c_i integers mod p, form a complete set of residues modd $p, f(x)$. If furthermore $f(x)$ is irreducible mod p, then these residues form the finite field $GF(p^r)$. As an example, $0, 1, x, x + 1 \;(\text{modd } 2, x^2 + x + 1)$ form the finite field $GF(2^2)$. An element y of $GF(p^r)$ is called a *primitive root* if $1, y, y^2, \ldots, y^{p^r-2}$ are all distinct and are the $p^r - 1$ elements of $GF(p^r)$, excluding 0. There is always a primitive root in $GF(p^r)$. [Indeed, there are $\phi(p^r - 1)$ primitive roots.]

Most constructions based on primes work equally well for finite fields. For example, if $p^r \equiv 3 \pmod 4$ then there are exactly $(p^r - 1)/2$ elements of $GF(p^r)$ that are nonzero squares, that is, of the form

$$u^2 \equiv a \not\equiv 0 \ [\text{modd } p, f(x)].$$

These may be used to construct an Hadamard matrix of order $p^r + 1$.

For example, $x^3 - x + 1$ is irreducible modulo 3, and the 27 residues

$$c_0 + c_1 x + c_2 x^2 \ (\text{modd } 3, x^3 - x + 1)$$

form the finite field $GF(3^3)$. In this field, x is a primitive root. The nonzero squares are

$$1, x^2, x^2 - x, x^2 + x + 1, -x^2 - 1, x^2 + x, x^2 - 1,$$
$$-x, -x + 1, x^2 - x + 1, -x^2 + x + 1, -x - 1, -x^2 - x + 1.$$

Let us number the 27 elements of $GF(3^3)$ in some way as

$$u_0 = 0, \qquad u_1, \ldots, u_{26} \qquad (\text{e.g., } u_i = x^{i-1}, i = 1, \ldots, 26).$$

We may use these squares in $GF(3^3)$ to construct an Hadamard matrix of order 28. We take a first row and column of $+1$'s. Numbering the remaining positions $0, \ldots, 26$ in row and column, we put $a_{ij} = +1$ if $u_i - u_j$ is a square, and $a_{ij} = -1$ otherwise. This rule yields a complete Hadamard square of order 28.

We can also use finite fields to construct orthogonal Latin squares. Let

$$n = p_1^{e_1} \cdots p_s^{e_s}$$

be the factorization of an integer n as a product of distinct prime powers. For simplicity, let us write

$$n_1 = p_1^{e_1}, \qquad n_2 = p_2^{e_2}, \ldots, n_s = p^{e_s}.$$

We form a system M of elements $(x_1, \ldots, x_s)$, $x_i \in GF(n_i)$, and define an addition and multiplication of elements of M by the rules

$$(x_1, x_2, \ldots, x_s) + (y_1, y_2, \ldots, y_s) = (x_1 + y_1, \ldots, x_s + y_s),$$
$$(x_1, x_2, \ldots, x_s)(y_1, \ldots, y_s) = (x_1 y_1, x_2 y_2, \ldots, x_s y_s).$$

Here $x_i + y_i$ and $x_i y_i$ are defined as elements of $GF(n_i)$. Let z_i be a primitive root of $GF(n_i)$, $i = 1, \ldots, s$, and put $z = (z_1, \ldots, z_s)$. If t is the minimum of $n_1 - 1, n_2 - 1, \ldots, n_s - 1$, then the elements z^j,

$j = 0, \ldots, t - 1$, are all different in each component $1, \ldots, s$. Hence the linear relations

$$y = z^j x + b, \qquad j = 0, 1, \ldots, t - 1,$$

with $x, y, b \in M$, are such that if

(13.30)
$$y = z^j x + b_1, \qquad 0 \le j \le t - 1,$$
$$y = z^k x + b_2, \qquad 0 \le k \le t - 1,$$

then z^j, z^k, b_1, b_2 determine x and y uniquely except when $j = k$. This

TABLE 13.6
Two Orthogonal 12 by 12 Latin Squares

Row number	Column number																							
	0	1	2	3	4	5	6	7	8	9	10	11	0	1	2	3	4	5	6	7	8	9	10	11
0	0	1	2	3	4	5	6	7	8	9	10	11	0	1	2	3	4	5	6	7	8	9	10	11
1	5	0	1	2	3	4	11	6	7	8	9	10	10	11	6	7	8	9	4	5	0	1	2	3
2	4	5	0	1	2	3	10	11	6	7	8	9	2	3	4	5	0	1	8	9	10	11	6	7
3	3	4	5	0	1	2	9	10	11	6	7	8	6	7	8	9	10	11	0	1	2	3	4	5
4	2	3	4	5	0	1	8	9	10	11	6	7	4	5	0	1	2	3	10	11	6	7	8	9
5	1	2	3	4	5	0	7	8	9	10	12	6	8	9	10	11	6	7	2	3	4	5	0	1
6	6	7	8	9	10	11	0	1	2	3	4	5	9	10	11	6	7	8	3	4	5	0	1	2
7	11	6	7	8	9	10	5	0	1	2	3	4	1	2	3	4	5	0	7	8	9	10	11	6
8	10	11	6	7	8	9	4	5	0	1	2	3	11	6	7	8	9	10	5	0	1	2	3	4
9	9	10	11	6	7	8	3	4	5	0	1	2	3	4	5	0	1	2	9	10	11	6	7	8
10	8	9	10	11	6	7	2	3	4	5	0	1	7	8	9	10	11	6	1	2	3	4	5	0
11	7	8	9	10	11	6	1	2	3	4	5	0	5	0	1	2	3	4	11	6	7	8	9	10

follows from the fact that (13.30) gives

(13.31) $$(z^j - z^k)x_1 = b_2 - b_1,$$

and unless $j = k$, the difference $z^j - z^k$ is distinct from 0 in each component; hence (13.31) determines x uniquely, and then either equation (13.30) determines y uniquely.

We may use the system M to determine a set of t mutually orthogonal n by n Latin squares. Designate the n elements of the system M by $a_0 = 0, a_1 = 1, \ldots, a_{n-1}$ in some manner. The cells of the squares are (x, y), where x, y are elements of M. In the jth square, $j = 0, \ldots, t - 1$, assign the number b to all cells (x, y) for which $y = z^j x + b$. It easily follows that each of $0, \ldots, n - 1$ occurs exactly once in each row and exactly once in each column of the jth square, and thus we have t Latin squares. The uniqueness of x and y in (13.30) when $j \ne k$ is precisely the condition needed for the t squares to be orthogonal.

We state this result as a theorem due to MacNeish [19].

Theorem 13.6. If $n = p_1^{e_1} p_2^{e_2} \cdots p_s^{e_s}$ is the factorization of n into prime powers, and if $v(n) = \min{(p_i^{e_i} - 1)}$, then there exist $v(n)$ mutually orthogonal Latin squares of order n.

We can illustrate this for $n = 12 = 2^2 \cdot 3$ by taking the elements of $GF(4)$ as $0, 1, x, x + 1$ (modd $2, x^2 + x + 1$), of $GF(3)$ as $0, 1, -1$ (modd $2, x^2 + x + 1$), and of $GF(3)$ as $0, 1, -1$ (mod 3). We have $z_1 = x$, $z_2 = -1$ as primitive roots. The elements of M may be numbered as follows:

$$
\begin{array}{lll}
0 = (0, 0), & 5 = (1, -1), & 9 = (x + 1, 0), \\
1 = (1, 1) = z^0, & 6 = (x, 0), & 10 = (x, 1), \\
(13.32) \quad 2 = (0, -1), & 7 = (x + 1, 1), & 11 = (x + 1, -1). \\
3 = (1, 0), & 8 = (x, -1) = z^1, & \\
4 = (0, 1), & &
\end{array}
$$

We obtain the two orthogonal 12 by 12 squares shown in Table 13.6.

Group-theoretical methods

The following method gives

$$v(n) = \min_i{(p_i^{e_i} - 1)}$$

mutually orthogonal squares of order n when

$$n = p_1^{e_1} p_2^{e_2} \cdots p_s^{e_s}$$

is the factorization of n into prime powers. For a long time it was thought that this was the maximum possible number. Here for an $n = 4k + 2$ the value of $v(n)$ is 1, and this supported the Euler conjecture that for $n = 4k + 2$ no orthogonal pair exists. Recently, by means of constructive methods in part recursive and in part group theoretical, both conjectures have been shown to be false. A pair of 10 by 10 orthogonal squares is given in Table 13.8, on page 400.

By *group-theoretical* methods we shall mean those that presuppose a certain group of automorphisms for the design.

For example, to construct a Steiner triple system of order $v = 6t + 3$ with a cyclic automorphism group of order $2t + 1$, we proceed as follows: The parameters are

$$b = (3t + 1)(2t + 1), \, v = 6t + 3, \, r = 3t + 1, \, k = 3, \, \lambda = 1.$$

Our objects will be in three sets of $2t + 1$ each:

$$A_0, A_1, \ldots, A_{2t}; \quad B_0, B_1, \ldots, B_{2t}; \quad C_0, C_1, \ldots, C_{2t}.$$

Each set is permuted cyclically by the generating automorphism of order $2t + 1$.

We observe that the t pairs $(1, 2t)$, $(2, 2t - 1)$ ..., $(t, t + 1)$, regarded as residues modulo $2t + 1$, have as differences $\pm(2t - 1)$, $\pm(2t - 3)$, ..., ± 1; differences are taken in both orders, so that every nonzero residue modulo $2t + 1$ appears exactly once in the list. Consequently, the $3t + 1$ triples,

(13.33)
$$(A_1, A_{2t}, B_0), (A_2, A_{2t-1}, B_0), \ldots, (A_t, A_{t+1}, B_0),$$
$$(B_1, B_{2t}, C_0), (B_2, B_{2t-1}, C_0), \ldots, (B_t, B_{t+1}, C_0),$$
$$(C_1, C_{2t}, A_0), (C_2, C_{2t-1}, A_0), \ldots, (C_t, C_{t+1}, A_0),$$
$$(A_0, B_0, C_0),$$

are such that under application of the generating automorphism α,

$$\alpha = (A_0, A_1, \ldots, A_{2t})(B_0, B_1, \ldots, B_{2t})(C_0, C_1, \ldots, C_{2t}),$$

we have $b = (3t + 1)(2t + 1)$ triples. We now verify that these triples form a Steiner triple system on the

$$v = 3(2t + 1) = 6t + 3$$

objects. It is sufficient to show that each pair of distinct objects occurs together in exactly one triple. A pair A_r, A_s, $r \neq s$, occurs in the triple $(A_{j+u}, A_{2t+1-j+u}, B_u)$, where

$$r - s \equiv \pm[j - (2t + 1) + j] \pmod{2t + 1},$$

and u is chosen to make $j + u$ and $2t + 1 - j + u$ equal to r and s, or s and r, respectively. Similarly, pairs B_r, B_s, and C_r, C_s arise from triples in the second or third rows of (13.33). In the first row of (13.33), B_0 occurs with each A_r except A_0. Hence an automorphism gives us a triple with a prescribed pair B_i, A_j, $j \neq i$. Similarly, the second row under the automorphism gives prescribed pairs C_i, B_j, $j \neq i$; and the third gives A_i, C_j, $j \neq i$. The last triple (A_0, B_0, C_0) under the automorphisms gives all pairs A_i, B_i; A_i, C_i; and B_i, C_i. Since each pair of the $6t + 3$ objects occurs once in the $(2t + 1)(3t + 1)$ triples, a count shows that each pair of distinct objects occurs exactly once. Thus the triples of (13.33) and their images under powers of the automorphism α give the desired Steiner triple system.

An examination of the two orthogonal squares of order 12 in Table 13.6 shows that the second may be obtained from the first by merely permuting the rows. This can also be expressed by saying that if the elements of the column headed 0 are the elements $0 = x_0, x_1, \ldots, x_{11}$ of M, then the column headed b_i, $i = 1, \ldots, 11$, has for its elements b_i, $b_i + x_1, \ldots, b_i + x_{11}$.

In other words, the squares are invariant under the additive group of the system M. This property has been taken by A. L. Dulmage, D. M. Johnson, and N. S. Mendelsohn [12] as the basis of a search on high-speed computers, and they have found families of five mutually orthogonal Latin squares of order 12. In Table 13.7, we give the first columns of five mutually orthogonal squares.

TABLE 13.7

First Columns of Five Mutually
Orthogonal 12 by 12 Latin Squares

Row number	Square number				
	1	2	3	4	5
0	0	0	0	0	0
1	5	1	11	9	7
2	4	7	9	5	2
3	3	2	1	11	10
4	2	8	6	1	11
5	1	6	3	8	4
6	6	9	2	3	8
7	11	3	10	4	5
8	10	5	7	6	3
9	9	10	4	7	1
10	8	4	5	2	9
11	7	11	8	10	6

The rows are the same as those in Table 13.6, and thus the first square in Table 13.7 is identical with the first square in Table 13.6. Exhaustive machine search has shown that a number of families of five mutually orthogonal squares of this type exist, but none exists with six mutually orthogonal squares of this type.

Using a combination of arithmetical and group theoretical methods, E. T. Parker [23] has constructed a family of pairs of orthogonal squares of order $(3q - 1)/2$, where q is any prime power such that $q \equiv 3 \pmod 4$; in particular, with $q = 7$, he obtained the first known pair of orthogonal 10 by 10 squares. This, and additional work with R. C. Bose and S. S. Shirkhande [5], completely demolished the Euler conjecture.

Let t be a primitive root of $GF(q)$, $q = p^s \equiv 3 \pmod 4$, and let

$$(13.34) \qquad x_i, \qquad i = 1, \ldots, \frac{(q-1)}{2},$$

be $(q-1)/2$ further symbols. We construct the ordered quadruples

$$
\begin{aligned}
&[x_i, b, t^{2i} + b, t^{2i}(t+1) + b],\\
&[t^{2i}(t+1) + b, x_i, b, t^{2i} + b],\\
&[t^{2i} + b, t^{2i}(t+1) + b, x_i, b],\\
&[b, t^{2i} + b, t^{2i}(t+1) + b, x_i].
\end{aligned}
$$

(13.35)

Here $i = 1, \ldots, (q-1)/2$, and b ranges over the elements of $GF(q)$. We have already observed, in Sec. 13.2, how a pair of orthogonal squares

TABLE 13.8
Two Orthogonal 10 by 10 Latin Squares

Row number	Column number																			
	0	1	2	3	4	5	6	7	8	9	0	1	2	3	4	5	6	7	8	9
0	0	4	1	7	2	9	8	3	6	5	0	7	8	6	9	3	5	4	1	2
1	8	1	5	2	7	3	9	4	0	6	6	1	7	8	0	9	4	5	2	3
2	9	8	2	6	3	7	4	5	1	0	5	0	2	7	8	1	9	6	3	4
3	5	9	8	3	0	4	7	6	2	1	9	6	1	3	7	8	2	0	4	5
4	7	6	9	8	4	1	5	0	3	2	3	9	0	2	4	7	8	1	5	6
5	6	7	0	9	8	5	2	1	4	3	8	4	9	1	3	5	7	2	6	0
6	3	0	7	1	9	8	6	2	5	4	7	8	5	9	2	4	6	3	0	1
7	1	2	3	4	5	6	0	7	8	9	4	5	6	0	1	2	3	7	8	9
8	2	3	4	5	6	0	1	8	9	7	1	2	3	4	5	6	0	9	7	8
9	4	5	6	0	1	2	3	9	7	8	2	3	4	5	6	0	1	8	9	7

corresponds to a set of ordered quadruples (r_i, c_j, f_k, s_l), giving for the cell in row r_i and column c_j the digit f_k of the first square and s_l of the second square. Since $(q-1)/2$ is odd, there exists a pair of orthogonal squares of this order; for example, such a pair can be constructed by the MacNeish technique. Let this be given in terms of ordered quadruples on the x's, say

(13.36) (x_i, x_j, x_k, x_l).

Parker showed next that the quadruples of (13.35) together with those of (13.36) yield a pair of n by n orthogonal squares with $n = (3q-1)/2$. Taking $q = 7$, $t = 3$, $n = 10$, and writing $x_1 = 7$, $x_2 = 8$, $x_3 = 9$, we have the result shown in Table 13.8.

A further constructive method is based solely on the requirement of a specified group of automorphisms. This cannot be treated here at the same length given the subject in [15]. One very special kind of automorphism, however, will now be considered.

Suppose D is a symmetric block design of parameters v, k, λ. We assume that there is an automorphism of D, of order v, cyclic on the objects and blocks of D. We can then take as our objects the residues $0, 1, \ldots, v - 1$ modulo v, and as the blocks B_0, B_1, $\ldots$, B_{v-1} with subscripts modulo v, and our automorphism will be

(13.37) $\qquad \alpha:\ i \to i + 1 \pmod{v},\ B_i \to B_{i+1}.$

We call D a *cyclic block design*. The block designs in Examples 13.1 and 13.3 are both of this kind. We define a *difference set* Δ of k residues modulo v to be a set $a_1, \ldots, a_k \pmod{v}$ for which each nonzero residue d mod v can be expressed as

(13.38) $\qquad d \equiv a_i = a_j \pmod{v},\ a_i, a_j \in \Delta$

in exactly λ ways, where $k(k - 1) = \lambda(v - 1)$. The theory of cyclic designs and the theory of difference sets are essentially the same thing, as the following theorem shows.

Theorem 13.7. If $a_1, \ldots, a_k$ are the residues modulo v that form a block B_j of a cyclic v, k, λ design, then they are a difference set modulo v. Conversely, if $a_1, \ldots, a_k$ are a difference set, then blocks

$$B_i:\quad a_1 + i, a_2 + i, \ldots, a_k + i, i = 0, \ldots, v - 1$$

of residues modulo v are the blocks of a cyclic v, k, λ design with automorphism $\alpha:\ i \to i + 1 \pmod{v}$, $B_i \to B_{i+1}$.

PROOF. Suppose the block B_j of the cyclic v, k, λ design D contains the residues $a_1, \ldots, a_k$. Then the block B_{j+s} contains the residues $a_1 + s$, $a_2 + s, \ldots, a_k + s$. Let d satisfy $d \not\equiv 0 \pmod{v}$. Then there are exactly λ different blocks that contain both d and 0. In other words, there are exactly λ choices of s, a_i, a_j such that $a_i + s \equiv d$, $a_j + s \equiv 0$. Otherwise expressed, there are λ choices of a_i and a_j such that

(13.39) $\qquad a_i - a_j \equiv d \pmod{v},$

since clearly $a_j + s \equiv 0$ determines $s \equiv -a_j$. But (13.39) asserts that $a_1, \ldots, a_k$ are a, v, k, λ difference set. We may easily reverse this argument, however, to show that if $a_1, \ldots, a_k$ is a v, k, λ difference set, then in the blocks

$$B_i:\quad a_1 + i, a_2 + i, \ldots, a_k + i, i = 0, \ldots, v - 1,$$

the pair of objects 0 and $d \not\equiv 0$ occur together exactly λ times. Then for any j, the values j and $j + d$ occur together exactly λ times; but this means that any two distinct objects r and s occur together exactly λ times (if we take $j \equiv r$ and $d \equiv s - r$). This shows that the B_i are blocks of a symmetric design D, and from their construction it is immediate that the α in the theorem, or in (13.37), is an automorphism of D.

In the definition of a cyclic block design D, it is not asserted that D has no automorphisms except the powers of the defining automorphism α. Indeed, every known cyclic block design has further automorphisms. We define a *multiplier* of a cyclic block design as a residue t modulo v, such that the mapping β,

$$(13.40) \qquad \beta: \quad i \to ti \;(\mathrm{mod}\; v), \qquad i = 0, \ldots, v-1,$$

is an automorphism of D. Since 1 $(\mathrm{mod}\; v)$ is an object of D, (13.40) automatically requires that $(t, v) = 1$; that is, it requires that t and v are relatively prime. The multipliers of D, if they exist, are a subgroup of the multiplicative group of residues modulo v relatively prime to v. Every known cyclic block design D has a multiplier $t \not\equiv 1 \;(\mathrm{mod}\; v)$. A basic theorem of Hall and Ryser [16] asserts the existence of a multiplier in many cases.

Theorem 13.8. Let $a_1, \ldots, a_k$ be a difference set modulo v. Suppose that p is a prime number dividing $n = k - \lambda$, that $(p, v) = 1$, and that $p > \lambda$. Then p is a multiplier of the block design determined by the difference set.

Space does not permit including the proof here. With the existence of a multiplier, however, for parameters of moderate size it usually is easy to construct a difference set if one exists, or else to prove that no difference set exists.

There always exists a difference set for the parameters

$$v = \frac{p^{(m+1)r}-1}{p^r-1}, \qquad k = \frac{p^{mr}-1}{p^r-1}, \qquad \lambda = \frac{p^{(m-1)r}-1}{p^r-1}$$

of the $(m-1)$-dimensional subspaces of $PG(m, p^r)$. This result is due to Singer [26]. The quadratic residues modulo p, where $p \equiv 3 \;(\mathrm{mod}\; 4)$ is a prime number, also form a difference set. This result is due to Paley [22], as remarked earlier.

A more unusual case is the following one, which was found by the author [14]. If p is a prime of the form $4x^2 + 27$, and if t is a primitive root of p chosen appropriately, then the $k = (p-1)/2$ distinct residues

$$(13.41) \qquad\qquad x^6, 3x^6, 27x^6 \;(\mathrm{mod}\; p), \qquad x \neq 0,$$

form a difference set. Here as x runs over the residues $1, \ldots, p-1$, the value x^6 takes on $(p-1)/6$ different values, each one six times. The first two instances of this are $p = 31$ and $p = 43$. The corresponding difference sets are

$$v = 31, k = 15, \lambda = 7, n = 8:$$

1, 2, 3, 4, 6, 8, 12, 15, 16, 17, 23, 24, 27, 29, 30 (mod 31)

and

$$v = 43, k = 21, \lambda = 10, n = 11:$$

1, 2, 3, 4, 5, 8, 11, 12, 16, 19, 20, 21, 22, 27, 32, 33,

35, 37, 39, 41, 42 (mod 43).

A further type is due to A. Brauer [6]. If p and $q = p + 2$ are both prime numbers, and so form a pair of "twin primes," then with $v = pq$ and $k = (pq - 1)/2$, the k residues mod $pq = v$, namely, the values a such that

$$\left(\frac{a}{pq}\right) = +1, \quad \text{and} \quad 0, q, 2q, \ldots, (p - 1)q \text{ (mod } pq),$$

form a difference set mod $v = pq$. Here by the notation

$$\left(\frac{a}{pq}\right) = +1$$

we mean that either both congruences $x^2 \equiv a$ (mod p) and $y^2 \equiv a$ (mod q) have solutions or neither does. An instance of this theorem is the case $p = 5, q = 7, v = 35, k = 17, \lambda = 8$, given by

0, 1, 3, 4, 7, 9, 11, 12, 13, 14, 16, 17, 21, 27, 28, 29, 33 (mod 35).

Other cases arise; for example, with $v = 37$, $k = 9$, $\lambda = 2$, we have $n = 7$, and Theorem 13.8 gives 7 as a multiplier. The solution is

1, 7, 9, 10, 12, 16, 26, 33, 34 (mod 37).

It is somewhat surprising that the apparently drastic conditions of Theorem 13.8 are so often satisfied. Again, the example in (13.41) shows that it is not easy to separate the arithmetical and group-theoretical aspects of these construction methods. Difference sets corresponding to noncyclic groups have been studied by R. H. Bruck [7], but it appears to be true that there are far more designs corresponding to the cyclic groups than to other groups.

MULTIPLE-CHOICE REVIEW PROBLEMS

1. There is at least one Steiner triple system with v objects for

   (a) $v = 95$,
   (b) $v = 100$,
   (c) $v = 105$,
   (d) none of the above.

2. The number of symmetric block designs with $v = 22$, $k = 7$, $\lambda = 2$ is

(a) exactly two,
(b) exactly one,
(c) none,
(d) infinitely many.

3. Five of the blocks of a Steiner triple system are 123, 145, 167, 256, 247. The remaining two blocks are

(a) 346, 357,
(b) 345, 367,
(c) 456, 123,
(d) 234, 567.

4. A binary code contains 144 words of 72 bits each. Any two words differ by 36 or 72 bits. This will correct as many as but not more than

(a) 9 errors,
(b) 17 errors,
(c) 18 errors,
(d) 36 errors.

5. The digits $1, \ldots, 9$ are arranged in a Latin rectangle with 8 rows and 9 columns. We can add a further row to form a Latin square

(a) always in at least two ways,
(b) always in exactly one way,
(c) always in no way,
(d) in a variable number of ways, depending on the particular rectangle.

REFERENCES

1. Baumert, L., S. W. Golomb, and M. Hall, Jr., "Discovery of an Hadamard Matrix of Order 92," *Bull. Amer. Math. Soc.* **68** (1962), 237–238.
2. Bhattacharya, K. N., "A New Balanced Incomplete Block Design," *Science and Culture* **9** (1944), 508.
3. Bhattacharya, K. N., "On a New Symmetrical Balanced Incomplete Block Design," *Bull. Calcutta Math. Soc.* **36** (1945) 91–96.
4. Bose, R. C., "On the Construction of Balanced Incomplete Block Designs," *Ann. Eugenics* **9** (1939), 353–399.
5. Bose, R. C., S. S. Shrikhande, and E. T. Parker, "Further Results on the Construction of Mutually Orthogonal Latin Squares and the Falsity of Euler's Conjecture," *Canadian J. Math.* **12** (1960), 189–203.
6. Brauer, A., "On a New Class of Hadamard Determinants," *Math. Z.* **58** (1953) 219–225.
7. Bruck, R. H., "Difference Sets in a Finite Group," *Trans. Amer. Math. Soc.* **78** (1955) 464–481.
8. Bruck, R. H., and H. J. Ryser, "The Nonexistence of Certain Finite Projective Planes," *Canadian J. Math.* **1** (1949) 88–93.
9. Chowla, S. and H. J. Ryser, "Combinatorial Problems," *Canadian J. Math.* **2** (1950) 93–99.

10. Connor, V. S., "On the Structure of Balanced Incomplete Block Designs," *Ann. Math. Stat.* **23** (1952) 57–71.

11. Connor, W. S., and M. Hall, Jr., "An Embedding Theorem for Balanced Incomplete Block Designs," *Canadian J. Math.* **6** (1953), 35–41.

12. Dulmage, A. L., D. M. Johnson, and N. S. Mendelsohn, "Orthomorphisms of Groups and Orthogonal Latin Squares, I." *Canadian J. Math.* **13** (1961) 356–372.

13. Euler, L., *Commentationes Arithmetical*, II, 302–361, Petrograd, 1849.

14. Hall, Marshall, Jr., "A Survey of Difference Sets," *Proc. Amer. Math. Soc.* **7** (1956), 975–986.

15. Hall, Marshall, Jr., *Combinatorial Analysis*, Ginn and Company, New York, to appear.

16. Hall, Marshall, Jr., and H. J. Ryser, "Cyclic Incidence Matrices," *Canadian J. Math.* **3** (1951) 495–502.

17. Hanani, H., "On Quadruple Systems," *Canadian J. Math.* **12** (1960) 145–157.

18. Hardy, G. H., and E. M. Wright, *An Introduction to the Theory of Numbers*, Oxford University Press, New York, 1938.

19. MacNeish, H. F., "Euler Squares," *Ann. Math.* (2), **23** (1922), 221–227.

20. Mann, Henry B. *Analysis and Design of Experiments*, Dover Publications, New York, 1949.

21. Moore, E. H., "Concerning Triple Systems," *Math. Ann.* **43** (1893) 271–285.

22. Paley, R. E. A. C., "On Orthogonal Matrices," *J. Math. Phys.* **12** (1933), 311–320.

23. Parker, E. T., "Orthogonal Latin Squares," *Proc. Nat. Acad. Sci.* **45** (1959), 859–862.

24. Reisz, M., "Uber eine Steinersche Combinatorische Aufgabe welche in 45 st Band dreses Journals, Seite 181, gestellt worden ist," *J. Reine Angew. Math.* **56** (1859), 326–344.

25. Ryser, H. J., "A Note on a Combinatorial Problem," *Proc. Amer. Math. Soc.* **1** (1950) 422–424.

26. Singer, J., "A Theorem in Finite Projective Geometry and Some Applications to Number Theory," *Trans. Amer. Math. Soc.* **43** (1938) 377–385.

27. Steiner, J., "Combinatorische Aufgabe," *J. Reine Angew. Math.* **45** (1853) 181–182.

CHAPTER 14

Introduction to Information Theory†

JACOB WOLFOWITZ

14.1 INTRODUCTION

The term *information theory* has been used by Shannon [1], the founder of the theory, to denote the theory of coding. This chapter is devoted to an introduction to the existence theorems of such a theory. For proofs of the theorems described here and for additional results the reader is referred to [2].

14.2 CHANNELS

In this section we shall describe finite channels. Let $A^* = \{1, \ldots, a\}$ and $B^* = \{1, \ldots, b\}$ be, respectively, the input and output alphabets. The *alphabet* that we use in everyday life consists of 26 Latin letters, 10 numerical symbols, various punctuation marks, and a space between words, which is itself a punctuation mark. The alphabets A^* and B^* are essentially no different and no less general. To avoid the trivial, we assume that both a and b are greater than 1.

Any sequence of n letters, or *elements*, from A^* (respectively, from B^*) is called a *transmitted* or *sent n-sequence* (respectively, a *received n-sequence*). In any one discussion, n will be fixed. The sender transmits n-sequences over a *channel*. When he sends such a sequence, say u_0, the receiver receives a *chance* received n-sequence; that is, the sequence received depends upon chance. Call the chance received n-sequence $v(u_0)$. Its distribution depends on u_0 and the channel. In fact, for mathematical

† Work under contract with the U.S. Air Force.

purposes the channel is simply the function

$$(14.1) \qquad P\{v(u_0) = v_0\},$$

that is, the probability that, when the n-sequence u_0 is sent, the chance received sequence should be v_0; this function is defined for any transmitted n-sequence u_0 and any received n-sequence v_0. When necessary to avoid confusion, dependence on n should be indicated. Usually the function (14.1) is defined for every n.

One of the simplest and most important of all channels is the *discrete memoryless channel* (dmc). It is described by means of a *channel probability function* (cpf) $w(j \mid i)$, defined for every $i \in A^*$ and every $j \in B^*$. This can be any function for which always $w(j \mid i) \geq 0$ and

$$\sum_{j \in B^*} w(j \mid i) = 1, \qquad i \in A^*.$$

Different functions w define different dmc's. Let

$$(14.2) \qquad u_0 = (a_1, a_2, \ldots, a_n),$$

$$(14.3) \qquad v_0 = (b_1, b_2, \ldots, b_n).$$

Then

$$(14.4) \qquad P\{v(u_0) = v_0\} = \prod_{k=1}^{n} w(b_k \mid a_k).$$

We see that $w(j \mid i)$ can be regarded as the probability that, when the letter i is sent, the letter j is received. In that case, the individual letters received are independently distributed.

14.3 CODES

Codes are the primary objects of our study. A *code* (n, N, λ), where n is the *length of each word*, N is the *length of the code*, and λ is the *maximum probability of error*, is a system

$$(14.5) \qquad \{(u_1, A_1), \ldots, (u_N, A_N)\},$$

where $u_1, \ldots, u_N$ are transmitted n-sequences, $A_1, \ldots, A_N$ are *disjoint* sets of received n-sequences, and

$$(14.6) \qquad P\{v(u_i) \in A_i\} \geq 1 - \lambda, \qquad i = 1, \ldots, N.$$

A code is used as follows: When the sender wishes to send the ith message, he sends u_i. When the message received lies in A_j, the receiver concludes that the jth message was sent. If the message received does not lie in any A_j, he may draw any conclusion he wishes about the

message that has been sent. The probability that any message sent will
be correctly understood by the receiver is at least $1 - \lambda$.

Our general problem is this: For various channels of interest, given n
and λ, $0 < \lambda < 1$, how big can N be? Our results will be asymptotic in n.

The closely related problem of constructing the codes whose existence
is guaranteed by the theorems that will be cited in Sec. 14.5 is as yet
only partially solved.

The question may be raised as to why all the words of a code should
be of the same length n. The answer is that, since our results are
asymptotic in n, this will hardly matter, and that it is very difficult to
treat the case in which the words are of unequal length (see [2], p. 1).

For a heuristic, semi-intuitive discussion of the dmc, see [2], Chapter 1.

14.4 ENTROPY; GENERATED SEQUENCES

Any vector with nonnegative components that add to 1 may be called
a *probability distribution*. A probability distribution on A^* (respectively
on B^*) will have a (respectively b) components.

The "entropy" of a probability distribution π,

(14.7) $$\pi = (\pi_1, \ldots, \pi_c),$$

is defined to be

(14.8) $$H(\pi) = -\sum_{i=1}^{c} \pi_i \log_2 \pi_i.$$

Logarithms to the base 2 are used for historical reasons only, and any
other base would do as well. If $\pi_i = 0$, the ith term of the right-hand
member of (14.8) is defined to be 0. This last convention always applies.

Many combinatorial properties of entropy are known and used
([2], Sec. 2.2). The significance of this function and some of the reasons
for introducing it will shortly become apparent.

In the remainder of this section we shall assume that we are dealing
with the dmc because (1) this is the simplest channel on which to illustrate
the ideas to be described, (2) it is an important channel *per se*, and (3) the
treatment of other channels can sometimes be reduced to a treatment of
the dmc.

Fix i at any value in A^*, and consider $w(\cdot \mid i)$ as a function of its first
argument. This is a probability distribution on B^*. We write its entropy
as $H[w(\cdot \mid i)]$.

Let u_0 be any n-sequence, sent or received. We define $N(i \mid u_0)$ as the
number of elements i in u_0. (If u_0 is a transmitted sequence then i ranges
over the set A^*, but if u_0 is a received sequence then i ranges over B^*.) Let
u_0 and v_0 be, respectively, a transmitted n-sequence and a received n-
sequence. We define $N(i, j \mid u_0, v_0)$, $i = 1, \ldots, a$ and $j = 1, \ldots, b$, as

the number of values k, $k = 1, \ldots, n$, such that the kth element of u_0 is i and the kth element of v_0 is j.

Let $\pi = (\pi_1, \ldots, \pi_a)$ be a probability distribution on A^*. A transmitted n-sequence u_0 will be called a π-*sequence* or a πn-*sequence* if

$$|N(i \mid u_0) - n\pi_i| \leq 2\sqrt{an\pi_i(1 - \pi_i)}, \qquad i = 1, \ldots, a.$$

Let $\delta > 2 \max (a, b)$ be a number that is determined appropriately in the course of the various proofs. A received n-sequence v_0 is said to be generated by an n-sequence u_0 if

$$|N(i,j \mid u_0, v_0) - N(i \mid u_0)w(j \mid i)| \leq \delta\{N(i \mid u_0)w(j \mid i)[1 - w(j \mid i)]\}^{1/2}$$

for $i = 1, \ldots, a$ and $j = 1, \ldots, b$. (Readers familiar with the binomial distribution will recognize the significance of this.)

Let $\pi' = (\pi'_1, \ldots, \pi'_b)$ be a distribution on B^* defined by

$$\pi'_k = \sum_{i=1}^{a} \pi_i w(k \mid i), \qquad k = 1, \ldots, b.$$

Let $H(\pi')$, as usual, be its entropy.

The following lemmas are easy to prove; their proofs can be found in [2], Sec. 2.1.

Lemma 14.1. Let $\epsilon > 0$ be any number and let δ be sufficiently large (δ is a function of ϵ). Let u_0 be any transmitted n-sequence. Then

(14.9) $P\{v(u_0)$ is generated by $u_0\} > 1 - \epsilon$.

Lemma 14.2. There is a constant $D_1 > 0$ such that the number of n-sequences generated by any one transmitted πn-sequence is less than

(14.10) $$\exp_2 \left\{ n \sum_{i=1}^{a} \pi_i H[w(\cdot \mid i)] + \sqrt{n}\, D_1 \right\}.$$

Lemma 14.3. Let u be a *chance transmitted* n-sequence, whose elements are independently distributed with the same distribution π. Let $v(u)$ be the chance received n-sequence. Let v_0 be a received n-sequence that is generated by any πn-sequence. There exists a constant $D_2 > 0$, which does not depend on v_0, such that

(14.11) $P\{v(u) = v_0\} < \exp_2 [-nH(\pi') + \sqrt{n}D_2]$.

Lemma 14.4. The number of n-sequences generated by any one transmitted πn-sequence is greater than

(14.12) $$\exp_2 \left\{ n \sum_{i=1}^{a} \pi_i H[w(\cdot \mid i)] - \sqrt{n}\, D_1 \right\}.$$

Lemma 14.5. The number of n-sequences generated by *all* transmitted πn-sequences is less than

$$\exp_2[nH(\pi') + \sqrt{n}D_3],$$

where $D_3 > 0$ is a constant.

14.5 THE DISCRETE MEMORYLESS CHANNEL

Consider a dmc with cpf w. The quantity

$$(14.13) \qquad C = \max_\pi \left\{ H(\pi') - \sum_{i=1}^a \pi_i H[w(\cdot \mid i)] \right\}$$

is called the *capacity* of the channel. The maximum in (14.13) is with respect to all probability distributions on A^*. The justification for the name *capacity* is the following two theorems.

Theorem 14.1. (Coding Theorem.) Let $\lambda, 0 < \lambda \leq 1$, be arbitrary. There is a positive function of λ, say $K(\lambda)$, such that, for any n, there exists a code

$$(14.14) \qquad (n, 2^{nC - \sqrt{n}K(\lambda)}, \lambda)$$

for the dmc.

(If the quantity $2^{nC - \sqrt{n}K(\lambda)}$ is not an integer, let it be replaced by its integral part. This convention will be followed throughout the rest of this chapter.)

Theorem 14.2. (Strong Converse of the Coding Theorem.) Let $\lambda, 0 \leq \lambda < 1$, be arbitrary. There is a positive function of λ, say $K'(\lambda)$, such that, for no n, does there exist a code

$$(14.15) \qquad (n, 2^{nC + \sqrt{n}K'(\lambda)}, \lambda).$$

(This theorem should not be confused with what is often called in the literature the converse of the coding theorem, where the conclusion of Theorem 14.2 is asserted *for sufficiently small* λ. The latter result is the *weak converse*; see [2], Sec. 7.4.)

Theorems 14.1 and 14.2 are proved in [2], Chapter 3. We sketch below the fundamental ideas of the proofs.

SKETCH OF THE PROOF OF THEOREM 14.2:

(a) Suppose first that the u's in (14.5) are all πn-sequences. By application of Lemma 14.1 we show that it is sufficient to assume that A_i, $i = 1, \ldots, N$, contains only sequences generated by u_i. We now suppose that this is the case.

(b) Since the A_i are disjoint, the total number of sequences in all the A_i (in $A_1 \cup \cdots \cup A_N$) is, by Lemma 14.4, greater than

$$(14.16) \qquad N \cdot \exp_2 \left\{ n \sum_{i=1}^{a} \pi_i H[w(\cdot \mid i)] - \sqrt{n}\, D_1 \right\}.$$

(c) By Lemma 14.5, the total number of sequences in all the A_i is less than

$$\exp_2 \{ nH(\pi') + \sqrt{n} D_3 \}.$$

(d) From (b) and (c) it follows that

$$\exp_2 \{ nH(\pi') + \sqrt{n}\, D_3 \} > N \exp_2 \left\{ n \sum_{i=1}^{a} \pi_i H[w(\cdot \mid i)] - \sqrt{n}\, D_1 \right\},$$

and therefore

$$(14.17) \qquad N < \exp_2 \{ nC + \sqrt{n}(D_1 + D_3) \}.$$

(e) Now drop the assumption that all the u's in (14.5) are πn-sequences. Divide the code (14.5) into subcodes in the following way: All the u's in each subcode have the same numbers of elements i, for $i = 1, \ldots, a$. It follows that we can divide the original code into fewer than $(n + 1)^a$ subcodes, because the number of elements i, $i = 1, \ldots, a$, in any transmitted n-sequence can be only $0, 1, 2, \ldots,$ or n.

(f) Each subcode has its length bounded by the right-hand member of (14.17). There are fewer than $(n + 1)^a$ subcodes. Hence the length of the entire code is less than

$$(14.18) \quad (n + 1)^a \exp_2 \{ nC + \sqrt{n}(D_1 + D_3) \} < \exp_2 \{ nC + \sqrt{n}K \},$$

for suitable K. This completes the argument.

SKETCH OF THE PROOF OF THEOREM 14.1:

(a) Let $\bar{\pi}$ be a distribution on A^* such that

$$(14.19) \qquad \left\{ H(\bar{\pi}') - \sum_{i=1}^{a} \bar{\pi}_i H[w(\cdot \mid i)] \right\} = \max_{\pi} \left\{ H(\pi') - \sum_{i=1}^{a} \pi_i H[w(\cdot \mid i)] \right\}.$$

Take δ so large that (14.9) holds for an $\epsilon < \lambda/2$.

(b) Construct a code as follows: u_1 is any $\bar{\pi}n$-sequence and A_1 consists of all sequences generated by u_1. Let u_2 be any $\bar{\pi}n$-sequence such that, if A_2 is defined to be the set of all sequences generated by u_2 and not in A_1, we have

$$P\{ v(u_2) \in A_2 \} \geq 1 - \lambda.$$

Continue the construction in this manner. At the kth step, choose any $\bar{\pi}n$-sequence to be u_k if, when A_k is defined to be the set of all sequences generated by u_k and not in any of $A_1, \ldots, A_{k-1}$, we have

$$(14.20) \qquad\qquad P\{v(u_k) \in A_k\} \geq 1 - \lambda.$$

Proceed in this manner *as long as continuation is possible*. Let (14.5) be the resulting code. We now have to prove that N is large enough.

(c) Let u_0 be any $\bar{\pi}$-sequence that is not one of $u_1, \ldots, u_N$. Then

$$(14.21) \quad P\{v(u_0) \text{ is generated by } u_0 \text{ and belongs to one of } A_1, \ldots, A_N\} \geq \frac{\lambda}{2},$$

since, if this were not so, we could prolong the code by adding (u_0, A_0) to it, where A_0 is the totality of all sequences generated by u_0 and not in any of $A_1, \ldots, A_N$. This would violate the fact that prolongation of the code is impossible.

(d) Since $\lambda \leq 1$, we have $1 - \lambda/2 \geq \lambda/2$. Hence, from the size of δ in (a), it follows that (14.21) holds even if u_0 is one of $u_1, \ldots, u_N$.

(e) Let u be a *chance* transmitted n-sequence, whose elements are independently distributed with common distribution $\bar{\pi}$. It is not difficult to show, from (d) and Lemma 14.1, that

$$(14.22) \qquad\qquad P\{v(u) \in [A_1 \cup \cdots \cup A_N]\} > \frac{\lambda}{8}.$$

(f) By a little argument, using (14.22) and Lemma 14.3, one can show that $[A_1 \cup \cdots \cup A_N]$ contains more than

$$(14.23) \qquad\qquad \exp_2 \{nH(\bar{\pi}') - \sqrt{n}D_4\}$$

sequences, where $D_4 > 0$ is a constant.

(g) By Lemma 14.2, the number of sequences in $[A_1 \cup \cdots \cup A_N]$ is less than

$$(14.24) \qquad N \cdot \exp_2 \left\{ n \sum_{i=1}^{a} \bar{\pi}_i H[w(\cdot \mid i)] + \sqrt{n}\, D_1 \right\}.$$

(h) Since the expression in (14.24) is greater than the expression in (14.23), the desired result follows.

14.6 COMPOUND CHANNELS

Consider now a dmc with this difference: Instead of a single cpf w there is given a set S^* of cpf's, say $S^* = \{w(\cdot \mid \cdot \mid s), s \in S\}$. Here the third index, s, distinguishes the cpf. The set S^* may have infinitely many elements. For each s, $w(j \mid i \mid s)$ is a cpf defined for $i = 1, \ldots, a$ and $j = 1, \ldots, b$. The compound channel *transmits* as follows: Each word of n letters (n-sequence) is transmitted according to some cpf in S^*; the cpf may vary *arbitrarily* in S^* from one such word to another.

Let P_s now denote probability according to the cpf $w(\cdot \mid \cdot \mid s)$. A code (n, N, λ) for the compound channel is a system (14.5) with all the requirements, except that (14.6) is replaced by the stronger requirement

$$(14.25) \qquad P_s\{v(u_i) \in A_i\} \geq 1 - \lambda, \qquad i = 1, \ldots, N; s \in S.$$

Thus, even if Maxwell's demon tried maliciously to vary the cpf so as to make things as difficult as possible, the probability that any word sent would be incorrectly understood by the receiver is $\leq \lambda$.

The question is, how long can codes be and still meet this stronger requirement (14.25)? It must be borne in mind that the cpf's in S^* may be very "antithetical" to each other. The fact is that theorems exactly like Theorems 14.1 and 14.2 hold for the compound channel (see [2], Sec. 4.3 and 4.4), but we shall not state them here. Thus we see that the maximum length of the code hinges on the capacity C_1 of the compound channel.

If C_1 were 0 in most cases, little could be done with a compound channel. Let $C(s)$ be the capacity of the dmc with the single cpf $w(\cdot \mid \cdot \mid s)$. Define

$$C_2 = \inf_{s \in S} C(s) = \inf_{s \in S} \max_{\pi} \left\{ H(\pi' \mid s) - \sum_{i=1}^{a} \pi_i H[w(\cdot \mid i \mid s)] \right\}.$$

Then obviously we have $C_1 \leq C_2$, for the demon could use the "worst" cpf for *every* word, that is, the one with the smallest capacity. (If S is an infinite set and there is no worst cpf, one uses a cpf with a capacity arbitrarily close to the infimum.) The fact is that

$$C_1 = \max_{\pi} \inf_{s \in S} \left\{ H(\pi' \mid s) - \sum_{i=1}^{a} \pi_i H[w(\cdot \mid i \mid s)] \right\},$$

and, surprisingly and pleasantly, C_1 is not 0 unless C_2 is 0. Thus C_1 is not 0 unless S^* contains a cpf whose capacity is 0 (or a sequence whose capacities approach 0).

Consider now a compound channel as above except that the receiver now knows which cpf is being used (but the sender does not). It has been shown that results like Theorems 14.1 and 14.2 hold and that the capacity of this channel is also C_1. Thus knowledge of the cpf by the receiver alone does not increase the capacity! (For an explanation of this fact and statements and proofs of the theorems, see [2], Sec. 4.5.)

Consider the compound channel as above, except that the cpf is now known to the sender but not to the receiver. The capacity is then C_2, which in general is greater than C_1.

14.7 OTHER CHANNELS

Many other interesting channels have been studied in addition to those discussed, and many more await the ingenious research worker. We conclude this chapter by mentioning a few miscellaneous results concerning some of them.

Discrete memoryless channel with feedback. Consider the dmc (with single cpf). Suppose the sender can look over the receiver's shoulder and see what the latter is receiving. The sender can choose subsequent letters to be sent in order to correct previous reception, but he can communicate with the receiver only over the channel. The capacity of this channel is the same as if there were no feedback!

Discrete channels with memory. In such channels the distribution of a letter received, say the kth, depends not only on the kth letter sent, but also on previous letters sent and/or received. The "memory" can be finite or not.

Semicontinuous channels, with or without memory. In such channels the input alphabet is finite, but the output alphabet is infinite (all the points of the real line, for example).

Continuous channels. In such channels both the input and output alphabets are infinite.

Message sequences in the interior or on the periphery of the n-sphere with center at the origin and radius $\sqrt{n}$; each coordinate subject to normal error with mean 0 and variance σ^2. The capacity of this channel is

$$\tfrac{1}{2} \log_2 \frac{1 + \sigma^2}{\sigma^2} .$$

If the message sequences are restricted to a thin shell at the periphery the capacity is unchanged!

Compound channels in which the cpf for each letter is determined by chance. Variations of such channels, in which the receiver, or sender, or both, know the stochastically determined cpf for each letter.

Compound channels without memory in which the cpf varies arbitrarily from letter to letter. Again the cpf may be known to the sender, or to the receiver, or to both.

MULTIPLE-CHOICE REVIEW PROBLEMS

1. Define $w(j \mid i)$ as the probability that the letter j is received when the letter i is sent. Consider a discrete memoryless channel, with input and output letters 1, 2, and

$$w(2 \mid 1) = w(1 \mid 2) = 1,$$
$$w(1 \mid 1) = w(2 \mid 2) = 0.$$

The capacity of this channel is

(a) 0, (b) 1/2, (c) 1, (d) 2.

2. Consider the same situation as in Problem 1, except that now

$$w(1 \mid 1) = w(1 \mid 2),$$
$$w(2 \mid 1) = w(2 \mid 2).$$

The capacity of this channel is

(a) 0, (b) 1/2, (c) 1, (d) 2.

3. Consider the same situation as in Problem 1, except that now

$$w(1 \mid 1) = w(2 \mid 2) = p,$$
$$w(2 \mid 1) = w(1 \mid 2) = 1 - p,$$

The capacity of this channel is

(a) 1, (b) 0,
(c) $p \log_2 p + (1 - p) \log_2 (1 - p)$,
(d) $1 + p \log_2 p + (1 - p) \log_2 (1 - p)$.

4. A channel has a input letters and b output letters. The maximum possible value of the capacity is

(a) 0, (b) 1, (c) $\log_2 a$, (d) $\log_2 b$,

5. In a general compound memoryless channel, the capacity is larger when

(a) neither the sender nor the receiver knows the channel probability function governing the transmission of any word,
(b) the sender but not the receiver knows the channel probability function governing the transmission of any word,
(c) the receiver but not the sender knows the channel probability function governing the transmission of any word,
(d) none of the above.

REFERENCES

1. Shannon, C. E., "A Mathematical Theory of Communication," *Bell System Tech. J.* **27** (1948), 379–423, 623–656.
2. Wolfowitz, J., *Coding Theorems of Information Theory*, Springer-Verlag, Heidelberg, Germany, 1961. (Also Prentice-Hall, Englewood Cliffs, New Jersey, 1961.)

CHAPTER 15

Sperner's Lemma and Some Extensions

CHARLES B. TOMPKINS

15.1 INTRODUCTION

We shall attack here a problem of proving explicitly some intuitively obvious facts. As an illustration, consider a ball B, that is, an ordinary sphere S and its interior in Euclidean space of three dimensions. Now suppose that this ball is subjected to a squeezing process, so that it becomes somewhat smaller and occupies a region R lying completely inside the ball B and having no points on the bounding sphere. We assume that there is no tearing involved in this deformation process, so that two points that were close together in B remain close together in R; but we admit collapse, so that two points that were far apart in B may be close together or even coincident in R. We shall speak of the points in B and their images, after the deformation, in R. Construct a vector field over the spherical boundary S of B by drawing the vector that joins each point of S with its image in R. Since R has no points in common with S, each of these vectors has a positive length; and since no tearing was allowed, this vector field is continuous. Furthermore, since R lies completely inside S, the vectors point generally inward. From this statement alone it is at least plausible, and, to some, obvious, that as S is traversed the directions of the vectors sweep through one complete solid angle. Furthermore, it is clear that similar statements can be made for spheres concentric with S and just slightly smaller than S in B. As we consider all spheres concentric with S and lying in B, we realize that this sweeping out of a complete solid angle must persist unless a vector can somehow manage to point in all directions simultaneously; this is possible in the limit of a continuous vector field only for the null vector—one of length 0. Thus we

416

conjecture, or see, that, under the conditions stated, one vector connecting a point of B to its image in R must have length 0 and that therefore after the deformation the image of this point is at exactly the position of the point; such a point is called a *fixed point*.

Although all this may be intuitively clear, and was stated by Gauss and probably even earlier workers, precise proofs are not immediately obvious. Sperner's lemma provides an arithmetization that is applicable to the proof of the assertions made. Furthermore, it illustrates the kind of elegant arithmetization that sometimes changes the course of mathematics and is so easily understood that the really powerful role of abstraction becomes apparent and easily appreciated. It should be noted in passing that the proof of Sperner's lemma in its original form illustrates a point that, as far as the author knows, was first stated by Hermann Weyl in a public lecture, an important point about proofs by mathematical induction: It is sometimes necessary to replace the desired proposition by a stronger one to carry out a proof by mathematical induction. In Sperner's lemma, we want to show that a configuration occurs at least once, but in the inductive proof we must show that it occurs an odd number of times. There is a delicate balance in inductive proofs: If the proposition stated is too strong, then the proof that it is valid for a value $k + 1$ of the index governing the induction, provided that it is valid for all lower positive integral values $1, \ldots, k$ of the index, becomes impossibly demanding; if the proposition stated is too weak, not enough information is furnished, by the assertion that it is valid for all values of the governing index through the value k, to provide means for proving its validity for the index value $k + 1$.

Finally, we shall extend the lemma and its applications. We could have started with the extensions (and indeed the reasoning used in the first paragraph to justify our belief in the existence of at least one fixed point is somewhat stronger than that intended in the original application of the lemma) and have skipped the lemma entirely, but this would entail several losses. First, the lemma illustrates the value of abstraction in research and in exposition. Secondly, there are extensions of the lemma in other directions that may be valuable in combinatorial problems; for some of these, see [14]. Finally, to put the lemma back into geometric context from its completely abstract form is in itself instructive. Somehow through mathematics there runs a theme, which—in the form to which the author responds most warmly—is that geometry is the mother of mathematical invention, that analysis and algebra are the necessary implements for statement of proofs of any complexity, and that geometry frequently provides understanding or meaningful context for concepts that are developed abstractly and ungeometrically either by happy observation or by extension of the ideas and operations of analysis and algebra. The author hopes that this overly simplified theme will not be lost during a coda in which he indicates how the geometrical statements presented during the main body can be arithmetized in a way to make them perfectly precise.

15.2 SPERNER'S LEMMA

Sperner's lemma is illustrated in Fig. 15.1, where a triangle of no particularly prescribed shape is depicted at the left, with its vertices labeled 0, 1, 2. To the right is the same triangle, but the triangle and its

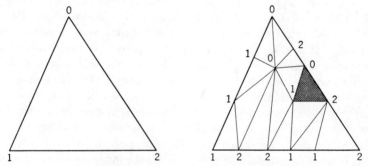

Fig. 15.1   Example of Sperner's lemma.

interior have been subdivided and covered with a set of smaller triangles. The vertices of the large triangle retain their old labels. The vertices that have been added receive labels from the same set of numbers, 0, 1, 2, subject only to the restriction that any vertex lying on a side of the large triangle must carry the label of one of the endpoints of that side. Vertices added inside the large triangle may carry any of the three labels.

The lemma states that at least one of the small triangles carries the full set of labels, 0, 1, and 2 on its three vertices, no matter how the labels are assigned, subject to the preceding rule. In Fig. 15.1, such a small triangle has been shaded; there are two others not shaded.

Before the reader seeks a counterexample, he should read at least far enough to know what is meant by a subdivision. If he continues and reads the proof, he will not waste time seeking the nonexistent counterexample.

Sperner's lemma applies not only to triangles but to figures, such as tetrahedra, that generalize triangles in higher dimensions. These figures are called *n-dimensional simplexes*; we usually write *n-simplex* rather than *n*-dimensional simplex. A 1-simplex is a pair of points and the line segment between them, to be thought of here as lying in a Euclidean plane or Euclidean space if the reader does not like to contemplate living on a line segment. A 2-simplex contains the vertices, sides, and interior of a triangle. A 3-simplex contains the vertices, edges, faces, and interior of a tetrahedron. An n-simplex is built on $n + 1$ vertices in n-dimensional space. The n-simplex contains all the points of the smallest convex region containing these vertices in just the way that the 3-simplex contains all the points of the smallest convex region containing its four vertices. We add, in order to save words in future remarks, that a 0-simplex is a single point and that a (-1)-simplex contains no points.

For elementary remarks concerning spaces of more than three dimensions, see the author's chapter in [3, pp. 448–479]. Some remarks at the end of the present chapter, however, will indicate a general method for precise handling of higher-dimensional simplexes that might console those who are not comfortable in more than three dimensions.

We shall now turn to some definitions that will be needed in the statement and the proof of the lemma. These definitions may be accepted in the spirit of the somewhat imprecise description given, but it should be noted that they are accurate in the language of the more precise style that was mentioned in the preceding paragraph.

The simplexes *naturally associated* with an n-simplex S are the simplexes of all dimensions, from -1 through n, having vertices that are vertices of S. There are 2^{n+1} of these, including the (-1)-simplex with no points. For example, for the tetrahedron they are the single (-1)-simplex, the four vertices as 0-simplexes, the six edges as 1-simplexes, the four faces as 2-simplexes, and the whole tetrahedron and its interior as a 3-simplex—a total of $16 = 2^4$.

The intersection of any two different simplexes each of which is naturally associated with S is a simplex naturally associated with S and with dimension no higher than that of either of the two simplexes; unless one of the two simplexes contains the other, the dimension of the intersection will be less than the dimension of either; if one simplex contains another, the dimension of the containing simplex is at least as great as that of the contained simplex, and if both are associated with S then either they are coincident or the containing simplex has higher dimension than the contained simplex. All these statements are as obvious as they sound.

If p is a point of an n-simplex S, then the *carrier of p in S* is the simplex of smallest dimension, naturally associated with S, that contains p. The carrier of any point p in S is well defined, for if p lies on any two different simplexes of the same dimension it also lies on their intersection, which is a simplex of lower dimension naturally associated with S, as was noted.

We shall use the notion of the carrier in the next definition and then provide an illustration of its meaning.

A set of points of an n-simplex S is *an admissibly labeled set* if the set contains all the vertices of S, if these vertices are labeled from 0 through n in any order so that each label is attached to exactly one vertex, and if every other point of the set has a label that is the label of one of the vertices of its carrier.

It is implied that the assignment of labels of points that are not vertices of S from among the labels of the vertices of the carriers is completely

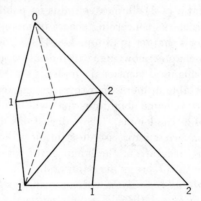

Fig. 15.2 Simplicial subdivision.

arbitrary, and the assignment to one point is completely independent of the assignment to any other. The definition of admissible labeling is a general definition that includes the special restriction imposed in the two-dimensional case, in the first paragraph of this section, in connection with the example illustrated by Fig. 15.1. In particular, in this example the sides of the large simplex are carriers of the vertices of the subdivision that lie on these sides.

A *subdivision of an n-simplex* S is a finite set $\{S_i\}$ of n-simplexes, each contained in S, such that (1) each point of S is contained in at least one simplex of S_i, (2) no two different simplexes S_j and S_k of $\{S_i\}$ intersect in an interior point of either, and (3) if p is a vertex of any simplex S_j of $\{S_i\}$, then it is a vertex of each simplex S_k of $\{S_i\}$ to which it is incident.

We shall first explain notation and then give examples. By a set $\{S_i\}$ we mean a set of n-simplexes that can be labeled by an index having generic symbol i; this index takes on values from some finite index set I, usually the first few positive integers. In referring to a simplex S_j of this subdivision, we imply that some particular value from I has been assigned to the index, and we denote this value by j; the symbol S_k has an analogous meaning.

The subdivision shown in Fig. 15.1 satisfies the foregoing definition. The large triangle and its interior are completely covered by the small 2-simplexes, and these have only boundary points in common. Condition (3) of the definition is also satisfied. In Fig. 15.2 we illustrate how condition (3) might not be satisfied. This figure depicts the four small 2-simplexes near the lower right-hand vertex of Fig. 15.1; the top simplex is the one that is shaded in Fig. 15.1. If further subdivision were attempted by inserting the dashed line from the lower left-hand vertex in Fig. 15.2 to the opposite edge of the small simplex, as shown, then a vertex would be formed that is not a vertex of the upper triangle, to which it is incident. Thus the upper dashed line (or some more complicated completion) would have to be added in order to satisfy requirement (3).

It is clear that the set of points that are vertices of the simplexes of an admissible subdivision of S is contained in S and that this set of vertices contains the vertices of S. We now state Sperner's lemma.

Sperner's Lemma. For any admissible labeling of the set of vertices of any subdivision of an n-simplex S, there exists at least one simplex S_j of the subdivision with vertices carrying a complete set of labels.

The proof is inductive, the induction being with respect to the number of dimensions n. In accordance with the remarks in the introduction, however, we find it convenient to prove a stronger proposition, that *the number of simplexes of the subdivision with vertices carrying a complete set of labels is odd.* Since 0 is not an odd number, this implies Sperner's lemma. We call this assertion the *strong assertion*.

For the one-dimensional case, the proposition involves a line segment with endpoints labeled 0 and 1, respectively. This segment is subdivided by punctuation with a finite set of points. Each of these points is arbitrarily labeled 0 or 1, independently of the labels of the other points. We choose from the short line segments (the ones containing no vertices of the subdivision on their interior) those with at least one endpoint labeled 0. The chosen segments are of two types: those with both ends labeled 0, and those with one end labeled 0 and one end labeled 1. Considering each such segment independently, we count the number of endpoints labeled 0. If there are a simplexes with two 0 labels, they will account for $2a$ endpoints labeled 0. If there are b simplexes each of which has a single 0 and a single 1 label, they will account for b endpoints labeled 0. Thus the number of segment endpoints labeled 0 is $2a + b$.

We now look at the line segment as a whole. One vertex with label 0 lies at the end of the segment, but only one. Every other vertex labeled 0, if any exists, lies on the interior of the large 1-simplex and is therefore incident to two simplexes of the subdivision; all these other vertices, then, are counted twice as endpoints of segments of the subdivision. Thus the total count of 0-labeled endpoints of simplexes of the subdivision must be twice the number of 0-labeled points on the interior of the original segment plus 1, corresponding to the original vertex labeled 0; this number is odd. Hence the number obtained at the end of the preceding paragraph, $2a + b$, is odd, and this implies that b is odd. Since b is just the number of simplexes of the subdivision carrying a complete set of labels 0 and 1, this proves the strong assertion for the one-dimensional case.

This same counting procedure will work for all dimensions, but it might be well, before proceeding to the general inductive step, once more to consider Fig. 15.1. Here we shall look at the (small) 2-simplexes of the subdivision, and for each, independently of all others, we shall count the number of sides that carry the label 0 at one end and the label 1 at the other end. The shaded triangle in the subdivision carries one such side, and indeed any triangle carrying all three of the labels will have one such side. The left-hand simplex of the subdivision at the very top of the figure has two vertices labeled 0 and one vertex labeled 1; it has two sides carrying labels 0 and 1. Proceeding down and to the left, we see that the next triangle has two vertices labeled 1 and a single vertex labeled 0; it also has two sides carrying the labels 0 and 1. These are the only types of labeled simplexes that can possibly have any sides with one end labeled 0 and the other labeled 1. Denote by a_1 the number of simplexes of the subdivision with two vertices labeled 0 and one vertex labeled 1, and by a_2 the number with two vertices labeled 1 and one vertex labeled 0; set $a = a_1 + a_2$. Then a is the number of simplexes of the subdivision that carry both labels 0 and 1 but not label 2 on their vertices. Let b denote the number of simplexes of the subdivision that carry all three labels. Then the total count of bounding sides with one end labeled 0 and the other labeled 1 is $2a + b$.

Some of the sides, like the one separating the two top triangles incident to the left-hand side of the original simplex in Fig. 15.1, lie between two simplexes of the subdivision; these, of course, occurred twice in the preceding count. Others, like the top segment of the left-hand side of the big simplex, lie on the edge of the big simplex; these were included just once in the count. All the subdivision sides that lie on the edge of the big simplex and that are labeled 0 at one end and 1 at the other end must lie on the side of the large simplex having vertices labeled 0 and 1, for one of these two labels is inadmissible on each of the other two sides of the big simplex. This side is a 1-simplex for which the strong assertion has already been proved. Hence the number of sides counted once is odd, and the total count is this odd number plus twice the number of sides counted twice; this again is an odd number. Thus again we have the result that $2a + b$ is an odd number, and it follows that b is odd. This proves the strong assertion for the case $n = 2$.

We shall now use this same proof to show that if the strong assertion is valid for any subdivision of any simplex of dimension k, then it is valid for any subdivision of any simplex of dimension $k + 1$. Consider a subdivision of a $(k + 1)$-simplex with the vertices admissibly labeled. Let a be the number of simplexes of the subdivision with vertices carrying all the labels 0 through k but not the label $k + 1$. Since any $(k + 1)$-simplex has exactly $k + 2$ vertices, each of these simplexes has some one label occurring at two vertices and all the others (except, of course, $k + 1$) at exactly one vertex. There are, then, exactly two ways in each of these simplexes in which a k-simplex naturally associated with it and having vertices carrying all the labels 0 through k can be chosen. Therefore the number of k-faces carrying all the labels 0 through k in this set of simplexes of the subdivision is exactly $2a$. Similarly, if there are b simplexes of the subdivision with vertices carrying all the labels 0 through $k + 1$, then each of these has a single k-face carrying all the labels 0 through k, and the number of such k-faces in these simplexes of the subdivision is b. Thus the total number of k-faces of simplexes of the subdivision carrying all the labels 0 through k is $2a + b$.

As before, some of these faces lie inside the original simplex and some lie on the boundary. Those that lie inside must be incident to exactly two simplexes of the subdivision, one on each side of the face. Those that lie on the boundary are incident to a single simplex of the subdivision, and all these lie on the k-simplex naturally associated with the original $(k + 1)$-simplex with vertices carrying the labels 0 through k [since every other k-face of the original $(k + 1)$-simplex excludes one of these labels]. The inductive assumption is that the strong assertion is valid for simplexes of dimension k; hence the number of k-faces of the subdivision counted once is odd. The total count, as before, is this odd number plus twice the number of k-faces counted twice, and this is again an odd number. Thus, as always, $2a + b$ is an odd number, and this implies that b is odd. Since b is the number of simplexes of the subdivision with vertices carrying all the labels 0 through $k + 1$, this proves the strong assertion for each admissible labeling of any subdivision of an arbitrary $(k + 1)$-simplex. Thus we have completed the inductive proof, for we have shown that the set of numbers n for which the strong assertion, and hence Sperner's lemma, is valid includes the number 1 and that if it includes any number k then it includes the next higher number, $k + 1$.

15.3 BROUWER'S FIXED-POINT THEOREM

The most immediate application of Sperner's lemma is a famous fixed-point theorem due to L. E. J. Brouwer. This theorem pertains, in the form in which it will be stated later, to n-simplexes, rather than n-balls, but otherwise it is illustrated by the discussion given in Sec. 15.1 about a continuous crushing of a ball into a subset of its interior. Actually, Brouwer's theorem is stronger, for it does not require that the image be interior to the simplex—boundary points may be included in the set of image points. As far as the role of the n-simplex versus that of the n-ball is concerned, this is unimportant, for it is easy to use either as a completely faithful map of the other, in the sense that each point of one corresponds to a single point of the other and that neighboring points of one correspond to neighboring points of the other; any distortion introduced in the mapping operation is of no importance in the application of the theorem. The argument outlined in Sec. 15.1 is somewhat more widely applicable, however, and hence is somewhat stronger, than the argument that will be used here to prove Brouwer's fixed-point theorem. The stronger argument will subsequently be developed and related, as indicated in Sec. 15.1, to Sperner's lemma. We now state Brouwer's fixed-point theorem, but we shall defer the proof until some algebraic preliminaries have been developed.

Brouwer's Fixed-Point Theorem. Any continuous function from an n-simplex to itself leaves at least one point fixed.

In saying that the function leaves at least one point fixed, we mean that there is at least one point of the n-simplex that is its own image. A function from an n-simplex to itself is simply a rule assigning to each point of the n-simplex an image that is a point of the same n-simplex; it is not required that every point of the n-simplex be the image of some point under the function, nor is it required that every pair of points have different images. A function is continuous at a point p if, given any neighborhood of the image q of p under the mapping, there is a neighborhood of p such that the image of each point of this neighborhood lies in the chosen neighborhood of q. (In this context, a neighborhood of p or q may be considered to be the interior of an n-ball—but not the bounding sphere of the n-ball—centered at p or q, respectively. If the n-ball has points that are not points of the n-simplex, then the neighborhood is the intersection of its interior with the n-simplex.) For purposes of testing continuity, the neighborhoods that are considered are usually small ones; the definition is simply the analyst's way of saying that the deformation described by the function does not permit tearing, but it does permit collapse.

In order to prove Brouwer's fixed-point theorem using Sperner's lemma, we shall have to provide a systematic way of labeling points of a subdivision. This could be done geometrically, but the presentation is more compact and more easily followed if it is done algebraically in terms of coordinates. We now turn to the algebraic preliminaries that were mentioned previously.

One of the easiest ways of looking at n-dimensional geometry is in terms of coordinate axes similar to those used in coordinate or analytic geometry. These coordinates are considered to be independent; that is, if there are n coordinates, then each is supposed to lie outside the flat space determined by the rest. In many applications it is convenient to consider these coordinate axes to be mutually perpendicular and to consider coordinates measured along them as being measured in terms of distance. If this is done, then the ordinary Pythagorean theorem for distance, directly generalized from geometry of two and three dimensions, holds: The distance between two points, one with coordinates u_i and the other with coordinates v_i, is given by the formula

$$\left[\sum_{i=1}^{n}(u_i - v_i)^2\right]^{1/2}$$

For our application, it is best to ignore the Pythagorean aspects of the geometry and to seek a symmetric coordinate description of an n-simplex. Ignoring the Pythagorean theorem amounts to ignoring the concept of perpendicularity, but the concept of parallelism remains. We take coordinate axes; it will be convenient to take $n + 1$ of them, which are independent in the foregoing sense, but we do not notice the angles between them, and we are not concerned with distances. Indeed, we take as origin of coordinates some point that is not a vertex of the n-simplex we want to describe, and that is otherwise arbitrary except for the requirement that the rays from the origin through the vertices of the n-simplex (there are, of course, just $n + 1$ of these rays) are independent. We choose these rays to be coordinate axes, and the coordinates of each vertex are taken to be all 0 except for the coordinate pertaining to the axis passing through that vertex, which is taken to have value 1.

Although we shall not go into the axioms of multidimensional geometry here, it is assumed that steps of equal length can be taken along any coordinate axis. Furthermore it is assumed that steps of equal length can be taken along any two parallel directions. Then the point with coordinates u_i is obtained by moving to the point on the first coordinate axis with coordinate u_1, then parallel to the u_2 axis a length equal to that from the origin to the point on the second coordinate axis with coordinate u_2, etc. Under these assumptions, the points of the n-simplex that has vertices with coordinates of value 0 except for a single coordinate of value 1 are those points with nonnegative coordinates having sum equal to 1. (An example is available in any room: Take the vertical line and the two horizontal lines at any corner of the floor as coordinate axes, choose points on each one yard or so from the corner, and fill in the triangle between these points. This is a 2-simplex in the 3-space. If we substitute a step for a yard, then there might be difficulty in stepping the same length in all directions—particularly upward. If, in addition, the room is old and has settled so that the corner is not square, then the notion of perpendicularity has been eliminated.) The $n + 1$ nonnegative coordinates summing to 1 and representing a point in an n-simplex constructed in this way are called *barycentric coordinates*.

Any vertex has barycentric coordinates that are all 0 except for a single coordinate, which has value 1. Any (one-dimensional) edge is composed of points such that all their barycentric coordinates have value 0 except for the two corresponding to the coordinates that are not 0 for the two vertices of the edge. Remembering the definition of the carrier of a point, we can assert that *the only positive barycentric coordinates of a point of the n-simplex are those that have value 1 at the vertices of the carrier of the point in the simplex.*

Since the sum of the barycentric coordinates of a point is 1, it follows that as a point leaves its carrier its positive barycentric coordinates must generally decrease as one or more of its zero coordinates become positive. We shall have to look later at this phenomenon somewhat more carefully. First, however, we give a labeling rule associated with a function from an n-simplex with barycentric coordinates to itself.

The label assigned by a continuous function from an n-simplex S to itself is defined as follows: If a point p of S with image q is labeled and if the barycentric coordinates of the points are respectively p_i and q_i, then the *label* of p is j if j is the smallest index value for which $q_j \leq p_j \neq 0$.

For "general functions" and "general points," both phrases being used implicitly, it will frequently be true that there is only one index for which the barycentric coordinates satisfy the inequalities that must be satisfied by the q_j and p_j, and the phrase "smallest index value" is inserted only as a rule to break ties. The rule in principle is classical; this compact form was copied from H. W. Kuhn [14].

We point out first that this rule satisfies the conditions imposed previously for the admissible labeling of a set of points in S for any finite set containing all the vertices of S. Since the only nonzero coordinates of the vertices are the individual ones corresponding to the individual coordinate axes through the vertices, we see that each vertex will carry a different label by virtue of the inequality $p_j \neq 0$. This same inequality and the statement already made about the relation between positive coordinates of a point and the vertices of its carrier assure that the other condition for admissible labeling is satisfied. Thus it is true that *if the vertices of any subdivision of an n-simplex S are labeled in accordance with the label assigned by a continuous function from S to itself, then this subdivision and its labeling will satisfy the condition in the conclusion of Sperner's lemma.*

We now need one additional assumption, which is plausible enough to allow deferment of its justification: Given any simplex S, there are subdivisions that are arbitrarily fine; that is, there are subdivisions such that the length of the longest one-dimensional edge of a simplex of the subdivision is arbitrarily small. This is clearly true if, as seems evident, the process of subdivision can be continued indefinitely and if subdivision generally replaces a simplex by a set of smaller ones.

Making this assumption, we can now finish the proof of Brouwer's fixed-point theorem. We take an infinite sequence of subdivisions of S with *mesh*, that is, length of the longest one-dimensional edge, approaching 0. From each subdivision, we choose one simplex that carries all labels, and in this simplex we choose a single point. We thus have an infinite sequence of points in the original simplex S, and we can choose a subsequence that converges to a single point. This point clearly is the limit point of the sequences of all vertices of all the simplexes from which the points of the convergent subsequence were originally chosen. Since, according to our labeling rule, for one of these vertices we have $q_j \leq p_j$ for each j, and since such inequalities are clearly preserved in the limit, it is true that no barycentric coordinate of the image of the limit point of the chosen subsequence exceeds the corresponding barycentric coordinate of the limit point itself. Hence, since the barycentric coordinates must sum to 1, the barycentric coordinates of the image of this limit point must be exactly the same as the barycentric coordinates of the limit point itself, and—since these are coordinates in a coordinate space of $n + 1$ dimensions—the points must coincide.

This completes the proof of Brouwer's fixed point theorem.

15.4 SUBDIVISIONS OF A SIMPLEX

Here we face more squarely a problem evaded in the preceding sections, that of whether there is any way of successively subdividing an n-simplex in accordance with the restrictions imposed on a subdivision, and, in particular, whether this can be done so that the mesh of the subdivisions approaches zero. We shall consider this last part to be reasonably obvious once the first part has been treated, and so we shall not belabor that point.

Although it is easy to see many ways of subdividing a 2-simplex, the successive subdivision of even a 3-simplex requires some care.

We recall the following facts: (1) we used the assumption that interior faces of a subdivision are incident to exactly two simplexes of the subdivision, and (2) we required that each vertex of any simplex of a subdivision be a vertex of each simplex of the subdivision to which it is incident. We shall present two schemes of subdivision for which these properties are clearly hereditary. In order to do this, however, we shall assume that the vertices of the simplex to be subdivided are given in terms of barycentric coordinates of some containing simplex, and we shall work in terms of these barycentric coordinates.

The property (2) required of vertices will be guaranteed by having all new vertices depend only on the barycentric coordinates of the vertices of the carriers of these new vertices in the simplexes to be subdivided. If two simplexes that are to be subdivided have naturally associated simplexes in common, then these naturally associated simplexes have the same vertices, and any new vertices that are added in the common naturally associated simplexes are in positions that are independent of the parent simplex and hence are coincident.

The first subdivision that will be discussed is *barycentric subdivision*. Given any k-simplex of which the vertices are points of an n-simplex S with barycentric coordinates, the *barycenter* of the k-simplex is the point with barycentric coordinates having values that are the arithmetic means of the corresponding coordinates of all the vertices of the k-simplex. For a barycentric subdivision of an n-simplex, which is presumably one simplex of a subdivision of S, construct the barycenters of the n-simplex itself and of all the simplexes of lower dimension, except the (-1)-simplex, naturally associated with it. Each simplex of the barycentric subdivision has as vertices a set of $n + 1$ points that are barycenters of simplexes of this naturally associated set, the simplexes having all dimensions from n through 0; thus each dimension appears once. Furthermore, if a vertex v_2 is the barycenter of a simplex S_2 of lower dimension than the simplex S_1 of which v_1 is the barycenter, and if both v_2 and v_1 are vertices of a simplex of the subdivision, then S_2 is required to be naturally associated with S_1. Thus, for example, the vertices of one of the simplexes of the barycentric subdivision of a 3-simplex (that is, a tetrahedron and its interior) would include the centroid of the tetrahedron (that is, the barycenter of the 3-simplex), the centroid of one of its faces, the center point of one side of that face, and an endpoint of that side. A little pictorial and arithmetic experimentation will convince the reader that barycentric subdivision does provide a means of successive subdivision meeting all requirements stated.

This experimentation will also lead to the observation that the simplexes of successive barycentric subdivisions become more and more elongated relative to their width, although their mesh goes to zero, and this is not always an acceptable property. We shall therefore present in outline another method of subdivision, due to H. Whitney [22, pp. 358–360]. This method of subdivision has the property that the simplexes produced under successive subdivision take only a finite number of shapes.

In Whitney's scheme, the vertices of the subdivision of an n-simplex are all either vertices of the simplex to be subdivided or barycenters (midpoints) of one-dimensional edges of the simplex that is to be subdivided. Furthermore, the midpoint of each edge is used as a vertex of each simplex of the subdivision to which it is incident, and thus the foregoing vertex requirement (2) is inherited.

To carry out the subdivision, label the vertices of an n-simplex v_i, with i running from 0 through n. We now define a set of points v_{ij} for $i \leq j$, with both indices running from 0 through n (or more precisely, with i running from 0 through n and j running from i through n). We set $v_{ii} = v_i$, and we take v_{ij}, $i \neq j$, to be the midpoint of the edge with endpoints v_i and v_j. We now assign a partial order to the points v_{ij}. One such point precedes a second if and only if the first index of the first point is at least as great as the first index of the second point and the second index of the first point is at most as great as the second index of the second point. Thus we can pass from a point to an immediately following point by holding the first index fixed and increasing the second index (if possible) or by holding the second index fixed and decreasing the first index (if possible), the increment in each case being 1. It is clear that no point can precede a point of type v_{jj}, since the second index is required to be at least as large as the first, and that no point can follow a point of type v_{0n}. From these observations, it is clear that maximal sets of ordered points will all contain $n + 1$ points and that there will be 2^n such sets. Rather than setting down a somewhat tedious proof of this assertion, we shall give the complete set of maximal sets of ordered points for $n = 3$:

$$v_{00}, v_{01}, v_{02}, v_{03} \qquad v_{11}, v_{12}, v_{13}, v_{03}$$

$$v_{11}, v_{01}, v_{02}, v_{03} \qquad v_{22}, v_{12}, v_{13}, v_{03}$$

$$v_{11}, v_{12}, v_{02}, v_{03} \qquad v_{22}, v_{23}, v_{13}, v_{03}$$

$$v_{22}, v_{12}, v_{02}, v_{03} \qquad v_{33}, v_{23}, v_{13}, v_{03}$$

These maximal sets of ordered points displayed are arranged so that the four maximal sets of ordered points for $n = 2$ can be obtained from the left-hand column by omitting the last point of each set. The count of the number of such sets is apparent; starting with v_{0n} at the right, there are two ways to choose the next element to the left at each step.

The subdivision consists of simplexes having as vertices the points of these maximal sets of ordered points. Experiment will lead almost immediately to an understanding and an inductive proof that the sub-division has the properties (1) and (2), as claimed. An example in two dimensions is easily constructed from the table and the remarks given earlier, and the course of the induction will be perceived if the reader tries to extend this example to three dimensions.

15.5　REVIEW OF POSITION—SECOND INTRODUCTION

We have now proved Sperner's lemma, and we have deduced the Brouwer fixed-point theorem. The lemma may be amusing; the fixed-point theorem may not have obvious significance to an inexperienced reader, but it is one of the famous and exciting theorems of topology. In this section, we shall review our position with regard to these two accomplish-ments, we shall note that we are not obviously close to the goal we set in the first introductory section, and we shall outline our procedure for getting to this goal.

The power of the Brouwer fixed-point theorem lies partly in the fact that it is not restricted to simplexes, as we shall note immediately, and partly in several famous extensions, some of which will be mentioned briefly.

We first concern ourselves with extensions of the Brouwer fixed-point theorem to domains that are not simplexes. To do this, we shall note an obvious generalization of the classical function of a map: A map presents a continuous, although sometimes distorted, image of a mapped domain; it has the properties that each point of the mapped domain goes into a single point on the map, each point on the map is the image of a single point of the mapped domain, and the functions from domain to map and map to domain are both continuous.

The word *map* may bring to the mind of the reader a piece of paper on which is depicted a portion of the surface of the earth. Distortion is necessary if the piece of paper is flat, for it is impossible to map a desired portion of a sphere onto a flat sheet without distortion. Some maps, for example, gnomonic projections, have the property that the images of great circles on the sphere are straight lines on the map; in these projec-tions, however, angles are distorted. Some maps, such as Mercator's projection, which takes meridians of longitude and parallels of latitude to straight lines, preserve angles but distort distances (the gnomonic projection also distorts distances). For example, as one approaches the poles on a Mercator projection, distances on the map become unrealistically magnified relative to distances nearer the equator, so that many schoolchildren have distorted ideas concerning the size of geographical objects in the far north and the far south.

Despite such distortion, since the domain mapped is limited to one in which the properties listed are applicable, it is still true that the map introduces only stretching and shrinking operations, and, of particular importance in the application here, it is true that *a continuous function from the region mapped to itself induces a continuous function from the map to itself.*

It is not hard to visualize an extension of this mapping idea to other dimensions. Trivially, the ordinary globe or, to the extent that it is valid, a model may be considered to be a three-dimensional mapping. The globe could be imagined to map the surface of the earth and the interior of the earth into an object with similar shape but much smaller size, so that convenient examination is possible. The same is true for models. Occasionally, for models, the size may be larger—models of biological cells, molecules, crystal cells, etc. Going further, we can easily construct a mapping from the points of a 3-simplex to the points of a three-dimensional ball, that is, from the points of a tetrahedron and its interior to the points of a sphere and its interior. For example, if the tetrahedron is inscribed in the sphere, segments from the center of the sphere to the boundary tetrahedron of the simplex may be mapped onto radii of the sphere by simple expansion from the center of the sphere; the expansion may be uniform along each of these segments and just sufficient to take the segment into a radius. Such a mapping is completely faithful for purposes of Brouwer's fixed-point theorem; that is, if we consider any continuous function from the sphere to itself, then the naturally induced function from the simplex to itself is continuous, and hence has a fixed point. The point of the sphere with image fixed under the induced function must be a fixed point under the original function.

A mapping with the properties we have listed is called a *homeomorphism*, and the mapped domain and its map are called *homeomorphs*. The pertinent properties are the following: (1) every point of the mapped domain has a single image on the map; (2) every point on the map is the image of exactly one point of the mapped domain; (3) the mapping function taking the domain onto the map is continuous; and (4) the inverse function assigning to each point on the map the point of the mapped domain of which it is the image is continuous. We shall not go into extensive discussion here, but it is obvious that there are many homeomorphs of simplexes of all dimensions, and that the Brouwer fixed-point theorem must be valid for such homeomorphs.

Brouwer's Fixed-Point Theorem. A continuous function from any homeomorph of an n-simplex to itself leaves at least one point fixed.

In particular, the admissible space of strategies for a two-person, zero-sum game is homeomorphic to a simplex of the correct dimension. For elementary discussions of the theory of games, see [3, pp. 191–210] or [23], and for more advanced discussions, see the literature cited in these places and the plentiful recent literature. The classical reference to the theory of games and its most direct applications is the famous book by J. von Neumann and O. Morgenstern [18].

S. Kakutani noted that, although the Brouwer fixed-point theorem cannot be applied directly to prove the fundamental theorem of the theory of games, a fairly straightforward extension can be used. In this extension, there is a function from a simplex to sets of points of the simplex such that the image of each point is a convex set of points containing at least one point. Since the number of points may vary among the images of various points of the simplex, continuity is difficult to define, but *upper semicontinuity* is meaningful. Under this condition, Kakutani [13] was able to use a simple limiting argument to prove the Kakutani fixed-point theorem, which has the fundamental theorem of the theory of games as an immediate corollary.

Other fairly elementary but interesting applications of results of this type have been presented by A. W. Tucker [20] and by Ky Fan [7].

Another type of limiting argument was used by J. Schauder to apply to spaces with infinitely many dimensions. In this case, it is difficult to use Sperner's lemma directly because of the elusiveness of the individual elements of infinite sequences, and Schauder extended Brouwer's fixed-point theorem to apply to convex bodies in Banach spaces [19]. Banach spaces are those that may have infinitely many dimensions, in which the notions of line segment and vector are preserved and that are well enough behaved to permit many of the properties of finite-dimensional Euclidean spaces to be preserved. A clear exposition of the elementary Schauder theory and some of its direct application may be found in [4, Chapter 15] and other works on topology. The Schauder theory has been applied both to prove the existence of solutions for elliptic partial differential equations, as in [19], and to give a proof of the existence of a solution to the standard ordinary differential-equation problem

$$y' = f(x, y), \qquad y(x_0) = y_0,$$

where x is a real variable, y is an n-tuple of reals, and x_0 and y_0 are real constants. Most problems of ordinary differential equations involving initial conditions that are now treated can be put into this form. The Schauder theory yields solutions under attractive degrees of generality for the function f.

We now discuss the steps that will enable us to attain the goal set in the first introduction. In this connection, we note that all arguments to be used in the following sections are insensitive to homeomorphisms and hence that the domain of applicability of the results we attain is considerably wider than that actually used in attaining the results.

The crucial point that will permit attaining our original goal is to notice that the labeling of the vertices of a subdivision is equivalent to providing a function from each simplex of the subdivision (domain simplex) to a single image simplex (which is not part of the subdivision). This function can be considered to be linear, and it is constructed for any domain simplex of the subdivision by labeling the vertices of the image simplex, assigning vertices of the image simplex as images of the vertices of the domain simplex in accordance with the labels of the domain simplex and interpolating linearly for the other points of the domain simplex. We shall describe such a mapping more precisely later. We shall need to keep track of a signed number of times that the image is completely covered by one of the simplexes of the subdivision; that is, we shall have to prescribe a sign to each labeling in which all the labels appear. Thus we shall end with an integer, the sum of all these signed numbers, associated with the image simplex.

It is just as easy to develop a reasonably general *simplicial homology theory*, which is basic to algebraic topology, as it is to carry out this analysis in any other way. To this end, in Sec. 15.6 we shall discuss oriented simplexes, their oriented boundaries, and oriented simplicial mappings. In Sec. 15.7 we shall discuss k-chains, cycles, and bounding cycles. Generally speaking, a cycle is an object with no boundary, such as a sphere or a torus for two dimensions, a circle for one dimension, etc.; noncycles are the circle and its interior in two dimensions, a line segment in one dimension, etc. It is clear that all the $(n-1)$-faces of an n-simplex form a cycle, and it is the only cycle that can be built from the simplexes of dimension less than n naturally associated with the n-simplex and that is nonbounding among these lower-dimensional simplexes; that is, it is not the boundary of some other geometrical object that can be built from these simplexes. This is the property we shall eventually exploit, and in Sec. 15.8 we shall prove a fundamental index theorem that will permit us to attain the goals set in Sec. 15.1. In Sec. 15.9 we shall run quickly through a few examples, omitting details; these will include the fundamental theorem of algebra (that every polynomial of positive degree has at least one root in the complex plane).

Finally, in Sec. 15.10 we shall confess to some of our sins and suggest references that may be used to obtain greater precision.

15.6 ORIENTED SIMPLEXES, THEIR BOUNDARIES, AND ORIENTED SIMPLICIAL MAPPINGS

The idea of orientation is not necessary in algebraic topology, but we can introduce it to obtain results that are not otherwise as easily obtained. The idea of orientation is a direct extension of the idea of clockwise and counterclockwise rotations. Thus an oriented 2-simplex is not only a set of points, a triangle and its interior, but also a cyclic direction of progression around its bounding triangle. If two 2-simplexes contain the same points, that is, if they are built on the same three vertices, but if the assigned directions of progression around the boundary are different, they are considered to be different oriented 2-simplexes; they are related, and we usually refer to one as the negative of the other. Thus, if we have one oriented two-simplex and wish to denote the oriented 2-simplex with the same points but opposite orientation, we simply adjoin to a symbol denoting the first simplex the coefficient -1. We shall extend this idea in two ways: (1) we shall provide a definition of orientation of simplexes of any number of dimensions and (2) we shall extend the set of coefficients that may be adjoined to any symbol denoting a simplex to include all integers. (Various other sets of coefficients are admissible in different versions of algebraic topology, but discussion of these would divert us here. A lucid exposition at a reasonably elementary level can be found in [15]; a more formal development appears in [6]. In particular, [15] is recommended as a source of expanded remarks along the lines made here and also of pertinent related topics. See also the more recent [17].

We continue to think of simplexes as embedded in euclidean space of comfortably high dimension, although [6] relaxes this restriction in a way that will be described superficially in Sec. 15.10. The points of an n-simplex are determined by specifying its $n + 1$ vertices. These should be so situated that they do not all lie in an $(n - 1)$-plane. The oriented n-simplex (points and orientation) may be specified by specifying the $n + 1$ vertices and assigning a linear order to them. Several different orders, however, will lead to the same oriented n-simplex. For the familiar case in which $n = 2$, let the three vertices be denoted by a, b, and c; then the orders (a, b, c), (b, c, a), and (c, a, b) are all equivalent and correspond to a progression around the bounding triangle in the same direction. The oppositely oriented simplex may be specified by any of the orders (b, a, c), (a, c, b) or (c, b, a).

Note that these six orders are just the six different permutations on the three elements a, b, and c. In order to introduce a compact algebraic procedure for determining relative orientations of simplexes, we shall discuss the parity of permutations. This is a well-known and well-understood concept, but we shall give precise definitions; generally, the parity of a permutation is even, referred to some preferred linear order of the elements, if the number of successive transpositions of pairs of elements required to restore the preferred order from the permuted order is even, and the parity is odd if the number of transpositions required to restore the preferred order is odd. Note that the equivalent orders in the examples of the oriented 2-simplexes are all even permutations of each other, and that the permutation of the order of the vertices from one oriented simplex to an oppositely oriented one in the example is odd. This will be taken as the basic rule for assignment of relative orientations.

We now take a set of $n + 1$ marks in preferred order and replace them by the set $0, 1, \ldots, n$. Consider any permutation P of the original marks; by making the substitution performed on the marks of the original permutation, we can replace P by an ordered set of marks $p_0, p_1, \ldots, p_n$; in this set, each mark p_i is one of the integers between 0 and n, inclusive, and no integer is used twice. In order to define the parity of the permutation P relative to the preferred order, we define the oddity of any element p_i of the induced permutation:

The *oddity* of the ith element (starting with the 0th element) p_i is the remainder after division by two of the number of values of j with the properties that $j > i$ and $p_j < p_i$.

The *parity* of a permutation is the parity (even or odd) of the sum of the oddities of the elements of the permutation.

The reader is advised to check these definitions for the permutations on three objects in the preceding examples. It is also advisable to prove the easy theorem that transposition of any two elements of a permutation reverses its parity—a theorem mentioned previously. Finally, if P_1, P_2, and P_3 are permutations on n objects, if P_2 is of even parity with respect to P_1, and P_3 is of even parity with respect to P_2, then any of the permutations is of even parity with respect to any other.

We now consider an n-simplex with vertices $a_0, a_1, \ldots, a_n$, and we assign the linear order in which they were written to these vertices. This is a realization of an oriented n-simplex, and all other realizations of this n-simplex through the specification of its vertices and a linear order must specify the same vertices and a linear order that is an even permutation of this order.

An *oriented n-simplex* is an n-simplex together with a maximal set of linear orderings of its vertices, such that each ordering of the set is an even permutation of the others.

It is also true that the orientation assigned to an n-simplex introduces a natural orientation to the boundary simplexes, or naturally associated simplexes, of $n - 1$ dimensions. This is clearly illustrated in the case $n = 2$, already described. The 1-simplexes on the boundary are traversed consistently with the orientation, and this implies that one 0-simplex of each is the initial vertex and the other is the terminal vertex. Indeed, the orientation of an n-simplex may be directly related to that of its boundary $(n - 1)$-simplexes by agreeing that it describes the orientation of each boundary $(n - 1)$-simplex as seen from the vertex of the n-simplex that is not a vertex of the $(n - 1)$-simplex. Before we present a precise definition, it will be convenient to record a definition of oppositely oriented simplexes.

Two simplexes S_1 and S_2 are *equal but oppositely oriented* if they have the same set of vertices and the orders assigned to the vertices of S_1 are odd permutations of those assigned to the vertices of S_2.

As we noted at the beginning of this section, if S_1 and S_2 are equal but oppositely oriented, we write $S_2 = -S_1$.

Let S be an oriented n-simplex with vertices $a_0, a_1, \ldots, a_n$, admissible in the order written; *the boundary oriented $(n - 1)$-simplexes of S* are the simplexes $(-1)^k S_k$, where S_k is an oriented $(n - 1)$-simplex with vertices that include all the vertices of S except a_k, and that may be written in the order above, with the omission of a_k.

In order to prove that this definition is consistent, one should prove that the same boundary $(n - 1)$-simplexes are generated for any admissible ordering of the vertices of the n-simplex, but this is trivial and will not be included here.

We shall introduce addition and cancellation of coefficients, such as the -1 in the definition immediately above, in the next section. It will turn out that each $(n - 2)$-simplex naturally associated with an n-simplex occurs on the boundary of two naturally associated $(n - 1)$-simplexes but with opposite orientations; thus the cancellation will leave a net number of 0 occurrences of any $(n - 2)$-simplex, and therefore of any lesser-dimensional simplex, on the boundary of an n-simplex, and they can be ignored. This statement is easily illustrated for the cases $n = 2$ and $n = 3$, and the reader is encouraged to produce these illustrations for himself.

We finally turn to simplicial maps. A simplicial map is a map from an oriented domain simplex S_d to a range simplex S_r; the domain simplex and the range simplex may for our purposes be restricted to have the same dimension n.

The simplicial mapping is determined completely by the images of the vertices of S_d, and it is required that these images be vertices, not necessarily all different, of S_r. The images of the 1-simplexes naturally associated with S_d are simplexes naturally associated with S_r, and of dimension no greater than 1. The mapping, which may not be a homeomorphism, is determined by linear interpolation between the images of the vertices. This process of linear interpolation may be continued inductively, but it is somewhat easier to think in terms of barycentric coordinates, which were introduced in Sec. 15.3. Let p_i be the barycentric coordinates of a point P of S_d, and let q_σ be the barycentric coordinates of its image Q, in some simplex naturally associated with S_r, with vertices the images of all the vertices of S_d; this may be S_r itself, and for oriented simplexes we must always keep in mind the equal simplexes oppositely oriented. Then for each σ the value of q_σ is the sum of the values of p_i over the set I of vertices that are mapped into the σth vertex. This gives a well-defined linear interpolation.

We shall be particularly interested in simplicial mappings in which each vertex of S_r is the image of one vertex of S_d. These mappings cover S_r, and the orientations of S_d and S_r may be consistent or inconsistent; that is, for an admissible linear ordering of the vertices of the oriented simplex S_d, the consistently ordered images may or may not be an admissible linear ordering of the vertices of the oriented simplex S_r.

Without pursuing details here, we note that we shall relate the number of consistent covering mappings on S_r minus the number of inconsistent ones with a similar difference for one of the oriented faces, where S_r will be a fixed simplex and S_d will be allowed to vary over all the simplexes of a subdivision. To this end, we shall present some of the elementary aspects of simplicial *homology theory* in the next section.

15.7 HOMOLOGY—k-CHAINS, CYCLES, BOUNDING CYCLES

We have remarked earlier that we shall be interested in the net signed number of coverings of a range simplex by simplical mappings of the domain simplexes of a subdivision. We have also remarked that we shall use the coefficient -1 adjoined to a symbol denoting an n-simplex to denote the equal but oppositely oriented simplex. We then suggested the natural extension that would permit any integer to be used as a coefficient to the symbol denoting any simplex; this has no obvious merit for the simplexes of the subdivision, but it has obvious implications in connection with our avowed purpose of counting algebraically the number of coverings of the range simplex. It will be convenient, however, to admit the coefficient 0 for the domain simplexes of the subdivision, and no particular difficulty will ensue if we admit any other integer, positive or negative (a negative integer having implications of orientation opposite to that of the oriented simplex for which it is a coefficient), and by admitting such coefficients we shall be able to develop a part of the classical *homology theory* of classical algebraic topology; we repeat that this theory is developed more lucidly and more generally in [15] and in other topology texts.

We shall need to introduce the idea of a *simplicial k-chain*. This is an abstract object built on a finite number of k-simplexes to each of which is adjoined an integer coefficient. In subdivisions of a simplex discussed in connection with Sperner's lemma and Brouwer's fixed-point theorem, an orientation would be assigned to the original simplex, parallel orientations would be assigned to each simplex of the subdivision, and the chain would be built on all these oriented simplexes of the subdivision; the coefficient 1 would be adjoined to each.

We shall find that we need to apply the homology theory to each subdivision independently, so that there is at any time only a finite number of simplexes, those of the subdivision being considered, available upon which to build a chain. It is usually convenient in our applications to orient the simplexes in a particular way, corresponding with an orientation in the original problem presented, but this is not essential to the study. We base our study on an arbitrary finite set of simplexes, subject to restrictions that all naturally associated simplexes of lower dimension are also available for building chains; the simplexes of the set are assigned orientations, either naturally or arbitrarily, and the geometrical object created is called a complex.

A *complex* is a finite set of oriented simplexes with the property that the set contains every simplex (with some orientation assigned) naturally associated with every simplex of the set.

It should be noted that the simplexes are not all of equal dimension, and that orientations are assigned to all simplexes in the complex; if there is a natural orientation introduced to a boundary simplex, the orientation assigned may or may not agree with it. In case S is an n-simplex in the complex, the $(n - 1)$-simplexes on the boundary of S have a naturally assigned orientation, but it turns out that the lower-dimensional simplexes naturally associated with S do not; orientations are in some way assigned to these simplexes.

Since in our applications, and in all the simplest applications of homology theory, the simplexes that can be used to build a k-chain are from a fixed finite set, it follows that we can attain some simplification of description by assuming that this finite set is the set of k-simplexes of some complex with the assigned orientation. If one of these simplexes actually is not used, then we simply assign a 0 coefficient to it. We may formalize this, with the understanding of the meaning of a 0 coefficient, in the following definition.

A *k-chain over a complex K* is the set of k-dimensional oriented simplexes of K together with an integer coefficient adjoined to each simplex.

We suggested one natural example of a k-chain—all the simplexes of a subdivision, consistently oriented, with coefficients 1. Another natural example has already been introduced; it is the set of bounding k-simplexes of a $(k + 1)$-simplex. To be more specific, let S be an oriented $(k + 1)$-simplex. It can be made into a $(k + 1)$-chain by adjoining the coefficient 1; this is irrelevant now, but the relevance of this conversion will immediately become apparent. The simplex S has $k + 2$ naturally associated k-simplexes on its boundary, and these may be oriented and taken together with the correct coefficients 1 or -1, as indicated in the preceding section, to form a k-chain, which is the oriented *boundary* of S.

Thus each oriented simplex has a boundary that is a chain of dimension one less than the dimension of the simplex. Implicitly, this chain is generated when the simplex is thought of as being taken once (being the only significant simplex in a chain and having coefficient 1), and we should be inclined to double the coefficients of the boundary simplexes if we considered the simplex to be taken twice (to be the only significant simplex in a chain and to have coefficient 2). This doubling situation could arise naturally through a limiting process in which two simplexes depending, say, on time drift closer together, along with their boundaries, until in the limit they coincide. Since the applications will include this type of argument, we can see our way to the definition of the boundary of a k-chain. This will immediately be formally presented in three steps.

For $k > 0$, the *boundary of an oriented k-simplex S* is a $(k - 1)$-chain of which the simplexes are the boundary $(k - 1)$-simplexes of S, each oriented consistently with S in the sense of the preceding section, and to each of which is adjoined the coefficient 1. The boundary of a 0-simplex is vacuous.

We next introduce the idea of the sum of two or more k-chains. This idea is required in our applications in two places, one of which has already been mentioned—the algebraic net count of the number of coverings of the range simplex in our principal application. The second main application is in a lemma stated later—that the boundary of a boundary is a chain with all coefficients having value 0. The sum of k-chains is a k-chain with coefficients that are the sums of the coefficients of chains to be summed.

A final formality remains before we define the boundary of a chain. The chain may be built on simplexes that share faces, so one of the $(k - 1)$-simplexes in the boundary of a k-chain may receive contributions from two sources. The resolution of this difficulty is trivial, and we proceed to it without further discussion by defining the sum of k-chains.

The *sum* of a finite number of k-chains of a complex K is a k-chain of K

with coefficients that are the sums of the corresponding coefficients of the summand k-chains.

The definition of the boundary of a k-chain is now immediate. It is, of course, always essential for us to remember that the faces of any simplex in a complex K are also simplexes, with possibly opposed orientations, of K.

The *boundary* of a k-chain of a complex K is vacuous if $k = 0$, and is otherwise a (possibly vacuous) $(k - 1)$-chain. This boundary is the sum of the boundaries of the k-simplexes of K, each boundary being taken a number of times equal to the coefficient of that simplex in the chain, negative coefficients implying opposite orientations.

By a *vacuous $(k - 1)$-chain*, we mean one with coefficients all equal to zero. That this is possible is illustrated by the simple, but fundamental, example presented in the following lemma.

Lemma. The boundary of the boundary of a k-chain is vacuous.

The proof will be largely omitted here; it is an exercise in algebra. The lemma is true if and only if the boundary of the boundary of an oriented simplex is vacuous. The rule for generating the oriented boundary of an oriented simplex leads immediately to this conclusion. We shall not present the algebraic proof here, but the reader is advised to illustrate the lemma for the 2-simplex and the 3-simplex, finding cancellation in the 0-simplexes and the 1-simplexes, respectively.

Although the boundary of a boundary is vacuous, a boundary is not necessarily the only type of chain with a vacuous boundary. For example, consider a complex composed of the faces of a tetrahedron (but not its interior) all consistently oriented, and, of course, the edges and vertices, which are required to be part of the complex by the definition of a complex given earlier in this section.

The 2-chain built on all these faces, each with coefficient 1, has a vacuous boundary, for it would be the boundary of the 3-simplex if one were present. It is clearly not the boundary of anything in the complex chosen for the example, however, for there are no 3-simplexes. Thus there are two kinds of chains with vacuous boundaries—those that are themselves the boundaries of chains of one higher dimension and those that are not. These are called, respectively, bounding cycles and nonbounding cycles. We supply formal definitions:

A chain with a vacuous boundary is called a *cycle*.

A k-cycle of K that is the boundary of some $(k + 1)$-cycle of K is called a *bounding k-cycle of the complex K.*

A k-cycle of K that is not the boundary of any $(k + 1)$-cycle of K is called a *nonbounding k-cycle of the complex K.*

Homology theory is the study of the cycles of complexes, with the understanding that any two cycles of a complex will be considered to be *equivalent* if their difference (as the difference of two chains) is a bounding cycle. In some texts, considerable effort is expended to deal with degenerate cycles, in which vertices may coincide; such cycles, however, may be ignored by noting (as A. W. Tucker did several years ago) that they bound.

In modern topology, *homotopy theory* is used to present results that are frequently stronger than those of homology theory; it is also more difficult. Furthermore, restriction to simplicial complexes is usually relaxed; J. H. C. Whitehead [21] introduced a satisfying general geometrical object for extensive development of homotopy theory and called it the *CW-complex*. A lucid exposition of many of the parts of topology related to homotopy, including a means for deriving an understanding of the role of CW-complexes, may be found in the pamphlet by P. J. Hilton [10], where CW-complexes are treated in Chapter 7, pp. 95–113.

In our own examples here, the complexes are all simple and easily described, and the homology theory is almost trivial. We shall not need to pursue the details of algebraic topology further to finish our arguments.

15.8   THE FUNDAMENTAL INDEX THEOREM

We shall now turn to a fundamental index theorem that generalizes Sperner's lemma suitably for our purposes. It is based in a nondemanding way on the homology theory outlined in the preceding section; it uses practically no machinery except the ideas of orientation, boundary, and the adjunction of a coefficient to a simplex. The theorem has been known in essence for about a century, but the presentation given here is not quite in the classical mold. The proof is probably due essentially to Lefschetz, who first presented many of the fixed-point and coincident-point theorems of topology. Some later applications to cases that may be more restrictive than those in which our ultimate interest may lie have been presented in [2] and [5]; an exposition that does not differ remarkably from the one presented here may be found in [4, p. 57].

We shall first state some definitions and the general theorem, and immediately afterward present the application of interest to us (in terms of simplicial subdivisions).

Let C^k be an oriented simplicial k-chain, S_r^k be an oriented k-simplex, and f be a mapping of the vertices of the simplexes of C^k to the vertices of S_r^k. (Equivalently, f is a simplicial mapping of every simplex of C^k to S_r^k.) Consider the set of simplexes of C^k with vertices mapped by f onto the complete set of vertices of S_r^k; index this set of simplexes arbitrarily so that S_i^k denotes the ith such simplex of C^k, and denote by c_i the coefficient of S_i^k in C^k. Finally, write $h_i = 1$ if f takes S_i^k to S_r^k with consistent orientation, and write $h_i = -1$ if f takes S_i^k to S_r^k with opposed orientation. Then the *index* of f over C^k is

$$\sum_i h_i c_i.$$

We now construct a special chain that depends on a k-chain C^k, a k-simplex S_r^k, and a mapping f^k of the vertices of the k-chain to the vertices of the simplex. The chain and the simplex are oriented, and this mapping has an index in accordance with the preceding definition. We shall denote the boundary of C^k by $\delta(C^k)$; this is, of course, a $(k-1)$-chain. We shall define the *reduced boundary under* f^k by modifying this chain; we shall denote this reduced boundary by $\delta^-(C^k; f^k, a)$. The reduced boundary is the old boundary, from which all terms involving boundary $(k-1)$-simplexes with one or more vertices mapped to one distinguished vertex a of S_r^k have been removed. The choice of this distinguished vertex is arbitrary, and it is most easily taken to be the initial one in some ordering of the vertices of S_r^k that is consistent with its orientation. (In notation, it is most convenient to take the final vertex, but this leads to a bit of detail concerning orientations in case k is odd.) We shall also define the *reduced mapping* f^{k-1}, seemingly depending on this distinguished vertex, to be the mapping f restricted to the vertices of the reduced boundary $\delta^-(C^k; f^k, a)$. Our principal theorem will be that the index K of the mapping f^{k-1} over $\delta^-(C^k; f^k, a)$ is equal to the index of f^k over C^k. We should note that the boundary $\delta(C^k)$ is a cycle, by the lemma of the preceding section, but generally the reduced boundary is not a cycle. Indeed, the principal theorem can be extended meaningfully through a decreasing set of dimensions by reapplying it to each reduced boundary generated. This, however, is not usually a rewarding process.

Let C^k be an oriented simplicial k-chain, S_r^k an oriented k-simplex, f^k a mapping of the vertices of the simplexes of C^k to those of S_r^k, and a a vertex of S_r^k. The *reduced boundary* of C^k relative to f^k and the distinguished vertex a is a $(k-1)$-chain, $\delta^-(C^k; f^k, a)$, on the same simplexes as the boundary $\delta(C^k)$, having the same coefficients except that the coefficients in the chain $\delta^-(C^k; f^k, a)$ of all simplexes having at least one vertex mapped to a by f^k are set to the value 0.

Let f^k be a mapping of the vertices of the simplexes of an oriented simplicial k-chain, C^k, to the vertices of an oriented simplex S_r^k, and let a be a distinguished vertex of S_r^k. The *reduced mapping* of f^k relative to a is the mapping f^{k-1} obtained by restricting the domain of f^k to vertices that are not mapped to the vertex a and that are, moreover, incident to a $(k-1)$-simplex of $\delta(C^k)$, none of whose vertices are mapped to a.

Principal Index Theorem. If C^k is an oriented simplicial k-chain, S_r^k is an oriented simplex, f^k is a mapping of the vertices of the simplexes of C^k to the vertices of S_r^k, a is a distinguished vertex of S_r^k, $\delta^-(C^k; f^k, a)$ is the reduced boundary of C^k relative to f^k and a, and f^{k-1} is the reduced mapping of f^k relative to a, then the (k-dimensional) index L of f^k over C^k is equal to the [($k-1$)-dimensional] index K of f^{k-1} over $\delta^-(C^k; f^k, a)$:

$$L = K.$$

Before we prove this theorem, we shall outline its application to our problems. To construct the k-chain, we must have a complex with simplexes of at least k dimensions. This complex in our applications will always be a simplicial subdivision of some k-dimensional region. In the application to the Brouwer fixed-point theorem, the region will be a tetrahedron and its interior, or any homeomorph thereof, or more generally the points of a k-simplex. In an application to prove the fundamental theorem of algebra, the region will be a large circular disc in the plane of complex numbers with center at the point corresponding to the number 0. In an example pertaining to tangent vector fields to a sphere, the region will be all the surface of the sphere except for small neighborhoods of two antipodal points, say, the north and south poles. In still another example pertaining to a particular method for solving linear inequalities (and equalities), the region will be somewhat more abstract, but it will be built on a closed ball in Euclidean space of n dimensions.

In each case, subdivisions of arbitrarily fine mesh can be produced to furnish a complex of k dimensions, with k-simplexes serving as a basis for a chain to be analyzed in the light of the foregoing principal index theorem. The k-simplexes will have a natural orientation, and the chains will invariably be constructed by adjoining the coefficient 1 to each of these naturally oriented simplexes. The boundary of each chain studied will invariably be constructed on simplexes that lie on the boundary of the basic region under study. The conditions of turning outlined in the first introduction will imply that the number K of the principal theorem differs from 0. Thus the principal theorem will imply that the number L differs from 0, and a limiting argument like that used in the proof of the Brouwer fixed-point theorem will establish the existence of a fixed point; the existence of this point will, in turn, establish the desired result. The generation of the simplicial mapping f^k in each case will be through a mapping of the k-dimensional region of interest to a Euclidean k-space in which the range simplex S_r^k lies. Again, a rule for assigning a label to each point of the mapped region will be supplied; roughly, the label assigned will be the number of the vertex of the range simplex closest to the image of the point under the mapping from the region of interest to the Euclidean space. This assignment is the one used in connection with the proof of the Brouwer fixed-point theorem. For any point of the region of interest that is used as a vertex of a subdivision, the generated mapping will assign this vertex to the vertex of the range simplex having the same label.

The implied examples will be presented more fully in the next section.

We now turn to the proof of the principal theorem. We might note first, and unimportantly except in trying to create an understanding of the arithmetical proof in the mind of the reader, that the index K is a measure of the number of times that the boundary of the range simplex is swept out, for each complete sweeping uses each oriented face exactly once, with sign taking care of orientation. This is related to the statement made earlier that the only $(k-1)$-cycle that is nonbounding on the complex constructed of the simplexes naturally associated with S_r^k, excluding S_r^k itself, is the cycle composed of the bounding faces, each taken once and each orientated consistently with S_r^k. Now, however, we must modify this statement slightly, for we have learned about chains since the statement was introduced. Other nonbounding cycles occur, but they are all integer multiples of the nonbounding cycle already mentioned.

We now continue the proof, which is perfectly straightforward (the ingenuity having been provided by the mathematicians who invented the essence of the statement of the theorem). We consider those simplexes of C^k with vertices mapped onto the complete set of vertices of S_r^k. These contribute to the index L by definition, and they are the only simplexes that do contribute to L. Each has exactly one face mapped onto the face of S_r^k, from which the distinguished vertex is excluded. Orientations, if checked carefully, will be observed to agree or disagree on this face in accordance with their agreement or disagreement for the mapping of the whole simplex. Thus the portion of the number K contributed by simplexes of C^k having vertices mapped onto the complete set of vertices of the k-simplex S_r^k is exactly equal to the number L of the theorem.

We now consider simplexes of C^k with vertices mapped onto the whole set of vertices of S_r^k, excluding the distinguished vertex a. We observe, as we did in the proof of Sperner's lemma, that one vertex of S_r^k must be the image of two vertices of any such simplex. Thus two faces of each such simplex are mapped onto the face of S_r^k that excludes the distinguished vertex, and it is easy to confirm that these faces have opposite orientations. Thus each such simplex contributes nothing to the number K.

Finally, consider simplexes of C^k with vertices having images that omit at least one vertex of S_r^k other than a. These contribute nothing to K. This completes the proof of the principal theorem.

15.9 SOME APPLICATIONS

In this section we conclude the main presentation with a set of three actual applications of the results obtained so far, a further example of a slight extension (proved in the literature but requiring a little more examination, although no more tools, than have been provided here), an announcement of an extension due to Schauder and Leray [1] analogous to the extension by Schauder [19] of the Brouwer fixed-point theorem, and a note concerning another method of arriving at these same applications by integrating solid angles.

The applications will include a re-examination of the Brouwer fixed-point theorem from the point of view of the first introduction, an outline of a proof of the fundamental theorem of algebra (that every polynomial of positive degree has at least one zero in the complex plane), and an outline of a proof of a theorem that every continuous field of vectors tangent to a 2-sphere has at least one vector of length 0. From this last theorem it can be deduced that every continuous mapping from a 2-sphere to itself has either a point that is fixed or a point of which the image is its antipode.

The re-examination of the Brouwer fixed-point theorem will deviate only slightly from the proof given in Sec. 15.3. We retain the sequence of subdivisions with mesh approaching zero, and we retain the labeling rule stated in connection with our earlier proof. We take the simplex S_r^k to be the original simplex. The inductive step used in connection with the proof of Sperner's lemma in Sec. 15.2 is replaced by a step based on a slightly stronger hypothesis, the validity of which will immediately be established.

We start, then, with a continuous mapping from a k-simplex to itself. We produce a sequence of subdivisions of the simplex with mesh approaching zero. We shall consider each of these subdivisions to define a simplicial complex, and we shall build a k-chain on this complex. In particular, to each k-simplex of the subdivision we assign an orientation consistent with that of the original k-simplex. (This is easy to define because the simplexes of the subdivision lie in the k-plane containing the original k-simplex.) The k-chain C^k is one that adjoins to each oriented simplex of the subdivision the coefficient 1. The boundary $\delta(C^k)$ consists of $(k-1)$-simplexes that make up the boundary of the original k-simplex, all internal $(k-1)$-simplexes having a coefficient 0 because of the cancellation rule introduced in connection with the arithmetic operations to be performed on chains.

The original k-simplex is taken as the range simplex S_r^k for the simplicial mapping to be introduced. The image of each vertex of any subdivision is defined exactly as it was in the proof of the Brouwer fixed-point theorem in Sec. 15.2. This mapping is admissible for the fundamental index theorem proved in Sec. 15.8. We now introduce a somewhat stronger hypothesis than the hypothesis concerning parity that was used to prove Sperner's lemma in Sec. 15.2. The hypothesis that will be used here is that the index of any such mapping of vertices over the subdivision complex is 1; that is, $L = 1$. We proceed to establish the validity of this hypothesis by induction over very nearly the same route as was used in the proof of Sperner's lemma.

First, we must establish the validity of the hypothesis for the case $k = 1$. This is trivial and follows exactly the same argument used for the 1-dimensional case in the proof of Sperner's lemma. We must next show that if the hypothesis is valid for mappings of $(k - 1)$-dimensional simplexes, then it is valid for mappings of k-dimensional simplexes. To do this, we again note that the labeling of the distinguished face, which does not contain the distinguished vertex of S_r^k, obeys the rules established in connection with the proof of the Brouwer fixed-point theorem; no vertex of the distinguished face is mapped to the distinguished vertex. Thus the labeling describes a mapping from the vertices of the subdivision of the distinguished face to the vertices of the distinguished face. We call this face S_r^{k-1} and the mapping f^{k-1}. We note that this mapping is a continuous mapping from the distinguished face to itself, and the validity of the hypothesis would require that the index or this mapping over the distinguished face be 1. This, however, is the index K of the mapping from the k-simplex to itself, and by the fundamental index theorem for any such mapping we have $L = K$. Thus we have shown that if the hypothesis is valid for dimension $k - 1$, it is valid for dimension k, and this completes the inductive argument.

This completes the proof of the Brouwer fixed-point theorem by the means described in the first introduction. The theorem, by giving an explicit value 1 for the index of the mapping, is somewhat stronger than the Brouwer fixed-point theorem, just as the index theorem is somewhat stronger than Sperner's lemma. The stronger version has been well known for some time, but so far as the author knows it was first printed explicitly in [5]. It is implied in [2] and is more or less explicitly stated in the portion of [4] dealing with Sperner's lemma.

As an exercise, the reader might prove by the induction used in the proof of Sperner's lemma that the number of simplexes of the subdivision carrying all labels in an order consistent with that of the original simplex exceeds by 1 the number with the opposite orientation.

We now turn to a proof of the fundamental theorem of algebra: Every polynomial of degree at least 1 over the field of complex numbers has a zero in the complex plane. For this, we consider a polynomial P mapping the complex z-plane to the complex w-plane. That is, we write $w = P(z)$, where z and w are complex numbers and

$$P(z) = a_n z^n + a_{n-1} z^{n-1} + \cdots + a_1 z + a_0.$$

We assume that the a_j are (real or) complex numbers, and that $a_n \neq 0$.

Again, we must furnish a chain, a simplicial mapping of this chain to a range simplex, a reduced mapping of the reduced boundary, and a proof that this index differs from zero, as it will turn out. The chain is built on a complex that is a subdivision of a large circular disc in the z-plane with center at the point corresponding to $z = 0$. It is clear that such a subdivision can be made arbitrarily fine, and we fix the disc, in a manner to be divulged shortly, and choose a sequence of simplicial subdivisions with mesh approaching zero. We shall use a limiting argument like that used in the first proof of the Brouwer fixed-point theorem, and note here that such an argument will certainly work if we can prove that the index of a mapping of a chain built on this subdivion, acting as the complex carrying the chain, has index different from zero. If this is so, the number of simplexes with vertices mapped to three distinct points of the range simplex is certainly not zero, and the limiting argument will work.

We take the range simplex S_r^2 to lie in the w-plane, and we may as well be specific about its vertices. We shall take them to be the points $w = 1$, $w = i$ (where i, of course, is the square root of -1), and $w = -1 - i$. Any other simplex containing the point $w = 0$ in its interior would do just as well. We now provide a rule for mapping any point of the z-plane to one of the vertices of the simplex S_r^2. Roughly, this rule is that the vertex lying closest to the image $P(z)$ of the point z is taken as the image of z. More specifically, and only approximately agreeing with this rough description, we take the following rule: If a point z is to be mapped to a vertex of S_r^2, it is mapped to the vertex $w = 1$ if and only if the real part of $P(z)$ is not negative and the imaginary part does not exceed the real part (signs being considered); it is mapped to the vertex $w = i$ if and only if the imaginary part is not negative and the imaginary part exceeds the real part; otherwise, it is mapped to the vertex $-1 - i$. It is immediately verified that a point z that is a limit point of points mapped to all three vertices must be mapped to the point $w = 0$, and hence must be a root of P, thus establishing the theorem. Without further comment, we assume that the reader will extend the limiting argument of Sec. 15.3 to establish the existence of this root, once we establish that for every sufficiently fine subdivision there exists a simplex of the subdivision with vertices mapped onto all three vertices of S_r^2. We now turn to that task.

A more explicit description of the disc to be used is the following. Since the coefficient a_n of the polynomial differs from zero, it is true that the term $a_n z^n$ will dominate the polynomial on the boundary if the disc is sufficiently large. That is, if we desire that the first term be at least a million times as large as the sum of all the other terms in the polynomial in absolute value on the boundary of the disc, all we need to do is to specify that the disc be sufficiently large, the size, of course, depending on the particular polynomial considered. Now it is well known that the image of $a_n z^n$ in the w-plane describes a circle taken n times as z describes a circle taken one time, both circles having center at the 0-points of their respective planes. Therefore it follows immediately that the index K will be n if the disc is taken to be sufficiently large, and if a subdivision of the z-circle and its interior with consistent orientations of the 2-simplexes (to each of which is adjoined the coefficient 1) is taken as a chain over which indices are to be computed.

The rest of the proof is immediate, and it is left to the reader. The continuation of the theorem in algebra depends on dividing out any selected root and reapplying the theorem to find that there are n roots to an nth degree polynomial, although some of the roots may have to be counted multiply. We shall mention such multiple counting again in connection with the next example.

Our final direct example pertains to vector fields tangent to an ordinary 2-sphere. We shall show that such a vector field has a net total of two singularities—points where the vector has length 0. There may be many more than two singularities, but these are opposing in sense, and in this way the theorem is more complicated than the fundamental theorem of algebra, for a polynomial of degree n never has more than n roots. On the other hand, there may be only one singularity that is to be counted twice as being two coincident simple singularities; this is completely analogous to the fundamental theorem of algebra, since a polynomial of degree n may have fewer than n roots, some of which are to be counted multiply.

We carry through our usual recipe. We first assume that there exists a nonsingular vector field tangent to the sphere; this means that there is a vector, with length bounded away from 0, tangent to the sphere at each point, and that the vectors are continuous functions of their argument points. We then choose two antipodal points, which we call poles, north and south, and we carry out our argument with these two distinguished points. We delete from the sphere the interior of a small circle (of latitude) around each of these points, and we construct a complex on the remainder of the sphere by subdivision. As usual, we require a sequence of subdivisions with mesh converging to zero. The orientation of the simplexes of the subdivision should be taken to be consistent, so that the boundary of the usual chain, obtained by adjoining the coefficient 1 to each simplex of the subdivision, is a chain with simplexes on the two boundary circles.

We assign an orientation to each circle of latitude consistent, say, with the orientation of the portion of boundary of the complex surrounding the north pole. We assign two vectors to each point of the sphere, one tangent to the circle of latitude through the point and pointing in the direction of the positive orientation assigned to this circle, and the other pointing along the meridian circle toward the north pole. Each of these vectors is to be considered to have unit length, and of course they are perpendicular. We must now remember that there is a single vector at the south pole, and we assume that it has a length different from 0. The vectors assigned to the points on a very small circle around the south pole all have essentially the same direction. Near the point where the south pole's vector would project on this circle, these vectors must point very nearly north, and at the diametrically opposite point (of the circle, not the sphere) and neighboring it they must point very nearly south. Thus it is not hard to see that the vector, which from our description is fixed, makes a full rotation with respect to our moving coordinate axes as these axes make a rotation around the circle.

To be more precise, then, we map these two axes to the coordinate axes of a plane. We construct a simplex in this plane, just as we did for the proof of the fundamental theorem of algebra, and we construct the image of any desired vector of the sphere by assigning it plane components relative to the axes of the plane equal to its components on the sphere relative to the two reference vectors constructed through its point of origin. We assign a vertex of the simplex to any point in accordance with the rule set up for the proof of the fundamental theorem of algebra, using the image of the vector in just the way we used the w-image there.

Exactly the argument used to show an index of ± 1 around the south pole boundary shows an analogous index around the north pole. It is important, however, to note that the indices differ in sign as a result of the assigned orientation of the simplexes involved. Thus the total index K is ± 2; we shall not take the trouble here to determine the sign precisely, but the reader should try. (We note that the portion of the sphere retained for the complex may conveniently be mapped on a flat piece of paper using a Mercator projection, mentioned earlier, but that the vectors assigned to the points on the top and bottom boundary circles seem to make a complete rotation as one moves from the west edge of the map to the east edge. The determination of the value of the index K depends on a meticulous assessment of the directions of these rotations.) The index L for each subdivision must then take the value ± 2 (again the sign can be determined precisely by the study suggested), and the usual limiting argument shows that there must be at least one point where the vector has length 0. This completes the proof, but we shall discuss this proposition a little further.

Although the fundamental theorem of algebra gives an index of n for a polynomial of degree n, and it is true that there can be no more than n roots of a polynomial, it is obvious that there can be as many points as are desired with vectors of length 0. Indeed, for example, the whole equator or any reasonable set of points on the equator may be assigned vectors of length 0, vectors of length 0 may be assigned to the north and south poles, and, say, northerly directed vectors with positive lengths may be assigned to all other points of the southern hemisphere, and southerly directed vectors with positive lengths may be assigned to all other points of the northern hemisphere. Whenever the computation of an index makes sense, the index is 2. The meaning of this statement in geometric terms is important, but it will not be expounded here. The reader is referred to Lefschetz [15] or Alexandroff and Hopf [1].

We also note in passing that a phenomenon analogous to the phenomenon of multiple roots of a polynomial can occur in connection with the vectors of length 0 on the sphere. Although we expect a net total of 2 in accordance with the index, these two may coalesce. An example is easily constructed by considering the sphere to be split by a plane of the 0th and 180th meridian (in degrees). If the sphere is examined from a distant point on the ray from the center of the sphere through the point on the equator with longitude 90 degrees east, the sphere will resemble a circular disc (to a person with poor depth perception). The south pole is at the bottom, by custom. We consider that the vector through the south pole has 0 length, and we consider circles on the circular disc through the south pole and tangent there to the boundary of the disc. The vectors on each circle are all oriented in the same direction relative to that circle, becoming small near the south pole and large at the highest point of each circle. See Fig. 15.3, which illustrates the eastern hemisphere. The western hemisphere is covered by vectors that are parallel with their mirror images in the eastern hemisphere, with respect to the cutting plane through the 0th meridian. This example is unimportant, perhaps, but it may save some readers from a few sleepless nights trying to prove that there are at least two points with vectors of length 0.

We finally note that if we consider a mapping from a sphere to itself, we can draw a vector from each point to its image. If this is projected on its tangent plane at its point of origin, then a field of vectors tangent to the sphere results. The only zero vectors are those that go from a point to its antipodal point or that were zero vectors in the first place. Thus we have the following famous corollary:

Corollary. A continuous mapping from a 2-sphere to itself either has a fixed point or takes at least one point to its antipode.

We now turn to our last example, which will be given without detail. This concerns the method of Kaczmarz [12] for the solution of a set of linear equations (or inequalities), treated in [3, pp. 454–455]. We shall present and describe the case in which there is no solution, and we shall discuss procedures of the Kaczmarz method.

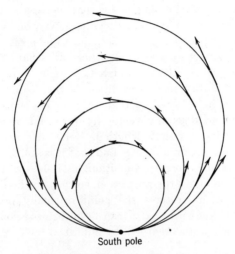

South pole

Fig. 15.3 Vectors on the sphere.

Let the equations have the form

$$\sum a_{ij}x_j = b_i,$$

and suppose, say, that the index i has a larger range than j. (Inequalities will serve just as well as equalities.) Suppose further that the system has no solution. The method of Kaczmarz is a relaxation method for approximating any solution that exists. To carry out the calculation, an approximate solution is guessed, and the equations are considered in cyclic order with respect to the index i; that is, the ith equation is considered after the $(i - 1)$-st, etc. An approximate solution is possible before considering any equation. It is modified, in the light of the equation considered, by treating the graph of that equation as a hyperplane in the space of the variables x, treated as coordinates, and by replacing the approximate solution (a point in the x-space) by its perpendicular projection on this hyperplane. It is easy to prove that this method converges to the closest solution when a solution exists. The speed of convergence is not guaranteed.

The question of determining what happens when there is no solution is not answered in the literature. It can, however, be treated without difficulty by index methods. To do this, consider the related set of equations.

$$\sum a_{ij}x_j = 0.$$

This set certainly has a solution, the point with coordinates $x_j = 0$ for every j. The studies of the Kaczmarz method cited show that the index of a Kaczmarz mapping from a large sphere with center at the origin to itself is 1. The Kaczmarz mapping is one that replaces each point on the sphere by the point obtained by projecting successively on the hyperplanes, the image from one projection being the point to be projected next.

Now consider a viewer at a great distance. It is hard for him to tell whether the hyperplanes all go through the origin or not; indeed, by moving the planes a very small percentage of the distance from viewer to origin, they can be restored to their original positions. At no time during this movement does any vector from a point on a huge sphere around the origin to its image have length 0. Hence we conclude, and could prove, that the index does not change and that it is still 1. This does not prove that the set of equations has a solution, but it does prove convergence to an invariant cycle under the projection, so that, in the limit, the complete set of projections restores the point originally projected. This situation, once it is known to exist, can be recognized computationally. The proof of this whole theorem will be published by T. S. Motzkin and the author.

The outline given, relating to the Kaczmarz projection method, used another theorem of great importance, the invariance of index under homotopy. For full discussion of this, see [10], [15] and also the more recent [11].

We should also note in this section that all the proofs could have been carried out by integration, avoiding reference to combinatorial theorems. Indeed, these integration proofs were the earliest ones. The integration computes the net solid angle swept out by a vector of which the index is being computed. The total angle swept out by the vector $w = P(z)$, where P is a polynomial of degree n, as z sweeps out a complete circle of large radius, is n, whereas the net total angle swept out as z transverses a very small circle is 0. Similarly, the total solid angle swept by the vector from the boundary points of a simplex to their images under a mapping of the simplex to itself is one sphere of the proper dimensionality. The vectors discussed in connection with the vector field on the sphere sweep out two complete circles as the origin of the vectors traverses the whole boundary, with proper regard being given to orientation.

These results are established in a remarkable paper by Hadamard [8] and in a more recent paper by Heinz [9], as well as in many other places.

A discussion of some of this work, particularly the early work involving integrals, may be found in the introductory survey of [15].

Finally, we note that index theorems have been established in Banach spaces of infinitely many dimensions in a manner analogous to Schauder's extension of the Brouwer fixed-point theorem. For such an extension, see Schauder and Leray [16] or the portion of [4] devoted to index theorems in Banach space. Such theorems have been applied to give the best known results concerning solution of some complicated problems in partial differential equations.

15.10 TOWARD GREATER RIGOR

The argument here has been largely geometric. It was not always thus, but the struggle between clarity and rigor is a bitter one in an expository presentation. Thus, in attempting to make a paper almost understandable, we resort to implicit arguments that do not bear close examination. This has happened in the final draft of this chapter; earlier drafts repaired the deficient arguments, but left an unattractive mess.

We note here briefly that mathematics must allow complete symbolization in order to be trusted. This means, for example, that our glib statements that two simplexes of a subdivision lie on different sides of a shared face and that, therefore, the face vanishes from any boundary chain of the chains we used must be examined and reduced to compact unmistakable notation. We have not carried out such a reduction here.

We note the existence of [6], which contains all the material necessary for such a reduction. We quote a basic definition from [6, p. 54]: "An *n-simplex s* is a set of $n + 1$ objects called vertices, usually denoted by $\{A\}$, together with the set of all real-valued functions α defined on $\{A\}$ satisfying

$$\sum_A \alpha(A) = 1, \qquad \alpha(A) \geq 0."$$

Any of these functions is a point of s, and the values taken by the function for the various vertices make up the barycentric coordinates of the point.

These points must be used as vertices in subdivisions, and the relations between the points of a subsimplex and those of the original simplex must be clarified. Again, this is easy with the algebraic tools that have now been employed in topology. The whole process, however, tends to hide the real geometric ideas present, at least during an initial reading.

More accurately, of course, the correctly abstracted and formalized proofs provide the only presentation that is completely understandable and that therefore does not hide the real meaning. We end with the expositor's perpetual dilemma: He must stop at such a level of precision that all his readers except those with excessive talent understand, or believe that they understand. The author takes the experienced and perplexed expositor's way out. He notes that the material in [6] is completely adequate for a formalization of the whole theory, and he recommends this to the ambitious reader as a wholesome mathematical exercise.

MULTIPLE-CHOICE REVIEW PROBLEMS

1. If the barycentric coordinates of the vertices A_0, A_1, A_2, A_3 of a 3-simplex are (1, 0, 0, 0), (0, 1, 0, 0), (0,0, 1, 0), (0, 0, 0, 1), respectively, then the barycentric coordinates of the midpoint of the edge joining A_1 and A_3 are

    (a) (0, 1/2, 0, 1/2),
    (b) (0, −1, 0, 1),
    (c) (0, 1, 0, 1),
    (d) (0, 1/3, 1/3, 1/3).

2. The oriented 2-simplex with vertices A_0, A_1, A_2, in that order, is the same as the one with vertices in the order

    (a) A_0, A_2, A_1,
    (b) A_1, A_2, A_0,
    (c) A_2, A_1, A_0,
    (d) A_1, A_0, A_2.

3. Under the labeling rule, "Assign the label j to a point $x = (x_0, x_1, x_2, x_3)$ under a transformation T with $Tx = y$ if j is the least number such that $y_j \le x_j \ne 0$," the label assigned to the centroid of the face with vertices (0, 1, 0, 0), (0, 0, 1, 0), (0, 0, 0, 1) under the transformation $Tx = (1/4, 1/4, 1/4, 1/4,)$, for every x, is

    (a) 0,   (b) 1,   (c) 2,   (d) 3.

4. The fixed point in the mapping of Problem 3 is

    (a) (0, 1/3, 1/3, 1/3),
    (b) (1, 0, 0, 0),
    (c) (1/2, 1/2, 0, 0),
    (d) (1/4, 1/4, 1/4, 1/4,)

5. For the accompanying figure,

    (a) the labels are assigned in a way that violates the conditions of Sperner's lemma,
    (b) the subdivision is not regular,
    (c) the topological index of the mapping is 0,
    (d) Sperner's lemma states only that there will be a side labeled 1, 0.

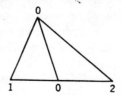

REFERENCES

1. Alexandroff, P., and H. Hopf, *Topologie I*, J. Springer Verlag, Berlin, 1935 (Grundlehren der mathematischen Wissenschaften, Vol. 45).

2. Bagemihl, F., "An Extension of Sperner's Lemma, with Applications to Closed Set Coverings and Fixed Points," *Fund. Math.* **40** (1953), 3–12 and an Addendum **41** (1955), 351.

3. Beckenbach, E. F. (editor), *Modern Mathematics for the Engineer*, McGraw-Hill Book Company, New York, 1956.

4. Bers, L., "Topology," Lecture notes from New York University, 1956–57.

5. Brown, A. B., and Stewart S. Cairns, "Strengthening of Sperner's Lemma Applied to Homology Theory," *Proc. Nat. Acad. Sci. U.S.A.* **47** (1961), 113–114.

6. Eilenberg, S., and N. Steenrod, *Foundations of Algebraic Topology*, Princeton University Press, Princeton, New Jersey, 1952.

7. Fan, Ky, "A Generalization of Tucker's Combinatorial Lemma with Topological Applications," *Ann. of Math.* (2), **56** (1952), 431–437.

8. Hadamard, J., "Note sur applications de l'indice de Kronecker," an appendix to J. Tannery, *Introduction à la théorie des fonctions d'une variable*, 2nd edition, Vol. 2, pp. 437–477, Paris, 1910.

9. Heinz, Erhard, "An Elementary Analytic Theory of the Degree of Mapping in n-dimensional Space," *J. Math. Mech.* **8** (1959), 231–247.

10. Hilton, P. J., *An Introduction to Homotopy Theory*, Cambridge Tracts in Mathematics and Mathematical Physics, No. 43, Cambridge, England, 1961 (revised edition).

11. Hu, S. T., *Elements of General Topology*, Holden-Day, Inc., San Francisco, 1964.

12. Kaczmarz, S., "Angenäherte Auflösungen von Systemen linearer Gleichungen," *Bull. International Acad. Polon. Sci., Cl. Sci. Math. Nat. Series A*, 355–357, 1937.

13. Kakutani, S., "A Generalization of Brouwer's Fixed Point Theorem," *Duke Math. J.* **8** (1941), 457–459.

14. Kuhn, H. W., "Some Combinatorial Lemmas in Topology," *IBM J. Res. and Dev.* **4** (1960), 518–524.

15. Lefschetz, Solomon, *Introduction to Topology*, Princeton University Press, Princeton, New Jersey, 1949.

16. Leray, Jean, and Jules Schauder, "Topologie et équations fonctionelles," *Ann. École Normale Supérieure* (3), **51** (1934), 45–78.

17. MacLane, S., *Homology*, Academic Press, New York, 1963.

18. von Neumann, J., and O. Morgenstern, *Theory of Games and Economic Behavior*, 2d edition, Princeton University Press, Princeton, New Jersey, 1947.

19. Schauder, J., "Über Linearen Elliptische Differentialgleichungen Zweiter Ordnung," *Math. Zeit.* **38** (1934), 257–282.

20. Tucker, A. W., "Some Topological Properties of Disk and Sphere," *Proc. First Canadian Math. Congress*, Montreal, Canada, 1945, pp. 285–309.

21. Whitehead, J. H. C., "Combinatorial Homotopy. I," *Bull. Amer. Math. Soc.* **55** (1949), 213–245.

22. Whitney, Hassler, *Geometric Integration Theory*, Princeton University Press, Princeton, New Jersey, 1957.

23. Williams, J. D., *The Compleat Strategyst*, McGraw-Hill Book Company, New York, 1954.

CHAPTER 16

Crystallography

KENNETH N. TRUEBLOOD

16.1 INTRODUCTION

Modern crystallography is concerned primarily with the atomic and molecular structure of solid substances possessing some degree of order, and with the manifestation of this structure and order in the macroscopic properties of the various substances. The borderline between crystallography and other branches of solid-state physics is not sharp, but solid-state phenomena that are primarily electronic are not usually considered within the bounds of crystallography and will not be discussed here.

Crystallography involves many aspects of mathematics that, at least in their beginnings, have been associated with combinatorial analysis, although some of them are now usually regarded as distinct from it— group theory, topology, probability, parquetry (the study of mosaics), and others. It is chiefly in this broad sense that crystallography fits into the present volume, although we shall also discuss briefly some other aspects of structure analysis that involve broad combinatorial considerations.

The development of modern crystallography began with the discovery of x-ray diffraction in 1912, for until that time crystallographers had to be content with studies of the external form and macroscopic properties of crystals. A first-hand historical account of this discovery has been given by M. von Laue [21]. Within a year after the discovery of the diffraction method, the atomic arrangements in many simple inorganic structures had been deduced. The power of diffraction methods for studying the structure of matter has grown at an accelerating pace since then, with the development of electron and neutron diffraction techniques, the application of new theoretical approaches, and especially the advent of high-speed computers.

Our present discussion is divided into four chief parts: (1) a consideration of the nature of the crystalline state; (2) a short exposition of the
principles of diffraction methods for studying crystal structures; (3) some
detailed approaches to solving the phase problem; and (4) a brief examination of a few representative atomic and molecular arrangements, chiefly
in solids. Those interested in a more detailed treatment, especially of the
basic principles and methods, should consult the general references
[8, 28, 30, 33, 46, 52] in the bibliography at the end of this chapter.

NATURE OF THE CRYSTALLINE STATE

The fundamental characteristic of crystals is a high degree of spatial
order; that is, the objects from which a crystal is composed are arranged
in a regular way with a precise spatial periodicity. Real crystals often
exhibit a great variety of imperfections—for example, short-range and
long-range disorder and dislocations, and other sorts of lattice defects.
Before these problems can be effectively treated, however, it is necessary to
consider the ideal crystal in which the order is perfect and three-dimensional. For many purposes this approximation to real crystals is a good
one, and we shall be primarily concerned with it here.

16.2 LATTICES

In the ideal crystal a basic pattern of structure, which is sometimes
called the *basis*, and which may be a single atom or a complex assemblage
of many atoms, is repeated indefinitely by translation along three non-
coplanar directions. The parallelepiped having as its sides the three unit
translations is termed the *unit cell*, and the aggregate of unit cells is the
crystal lattice. The lattice is thus an infinite set of points recurring
regularly in space, and it may be characterized fully by specifying the
three noncoplanar translation vectors $\mathbf{a}_i$ and then writing the general
translational symmetry operation

$$(16.1) \qquad \mathbf{T}_h = \sum_{i=1}^{3} h_i \mathbf{a}_i,$$

in which the h_i are integers. The lattice points may be considered as the
points generated when all combinations of h_i are used. There are just
fourteen different three-dimensional lattices (Table 16.1); they were
derived first by Frankenheim and Bravais in the first half of the nineteenth

TABLE 16.1
The Fourteen Bravais Lattices[1,2]

System	Type	Symbol	Lattice points in conventional unit		Conventional vector triple	
					Given by the symmetry[4]	To be specified
Cubic	Simple	P	000		$a = b = c$ $\alpha = \beta = \gamma = 90°$	a
	Face-centered	F	000 0½½ ½0½ ½½0			
	Body-centered	I	000 ½½½			
Tetragonal	Simple	P	000		$a = b$ $\alpha = \beta = \gamma = 90°$	a, c
	Body-centered	I	000 ½½½			
Hexagonal		P	000		$a = b$ $\alpha = \beta = 90°$ $\gamma = 120°$	a, c
Trigonal[3]		R	000		$a = b = c$ $\alpha = \beta = \gamma$	a α
Orthorhombic	Simple	P	000		$\alpha = \beta = \gamma = 90°$	a, b, c
	Side-centered	C	000 ½½0			
	Face-centered	F	000 0½½ ½0½ ½½0			
	Body-centered	I	000 ½½½			
Monoclinic	Simple	P	000		$\alpha = \gamma = 90°$	a, b, c β
	Side-centered	C	000 ½½0			
Triclinic		P	000			a, b, c α, β, γ

[1] It is conventional to label the edges of the unit cell a, b, c and the angles between them α, β, γ, with α the angle between b and c, etc. The positions of points within the unit cell are given in fractions of the translation along the corresponding edge.

[2] The symmetry of the lattice at each lattice point is that of the highest-symmetry point group for that system. For example, for the tetragonal lattice, it consists of a fourfold axis normal to a mirror plane, with two independent mirror planes parallel to and passing through the fourfold axis and at 45° to each other. The symbol is $4/mmm$. *Structures* based on the lattices may have lower symmetry.

[3] The lattice of the trigonal system is often called a *rhombohedral lattice* because of the shape of the unit cell. This system is sometimes considered a subdivision of the hexagonal system.

[4] Note that, in addition to the symmetry relations between the unit cell parameters, there may be *accidental* equalities as well, for example, an orthorhombic crystal with three axes equal to within a high degree of precision, or a monoclinic crystal with $\beta = 90.0°$. The space group of the crystal can have no higher symmetry, however, than that of the structure arranged on the lattice. For example, an essential characteristic of cubic symmetry is the presence of a threefold axis along the body diagonal of the unit cell; in the orthorhombic crystal with three equal axes, this threefold axis would not be present. If it *were* present, the crystal would be cubic.

century and are referred to as the *Bravais lattices*. These lattices are best classified in terms of the symmetry at each lattice point. With certain lattices the *primitive* unit cell described previously does not effectively display the lattice symmetry, and consequently the conventional unit cell is chosen in a *nonprimitive* way; that is, it is associated with more than one lattice point. This is illustrated in Fig. 16.1 for the unit cell of a face-centered lattice.

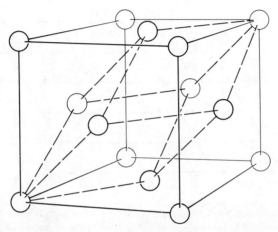

Fig. 16.1   A face-centered unit cell (solid lines) and the corresponding primitive unit cell (dashed lines). The larger cell has four times the volume of the smaller and thus has four lattice points associated with it: one-eighth of each of the eight corner points and one-half of each of the six face-centered points.

The lattice may then be regarded as an ordered three-dimensional array of points that can be used to pick out equivalent points in a crystal structure. The term *lattice* is sometimes misleadingly used to refer to a structure itself. The *structure* is distinct from the *lattice* in that it is an ordered array of objects (of any sort), rather than merely of points. These objects may themselves be totally asymmetric or may have any degree of symmetry; on the other hand, it is not meaningful to speak of the symmetry of a point, which by definition is dimensionless. In general, a symmetry operation may be regarded as any transformation of the space occupied by an object—for example, a crystal—that preserves linear dimensions and leaves invariant the properties of the object. For the moment we shall not be concerned with the purely operational problem of recognizing whether or not the properties really are invariant under the transformation; clearly in practice this will depend on the nature and sensitivity of whatever measuring devices are employed.

An ideal crystal with basis consisting of more than a single atom can be viewed as an indefinitely extended combination of several interpenetrating systems, each of which consists of identical atoms arranged on one of the fourteen Bravais lattices. With most crystals, the arrangement of matter within the unit cell itself has some symmetry (other than the purely translational symmetry that is associated with a nonprimitive unit cell)—an axis of symmetry, a mirror plane, or one of the other symmetry elements

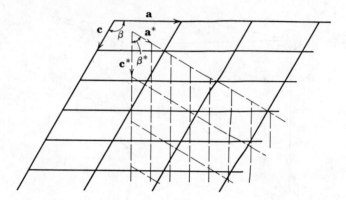

Fig. 16.2   Portions of a monoclinic net (solid lines) normal to **b**, and of a corresponding reciprocal lattice net (broken lines). Since the origin of each lattice is arbitrary, no relation between their origins is implied or intended, although in practice one normally chooses the lattice origins so that the lattice symmetry elements coincide with those of the crystal structure and thus with each other.

considered later. Under these conditions, the *asymmetric unit* of the structure is even smaller than the unit cell. The asymmetric unit is the smallest portion of the structure that must be specified, for if it is operated on by all of the translational and other symmetry elements present, then the entire structure is generated.

In addition to the lattice of the crystal structure in *real* or *crystal* space, which has already been described, there is a second lattice that can be defined in terms of the first and that plays an important role in many aspects of solid-state physics. This is the *reciprocal lattice,* and it is defined as follows, with unstarred quantities pertaining to the real lattice and starred ones to the reciprocal lattice:

$$V = \mathbf{a}_1 \cdot \mathbf{a}_2 \times \mathbf{a}_3,$$

(16.2) $$\mathbf{a}_1^* = \frac{\mathbf{a}_2 \times \mathbf{a}_3}{V}, \qquad \mathbf{a}_2^* = \frac{\mathbf{a}_3 \times \mathbf{a}_1}{V}, \qquad \mathbf{a}_3^* = \frac{\mathbf{a}_1 \times \mathbf{a}_2}{V}.$$

The vectors of the real lattice and the reciprocal lattice form an *adjoint set* in the sense in which this term is used in tensor calculus, for they satisfy the condition

$$\mathbf{a}_i \cdot \mathbf{a}_j^* = \delta_{ij},$$

where δ_{ij} is the Kronecker delta (1 if $i = j$; 0 otherwise). The symmetry of the two lattices is identical.

If the axes of either lattice are orthogonal, the relations between the lattices are simple: corresponding lattice vectors are parallel, and the lengths of those in one lattice are inversely proportional to the lengths of those in the other. With nonorthogonal axes, the relations are not hard to visualize geometrically; a two-dimensional example is given in Fig. 16.2, in which the axes are labeled $\mathbf{a}$ and $\mathbf{c}$ so that it can serve for the nets normal to $\mathbf{b}$ (or $\mathbf{b}^*$) in a monoclinic crystal. (The axial vectors are more commonly referred to by crystallographers as $\mathbf{a}$, $\mathbf{b}$, and $\mathbf{c}$ than as $\mathbf{a}_1$, $\mathbf{a}_2$, and $\mathbf{a}_3$, although both notations are used.) We shall see later that if a structure is arranged on a given lattice, then its diffraction pattern is necessarily arranged on the lattice reciprocal to the first.

16.3 POINT SYMMETRY

Any finite object, such as an isolated molecule or a real crystal, can possess only *point-symmetry*; that is, any symmetry operation must leave at least one point of the object invariant. On the other hand, an infinite array such as a lattice or an ideal unbounded crystal structure may have translational symmetry as well, since translation along any integral number of lattice vectors moves it into self-coincidence and thus is a symmetry operation. A translation operation leaves no point invariant since it moves all points equal distances in parallel directions; it is an example of a *space-symmetry* operation. Because most actual macroscopic crystals consist of many millions of unit cells, it is a fair approximation to regard the arrangement of atoms throughout most of a real crystal as possessing translational symmetry. Edge effects normally are small.

The geometrical requirements of lattices restrict the number of possible *rotational-symmetry* elements. If we define an axis of order n as an axis about which a rotation of $2\pi/n$ brings the object operated on into self-coincidence, it may readily be shown [46] that only axes of order 1, 2, 3, 4, and 6 are compatible with structures built on three-dimensional (or two-dimensional) lattices. Isolated molecules can, and sometimes do, possess symmetry axes of other orders. When, however, crystals are formed from a molecule with, for example, a fivefold axis, the fivefold axis cannot be a symmetry axis of the crystal as a whole. The molecule may still, within experimental error, retain its fivefold symmetry in the crystal, but it can

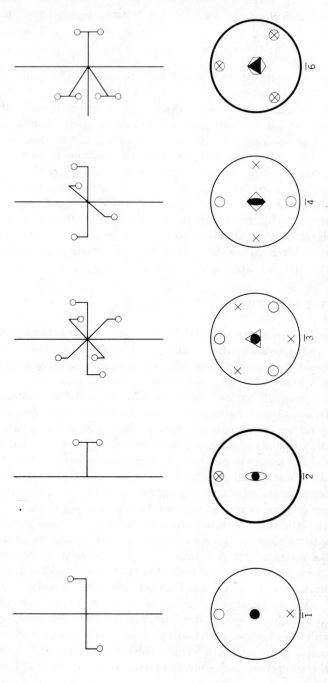

Fig. 16.3 The crystallographic rotatory-inversion axes (above) and their stereographic projections (below). In the projections, ○ and × are used to represent points on opposite sides of the median plane. The symbol in the center of each circle is that conventionally used to represent the axis in projection; for $\bar{2}$ and $\bar{6}$ a darkening of the outer circle is also used to represent the horizontal mirror plane.

never occur at a position such that this symmetry is a necessary consequence of a fivefold symmetry in the crystalline environment. In other words, if fivefold symmetry appears to persist in the molecule in the crystal, it is because the perturbations by the crystal field, which necessarily has some other symmetry, are too small to produce perceptible changes in whatever molecular properties are being studied.

An important point-symmetry operation is the *inversion* operation. If the center of inversion is at the origin, which is conventional, then every point (x, y, z) is converted to $(-x, -y, -z)$. This operation is symbolized by $\bar{1}$. There are additional point-symmetry operations that may be regarded as combinations of pure rotations with inversions; they are termed *rotatory-inversion* axes and are symbolized by $\bar{n}$, where the integer n may have only the values 2, 3, 4, and 6 (in addition to 1) in the point groups pertinent to crystals. The operation $\bar{2}$ is equivalent to a mirror plane normal to the rotatory-inversion axis; the symbol m is used far more often than $\bar{2}$ for this symmetry element. All these operations are illustrated in Fig. 16.3. It is important to note that the rotatory-inversion operation is neither a pure rotation nor a pure inversion alone; the point groups $\bar{n}$ with n even do not contain as symmetry elements either the n-fold rotation operation or the inversion operation, although in fact if n is odd then both these elements are present.

The rotatory-inversion operations differ from the pure rotations in an important respect: they convert an object into its mirror image. Thus a pure rotation can convert a left hand only into a left hand; on the other hand, a rotatory-inversion axis will, on successive operations, convert a left hand into a right hand, then that back into a left hand, and so on. If we define the order of a symmetry operation as the number of times the operation must be performed before we return to the original configuration, then clearly the order of each rotatory-inversion operation must be *even*, since otherwise we would end with an object of the opposite hand from that with which we started. For these axes, the order is $2n$ if n is odd, although it is just n if n is even. Objects that cannot be superimposed on their mirror images cannot possess any element of rotatory-inversion symmetry.

The 10 point-symmetry operations that have been described can be combined together in just 32 ways in three dimensions to form the 32 three-dimensional crystallographic point groups. These have the properties characteristic of an abstract group: (1) they all possess the identity operation, here symbolized by 1; (2) the inverse A^{-1} of each operation A of the group is a member of the group, being defined by $A^{-1}A = 1$; (3) the *product* AB of any two operations A and B of the group is itself an operation of the group; and (4) the combination law is associative $A(BC) = (AB)C$, although not necessarily commutative since in general

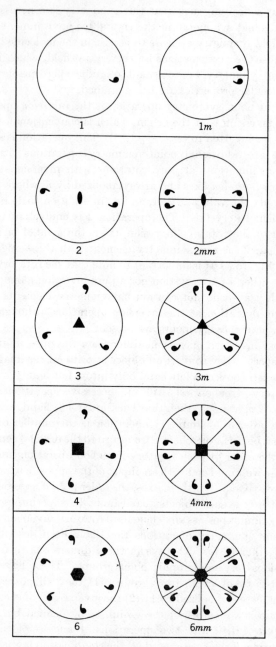

Fig. 16.4 The two-dimensional crystallographic point groups. Equivalent points are shown.

we do not have $AB = BA$. We shall not discuss the three-dimensional crystallographic point groups [21] here, but we shall rather restrict our attention to the less complicated two-dimensional ones, which are sufficiently illustrative for our purposes. There are just 10 of these, which may be formed by various combinations of 1-, 2-, 3-, 4-, and 6-fold rotations about a point with reflection in a mirror line. These are illustrated in Fig. 16.4, in which the graphical symbol at the center of each circle represents the order of the rotation axis in an obvious way, the straight lines designate mirror lines, and the commas represent points equivalent by symmetry. The numerical-literal symbols for these two-dimensional point groups are composed as follows: first comes the order of the rotation axis; next is indicated the presence of a unique mirror line, if any—other mirrors related by the rotation operation are also implied; and third is given a symbol for a second mirror line, if any, which is not equivalent by symmetry to the first—although in fact it is generated by the interaction of the rotation and the first mirror, as can be seen by studying the diagram for 2 *mm*, 4 *mm*, or 6 *mm*.

16.4 SPACE SYMMETRY

Combinations of the point-symmetry operations with translations give rise to various sorts of space-symmetry operations in addition to the

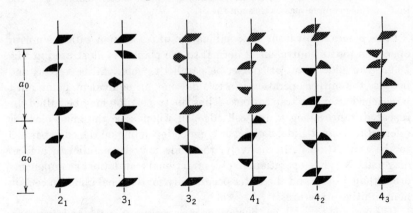

Fig. 16.5 Some crystallographic screw axes. Two identity periods of each are shown, as indicated on the left.

pure translations. *Screw axes*, some of which are illustrated in Fig. 16.5, result from the combination of translation and pure rotation. They are symbolized by n_m, which signifies a rotation of $2\pi/n$ followed by a translation parallel to the axis by the fraction m/n of the identity period along that axis. If $p = n - m$, then the axes n_m and n_p are *enantiomorphous*; that is, they are mirror images of one another. This can be seen by comparing 3_1 and 3_2, or 4_1 and 4_3. It is important, however, to note that it is only the axes that are enantiomorphous; structures built on them will not be enantiomorphous unless the objects in the structure are themselves enantiomorphous. Thus a *left* hand operated on by a 3_1 will give an arrangement that is the mirror image of that produced by the operation of a 3_2 on a *right* hand, but not, of course, the mirror image of that produced by the operation of a 3_2 on another left hand.

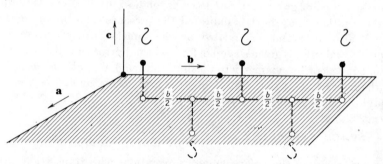

Fig. 16.6 The operation of a **b**-glide plane normal to **c**. The operation is a combination of a translation of **b**/2 and reflection in a plane that is normal to **c** and that (in this example) passes through the origin.

Glide planes result from the combination of translation with the mirror operation (or its equivalent, $\bar{2}$ normal to the plane), as illustrated in Fig. 16.6. The glide must, of course, be parallel to some lattice vector, and because the mirror operation is a twofold one, an equivalent point must be reached after two translations. Thus the translation may be half of the repeat distance along a unit-cell edge, in which case the glide plane is referred to as an **a**-glide, **b**-glide, or **c**-glide, depending on the edge parallel to the translation. Alternatively, the glide may be parallel to a face-diagonal. No glide operation involves fractional translational components other than $\frac{1}{2}$ or $\frac{1}{4}$, and the latter occurs only for diagonal glides in certain nonprimitive structures.

It is possible to combine the various pure rotations, rotatory inversions, screw axes, and glide planes in just 230 ways compatible with the geometrical requirements of three-dimensional lattices. There are thus 230 three-dimensional space groups, ranging from that with no symmetry other than the identity operation (symbolized by $P1$, the P implying primitive) to that with the highest symmetry, $Fm3m$, a face-centered cubic space group of order 192. We cannot here discuss the derivation, or demonstrate the uniqueness, of these space groups. Zachariasen [57] has given a detailed derivation of the crystallographic point groups and space groups, with the help of group theory and matrix algebra; Belov [2] has presented a somewhat briefer and less elegant but more geometric derivation. Much useful information about the crystallographic point groups and space groups is tabulated in [21].

It is interesting to note that these 230 unique three-dimensional combinations of the possible crystallographic symmetry elements were derived independently in the last two decades of the nineteenth century by Fedorov in Russia, Schönflies in Germany, and Barlow in England. It was not until several decades later that anything was known of the actual structure of even the simplest crystalline solid. Since the introduction of diffraction methods for studying the structure of crystals, the space groups of many thousands of crystals have been determined. Representatives of most, although not all, of the space groups have been found; a survey [12] of the literature through 1948 showed that certain space groups occur with high frequency. This is not surprising if one examines the structures built up by packing highly symmetrical objects such as spherical or nearly spherical atoms or ions—it is reasonable that these should form crystals of high symmetry. Among the more than 1200 organic crystals investigated through 1948, however, most of them involving molecules of irregular shape, more than 40% belonged to just three space groups. None of these is of high symmetry; two are monoclinic and one orthorhombic. Two-fold screw axes are present in all three space groups, and are the only symmetry element, other than the identity, present in two of them; in the third, a glide plane and center of symmetry are present as well. The crystallo-chemical implications of these facts have been considered by several people, for example, Nowacki [37] and Kitaigorodskii [26].

16.5 ANTISYMMETRY

Recently, the ideas of space-group theory have been extended to include an additional twofold symmetry operation in what may be regarded as a fourth dimension [23, 34]. This operation is one normally associated with the term *antisymmetry*; it might be described, for example, as a change of color (black to white), or of the sign of a function (plus to minus), or of the direction of spin (up to down). Thus a center of antisymmetry would change a black (or positive, or . . . .) right hand into a white (or negative or . . . .) left hand. The combination of this operation with the usual three-dimensional groups leads to 1651 possible "black and white" groups [3, 58]. They have been used in the discussion of the structures of ferromagnetic and ferroelectric materials and in the description of the relations between structure and twinning phenomena in crystals, as well as in certain aspects of Fourier analysis of electron density in crystals.

CRYSTAL STRUCTURE ANALYSIS BY X-RAY DIFFRACTION

16.6 DIFFRACTION

The wavelengths of easily accessible x-rays, that is, those produced by displacement of K-electrons from metals of intermediate atomic number, are of the order of one angstrom (10^0 Å). Because the repeat distances in most crystals are not greatly different from this value, usually being in the range $10^{1/2}$ to $10^{3/2}$ Å, the ordered array of atoms constituting an ideal crystal acts as a three-dimensional diffraction grating for x-rays. The scattering is primarily by the electrons in the atom, since that by the much heavier nucleus is negligible in comparison. Neutrons of comparable wavelength, for example thermal neutrons from a nuclear reactor, are also diffracted by crystals, being scattered by interaction with the nuclei of the atoms. Similarly, electrons of appropriate wavelength can be diffracted by crystals. We shall restrict our discussion, however, almost exclusively to x-ray diffraction.

Since a crystal may be regarded as a three-dimensional grating, the principles of crystal diffraction are closely analogous to those of optical diffraction phenomena, and may usefully be approached by means of these analogies. We shall begin by considering the scattering by a single scattering object or small group of such objects in an irregular array, and shall then consider the effect of arranging these objects in an ordered array on a lattice. Figure 16.7 shows the diffraction patterns from two slits of different width, the same light having been used to illuminate each slit. The effect of slit width is clear: the narrower the scattering region, the broader the resulting pattern. A similar effect is observed when one compares the scattering of x-rays by atoms of low and of high atomic number. Even if we consider the scattering per electron, to eliminate intensity differences arising from the difference in the number of electrons present, we find that the atom of high atomic number gives a significantly broader diffraction pattern. This occurs because the high nuclear charge concentrates the electrons, the atom of high atomic number thus having effectively a smaller volume per electron and thus being effectively a narrower slit. As thermal vibrations in a crystal increase and the atoms effectively occupy larger volumes, the scattering intensity at high angles falls off markedly. In accord with these generalizations, it is understandable that the scattering of neutrons by a stationary atomic nucleus would show almost no decrease with increasing scattering angle, for the wavelengths normally used, because nuclei are so much smaller than the usual neutron wavelengths; thermal vibration does, however, cause decreases with increasing angle by markedly increasing the effective volume of the nucleus.

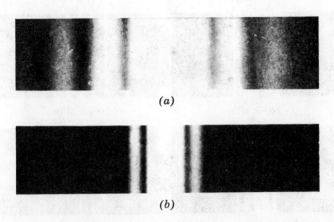

(a)

(b)

Fig. 16.7 Single-slit diffraction pattern. (a) Narrower slit. (b) Broader slit. (Reprinted by permission from Francis A. Jenkins and Harvey E. White, *Fundamentals of Physical Optics*, 3rd edition, p. 312, McGraw-Hill Book Co., New York, 1957.)

The illustrations in Fig. 16.7 might well represent illumination of the same slit by two different wavelengths; there is a direct relation between the width of the pattern and the wavelength used to form it.   This effect is sometimes important in x-ray work, when the use of too short a wavelength with a crystal having a large unit cell may give so concentrated a pattern that it is hard to resolve neighboring diffraction maxima.

Figure 16.8 shows how the diffraction pattern of a single slit is modified by interference effects from identical parallel slits when increasing numbers of slits are placed in regularly spaced linear array, that is, on a one-dimensional lattice. The important point to note is that *the diffraction pattern from a grating of slits represents a sampling of the single-slit pattern in narrow regions*; with even as few as 20 slits in the "grating," the small subsidiary maxima vanish almost completely. The spacing of the sampling regions (or "lines" for an infinite grating) is inversely related to the spacing of the slits. It can be seen that if the intensity of the single-slit pattern is zero or weak in a region where it is sampled by the multiple-slit pattern, then the latter is also zero or relatively weak there. This is illustrated in

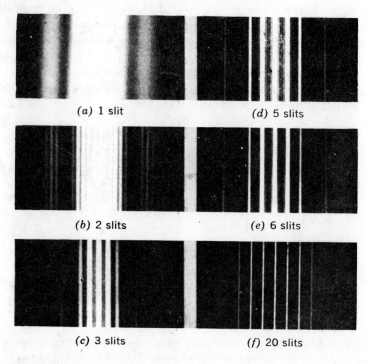

Fig. 16.8 Diffraction patterns from gratings consisting of increasing numbers of equidistant parallel slits from (a) 1 to (f) 20 slits. (Reprinted by permission from Francis A. Jenkins and Harvey E. White, *Fundamentals of Physical Optics*, 3rd edition, p. 329, McGraw-Hill Book Co., New York, 1957.)

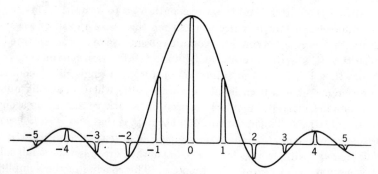

Fig. 16.9 Amplitudes of the diffraction maxima from a grating of coarse slits (individual peaks). The envelope is the (appropriately scaled) amplitude of diffraction from a single slit of the same size (see text).

Fig. 16.9; for convenience, the vertical scale of the single-slit curve has been multiplied by the number of elements in the grating so that it will be the exact envelope of the individual peaks from the grating. An intensity

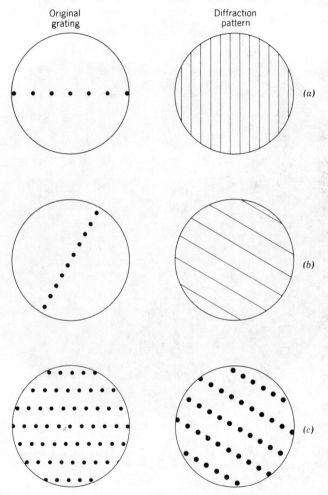

Original
grating

Diffraction
pattern

(a)

(b)

(c)

Fig. 16.10 Diffraction patterns from two different one-dimensional gratings *(a)* and *(b)*, and of the two-dimensional grating composed from them *(c)*.

curve would be similar in general appearance except that it would, of course, be everywhere positive, since the intensity of a wave disturbance is proportional to the square of its amplitude.

Figure 16.10 shows in a simple way how a two-dimensional array of simple scattering objects (here merely fine holes, that is, holes of diameter much less than the wavelength of the radiation used) produces a two-dimensional diffraction pattern. Each of the one-dimensional gratings of holes in Figs. 16.10*a* and 16.10*b* produces a diffraction pattern in which

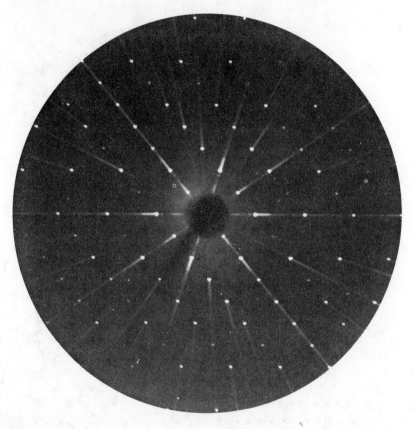

Fig. 16.11   X-ray diffraction photograph of ZrO_2, taken by the precession method, which gives an undistorted representation of the reciprocal lattice. The streaks are due to "white" (continuous) radiation.

interference effects have cut out all the intensity except at selected spots in directions *parallel* to the original lattice, but there has been no interference in directions normal to this line just because the original grating was one-dimensional. If, however, as in Fig. 16.10*c*, we have both sorts of one-dimensional gratings present at once, giving a regular two-dimensional grating, then each of the linear gratings may be considered to act independently, and the only regions in which there is no destructive interference are the intersections of the two sets of lines produced in Figs. 16.10a and 16.10b. The lattice of the diffraction pattern is necessarily then the reciprocal of the lattice of the original scattering objects, as was anticipated earlier.

The fact that the intensity is the same at all reciprocal lattice points in the diffraction pattern in Fig. 16.10c should not be thought to hold generally; it happens here because the scattering objects in the original "crystal lattice" are all particularly simple (just a single isotropic hole in each unit cell) and are much smaller than the wavelength of the radiation used in this hypothetical experiment. Consequently, the intensity of the diffraction pattern falls off only very slowly with the angle of scattering. Neither of these conditions is normally fulfilled in a real crystal: most unit cells contain a complex assembly of atoms, and each atom is comparable in linear dimensions to the wavelength of the radiation used. Figure 16.11 shows a typical x-ray diffraction photograph, taken by the "precession method," which records the diffraction pattern, and thus the reciprocal lattice, without distortion. Considerable variation in intensity at the different lattice points is evident. The analogy with Figs. 16.8, 16.9, and 16.10 holds; the x-ray photograph is merely a scaled-up sampling of the diffraction pattern of the contents of a single unit cell, with the sampling points arranged on a lattice reciprocal to the lattice of the crystal itself. These relations are illustrated [49] in Fig. 16.12, which shows (a) a schematic representation of a section of the reciprocal lattice of a crystal of phthalocyanine, with the size of each spot proportional to the intensity at that lattice point; (b) a mask representing the phthalocyanine molecule, viewed along the direction corresponding to the reciprocal lattice section in (a); (c) an optical diffraction pattern of the mask in (b), on the same scale as (a) and corresponding to the diffraction pattern of a single molecule of phthalocyanine in the orientation **b**; and (d) a superposition of some of the features of (c) on the weighted reciprocal lattice (a), showing the correspondence between them. It is noteworthy that even with the rather crude representation of the molecule by a mask containing appropriately placed identical holes (which, unlike atoms, all have the same scattering power and all have sharp edges), the agreement between the optical pattern and the crystal x-ray pattern is remarkably good. The diffraction pattern of the single molecule is a complicated one because of interference effects of the radiation scattered from the different "atoms" in the molecule; it reflects, however, the symmetry of that molecule.

The fundamental relationship between any object and its diffraction pattern, a relationship that underlies [54, 50, 31] all structure analysis by diffraction methods, is this: the diffraction pattern is the Fourier transform of the scattering object *and vice versa*. If the scattering object

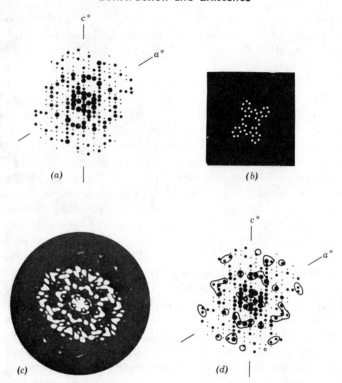

Fig. 16.12   (*a*) Weighted (*h0l*) reciprocal lattice section of a crystal of phthalo-cyanine. The size of each spot is proportional to the x-ray intensity at that lattice point. (*b*) A mask representing a projection of the molecule of phthalocyanine viewed down **b**. (*c*) Optical diffraction pattern of (*b*). (*d*) Superposition of the main features of (*c*) and (*a*). (Reprinted by permission from H. Lipson and W. Cochran, *Determination of Crystal Structures*, G. Bell and Sons, London, 1953.)

consists of molecules arranged on a crystal lattice, then its Fourier transform is the weighted reciprocal lattice, that is, the reciprocal lattice with appropriately varying intensity at the different lattice points.

16.7 FOURIER TRANSFORMS OF ATOMS AND GROUPS OF ATOMS

The one-dimensional Fourier transform $f(x)$ of a continuous function $g(u)$, such as the electron density distribution in an atom, is itself a continuous function; it is given by

$$(16.3) \qquad f(x) = (2\pi)^{-\frac{1}{2}} \int_{-\infty}^{\infty} g(u)e^{iux}\, du.$$

Conversely, $g(u)$ may be expressed as

$$(16.4) \qquad g(u) = (2\pi)^{-\frac{1}{2}} \int_{-\infty}^{\infty} f(x)e^{-iux}\, dx.$$

These expressions are easily generalized to the three-dimensional case. If the function g is spherically symmetrical, it may then be expressed as a radial distribution function; that is, integration over the angular variables may be separated. Since u or a radial variable r is a variable in real or crystal space, x must be a variable in reciprocal space, and the common variable used in diffraction theory is

$$S = \frac{2\sin\theta}{\lambda},$$

or quantities differing from this by a numerical factor. Here θ is half the angle between the incident and diffracted beams, and λ is the wavelength of the radiation used. If g is the electron distribution in the atom, then the corresponding f is called the *atomic scattering factor* or *atomic form factor* for x-rays [46, 50]. In practice, most atomic-scattering factors are derived from spherically symmetric electron distributions, which may have been derived by averaging over angular variables, and thus the scattering factors are themselves spherically symmetric. In a few instances, anisotropic atomic scattering factors have been derived—for example, for a carbon atom in a trigonal bonding environment. Proper use of this scattering factor requires knowledge of the orientation of the electronic bonding orbitals of this atom relative to the incident beam in the diffraction experiment; since this knowledge is usually the object of the study, one must use a spherically symmetric distribution at least in the first approximation. In fact, few structure analyses have yet been precise enough to require the use of scattering factors in which the intrinsic anisotropy of the electron distribution in the atom at rest has been taken into account. If there is significant anisotropy of this sort, then one can usually allow for it by an assumed anisotropy of thermal motion of the atom, at least to a fairly good approximation, although attempts have been made to distinguish these two effects.

The atomic scattering factor is usually expressed as the ratio of the scattering by the atom under given conditions to the scattering by a single free electron under the same conditions. At $\theta = 0°$, that is, for scattering in the forward direction, all electrons in the atom scatter in phase, and the atomic scattering factor equals the number of electrons present. Since atomic sizes are comparable to the wavelengths of x-rays normally used in diffraction experiments, atomic scattering factors fall off markedly with increasing angle of scatter because of interference between the radiation scattered from one part of the atom and that scattered from another. As mentioned earlier, the relative fall-off is smaller for atoms of high atomic number because their electrons are more effectively concentrated by the high nuclear charge.

In calculating the diffraction pattern to be expected from a group of atoms, the usual assumption is that the atoms are independent of one another. This would imply that the perturbation of the electron distribution of a given atom by the other atoms is negligible, that is, that the electronic rearrangements involved in bond formation are here unimportant. This assumption is a reasonable one in most studies because bonding electrons are necessarily "outer" electrons in an atom, and thus, as a consequence of the reciprocal relation between the dimensions of an object and those of its scattering pattern, the effects of these bonding electrons are predominantly at low scattering angles. Experimental errors are usually greatest in the comparatively few diffraction maxima observable in this region of the pattern, and accordingly minor variations in the distribution of bonding electrons are hard to detect. Bonding electrons have reportedly been detected in a few very precise studies of simple structures, but these cases are the exception.

If we can represent the group of atoms merely as a superposition of isolated atoms, each at a position in the unit cell given by a position vector $\mathbf{r}_j$ from the origin, then the scattering from the group of atoms can be represented [52, 54] merely as the sum of the scattering factors for the individual atoms, each multiplied by an appropriate phase factor:

$$(16.5) \qquad G = \sum_j g_j e^{2\pi i \mathbf{S} \cdot \mathbf{r}_j} = \sum_j f_j e^{-B(\mathbf{S})} e^{2\pi i \mathbf{S} \cdot \mathbf{r}_j},$$

in which G now represents the scattering from the entire group of atoms in a given direction, f_j represents the scattering factor of atom j in that same direction when the atom is at rest, $\mathbf{S}$ is a vector in reciprocal space that lies in the same plane as the incident and diffracted beams and is normal to the bisector of the angle between them, and $e^{-B(\mathbf{S})}$ is a *temperature factor* that takes account of thermal vibration of the atom. The magnitude of $\mathbf{S}$ is $(2/\lambda) \sin \theta$.

The function G as given by (16.5) is the Fourier transform of the electron distribution representing the group of atoms; in general, it is a continuous function. If this group of atoms is then repeated over and over periodically in three dimensions to form a crystal, the effect is to multiply G by a fringe function so that the transform is observed only at the points of the lattice reciprocal to the crystal lattice. Under these conditions, the vector $\mathbf{S}$ in (16.5) takes on only values corresponding to reciprocal lattice vectors, and the scalar product $\mathbf{S} \cdot \mathbf{r}_j$, satisfies

$$\mathbf{S} \cdot \mathbf{r}_j = hx_j + ky_j + lz_j.$$

The integers h, k, and l are the indices of the reciprocal lattice point at which diffraction is being observed, and x_j, y_j, and z_j are the positional parameters of atom j in the crystal lattice, expressed as fractions of the corresponding unit cell edges.

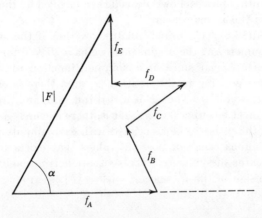

Fig. 16.13 Vector addition of atomic scattering factors with different phases to give an over-all structure factor with magnitude $|F|$ and phase angle α. The origin has arbitrarily been chosen at atom A and thus the phase of A is zero. This atom also has a scattering factor twice that of the other atoms. For this diffraction maximum the phase angles of the other atoms relative to this origin are B, 115°; C, 35°; D, 180°; E, 90°.

When the sum in (16.5) is then taken over the N atoms in the crystal unit cell, it is called the *structure factor* and is symbolized by F. Figure 16.13 is a simple vectorial illustration of the relation between F and the individual f's. The structure factor F is in general a complex quantity and may be written as

$$(16.6) \qquad F = A + iB = |F| e^{i\alpha},$$

with

$$(16.7) \qquad A = \sum_{j=1}^{N} f_j e^{-B(S)} \cos 2\pi(hx_j + ky_j + lz_j)$$

and

$$(16.8) \qquad B = \sum_{j=1}^{N} f_j e^{-B(S)} \sin 2\pi(hx_j + ky_j + lz_j).$$

It is clear that F has not only an amplitude, namely $(A^2 + B^2)^{1/2}$, but a phase as well, which may be defined in terms of the angle,

$$(16.9) \qquad\qquad \alpha = \cos^{-1} \frac{A}{(A^2 + B^2)^{1/2}} .$$

For any given crystal, values of $|F|$ may be observed at hundreds or even thousands of reciprocal lattice points (h, k, l); the phase of at least one of these F's may be given any arbitrary value by appropriate choice of the origin. The other phases, however, will then be fixed by the structure and the nature of the atoms present.

Equations (16.6), (16.7), and (16.8) show that if the structure has a center of symmetry at the origin, then we have $B = 0$ for all diffraction maxima (or *reflections*), since $\sin \theta$ is an odd function and for every term in (16.8) corresponding to the position (x, y, z) there is one of opposite sign for $(-x, -y, -z)$. Thus F is a real function, and, from (16.9), the phase angle must be either 0 or π; that is, there is merely a twofold ambiguity about the phase to be associated with each amplitude $|F|$, instead of the continuous range of possible values that exists for a structure without a center of symmetry. This situation is normally discussed in terms of the sign of the F, since $F = A = |F| \cos \alpha$, and $\cos \alpha$ must be either $+1$ or -1.

16.8 THE PHASE PROBLEM

The reciprocity inherent in (16.3) and (16.4) applies to structure factors as well. Just as the structure factors that make up the diffraction pattern represent the Fourier transform of the electron density of a group of atoms, so the Fourier transform of the group of structure factors comprising the diffraction pattern is the electron density of the scattering matter that gives rise to the diffraction pattern. If the latter transformation can be effected, then the electron distribution and thus the molecular structure of the scattering material can be determined and the primary goal of this sort of research achieved. To perform this Fourier transformation either physically or analytically, however, one must know the amplitude and phase of the diffraction pattern throughout the range in which it can be observed. The amplitude is readily measured since, except for certain easily calculated geometrical factors, it is proportional to the square root of the observed intensity, and the intensity can readily be measured with moderate precision. The phase, however, normally is not directly observable. Thus the central problem of structure analysis by diffraction methods is to determine the phases, or at least a sufficiently good approximation to them so that Fourier inversion will give a recognizable image of at least a portion of the scattering matter. The image may then be improved and the corresponding structure determined and refined by the methods discussed in Secs. 16.9 to 16.13.

The recombination of scattered radiation to give an image of the object that produced the scattering is, of course, just what a microscope does, and indeed Abbe's theory of image formation approaches the problem in terms of Fourier representations [6, 30]. This is illustrated in Fig. 16.14. The image formation can be considered to take place in two stages. In the first, the lens combines all the radiation scattered in a *given* direction to give *one* of the spectra S; note that each part of the scattering object contributes to each of these spectra. In the second stage, the radiation refracted by the lens travels on beyond the focal plane of the spectra until all the radiation scattered in all directions from a given *part* of the original object is focused at a point. This then produces an image of the scattering matter. There is no phase problem here because the relative phases of the individual scattered beams are preserved as they travel through the microscope, and thus when the beams recombine to form the image they still have the proper relative phases.

The diffraction maxima, or so-called *reflections* from a crystal, are precisely analogous to the spectra S except that, since x-rays cannot be focused, the lens is absent and the radiation scattered in a given direction will be resolved from that in a nearby direction only if the detector is placed sufficiently far away that the linear divergence of the beams has been greater than the dimensions of the scattering crystal. Since there is no focusing, the spread of each diffraction maximum is approximately the same as that of the cross-section of the crystal in the beam of radiation—assuming that the beam was collimated and not significantly divergent.

The second stage in x-ray diffraction must be done analytically because it cannot be done physically. If one knows the amplitude and phase of the radiation scattered in every direction, then it is possible to recombine these to give an image of the scattering matter, an operation analogous to that performed by the optical system of the microscope. In three dimensions, this involves a triple Fourier summation:

$$(16.10) \qquad \rho(x, y, z) = \frac{1}{V} \sum_{h=-\infty}^{\infty} \sum_{k=-\infty}^{\infty} \sum_{l=-\infty}^{\infty} F(h, k, l) e^{-2\pi i(hx+ky+lz)},$$

in which ρ is the electron density in the unit cell at a point (x, y, z), V is the volume of the unit cell, and F is the structure factor, which has both an amplitude and a phase, as given by (16.6). The phases, however, cannot be observed directly since they normally are lost in the process of measuring the intensity, which depends only on the amplitude. Hence the phases must somehow be deduced or approximated.

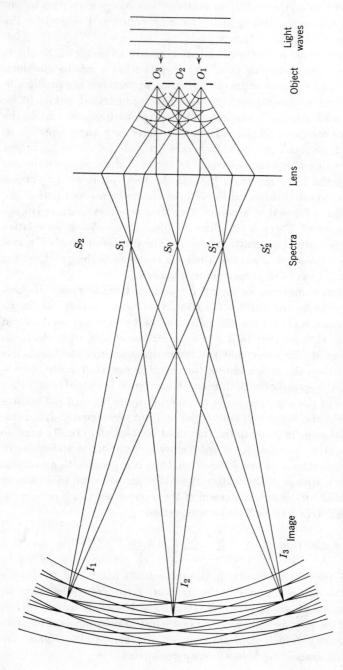

Fig. 16.14 Abbe's theory of image formation in the microscope. The object is a grating (only three slits are shown here). The objective lens is such that, for the radiation and grating spacing used, only the zeroth and first orders are intercepted and thus focused by the lens. [Reproduced by permission from W. L. Bragg, *Z. Kristallographie*, **70** (1929), p. 478.]

Figure 16.15 shows, in an oversimplified two-term one-dimensional example, the difficulty if one does not know the relative phases of the scattered beams that are to be recombined. In (b) the two waves are in phase, and in (a) they are out of phase by 40°; the results of summing them, shown in the lower halves of the drawings, are very different, with some of the peaks, which might represent the electron density in an atom, occurring at quite different places in the unit cell. Similarly, in the three-

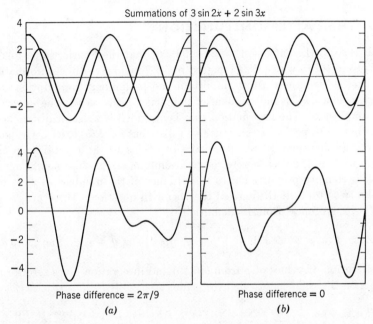

Summations of $3 \sin 2x + 2 \sin 3x$

Phase difference = $2\pi/9$ Phase difference = 0

(a)   (b)

Fig. 16.15 Summations of two terms of different frequency (and amplitude). (a) Phase difference 40°; (b) phase difference 0°. The character of the sum obtained varies markedly with variation in relative phase.

dimensional Fourier summations (16.10) by which the electron density in a crystal is reconstructed from the observed structure-factor amplitudes and the best available phases, a considerable amount of false detail will be present in the image if the phases are insufficiently accurate. Thus the first stage in the determination of a crystal structure is always an attempt to determine at least a portion of the phases. This can be done indirectly by deducing a "trial structure," that is, an approximation to the structure or to some part of it, from which approximate phases can be calculated either by (16.6), (16.7), and (16.8), or by some direct method that permits estimation of the phases with essentially no structural information inferred or assumed. In the next part of this chapter, we shall discuss briefly the most common and successful methods that are used for phase determination.

APPROACHES TO SOLVING THE PHASE PROBLEM

16.9 THE PATTERSON FUNCTION

In 1935, A. L. Patterson showed [38, 30] that the Fourier series analogous to (6.10) in which F is replaced by $|F|^2$ could give direct information about the structure. Most important, no preliminary assumptions are needed, since $|F|^2$ is independent of the phase, being merely the square of the amplitude. The information that is provided is a vector distribution function; the peaks in the Patterson distribution $P(u, v, w)$ correspond to vectors between all possible pairs of atoms in the crystal structure itself, with all the vectors referred to a common origin. The height of any peak is roughly proportional to the product of the numbers of electrons in the two atoms at the ends of the vector in question. These properties of $P(u, v, w)$ follow from its definition,

$$(16.11) \quad P(u, v, w) = V \int_0^1 \int_0^1 \int_0^1 \rho(x, y, z)\rho(x + u, y + v, z + w) \, dx \, dy \, dz.$$

Substitution of values of ρ from (16.10) and integration gives zero for all terms on the right-hand side except for

$$P(u, v, w) = \frac{1}{V} \sum_{h=-\infty}^{\infty} \sum_{k=-\infty}^{\infty} \sum_{l=-\infty}^{\infty} F(h, k, l)F(-h, -k, -l)e^{2\pi i(hu + kv + lw)}.$$

This leads quite immediately to the following (16.12) since, by (16.6), $F(h, k, l)$ and $F(-h, -k, -l)$ are complex conjugates:

$$(16.12) \quad P(u, v, w) = \frac{1}{V} \sum_{h=-\infty}^{\infty} \sum_{k=-\infty}^{\infty} \sum_{l=-\infty}^{\infty} |F|^2 (h, k, l)e^{2\pi i(hu + kv + lw)}.$$

The distribution represented by (16.12) can be considered to be the sum of N^2 individual terms, if N is the number of atoms in the unit cell:

$$(16.13) \quad \begin{aligned} P(u, v, w) &= \sum_{m=1}^{N} \sum_{n=1}^{N} P_{nm}(u, v, w), \\ P_{nm}(u, v, w) &= \frac{1}{V} \sum_{h=-\infty}^{\infty} \sum_{k=-\infty}^{\infty} \sum_{l=-\infty}^{\infty} f_n(h, k, l)f_m(h, k, l) \\ &\qquad \times e^{-B_n(S)}e^{-B_m(S)}e^{2\pi i(hu + kv + lw)}. \end{aligned}$$

Similarly, the expression (16.10) for the electron density can be considered to be the sum of N separate terms corresponding to the individual atoms, because of the relations given by (16.7) and (16.8).

Examination of (16.11) reveals why it corresponds to a vector distribution function. There is a contribution to the integral at every point (u, v, w) for which each of the density terms in the integrand is nonzero, and thus for each pair of atoms separated by (u, v, w), one atom at (x, y, z) and the other at $(x + u, y + v, z + w)$. For $u = v = w = 0$, we have a contribution from the interaction of each atom with itself. A simple one-dimensional example [7] is given in Fig. 16.16, which shows two unit cells of a simple three-atom structure, and two unit cells (in vector space) of the corresponding Patterson distribution. The Patterson function has peaks corresponding to the vectors between all pairs of atoms in the original structure, with weights approximately proportional to the products of the peak weights of the atomic distribution itself. For a unit cell with N atoms, the Patterson function has N^2 peaks, N of which coincide at the origin, corresponding to the self-interactions of the atoms, and some of the others of which may also coincide exactly because of symmetry.

In principle, it is possible to recover the fundamental set (the atomic distribution) from the vector set (the Patterson distribution), as first shown by Wrinch [53]. Although both real and hypothetical examples are known in which two distinct atomic distributions give rise to the same vector set [30], such homometric structures are exceedingly improbable with unit cells of even moderate complexity. The difficulties in solving structures from their Patterson distributions arise not from homometric ambiguities, but rather from the overlapping of peaks in Patterson space as a consequence of their significant width and high concentration. The individual peaks are somewhat wider than those of atoms in the corresponding atom space; this width depends on the radiation used and the limiting angle of observation of the diffraction pattern. In the usual x-ray experiments with copper or molybdenum K-radiation, individual *atoms* can be well resolved in three dimensions. Since, however, there are $N(N - 1)$ Patterson peaks (ignoring those at the origin, which give no useful information) in the same volume that would contain N atoms in the atom space, accidental overlaps of interatomic vectors become almost inevitable when there are even as few as 10 atoms in a unit cell; and with a moderately complex structure, with perhaps 40 atoms in the unit cell, most of the Patterson peaks will at least in part be composite. Techniques for sharpening and otherwise modifying the peak shape are normally used to improve resolution, but they still leave one far short of the ideal of point interactions, and many multiple overlaps invariably remain.

An essential early stage in the analysis of a Patterson distribution is, then, the identification of peaks, which may correspond to single inter-actions or to anticipated multiple interactions of vectors that either are expected, on the basis of the presumed molecular structure, to be at least

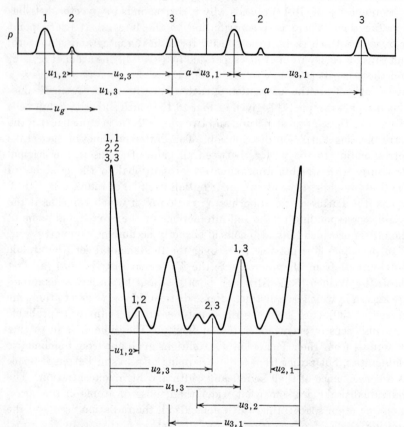

Fig. 16.16 Relation between the electron density (upper) and the corresponding interatomic vector map (lower). Two unit cells of a three-atom one-dimensional structure are shown. There are nine (3^2) peaks in the Patterson function; three coincide at the origin and the other six, distributed throughout the unit cell, fall into two groups of three related by centers of symmetry at the origin and halfway along the cell edge. (Reproduced by permission from M. J. Buerger, *Vector Space*, p. 14, John Wiley and Sons, 1959.)

approximately parallel and of comparable length, or are known from the unit-cell symmetry to have a certain multiplicity. For example, if we have a molecule containing a benzene ring, we know that there will be an approximately regular hexagonal array of carbon atoms in this ring, and these will give rise to a simple planar vector pattern with known relative weights of the different peaks. This pattern will normally be made more complex by overlaps of vectors from any additional atoms present, and possibly by overlap with its own image produced by reflection if mirror planes are present in the unit cell of the crystal; but if the unit cell contains a relatively small number of atoms, it may be possible to recognize in the Patterson distribution the pattern arising from the benzene ring. This sort of recognition of parts of structures with known stereochemistry has been an important factor in the solution of many structures of moderate complexity. Recently high-speed computer programs have been written to facilitate the searching of the Patterson function for regions of optimum fit [48].

It is impossible to discuss here all the methods that have been devised for analysis of Patterson distributions [7, 30]. It was early recognized by Harker that, in the presence of certain symmetry elements, specified sections of three-dimensional Patterson functions correspond in part to projections of the atomic distribution. Unfortunately, accidental overlaps considerably complicate the practical use of such *Harker sections* in most structures, although they are helpful in conjunction with other information. So-called *superposition* and related methods, which are really based on the ideas of Wrinch [53] although they were developed independently by various other investigators a decade or so later, involve the transcription of the Patterson distribution with its origin at one or more individual peaks and positions related to these by the symmetry of the atomic distribution. With the help of a high-speed computer for the tedious task of transcription and recognition of regions of significant overlap for which one searches, such methods can be quite powerful, but they depend again on the initial location of individual Patterson peaks, and in a complex structure overlaps make this difficult.

Despite these difficulties, however, analysis of Patterson interatomic vector distributions, supplemented if possible by chemical information about the shape and size of all or part of the molecule and any available physical information about the orientation of the molecules in the crystal, affords one of the most powerful tools available for deducing trial structures. The technique is especially helpful in the study of complex molecules when these contain some "heavy atoms," that is, one or a few atoms with atomic numbers considerably greater than those of most of the atoms present, as discussed in the following section.

16.10 HEAVY-ATOM METHODS

If the structure that is being investigated contains one or a few heavy atoms, the vectors between these heavy atoms (in the same and different asymmetric units in the unit cell) will give rise to much higher peaks in the Patterson function than will the light-atom-to-light-atom vectors. Consequently, these vectors can usually be identified, and the coordinates

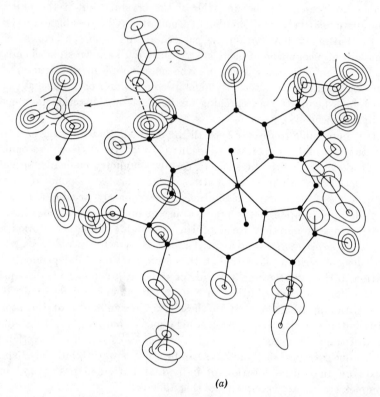

(a)

Fig. 16.17   Appearance of "side-chain" atoms in a vitamin B_{12} derivative at different stages in the structure analysis. Contour intervals are 1 $e/Å^3$. (*a*) Early approximation to the phases based on the positions of only 26 of the 73 atoms in the structure; the positions of these atoms are indicated by the black dots. (*b*) Nearly correct phases, based on the positions of 68 of the 73 atoms present. [Reproduced by permission from D. C. Hodgkin, J. Pickworth, J. H. Robertson, K. N. Trueblood, R. J. Prosen, and J. G. White, "The Crystal Structure of the Hexacarboxylic Acid Derived from B_{12} and the Molecular Structure of the Vitamin," *Nature* **176** (1955), p. 325.]

of the heavy atoms can be derived from them. In favorable circumstances it may be possible to locate not only the vectors between different heavy atoms in the unit cell but also the vectors between a given heavy atom and at least some of the lighter atoms in the molecule; this technique is especially promising when there is only one heavy atom in the asymmetric unit, and when the number of lighter atoms is not excessively large, for example, no more than thirty or forty. The image of the molecule, or of some part of it, so obtained can then be used in a trial structure for calculation of approximate phases, which, in conjunction with the observed amplitudes,

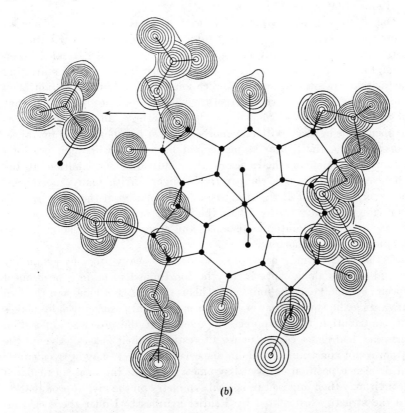

(b)

can be used to obtain a still better picture of the structure by Fourier synthesis according to (16.10). Even if none of the lighter atoms can be located from the vector distribution, the phases calculated from the position(s) of the heavy atom(s) alone often suffice to give a recognizable image of some or all of the structure after Fourier synthesis with the observed amplitudes. New phases, presumably closer to the true ones, can then be calculated, and the Fourier synthesis repeated. Illustrations of the operation of this process in the determination [22] of the structure of a vitamin B_{12} derivative are given in Fig. 16.17. There were 73 unique atoms to be found in this structure, including only one heavy atom, cobalt, with atomic number 27, about four times the average of the other atoms present. Figure 16.17a shows the electron density contours of the "side-chain" atoms attached to the ring system surrounding the cobalt atom, at an intermediate stage with phases based only on the positions of 26 of the atoms. Figure 16.17b illustrates the improvement in the appearance of these atoms when the phases are calculated with the positions of 68 of the 73 atoms present.

One drawback of the method is that when the heavy atom has an atomic number sufficiently high to dominate the vector distribution, it necessarily must dominate the x-ray scattering as well. It can be seen from Fig. 16.13 that if one atom has a much larger atomic scattering factor than the others, that is, a much longer vector in this diagram, then the phase angle for the whole structure will in general not be far from that of the single heavy atom alone, unless, of course, all the other atoms happen to be in phase with one another, a most improbable circumstance. Consequently, the lighter atoms cannot be located with high precision by any refinement method because their individual contributions to the observed intensities are comparatively slight. With many substances, however, the object of the study is a determination of the gross over-all molecular structure rather than the fine metric details, so that high precision in the interatomic distances and other aspects of the molecular geometry is less important.

Occasionally it may be possible to locate a heavy atom by symmetry considerations alone, since it may be constrained to lie in a position of special symmetry in the unit cell. This is more often a handicap than an advantage, however, because an atom in a special position usually makes no contribution to an appreciable fraction of the observed diffraction maxima, and hence cannot be used to estimate their phases. Even if the atom is not constrained to lie in a special position, it may by accident lie at or near a position of special symmetry, so that the phases calculated from it may then impose this special symmetry on the first approximation to the structure calculated by Fourier synthesis. Under these circumstances it is necessary to sort out the correct structure from its pseudo-symmetrical image, a procedure that in practice is not very difficult.

16.11 ISOMORPHOUS REPLACEMENT

Perhaps the most powerful method of phase determination is based on diffraction from a series of isomorphous crystals, that is, crystals with the same symmetry and essentially identical unit-cell dimensions and atomic arrangements but systematically varying identity of one or more of the atoms. The chemical problems in preparing such crystals normally are very difficult because even elements with similar chemical properties (for example, those in the same column of the periodic table) vary sufficiently in size that packing rearrangements often occur on substitution of one for another. The configuration corresponding to minimum energy is always extremely sensitive to subtle variations in molecular size and shape. The method has proved successful, however, in the analysis of the structure of a number of relatively small molecules, and it has recently provided the key to the remarkable determinations of the structures of the proteins myoglobin and hemoglobin, which contain more than 1000 atoms in the asymmetric unit [25, 44].

The principle of the method is relatively simple. The over-all structure factor F for any diffraction maximum may be considered to be the sum of a contribution F_i due to the atom or group of atoms with altered identity in the different isomorphous crystals, and a contribution F_R due to the remaining atoms, which are identical in nature and position in the different crystals. Thus considering also the components of F as defined in (16.6), (16.7), and (16.8), we have

$$F = F_i + F_R,$$

(16.14) $$A = A_i + A_R,$$

$$B = B_i + B_R.$$

Variations of F_i, that is, of A_i and, if the crystal is acentric, of B_i, in the different isomorphous crystals arise from differences in the scattering effects of the substituted atoms. Since the contribution of the other atoms F_R remains constant, the over-all F changes. In a centrosymmetric crystal, this is merely a change in amplitude, corresponding to a change in A_i, with $B \equiv 0$. If the position of the variable atom or group of atoms is known, for example, by analysis of a Patterson function, then not only the magnitude but also the sign of F_i (in this case A_i) can be calculated. Correlation with the observed changes in the amplitude of F then immediately permits prescribing a sign for each observed amplitude. For example, suppose that the observed magnitudes of a given reflection in two isomorphous compounds are 24 and 30, and the corresponding F_i are known to be -8 and -2. It is clear then that $F_R = +32$ and that the signs of these F's must be positive. It is often possible to deduce signs correctly by this method even when the data have not been put on an absolute scale. When the different substituents are known to be in the same position in the different crystals, unambiguous inferences can often be drawn, at least for the stronger reflections, even if one does not know ˎ the position of the substituted atoms, because one knows which substituted atom will have contributed the larger absolute value (corresponding to the heavier atom). In some cases, one of the isomorphous crystals may be unsubstituted, that is, have no atom at all in the position occupied by different atoms in the other isomorphous substances. Normally the varied atoms all occupy the same position in the asymmetric unit, although with very large molecules it is sometimes possible to alter the position of substituents without disturbing the arrangement of the unvarying major portion of the structure.

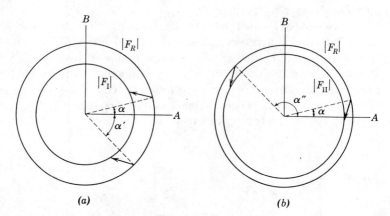

Fig. 16.18 Determination of phase angle from a triple of isomorphous noncentrosymmetric crystals. The heavy arrows correspond to the known amplitude and phase of the variable-atom contribution. The pairs illustrated separately in (a) and (b) each give rise to a twofold ambiguity in the phase angle for the unsubstituted crystal because there are two ways of placing the heavy arrows so that the vector sum of the unsubstituted contribution and that of the variable atom equals that of the substituted crystal. Only one of these angles (α), however, is the same in the two cases. Hence it must be the correct phase angle for F_R. [For simplicity it has been assumed here that the heavy atoms in I and II have the same scattering power but different positions and thus different phases. In (a) the phase angle of the heavy atom is about 170° and in (b) it is about 255°.]

With noncentrosymmetric crystals, it is necessary to have at least three isomorphous substances in order that the phase may be determined, since the phase angle has now a continuous range of possible values. This is illustrated in Fig. 16.18, in which for convenience it has been assumed that amplitude measurements have been made on the unsubstituted compound (R) as well as on crystals containing two different substituents (I and II). The amplitudes of the structure factors for I and R are represented by the radii of the circles in Fig. 16.18a, and those for II and R by the radii of the circles in Fig. 16.18b. The phases of I, II, and R are, however, initially indeterminate, since any combination of A and B is possible, consistent with the observed radius. If the positions of the variable atoms are known, both the amplitudes and phases of their contributions can be calculated; they are represented by the heavy arrows in Fig. 16.18. It can be seen that there are two, but only two, ways of placing the arrows consistent with each pair of measurements; by considering each pair in turn, however, this twofold ambiguity is removed and the phase of this reflection for any of the crystals (the illustration is for the unsubstituted one) can then be deduced.

In practice, because of experimental errors in the amplitude measurements and sometimes because of lack of perfect isomorphism as well, more than three isomorphous crystals are frequently used in order to increase the precision of measurement of the phase angles. In Kendrew's determination of the structure of myoglobin, six different isomorphous derivatives were used to establish the bulk of the phase angles out to a resolution of about 2 Å. The labor involved was staggering: quite apart from the years of effort involved in searching for the derivatives and in growing and photographing the crystals, about 10^4 intensity measurements had to be made on each of the derivatives, and then these intensities had to be processed to obtain the amplitudes and phases. This whole operation is now being extended to a resolution of about 1.5 Å, a process that involves just about double the labor but that may permit unambiguous characterization of almost all regions of this complex and biochemically fascinating structure.

16.12 DIRECT METHODS

Many attempts have been made to devise algebraic or statistical methods for the direct solution of crystal structures on the basis of intensity data, the unit-cell dimensions, and the chemical composition of the crystal, with no structural information of any sort [30, 55]. Most efforts so far have been directed to the solution of the problem in centrosymmetric crystals, and even there the success has been limited. Although there have been a few discussions of solutions in crystal space, that is, direct solutions for the atomic positional coordinates, the only really successful and practical approaches have involved calculations in reciprocal space, that is, calculations of phases, which in a centrosymmetric crystal amount to nothing more than signs to be associated with the amplitudes. The problem then consists of selecting the most appropriate combination of signs from among the 2^n possible combinations for a structure with n independent diffraction maxima; although a few signs can always be chosen arbitrarily, the number depending on the symmetry, this number is always negligible with respect to n, so this is a minor consideration in this general discussion.

These methods involve systematic examination of relationships—inequalities and equalities—among specified groups of the structure factors or their signs. Most of the methods depend either explicitly or implicitly on certain physical facts that impose limitations on the distribution of the magnitudes and phases of the structure factors, notably the fact that the electron density can never be negative and that it can, to a fair approximation, be considered as resulting from a superposition of spherically symmetric atoms of approximately the same shape. To the extent that different atoms do indeed have the same shape, we can write $f_i = Z_i \hat{f}$, with Z_i the atomic number of the ith atom and $\hat{f}$ a unitary atomic scattering factor. The relations between different structure factors are most effectively expressed in terms of unitary structure factors,

$$(16.15) \qquad U(h, k, l) = \frac{F(h, k, l)}{\sum_i f_i} = \frac{F(h, k, l)}{\hat{f} \sum_i Z_i} .$$

The magnitudes of the U's can never exceed unity, and they approach this value only when all atoms are almost in phase.

Progress to date with direct methods and prospects for their success in the near future have been summarized by Woolfson [55]. Perhaps the most elegant and comprehensive approach to the problem is that of Klug [29]. His analysis is based on the fact that the statistical problem in crystallography is one involving linear sums (the structure factors) of random variables (the atomic contributions to the structure factors), and that more or less standard methods exist in probability theory for dealing with this situation. Klug deals both with the determination of individual signs from intensity relations and with the derivation of relations between signs of different structure factors. Relations of the latter sort, which had been derived earlier in various ways, usually involve a joint probability distribution, that is, evaluation of the probability that a given structure factor has a particular sign when the signs of other specified structure factors are known. Before we discuss these relations, however, we shall consider the determination of individual signs from intensity relations because it can be presented more simply and was historically the first practical direct method, developed by Harker and Kasper in 1948 for their work on decaborane [20].

Their approach involves inequality relationships derivable in a variety of ways, for example, from Cauchy's inequality,

$$(16.16) \qquad \left| \sum_{j=1}^{N} a_j b_j \right|^2 \le \sum_{j=1}^{N} |a_j|^2 \sum_{j=1}^{N} |b_j|^2,$$

where the a_j and b_j may be real or complex. For convenience, we shall let H represent the triple (h, k, l), $\phi_j = 2\pi(hx_j + ky_j + lz_j)$, and $n_j = Z_j/\Sigma Z_i =$ fraction of the electrons associated with the jth atom. Then we have

$$U(H) = \frac{F(H)}{\hat{f}\sum\limits_{j=1}^{N} Z_j} = \frac{1}{\hat{f}\sum\limits_{j=1}^{N} Z_j} \sum\limits_{j=1}^{N} \hat{f}Z_j \cos \phi_j = \sum\limits_{j=1}^{N} n_j \cos \phi_j.$$

(Only cosine terms are involved, as in (16.7), because of the assumed center of symmetry at the origin.) If now we let $a_j = n_j^{1/2} \cos \phi_j$ and $b_j = n_j^{1/2}$ in Cauchy's inequality, and note that

$$\sum\limits_{j=1}^{N} n_j = 1,$$

then we obtain

$$\left| \sum\limits_{j=1}^{N} n_j \cos \phi_j \right|^2 \leq \sum\limits_{j=1}^{N} |n_j \cos^2 \phi_j| \sum\limits_{j=1}^{N} n_j \leq \frac{1}{2} \sum\limits_{j=1}^{N} n_j(1 + \cos 2\phi_j),$$

or

(16.17) $$U^2(H) \leq \tfrac{1}{2} + \tfrac{1}{2}U(2H).$$

It is sometimes possible with this relation to prove that $U(2H)$ is positive, but never that it is negative. For example, suppose we knew that $|U(h, k, l)| = 0.7$ and $|U(2h, 2k, 2l)| = 0.4$. Then (16.17) gives $0.49 \leq 0.5 \pm 0.2$, and clearly the plus sign must be chosen; that is, $U(2h, 2k, 2l)$ must be positive.

A great many other inequalities can be derived by making appropriate substitutions in (16.16); for example, we obtain

(16.18) $$|U(H) + U(H')|^2 \leq [1 + U(H + H')][1 + U(H - H')]$$

or, if there is a twofold screw axis parallel to $\mathbf{b}$,

(16.19) $$|U(h, k, l)|^2 \leq \tfrac{1}{2}[1 + (-1)^k U(2h, 0, 2l)].$$

Relation (16.18) is valid also if all the signs in it are reversed; it can be used, unlike (16.17) or (16.19), to establish the sign of a structure factor that does not have all even indices. Relations such as (16.19), or (16.18) with the signs changed, can be used to establish that a certain sign must be negative. The number of simple and useful inequality relations increases with increasing symmetry of the crystal. It is apparent that for practical application of these relations it is important that the magnitudes of the U's be as large as possible, since larger $|U|$ means a larger difference between $+U$ and $-U$, and thus a better chance of distinguishing between them analytically. Unfortunately, as the number of atoms in the structure increases, the chance that many of them will contribute in phase to a given diffraction maximum decreases, and consequently the average $|U|$ decreases. It is, in fact, equal to $1/\sqrt{N}$ in a structure with N equal atoms in the unit cell, and if N is sufficiently large then the values of $|U|$ will be distributed approximately in a Gaussian manner; for example, only about 5% of them will be larger than $2.5/\sqrt{N}$. Considerations of this sort and practical experience both suggest that inequalities alone are generally insufficient for solving a structure with more than about 10 atoms in the asymmetric unit. Since the pioneering efforts of Harker and Kasper [20], other investigators have commented on and extended inequality methods; for example, more general inequalities have been derived [24] by application of Herglotz's theorem, which expresses certain determinantal relations between the Fourier coefficients of a series that is known to represent a nonnegative function. The Harker-Kasper inequalities, however, are about as convenient and effective as any of the more extended ones [55].

Several different approaches have been used in the derivation of equality relationships between structure factors [27, 29, 30, 55]. The most useful of these, presented originally by Sayre [47] and developed since by many others [55], can be used to demonstrate that certain relationships must hold between the signs of different structure factors. If we designate the sign of a given structure factor $F(H)$ by $s(H)$, then it can be proved that, for example, $s(H)s(H') = s(H + H')$ if the corresponding unitary structure factors are large enough. Even if they are not sufficiently large for this relationship to be *necessarily* true, the *probability* that it is true can be calculated by a very simple formula involving only the magnitudes of the unitary structure factors and the scattering factors of the atoms present. Other sign relationships of a similar sort have been derived and found useful; in the presence of other symmetry elements, in addition to a center of symmetry, more specialized formulas are also applicable.

Most commonly in current practice with direct methods, a combined application of inequalities and sign relationships is made. The sign relationships, applied systematically and usually now with the help of a high-speed computer, are used to express the probable signs of a significant fraction of the larger unitary structure factors in terms of the signs of a very few of them. Under favorable circumstances the latter may be surmised from inequalities or physical considerations, for example, the impossibility of the molecule lying in certain positions in the unit cell; otherwise, trial-and-error methods can be used. Throughout the sign-determining process, an attempt is made to obtain several independent checks on a given sign by using as many different relations as can be found. This overdetermination to test for consistency is essential because of experimental errors in the observed amplitudes or in their scale factor, since theoretical conditions are never quite met (for example, different atoms do not in fact have quite the same shape), and since some of the sign-determining relationships that have probabilities smaller than unity will in fact be wrong. If even a relatively small fraction of the signs can be deduced correctly, for example, perhaps 15 % of them, including those to be associated with most of .the largest amplitudes, a recognizable Fourier map of the structure can often be obtained. Many detailed examples of the application of different methods are given by Woolfson [55].

Although comparatively little of the work with direct methods to date has taken full advantage of the potentialities of the present generation of high-speed computers, a promising beginning has been made [9, 10, 56]. The programs so far described are designed only for two-dimensional data. The more elaborate of them [56] starts with the raw crystallographic data, calculates a number of the most probable sets of signs with the help of various special tests and iterative procedures to screen out and correct the occasional sign that has first been chosen incorrectly, and then prints out approximate electron density maps calculated with as many as desired of the most plausible sets of signs. These maps are, of course, not highly precise, because they include contributions from only the stronger diffraction maxima, the signs of which are easier to fix, and because the methods are in essence statistical, so that the best trial model might correspond to any of the more probable sets of signs deduced. At this stage the judgment of the investigator must still be used in discriminating among the various Fourier maps to find which are most promising for refinement. Direct methods of this sort are still in their infancy and have thus far been used only in a comparatively small number of actual structural analyses. No doubt, however, they will play an increasingly significant role, particularly in the solution of structures of no more than moderate complexity. Woolfson has suggested [55] that this sort of method should be useful for structures with no more than about 50 or 60 atoms in the asymmetric unit. The most complex structure solved by direct methods to date has fewer than half that many unique atoms.

16.13 REFINEMENT

After approximate positions have been determined for most, if not all, of the atoms, refinement of the structure can be started. In this process the atomic parameters are varied systematically so as to give the best possible agreement of the observed structure-factor amplitudes and those calculated for the proposed structure. Since most crystal-structure problems involve many parameters, many successive refinement cycles are usually needed before the structure converges to the stage at which shifts in the atomic parameters from cycle to cycle are negligible with respect to the expected experimental errors. There are two common refinement techniques, one involving Fourier syntheses and the other a least-squares process; although they have been shown formally to be nearly equivalent, differing chiefly in the weighting attached to the experimental observations, they differ considerably in manipulative details, and we shall discuss them separately here.

Fourier methods

In each cycle of Fourier refinement, one first calculates phases, by means of (16.6) to (16.9), for the proposed structure as it exists at that stage in the analysis. These phases are used with the *experimental* amplitudes in a Fourier synthesis of an approximation to the electron density by means of (16.10) or by means of an analogous equation for fewer dimensions. The positions and shapes of the maxima in this synthesis then serve for improvement of the atomic parameters, which in turn are used for calculation of an improved set of phases. Alternate cycles of Fourier synthesis and phase calculation should be continued until there are no significant changes in the derived structure, or in the phases, from one cycle to the next. In practice, until the advent of modern computers, many analyses were stopped short of complete convergence because of the labor involved in the calculations. For example, calculation of electron density at intervals of 0.2 Å for a typical asymmetric unit 5 by 8 by 10 Å might involve some 50,000 separate summations by (16.10), each summation being over several thousand different terms that must previously have been calculated. In practice, the labor can be significantly reduced by proper combination of terms in the summation but even then it is still so great that three-dimensional structure determinations were rare before the days of electronic computers.

Fourier refinement can be accelerated, particularly in the late stages, by means of "difference" syntheses. The amplitudes used in such a synthesis are the differences of the experimental amplitudes and those calculated for the structure at that stage according to (16.6) to (16.8). The resulting maps are automatically corrected for *finite series* errors, which arise because one is seldom able to measure experimentally all the nonzero amplitudes of diffraction maxima, and consequently one seldom has the ideal situation of a convergent series, which was assumed in deriving (16.10). If amplitude differences are used as Fourier coefficients, the effects of ignoring any inaccessible diffraction maxima will be much smaller because their calculated amplitudes presumably are similar to their unmeasured experimental values, at least to the extent that the proposed structure is a good model for the actual structure. Consequently, the terms omitted from the summation because one has a finite number of observations are much smaller than before. The ideal difference map has zero curvature, zero slope, and even zero density everywhere, within experimental error. Deviations from these conditions in the intermediate stages of refinement serve to indicate the changes needed in the proposed model of the structure, either in the atomic positions or in temperature factors [see (16.5) to (16.8)] or occasionally even in atomic identity. Difference maps have proven very useful in locating relatively light atoms, for example, hydrogen atoms in organic structures or carbon atoms in the presence of many atoms of high atomic number. These atoms are seldom locatable until the parameters of the heavy atoms present have been well refined.

The process of Fourier refinement can be adapted for automatic operation with a high-speed computer. Instead of evaluating the electron density on a fixed lattice of points as in (16.10), we calculate it, together with its first and second derivatives, at the positions assumed for the atomic centers at this stage. The shifts in the atomic parameters can then be derived from the slopes and the curvatures in different directions. Normally we apply this *differential-synthesis* method to the difference density in order to avoid finite-series errors. In fact, however, the method is used much less extensively than least-squares refinement, for the latter is somewhat more convenient for computer application and has the advantage of a statistically sounder weighting scheme for the experimental observations.

One question that always arises in discussions of Fourier refinement is, "How good must the trial structure be, or how nearly correct must the phases be, for the process to converge?" This question cannot be answered precisely. When the initial phases are poor, the first approximations to the electron density will contain much false detail, together with peaks at or near the correct atomic positions. The sorting of the real from the specious is difficult; experience, a sound knowledge of the principles of structural chemistry, and a good deal of caution are necessary. At this stage it is often imperative, and always desirable, that contour plots of the approximate electron density or difference density be studied carefully. If most of the atoms included in the phasing are no more than about one-half their radius (0.2 to 0.3 Å) from their correct sites, then a few that are farther away and even one or two that may be wholly spurious can be tolerated. A very astute or fortunate investigator may be able to recognize portions of a molecule of known structure in a map produced from a trial structure that is appreciably worse than that described, but such perspicacity is uncommon.

Least-squares refinement

The most widely used technique for the rapid and routine refinement of crystal structures, almost invariably now with the help of a computer, is based on the method of least squares [11, 30]. If we let $\Delta\,|F|$ be the difference in the amplitudes of the observed (F_0) and calculated (F_c) structure factors, and let the standard deviation of the experimental value of $F_0(h, k, l)$ be $[w(h, k, l)]^{-\frac{1}{2}}$, then, according to the theory of errors, the best parameters of the structure are those corresponding to the minimum value of the quantity

$$(16.20) \qquad Q = \Sigma w(h, k, l)[\Delta\,|F(h, k, l)|]^2,$$

in which the sum is taken over all independent observations, that is, the unique diffraction maxima. The variable parameters that can be used in the minimization of Q normally include the atomic position parameters x_j, y_j, and z_j for each atom, although occasionally one or more of these may be fixed by symmetry, the temperature parameters for each atom, which may number as many as six if the vibration is represented by a general ellipsoid not constrained by symmetry, and an over-all scale factor for the experimental observations. Thus in a general case there may be as many as $9N + 1$ parameters to be refined for a structure with N independent atoms. In special situations, there may also be other parameters, for example, *site-occupancy* factors in a partially disordered structure in which some positions are not entirely equivalent in every unit cell.

If the total number of parameters to be refined is p, then the minimization of (16.20) in this p-dimensional space involves setting the derivative of Q with respect to *each* of these parameters equal to 0. This gives p independent simultaneous equations; the weights w and the observed amplitudes are independent of the variable parameters, whereas the calculated amplitudes depend on each of the parameters in an explicit manner given by the equations defining the structure factor. Thus the derivatives of Q are readily evaluated. Clearly, at least p experimental observations are needed to define the p parameters, but, in fact, since the observations usually have significant experimental uncertainty, it is desirable that the number m of observations exceeds the number of variables by an appreciable amount. In most practical cases with three-dimensional x-ray data, m/p is of the order of 5 to 10, so that the equations derived from (16.20) are greatly overdetermined.

Unfortunately, the equations derived from (16.20) are by no means linear in the parameters; it is clear from (16.6) to (16.8) that they involve various transcendental functions. The straightforward application of the method of least squares requires a set of linear equations. *If* a reasonable trial structure is available, then it is possible to derive a set of linear equations in which the variables are the *shifts* from the trial parameters, rather than the parameters themselves. This is done by expanding in a Taylor's series about the· trial parameters, retaining only the first-derivative terms on the assumption that the shifts needed are sufficiently small that the terms involving second- and higher-order derivatives are negligible:

$$(16.21) \qquad \Delta |F| = \sum_{i=1}^{p} \frac{\partial F}{\partial \xi_i} \Delta \xi_i,$$

with the ξ_i representing the parameters being refined. Obviously, the validity of the assumption depends on the closeness of the trial structure to the correct structure as well as on the shape of the p-dimensional surface corresponding to Q in the region of the trial structure. If conditions are unfavorable, the process may well converge to a false minimum in this p-dimensional space rather than to the deeper true minimum corresponding to the correct solution. Thus this method of refinement also depends for its success on the availability at the start of a reasonably good set of phases, that is, a good *trial structure*.

If the trial structure is sufficiently good, then the m observational equations of the form (16.21), one for each of the m independent observations, are appropriate for the usual least-squares treatment since they are linear in the parameters to be refined. From them are formed, in the usual manner, a set of p normal equations, each linear in the p parameters and each contributed to by each of the m observations in proportion appropriate to its statistical weight. Solution of these p simultaneous equations involves manipulation of a p-by-p matrix, a process that is straightforward with a modern computer even when p exceeds 100. Actually, in many least-squares refinements of crystal structures, the entire matrix is not calculated because it has been found by experience that cross terms between different atoms are usually negligible with respect to intra-atomic terms. This means that no matrix larger than 9 by 9 or 10 by 10 need be calculated, and consequently that a larger total number of parameters can simultaneously be refined without overtaxing the storage capacity of the computer.

Several refinement cycles are needed to reach convergence in the least-squares method because of the approximations involved in deriving (16.21), and sometimes because some cross terms have been ignored in the matrix of the normal equations. The shifts in the parameters obtained from one cycle are used to give improved trial-structure parameters, and then the entire process is repeated until the shifts calculated in a given cycle are much smaller than the estimated standard deviations of the corresponding parameters. These standard deviations are readily obtained from the inverse of the matrix of the normal equations. The entire least-squares process is admirably suited to automatic high-speed computation, but the cautious investigator will arrange for regular checks on the chemical and physical plausibility of the results after each cycle. Difficulties can then readily be spotted and corrected. For example, if a false atom has inadvertently been included in the initial trial structure, then its behavior will soon appear anomalous: it may move to a chemically unreasonable position (for example, too close to another atom) and its temperature factor will usually increase strikingly to a value far higher than that normally encountered for any real atom. This corresponds physically to a very high vibration amplitude, that is, a smearing of the atom throughout the unit cell, an almost infallible sign that there is no atom in the correct structure at the position assumed in the trial structure.

At the conclusion of any least-squares refinement process, it is always wise to calculate a difference Fourier synthesis. If it is zero everywhere, within experimental error, then the least-squares solution is a reasonable one. If it is not, and the peaks in it cannot be explained, for example, as the result of light atoms not yet included in the structure-factor calculations, then it is likely that the least-squares procedure has converged to a false minimum because the initial approximation of the trial structure was not sufficiently good. Another plausible trial structure must be sought and refinement tried again.

SOME ASPECTS OF ATOMIC ARRANGEMENTS

To illustrate some of the applications of the ideas and techniques that have been described, we shall discuss a few topics from among the thousands of published crystallographic studies and some peripherally related investigations. Unfortunately, even these topics must perforce be treated briefly.

16.14 TOPOLOGY AND SHAPES OF MOLECULES AND ASSEMBLIES OF MOLECULES

Topological structural chemistry was chiefly a nineteenth-century development, with the endless variety of structural possibilities inherent in the compounds of carbon ("organic compounds") providing the impetus. For example, the rules for formulating the saturated hydrocarbons are simple: each carbon must be linked to four other atoms, each hydrogen to only one other atom. The resulting possible formulas, when written in the shorthand notation common today, are indistinguishable from the "tree" representations of combinatorial mathematics [45] (see Chapters 5 and 6 of the present book).

Crystallographic molecular structure determinations not only have confirmed the topologies deduced from reactivity considerations for a great variety of molecules but also have provided detailed information about the three-dimensional molecular geometry of molecules of both known and hitherto unknown atomic arrangement. They have, in fact, provided the chief, although by no means the only, source of such information; for molecules with more than 15 or 20 atoms, no other method of structure determination is applicable, and for very large molecules (upward of 100 atoms) it is often simpler to establish the detailed molecular structure crystallographically than to deduce the connectivity of the atoms by more traditional chemical methods.

The most dramatic triumph of structural crystallography has been the recent determination of the detailed structure of the protein myoglobin [25], which serves for the storage of oxygen in muscle tissue. This molecule, with more than 1000 atoms (not including the hydrogen atoms, undetectable in this experiment), consists primarily of a topologically linear sequence of about 150 simpler units, amino acids, bonded together in a familiar manner. This single chain is folded in an unexpected way, which might be considered merely "random," except for the fact that the biochemically related hemoglobin molecule is folded in a remarkably similar fashion [44]; see Fig. 16.19. Hemoglobin is responsible for the storage and transport of oxygen in human and other mammalian blood. Although it has not yet been studied at as high a resolution as myoglobin, at 6 Å resolution Fig. 16.19 shows that the two chemically distinct (although related) chains that comprise the asymmetric half of the hemoglobin molecule are each folded in a fashion scarcely distinguishable from that of the myoglobin chain. It is tempting to speculate on the evolutionary significance of this finding, especially since the myoglobin sample was taken from a sperm whale (the high concentration of myoglobin in this animal is presumably responsible for its ability to stay submerged for long periods) and the hemoglobin was equine. Actually, it has been shown that samples of these proteins from different species differ only in minor respects.

In any event, whatever its significance, the folding clearly is highly specific and reproducible. It is not yet clear whether some new principle that might account for the manner of folding remains to be discovered or whether it can be discussed entirely in terms of the myriad of specific interactions between atoms in nearby parts of the chains and the deformation of bond angles and dihedral angles. Interactions between atoms in different regions of the same molecule, or in distinct molecules, are extremely sensitive to small variations in molecular size and shape because

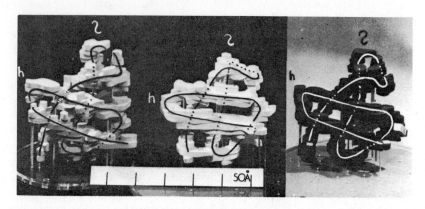

Fig. 16.19   Models of the appearance (at 6 Å resolution) of the polypeptide chains in myoglobin (left) and hemoglobin (center and right). The two crystallographically (and chemically) distinct chains in hemoglobin are here given different colors for convenience. [Reproduced by permission from M. F. Perutz, M. G. Rossman, A. F. Cullis, H. Muirhead, G. Will, A. T. C. North, "Structure of Haemoglobin," *Nature*, **195** (1960), p. 418.]

the attractive energies of such interactions vary about as $1/r^6$, and the repulsive energies as about $1/r^{12}$, where r is the distance of separation of the atoms.

When viewed at 2 Å or higher resolution, the atoms in the protein chain in most of the "straight" sections of the myoglobin molecule evident in Fig. 16.19 are found to be in a helical configuration with helix parameters almost indistinguishable from those postulated by Pauling, Corey, and Branson [42] in 1951 for the so-called α-helix. Helical configurations are very common for polymeric molecules, that is, molecules made by combining many similar smaller units according to some specific bonding scheme. Most natural and synthetic fibers and many biologically important macromolecules such as the nucleic acids (see Chapter 17) occur in at least partially helical configurations.

One of the important clarifications resulting from crystal-structure studies has been in the significance of inorganic formulas. For example, there are many different solid inorganic substances with empirical formula

XY_2. Before the structures of these materials were understood, they were all conventionally regarded as discrete molecules composed of one atom of X and two atoms of Y. Some, such as carbon dioxide, CO_2, or the nitrite ion, NO_2^-, are indeed discrete *monomeric* species. Others, however, are *polymeric*, that is, consist of a number of units hooked together so that the over-all composition is XY_2. For example, the ring illustrated in Fig. 16.20 has this composition, although each X is surrounded by three

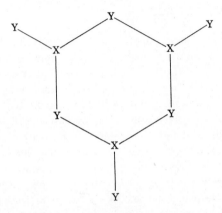

Fig. 16.20 A possible polymeric structure for a substance with composition represented by the formula XY_2.

Y's, and half the Y's are bonded to two X's; this is one of the forms of BO_2^-, which may also occur in rings of other sizes or in long chains—always, however, with the same environments for the individual atoms. Another possibility for XY_2 is an arrangement such that each X is surrounded by four Y's and each Y by two X's; this is the situation in the various forms of silica, SiO_2 (quartz, tridymite, cristobalite, and others). These consist of essentially infinite three-dimensional networks. Such structures exist for a great variety of different substances; many are topologically identical, differing in their geometrical details only because of differences in the electronic symmetry and size of the atoms involved.

The topology of structures built up from repeating units that can bond to one or more of their neighbors has been considered in some detail by Wells [51]. He classifies such structures in terms of the *coordination number* or *connectivity* of each structural unit, that is, the number of other such units to which it is linked. When the coordination number (CN) is 1, the structure necessarily consists of discrete objects each comprising a pair of the units. When the CN is 2, the structure must consist either of rings (of any size) or of infinite chains. When the CN is 3 or more, infinite networks are possible, and Wells has illustrated and discussed many of the different possibilities, showing how they are exemplified in actual structures, not only those built up from polyvalent atoms but also those involving larger units that can form a limited number of specific links, chiefly by hydrogen bonding.

16.15 PACKING OF SPHERES, MOLECULES, AND POLYHEDRA

The structures of many simple substances can also be profitably discussed in terms of the most effective ways of packing together spheres of various sizes [40, 51]. If all the spheres are identical, then there are two simple ways in which they can be packed in an ordered fashion so as to fill the maximum fraction, about 74%, of the available space. In each of these structures, any given sphere is surrounded by twelve equivalent spheres, six at the corners of a regular hexagon with the reference sphere at its center, three others above this plane, and three below it. One of these closest-packed arrays has hexagonal symmetry, the other cubic; it is, indeed, merely a face-centered cubic structure. About two-thirds of the elements crystallize with one or the other of these structures. The structures of the alkali halides and some other similar compounds were deduced by Pope and Barlow in the nineteenth century on the basis of considerations concerning the ways in which spheres of different sizes can be packed; this was long before there were any means to determine the actual atomic arrangements in crystals. The manner in which many large organic molecules crystallize can also be explained [26] in terms of the ideas of close packing, although the application of the principles is complicated because of the irregular shape of most such objects.

Still another illuminating way of describing and classifying many structures is in terms of the packing together of polyhedra of different sizes and shapes [40, 51]. The approach is not a new one; it has long been used, for example, in discussions of the coordination about different sorts of ions and in the analysis of energy levels in solids in terms of Brillouin zones. Recently, however, efforts have been made to systematize the description of many simple inorganic structures in terms of the way in which the different polyhedral holes present are filled, or remain unfilled. A modular algebra and a number of different sorts of models have been developed for the purpose [17, 32, 35]. In addition, many more complex structures, particularly those of alloys, have been analyzed in terms of the packing of spheres with preferred tetrahedral groupings, so that coordination polyhedra have triangular faces [15, 16]. In some of the more complex alloys of the transition metals, coordination polyhedra with 12, 14, 15, and 16 such faces are prominent, and the ways in which they can be combined to build up large structures have been deduced.

Many crystalline hydrates are known [14, 43, 51] to contain polyhedral frameworks of water molecules linked together by hydrogen bonds, which are primarily dipole-dipole attractions with only about 5 to 10% the strength of more usual chemical bonds, but still appreciably more strength

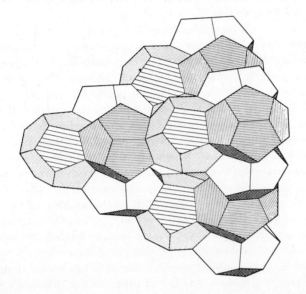

Fig. 16.21 The polyhedral structure of the 17 Å cubic gas hydrates. The structure may be regarded as built up from pentagonal dodecahedra and hexakaidecahedra (which have twelve pentagonal and four hexagonal faces). [Reproduced by permission from D. Feil and G. A. Jeffrey, "The Polyhedral Clathrate Hydrates, Part 2. Structure of the Hydrate of Tetra Iso-Amyl Ammonium Fluoride," *Journal of Chemical Physics*, **35** (1961), p. 1872.]

than most nonspecific nonbonded interatomic interactions. The polyhedral structure of the cubic gas hydrates with a 17 Å unit-cell edge is illustrated in Fig. 16.21. This structure [14] consists of slightly distorted pentagonal dodecahedra and hexakaidecahedra, which have 12 pentagonal faces and 4 hexagonal ones. Pauling has suggested [41] that these polyhedral arrays of water molecules may play a key role in general anesthesia induced by various comparatively unreactive gases such as nitrous oxide and cyclopropane. He has also suggested [39] that liquid water might contain a mixture of transient polyhedra of various sorts.

One of the most vexing and long-standing general problems remaining in structural physics and chemistry is that of finding a plausible and tractable model for the structure of a liquid. The problem is in significant part a combinatorial one because it is one of finding and enumerating the many different configurations of comparable energy that presumably may be present. Recently, Bernal [4, 5] has proposed a polyhedral model for simple liquids. He considers a general liquid composed of atoms or roughly spherical molecules that do not have strong specific interactions with one another, and starts with the premise that configurations of minimum energy will be characterized by approximately equal distances between the molecules because of the steepness of the intermolecular potentials. He then represents these intermolecular distances by the edges of five different sorts of polyhedra: the regular tetrahedron and octahedron, and semiregular equal-edged polyhedra with 12, 14, and 16 faces. The molecules are thus assumed to be at the vertices of the polyhedra. Small distortions, up to 10%, in a few edge lengths are necessary if all space is to be filled with these polyhedra (other than a 2:1 mixture of tetrahedra and octahedra). The model is a novel and provocative one, although much remains to be done with it before it can be called successful. Bernal's model is in some ways a special case of the more analytical treatment recently given by Gilbert [19] to the problem of the random subdivision of Euclidean space into disjoint regions. Gilbert has applied his analysis as a model for crystal grain boundaries and has used it for estimation of various geometrical parameters of such grains.

Before we leave the subject of polyhedra, it is interesting to note that they have cropped up in a number of other widely differing structural contexts. The boron atoms in some forms of elementary boron and in many of the compounds of boron with hydrogen and some other elements are arranged in nearly regular polyhedral frameworks, or parts of such frameworks; for example, in many of the boron hydrides the boron atoms are so arranged that they occupy the corners of an icosahedral fragment. At the other end of the scale of molecular sizes, many of the spherical viruses, which contain hundreds of thousands or even millions of atoms, have been found to consist of polyhedral arrangements of smaller presumably identical subunits. Evidence for these arrangements comes both from x-ray diffraction and from electron-microscope studies. For example, turnip yellow mosaic virus [36] consists of 32 subunits, 20 of which are at the 20 vertices of a pentagonal dodecahedron, and the other 12 of which are at the centers of the 12 faces of this polyhedron. The latter thus are at positions of fivefold symmetry in this virus particle (although, of course, as stressed earlier, these cannot be positions of fivefold crystal symmetry).

16.16 DISORDERED STRUCTURES AND ORDER-DISORDER PHENOMENA

Throughout this entire discussion we have thus far assumed that the crystals considered are ideal, whereas, in fact, most real crystals contain a variety of imperfections. We shall consider just one such defect, the

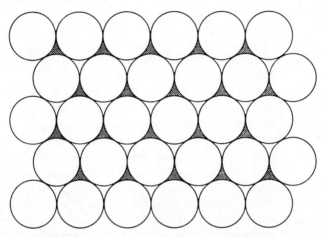

Fig. 16.22 A close-packed layer. The two sets of interstices (dark and light) represent respectively the two possible sets of projected positions for the centers of spheres in adjacent close-stacked layers. Each set necessarily forms a close-packed layer identical with (although displaced from) the original one.

presence of a degree of disorder, that is, lack of complete equivalence of adjacent unit cells. Such disorder necessarily is nonperiodic, since if it were periodic the crystal would be described in terms of the larger, and ordered, unit cell. The disorder may be one-, two-, or three-dimensional.

For simplicity, let us consider first a one-dimensional example, perfectly ordered layers stacked in a random fashion. In the simplest such case, the layers may all be identical; there are then two ways in which three layers of close-packed spheres can be stacked with successive layers in as close contact as possible. The important point is the relation of the outer two layers to the central layer. The spheres in the outer layers can nestle on the central layer in either of two sets of positions, which are represented as *dark* and *light* in Fig. 16.22. If both outer layers occupy the same set of projected positions when viewed normal to the layers (for example, the dark set in Fig. 16.22) then the central layer is designated as an H-layer for reasons that will shortly be made clear. Alternatively, the outer layers may occupy different projected positions, that is, in Fig. 16.22, the dark set on one side and the light set on the other, in which case the central layer is referred to as a C-layer. The two simplest ordered structures are those in which each layer is an H-layer or a C-layer; these are, respectively, the hexagonal and cubic closest-packed structures, with two-layer and three-layer repeating units. Other ordered structures with repeating units of four (HC), six (CCH), nine (HHC), and more layers are known, as well as substances in which the layers are stacked in a disordered manner.

With layers that are distinct from one another, either in structure or in composition, disordered arrangements are even easier to conceive, although, in fact, they may occur less frequently because they are energetically unfavorable. In general, disorder is most likely when the energy difference between the ordered and disordered structures is small. Entropy effects always favor the disordered state, but they normally are over-balanced by the energetic advantage of the ordered structure; this effect, however, becomes less pronounced as the temperature increases.

Another sort of disorder exists in certain phases of the ammonium halides [51]. The ammonium ion, NH_4^+, has a tetrahedral structure with the nitrogen atom at the center, and in these phases it lies at the center of a cube, whose eight corners are occupied by halide ions. The hydrogen atoms of the ammonium ion, which have a slight positive charge, tend to be oriented toward the negative halide ions. Now a tetrahedron may point to the corners of a circumscribed cube in either of two ways (Fig. 16.23). At low temperatures, the structure is ordered, all ammonium ions being oriented in a parallel way; at higher temperatures, however, it becomes disordered, each ammonium ion randomly adopting either one of the two possible orientations, independent of what its neighbors do. Neutron-diffraction studies of such phases indicate the presence of one-half hydrogen atom in each of eight equivalent positions at the corners of a cube about the nitrogen; this is precisely what an average of the unit cells contains. There are also some other ammonium halide structures known—for example, one in which the ammonium ions in one column of unit cells through the structure have one orientation while those in the four surrounding columns have the opposite disposition.

Rotational disorder occurs in many organic crystals containing molecules with an approximately circular cross section and no highly specific intermolecular bonding forces. Molecules in different unit cells are oriented at different angles with respect to the axis normal to the circular cross section. This effect has vitiated efforts to determine the detailed structures of many such substances, because the lack of equivalence of different unit cells means that one finds only the rotationally averaged structure.

(Actually, in x-ray experiments at a single temperature, this effect cannot be distinguished from free rotation of the molecules about the same axis.)

Another sort of orientation disorder occurs in structures containing particles with spin. Such effects are undetectable with x-rays, but the magnetic moments associated with electronic or nuclear spin are detectable by neutron diffraction [1]. For example, manganous oxide, MnO, has the sodium chloride structure, with the ions lying on two interpenetrating

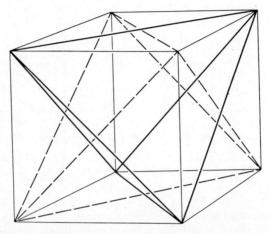

Fig. 16.23   The two ways of orienting a regular tetrahedron in a cube. The length of the edge of a tetrahedron is equal to that of the face diagonal of the cube.

face-centered cubic lattices. Since the manganous ion has unpaired electrons, it possesses a magnetic moment that can be either ordered or not in the structure. One of the ordered phases is the so-called antiferromagnetic phase, in which the individual moments are opposed in pairs, rather than being aligned parallel, as in a ferromagnetic substance. When the antiferromagnetic phase is examined with neutrons below the Curie temperature of 120°, the unit cell is found to be twice as large along each of the axes as in a comparable x-ray experiment. The reason is that in this antiferromagnetic phase alternate manganous ions have opposed spins; hence they are no longer equivalent in scattering the beam of neutrons despite the fact that they are indistinguishable with x-rays. The Curie temperature is the equilibrium temperature for the transition from the ordered phase to a disordered one. Ferroelectric materials, in which electric dipoles rather than magnetic ones are aligned, show similar effects; the transitions between ordered and disordered phases in these substances are frequently accompanied by detectable atomic movements, however, since the electric dipoles, unlike the magnetic ones, invariably involve two or more atoms.

Substitutional disorder is very common, particularly in alloys and related structures, and usually has marked effects on the mechanical, thermal, electrical, and magnetic properties of the materials. These effects have provided both impetus to and techniques for many experimental studies and theoretical analyses of order-disorder phenomena [13].

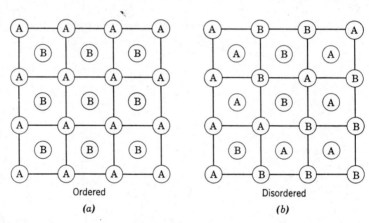

Ordered   Disordered
(a)   (b)

Fig. 16.24 Ordered (a) and disordered (b) arrangements of A and B atoms in the alloy AB. (Reproduced by permission from Charles Kittel, *Introduction to Solid State Physics*, 2nd edition, p. 337, John Wiley and Sons, New York, 1956.)

A simple two-dimensional example [28] of substitutional disorder is shown in Fig. 16.24. In the ordered state of this example, the nearest neighbors of any given atom of type A are atoms of type B, and vice versa. This situation results when the strongest interatomic attraction is between A and B rather than between like atoms; if the A-A and B-B attractions were appreciably stronger than the A-B, then separate phases of pure A and pure B would form.

Quantitative treatments of disorder phenomena necessitate a precise means of describing disorder. Perhaps the simplest of these is in terms of a long-range order parameter δ. Let us designate by a and b the lattices occupied respectively by the A atoms and the B atoms in the ordered state (Fig. 16.24a). Then we have

$$\delta = \frac{\text{(actual No. of A on } a) - \text{(No. of A on } a \text{ if random)}}{\text{(maximum possible No. of A on } a) - \text{(No. of A on } a \text{ if random)}}$$

$$= \frac{fN - N/2}{N - N/2} = 2f - 1,$$

where f is some fraction between 0.5 and 1 and N is the total number of atoms of A (or of B in this simple 1:1 example) present. If the structure is perfectly ordered, then $\delta = 1$. On the other hand, although when $\delta = 0$ the long-range disorder is complete in the sense that on the average throughout the structure there are equal numbers of A and B atoms on *each* of the lattices a and b, there may still be a considerable degree of *local* ordering. Consequently, additional parameters are needed to describe the situation. The most common of these is the short-range order parameter, of which the definition is analogous to that of δ except that what is considered is not the occupancy of an entire lattice but rather merely the occupancy of nearest-neighbor sites around a given atom by atoms of the opposite kind. In still more elaborate treatments, additional parameters are used to describe the ordering at next-nearest neighbor and surrounding sites. The application of these different order parameters in the interpretation of the variations of physical properties with the degree of ordering is beyond the scope of this discussion but clearly involves many combinatorial techniques, primarily, of course, those relevant to statistical mechanics.

MULTIPLE-CHOICE REVIEW PROBLEMS

1. A point-symmetry operation necessarily

- (*a*) involves a rotation,
- (*b*) leaves only one point of an object invariant,
- (*c*) leaves at least one point of an object invariant,
- (*d*) has none of the above restrictions.

2. In contrast to the point groups, the space groups

- (*a*) can involve no rotational symmetry element of order other than 1, 2, 3, 4, or 6.
- (*b*) can be extended by the idea of antisymmetry,
- (*c*) can involve rotatory-inversion operations,
- (*d*) are infinite in number.

3. The diffraction pattern of an array of objects on a lattice differs from that of a single object in

- (*a*) that it is a Fourier transform of the single-object pattern,
- (*b*) that it is a sampling of the single-object pattern at the points of a lattice identical in geometry and symmetry with that of the diffractor,
- (*c*) that the intensities and phases at the sampling points are reciprocally related to those of the single-object pattern at the same points,
- (*d*) none of the above ways.

4. If a structure has a center of symmetry, then

- (*a*) either a mirror plane or a twofold inversion axis must be present,

(b) the structure must be solved by the heavy-atom method,
(c) the diffracted waves are all necessarily in phase,
(d) none of the above is correct.

5. The interatomic vector map known as the "Patterson distribution" is a powerful aid, but, in general, it

(a) cannot be calculated without phase information,
(b) cannot be applied to noncentrosymmetric structures,
(c) does not permit resolution of peaks in a complex structure,
(d) is not limited in any of the above ways.

REFERENCES

1. Bacon, G. E., *Neutron Diffraction*, Oxford University Press, New York, 1955.
2. Belov, N. V., "A Classroom Method for the Derivation of the 230 Space Groups," *Proc. Leeds Phil. Lit. Soc., Sci. Sec.* **8** (1957), 1–46.
3. Belov, N. V., N. N. Neronova, and T. S. Smirnova, "Shubnikov Groups," *Akad. Nauk SSSR. Kristallografiya* **2** (1957) 315–325 (Russian).
4. Bernal, J. D., "A Geometrical Approach to the Structure of Liquids," *Nature* **183** (1959), 141–147.
5. Bernal, J. D., "The Structure of Liquids," *Scientific American* **203**, No. 2 (1960), 125–134.
6. Bragg, W. L., "Optical Representation of X-Ray Crystal Analysis Results," *Z. Krist.* **70** (1929), 475–492.
7. Buerger, M. J., *Vector Space*, John Wiley and Sons, New York, 1959.
8. Buerger, M. J., *Crystal Structure Analysis*, John Wiley and Sons, New York, 1960.
9. Cochran, W., and A. S. Douglas, "The Use of a High-Speed Digital Computer for the Direct Determination of Crystal Structures. I," *Proc. Roy. Soc. London, Series A* **227** (1955) 486–500.
10. Cochran, W., and A. S. Douglas, "The Use of a High-Speed Digital Computer for the Direct Determination of Crystal Structures. II," *Proc. Roy. Soc. London, Series A* **243** (1957), 281–288.
11. Cruickshank, D. W. J., "Fourier Synthesis and Structure Factors," Section 6 in J. S. Kasper and K. Lonsdale (editors), *International Tables for X-Ray Crystallography*, Vol. II, Kynoch Press, Birmingham, England, 1959.
12. Donnay, J. D. H., and W. Nowacki, *Crystal Data. Classification of Substances by Space Groups and Their Identification from Cell Dimensions*, Memoir 60, Geological Society of America, New York, 1954.
13. Elcock, E. W., *Order-Disorder Phenomena*, Methuen and Co., London, 1956.
14. Feil, D., and G. A. Jeffrey, "The Polyhedral Clathrate Hydrates, Part 2. Structure of the Hydrate of Tetra Iso-Amyl Ammonium Fluoride," *J. Chem. Phys.* **35** (1961), 1863–1873.
15. Frank, F. C., and J. S. Kasper, "Complex Alloy Structures Regarded as Sphere Packings. I. Definitions and Basic Principles," *Acta Cryst.* **11** (1958), 184–190.
16. Frank, F. C., and J. S. Kasper, "Complex Alloy Structures Regarded as Sphere Packings. II. Analysis and Classification of Representative Structures," *Acta Cryst.* **12** (1959), 483–499.
17. Gehman, W. G., "Standard Ionic Crystal Structures," *J. Chem. Ed.* **40** (1963), 54–60.

18. Gehman, W. G., *The Closest Packing of Spheres. A Unified Basis for Crystal Structures*, A. E. C. Report NAA-SR-6003, Atomics International, Canoga Park, Calif., 1961.

19. Gilbert, E. N., Unpublished manuscript submitted to *Ann. Math. Stat.*, 1961.

20. Harker, D., and J. S. Kasper, "Phases of Fourier Coefficients Directly from Crystal Diffraction Data," *Acta Cryst.* **1** (1948), 70–75.

21. Henry, N. F. M., and K. Lonsdale (editors) *International Tables for X-Ray Crystallography*, Vol. I, Kynoch Press, Birmingham, England, 1952.

22. Hodgkin, D. C., J. Pickworth, J. H. Robertson, K. N. Trueblood, R. J. Prosen, and J. G. White, "The Crystal Structure of the Hexacarboxylic Acid Derived from B_{12} and the Molecular Structure of the Vitamin," *Nature* **176** (1955), 325–328.

23. Holser, W. T., "Classification of Symmetry Groups," *Acta Cryst.* **14** (1961), 1236–1242, and references given there.

24. Karle, J., and H. Hauptmann, "The Phases and Magnitudes of the Structure Factors," *Acta Cryst.* **3** (1950), 181–187.

25. Kendrew, J. C., R. E. Dickerson, B. E. Strandberg, R. G. Hart, D. R. Davies, D. C. Phillips, and V. C. Shore, "Structure of Myoglobin," *Nature* **185** (1960), 422–427.

26. Kitaigorodskii, A. I., *Organic Chemical Crystallography*, translated from the Russian, Consultant's Bureau, New York, 1961.

27. Kitaigorodskii, A. I., *The Theory of Crystal Structure Analysis*, translated from the Russian, Consultant's Bureau, New York, 1961.

28. Kittel, C., *Introduction to Solid State Physics*, 2nd edition, John Wiley and Sons, New York, 1956.

29. Klug, A., "Joint Probability Distributions of Structure Factors and the Phase Problem," *Acta Cryst.* **11** (1958), 515–543.

30. Lipson, H., and W. Cochran, *The Determination of Crystal Structures*, George Bell and Sons, London, 1953.

31. Lipson, H., and C. A. Taylor, *Fourier Transforms and X-Ray Diffraction*, George Bell and Sons, London, 1958.

32. Loeb, A. L., "A Binary Algebra Describing Crystal Structures with Closely Packed Anions," *Acta Cryst.* **11** (1958), 469–476.

33. Lonsdale, K., *Crystals and X-Rays*, D. Van Nostrand Co., Princeton, New Jersey, 1949.

34. Mackay, A. L., "Extensions of Space-Group Theory," *Acta Cryst.* **10** (1957), 543–548.

35. Morris, I. L., and A. L. Loeb, "A Binary Algebra Describing Crystal Structures with Closely Packed Anions. II. A Common System of Reference for Cubic and Hexagonal Structures," *Acta Cryst*, **13** (1960), 434–443.

36. Nixon, H. L., and A. J. Gibbs, "Electron Microscope Observations of the Structure of Turnip Yellow Mosaic Virus," *J. Mol. Biol.* **2** (1960), 197–200.

37. Nowacki, W., "Symmetry and Physicochemical Properties of Crystallized Compounds. VI. The Distribution of Crystal Structures over the Space Groups and the General Structural Principles of Crystallized Organic Compounds," *Helv. Chim. Acta* **34** (1951), 1957–1962.

38. Patterson, A. L., "Direct Methods for the Determination of the Components of Interatomic Distances in Crystals," *Z. Krist.* **90** (1935), 517–542.

39. Pauling, L., "The Structure of Water," pp. 1–6 in D. Hadzi (editor), *Symposium on Hydrogen Bonding*, Pergamon Press, New York, 1959.

40. Pauling, L., *The Nature of the Chemical Bond*, 3rd edition, Cornell University Press, Ithaca, N.Y., 1960, especially Chapters 13 and 11.

41. Pauling, L., "A Molecular Theory of General Anesthesia," *Science* **134** (1961), 15–21.

42. Pauling L., R. B. Corey, and H. R. Branson, "The Structure of Proteins: Two Hydrogen-Bonded Helical Configurations of the Polypeptide Chain," *Proc. Nat. Acad. Sci. U.S.A.* **37** (1951), 205–211.

43. Pauling, L., and R. E. Marsh, "The Structure of Chlorine Hydrate," *Proc. Nat. Acad. Sci. U.S.A.* **38** (1952), 112–118.

44. Perutz, M. F., M. G. Rossman, A. F. Cullis, H. Muirhead, G. Will, and A. T. C. North, "Structure of Haemoglobin," *Nature* **185** (1960), 416–422.

45. Riordan, J., *An Introduction to Combinatorial Analysis*, John Wiley and Sons, New York, 1958.

46. Robertson, J. M., *Organic Crystals and Molecules*, Cornell University Press, Ithaca, New York, 1953.

47. Sayre, D., "The Squaring Method: A New Method for Phase Determination," *Acta Cryst.* **5** (1952), 60–65.

48. Sparks, R. A., American Crystallographic Association Meeting, Boulder, Colorado, 1961.

49. Taylor, C. A., and H. Lipson, "Optical Methods in Crystal-Structure Determination," *Nature* **167** (1951), 809–810.

50. Waser, J., and V. Schomaker, "The Fourier Inversion of Diffraction Data," *Rev. Mod. Phys.* **25** (1953), 671–690.

51. Wells, A. F., *Structural Inorganic Chemistry*, 3rd edition, Oxford University Press, New York, 1962, especially Chapter IV.

52. Wheatley, P. J., *The Determination of Molecular Structure*, Oxford University Press, New York, 1959.

53. Wrinch, D. M., "Geometry of Discrete Vector Maps," *Phil. Mag.* **27** (1939), 98–122.

54. Wrinch, D., *Fourier Transforms and Structure Factors*, Monograph 2, American Society for X-Ray and Electron Diffraction, Murray Printing Company, Cambridge, Mass., 1946.

55. Woolfson, M. M., *Direct Methods in Crystallography*, Oxford University Press, New York, 1961.

56. Woolfson, M. M., and R. A. Sparks, *Computer Determination of Crystal Structures*, Mathematics and Applications Department, International Business Machines Corporation, New York, 1961.

57. Zachariasen, W. H., *Theory of X-Ray Diffraction in Crystals*, John Wiley and Sons, New York, 1945.

58. Zamorzaev, A. M., Thesis, Leningrad, 1953.

CHAPTER 17

Combinatorial Principles in Genetics

GEORGE GAMOW

17.1 INTRODUCTION

A living cell always consists of two essential parts: (1) a *nucleus* containing *chromosomes*, which are the carriers of hereditary information; (2) *cytoplasm* forming the main body of the cell and consisting essentially of *enzymes*, which catalyze various kinds of biochemical reactions necessary for the development and survival of the organism. Comparing a living cell with a factory, we can say that its nucleus is the manager's office, the chromosomes serving as file cabinets for the storage of blueprints and production plans. Enzymes, which are also called *hormones* when they are secreted into the bloodstream, are the workers and the tools they use, carrying out various tasks according to the instructions obtained from the manager's office.

17.2 AMINO-ACID SEQUENCES IN PROTEINS

Chemically, enzymes and most of the hormones are long *protein* molecules. The study of proteins is well progressed today, and we know that these are linear sequences of comparatively simple molecules of various *amino acids* held together by so-called *peptide bonds*. Under the action of heat and mild acids, proteins break up into individual amino acids, which can be separated and their relative amounts measured.

Although an organic chemist can produce thousands of different amino acids simply by attaching a hydrogen atom, an amino group ($—NH_2$), and a hydroxyl group ($—COOH$) to *any* molecule, the breakup of protein molecules leads to *only twenty* different amino acids. These are called

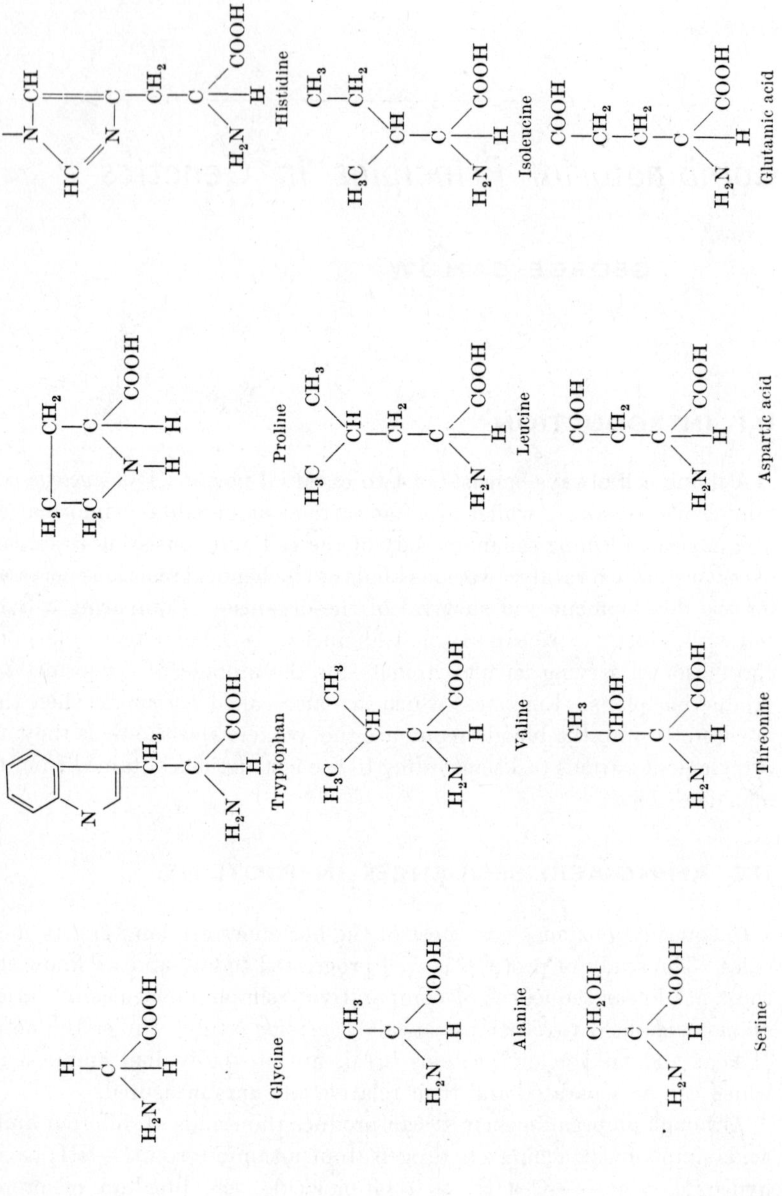

Histidine

Isoleucine

Glutamic acid

Proline

Leucine

Aspartic acid

Tryptophan

Valine

Threonine

Glycine

Alanine

Serine

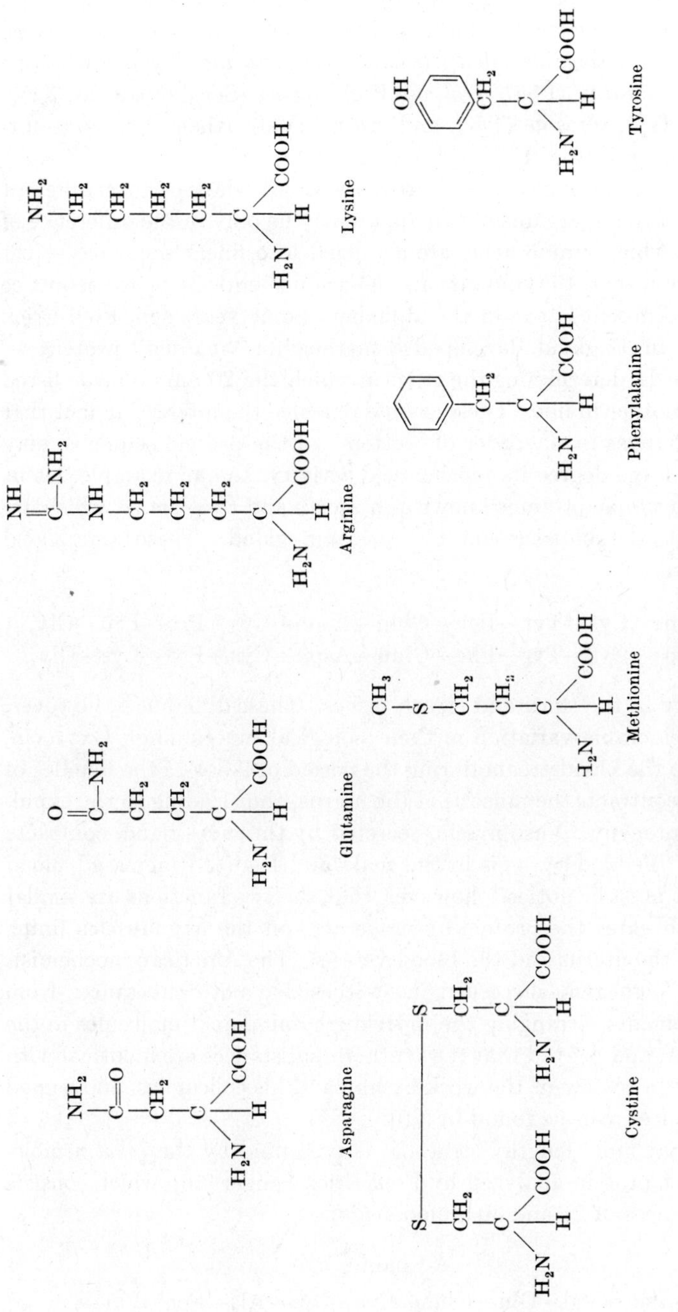

Fig. 17.1 The amino acids of proteins.

alanine (Ala), *arginine* (Arg), *asparagine* (Aspn), *aspartic acid* (Asp), *cysteine* (Cys), *glutamine* (Glun), *glutamic acid* (Glu), *glycine* (Gly), *histidine* (His), *isoleucine* (Ileu), *leucine* (Leu), *lysine* (Lys), *methionine* (Met), *phenylalanine* (Phe), *proline* (Pro), *serine* (Ser), *threonine* (Thr), *tryptophan* (Try), *tyrosine* (Tyr), and *valine* (Val); their structural formulas are shown in Fig. 17.1.

The differences among various protein molecules, which perform different functions in a living organism, can apparently be only in the difference of the *order* in which amino acids are arranged into linear sequences—just as, in a written text, the contents are determined entirely by the sequence of about 20 different letters of the alphabet. Some years ago, Fred Erick Sanger [16] in England developed a method for "reading" protein sequences, that is, determining the order in which the 20 amino acids listed follow one another in line. These studies revealed the interesting fact that just a few changes in the order of "letters" of the protein sequence may change to a large degree its biochemical activity. As an example, let us consider two simple proteins known as *oxytocin* and *vasopressin*, which are secreted into the bloodstream by a certain gland. These amino-acid sequences are

Oxytocin: Cys—Tyr—Ileu—Glun—Aspn—Cys—Pro—Leu—Gly,
Vasopressin: Cys—Tyr—Phe—Glun—Aspn—Cys—Pro—Lys—Gly,

differing only in the third and eighth places. These differences, however, cause a considerable variation in their biological functioning. Oxytocin, secreted into the bloodstream during the period of labor in the females of the species, contracts the muscles of the uterus, thus leading to the expulsion of the progeny. Vasopressin, secreted by the same gland, contracts the walls of the blood vessels in the male and leads to increased blood pressure. It must be noticed, however, that the two functions are similar since in both cases the protein hormone acts on the flat muscles lining the walls of the uterus and the blood vessels. The American biochemist, Vincent du Vigneaud, succeeded in synthesizing both substances from chemical elements, arranging the individual amino-acid molecules in the desired order, and proved that the synthetic substances are identical with natural ones; a review of the work by him and his colleagues, and a good list of references, can be found in [20].

A somewhat more lengthy structure is presented by the *insulin* molecule, the first protein analyzed by Fred Erick Sanger [16], which consists of two sequences of 21 and 30 amino acids:

A-Chain

Gly—Ileu—Val—Glu—Glun—Cys—Cys—Ala—Ser—Val—
—Cys—Ser—Leu—Tyr—Glun—Leu—Glu—Aspn—Tyr—Cys—Aspn

B-Chain

Phe—Val—Aspn—Glun—His—Leu—Cys—Gly—Ser—His—
—Leu—Val—Glu—Ala—Leu—Tyr—Leu—Val—Cys—Gly—
—Glu—Arg—Gly—Phe—Phe—Tyr—Thr—Pro—Lys—Ala

There is a cystine linkage (—S—S—) joining the cysteine pair (A-6, A-11), and also joining the cysteine pair (A-7, B-7) and the cysteine pair (A-20, B-19).

It is interesting to note that, although many amino acids in the sequence are essential for correct biological functioning of protein, others are apparently unimportant. Thus the insulins extracted from cattle, sheep, and hogs differ from one another by three adjacent amino acids in the middle of the sequence, and nevertheless insulins obtained from these animals are equally effective when introduced into a human patient. On the other hand, a single change in the essential part of the sequence may make the insulin completely ineffective.

The study of protein sequences is going ahead with giant strides. Recently, for example, Wendell M. Stanley et al. [19] in Berkeley determined a sequence of 158 amino acids in the tobacco mosaic virus.

The early stages of the work on protein decoding are described in [6], and more recent developments in [2]. These two articles contain 65 and 70 references, respectively, and in them the reader should be able to find most of the facts and references he might want on the subject.

17.3 DOUBLE-STRANDED SEQUENCES IN DNA

As was stated earlier, protein sequences in various enzymes and hormones are determined by the genetic material contained in chromosomes. This material is known as *nucleic acid*. Nucleic-acid molecules separated from cellular nuclei appear on the electromicrograms as long threads; these, however, are considerably thicker than those of the protein molecules. Chemical analysis indicates that these long nucleic-acid molecules represent sequences of comparatively simple atomic groups, which are known as *nucleotides*. In contrast to proteins, however, there are *only four* different types of nucleotides: *adenine* (Ad), *thymine* (Th), *guanine* (Gu), and *cytosine* (Cy). See Fig. 17.2. [Thymine, as we shall see later in this section, can be replaced by a slightly modified substitute structure known as *uracil* (Ur).] Each nucleotide consists of (1) the so-called *base*, which can appear in four different types as listed; (2) a sugar molecule, *deoxyribose*; and (3) a *phosphate group*, which, attaching itself to the sugar molecule in the next nucleotide, secures the formation of the nucleotide chain. The structural formula of a nucleic-acid sequence is shown in Fig. 17.3.

A further point is that in the nucleic acids forming the chromosomes we have actually a double-stranded chain in which the bases of one strand are attached to the bases of another by *double or triple hydrogen bonds*. Inspecting the chemical structure of the four bases, one finds that such double hydrogen bonding can exist only between adenine and thymine, and triple only between guanine and cytosine. Thus the sequence of

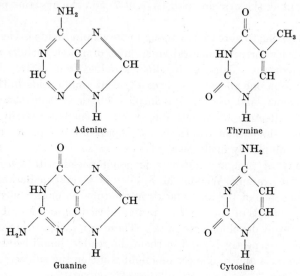

Fig. 17.2 The bases of the four nucleotides.

nucleotides in one of the two strands completely determines [22, pp. 174, 175] the sequence in the other, as in the following example (see Fig. 17.3):

$$—Cy—Ad—Ad—Th—Cy—Gu—\cdots$$
$$—Gu—Th—Th—Ad—Gu—Cy—\cdots$$

This double-strandedness of chromosomal nucleic acid, which is also known as DNA (for deoxyribonucleic acid), is essential for the process of replication. When a cell is preparing to divide, all DNA molecules in its nucleus apparently split into two single strands, and each of these strands is later regenerated into a double strand by catching the free nucleotides that are apparently present in large quantities in the surrounding medium. Thus one obtains two double-stranded DNA molecules, each of them being identical with the original one. If, as sometimes happens, a mistake occurs in the replication process, one or possibly both of the newly formed DNA molecules will differ in one or more nucleotides in the sequence, which leads to the changes in the progeny known as *mutations*.

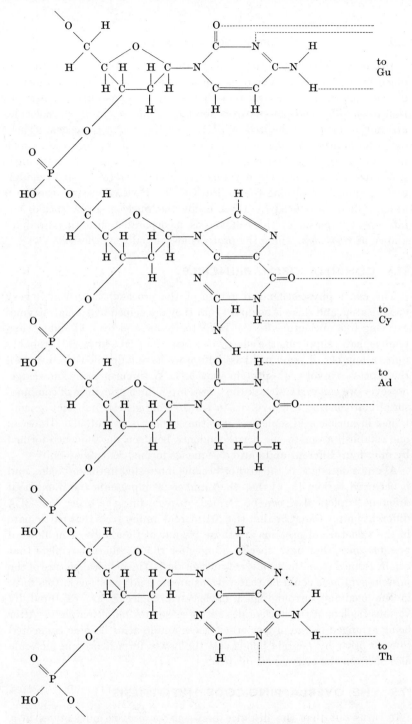

Fig. 17.3 DNA structure. The dashed lines indicate double or triple H-bonds between the bases.

In our example of a factory, the workers do not crowd into the manager's office for instructions; nor does the manager walk through the factory giving individual instructions to each worker. This passing on of information is the job of people known as foremen. Similarly, protein molecules are not directly synthesized by DNA in the cellular nucleus. There exists an intermediate agent known as RNA (for ribonucleic acid), which differs from DNA in three respects: (1) it is single-stranded; (2) its sugar molecules (ribose) carry one extra oxygen atom; and (3) by slight modification, namely, replacing the radical CH_3 by H, thymine is turned into uracil. Being produced by DNA inside the nucleus, RNA sneaks out into the cytoplasm of the cell and is incorporated into tiny granules known as *ribosomes*, where the proteins are actually synthesized.

17.4 COMBINATORIAL PRINCIPLES

The exact physicochemical nature of the processes described is very complicated, and there is no doubt that they are helped by special enzymes carrying free nucleotides to the RNA template, and by ATP (adenosine-triphosphate) supplying the energy necessary for that synthesis. Probably quite a number of years will pass before we have a detailed picture of all the processes involved in protein synthesis. With our present knowledge, however, we can certainly ask the question: Which principles of combinatorial mathematics are involved in transferring genetic information contained in nucleic acid sequences into the sequences of proteins? How can one establish a one-to-one correspondence between the sequence formed by only four different units into a sequence formed by twenty units?

About a decade ago, the author became interested in this problem, and it occurred to him [3, 4] that the number 20 represents the number of different triplets that can be formed, disregarding the order, out of 4 different units. Can it be that the 20 different amino acids that participate in the structure of proteins are those 20, out of thousands or millions of possible ones, that have specific affinity for the 20 different triplets that can be formed from the 4 different nucleotides?. The physical picture of the protein synthesis could be that various amino acids existing in free form in the medium surrounding a nucleic-acid molecule are captured by various triplets of the nucleotides acting as some kind of template. After being arranged in the proper order, these amino acids become connected to each other by peptide bonds and the newly formed protein molecule files off from the original template.

17.5 THE OVERLAPPING-CODE HYPOTHESIS

It turns out that the distance between consecutive nucleotides in a polynucleotide molecule is equal to the distance between consecutive amino acids in the molecule of a protein. Thus, if the nucleotides are arranged along a straight line, the captured amino acids will be separated from one another by distances three times as large as necessary for the formation of peptide bonds and will not stick together (see Fig. 17.4).

— Ad — Ad — Th — Gu — Th — Cy —

Val   Pro

Fig. 17.4 The nonoverlapping-code hypothesis.

It is natural to assume, as a first possibility, that we are dealing here with an *overlapping code*, in which the amino acids incorporated into the template share the two neighboring nucleotides, as shown in Fig. 17.5.

If this were the case, one would expect that the neighboring amino acids should show a certain correlation as a result of their having two

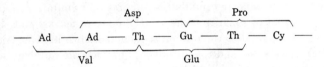

Fig. 17.5 The overlapping-code hypothesis.

nucleotides in common. Thus, if one of the amino acids in the sequence is determined by three Ad's, then each of its nearest neighbors must necessarily be determined by two Ad's and a third nucleotide (Ad Ad Ad, Ad Ad Th, Ad Ad Gu, or Ad Ad Cy). The question of whether or not such a correlation between the neighboring amino acids does really exist could be investigated in either a deterministic or a stochastic way.

The number of possible assignments of 20 amino acids to 20 nucleotide triplets is $20! = 3 \cdot 10^{17}$, which is equal to the age of our Universe expressed in seconds! Thus, the blind trial-and-error method of assignment is, of course, out of the question. As is true in any decoding operation, however, one can proceed in a more ingenious way by excluding various possibilities in specially selected stretches of protein sequences. Using this method, the author and his colleague Martynas Yčas [7] were able to find several stretches in the insulin sequence that definitely could *not* be interpreted in terms of the overlapping-code system for *any* possible assignment of amino acids to nucleotide triplets.

17.6 STATISTICAL INVESTIGATIONS

Another method of attacking the foregoing question is based on the statistics of various neighboring amino-acid pairs in the protein sequences.

To do this, we write an amino-acid sequence, for example,

$$\text{Ala—Val—Ser—Glu—Pro—Ala} \cdots,$$

and then writes the pairs which here are

$$(\text{Ala—Val}) \quad (\text{Val—Ser}) \quad (\text{Ser—Glu}) \quad (\text{Glu—Pro}) \quad (\text{Pro—Ala}) \cdots.$$

These pairs are marked in a 20×20 grid (400 squares), with the names of different amino acids written along the two axes. We now count the number of squares that remain unmarked, those that have only one mark, those with two marks, etc.

If the distribution is random, the result must agree with the well-known Poisson formula,

$$P(r) = N \frac{e^{-r}(\bar{n})^r}{r!},$$

where $P(r)$ is the expected number of squares that contain r marks, N is the total number of squares, n the total number of marks, and $\bar{n} = n/N$. Applying this method to the protein sequences known in 1956, Alexander Rich, Martynas Ycas, and the author [6] obtained the numbers shown in the second column of Table 17.1.

TABLE 17.1

Comparison of Frequency of Neighboring Amino-Acid Pairs in Protein Sequences, and of Letter Pairs in English and in Russian Texts, with Random (Poisson) Distribution.

Number of marks per square	Known protein sequence	Poisson distribution	*Paradise Lost*	*Dark Eyes*
0	264	264	305	361
1	103	116	55	32
2	27	26	23	4
3	4	4	7	2
4	2	0.4	3	1
5	0	0.04	3	0
6	0	0.003	2	0
7	0	0.0002	0	0
8	0	0.00001	2	0

The results agree extremely well with the expected random distribution shown in the third column. More recent studies, based on larger numbers of known protein sequences, make this agreement even better.

It is interesting to use the same method for analyzing the sequences of letters of the alphabet in an English text. In the fourth column of Table 17.1 are shown the results (normalized to 400 squares) obtained in the analysis of the opening lines of Milton's *Paradise Lost*. One observes here a considerable deviation from the Poisson formula, which is because in the English language certain letters have a strong tendency to pair off, as in the case of th, ck, ch, sh, etc.

The fifth column in Table 17.1 shows a similar analysis for the Russian song *Ochi Chornie* (*Dark Eyes*). Since Russian is more phonetic than English, the number of preferred letter pairs decreases, and the figures agree better with the Poisson distribution. No matter what the letters do in different human languages, however, the "protein language" behaves strictly in accordance with the random formula of Poisson. This proves definitely that the genetic code is *not* of the overlapping type.

17.7 GEOMETRIC IMPLICATIONS

Thus geometrically it is necessary to assume that, in the process of protein synthesis, the RNA molecule is probably coiled into the form of a solenoid, bringing the bases closer together. This would also explain why the order of the nucleotides in a triplet is nonessential, since indeed in this case the bases may form an equilateral triangle (Fig. 17.6).

An essential problem in the understanding of the process of protein synthesis is the question of how triplets are selected from a continuous sequence of bases. In fact, the same sequence can be read in three different ways:

```
..., Ad   Ad   Gu,  Cy   Gu   Ur,  Ad   Gu   Cy,  ...,
..., Ad,  Ad   Gu   Cy,  Gu   Ur   Ad,  Gu   Cy   ...,
..., Ad   Ad,  Gu   Cy   Gu,  Ur   Ad   Gu,  Cy   ...,
```

depending on how one puts the "commas" indicating which of the members of the sequence form the triplets of the template. Apparently one of the three readings must be correct, while the other two are incorrect. This problem was recently attacked by F. H. C. Crick et al. [1], who studied mutations in the bacteriophage known as T-4. The mutations may occur either by inserting one additional nucleotide into the RNA sequence (+) or by deleting one (−). It was found through complicated studies of mutant progeny that one single (+) or (−) mutation within a certain gene makes it inactive. This can be interpreted by saying that one single inserted or deleted member in the sequence makes the reading "nonsensical," by shifting all the commas beyond the mutant point one step to the right or to the left. The same is true for two (+) or two (−) mutations, which shift the commas by two steps. On the other hand, a pair of neigh-boring $(+)$ and $(-)$ mutations produces much smaller effect on the progeny, since in this case the commas are again in the same places beyond the second mutant point. Thus one has to assume that within each gene (i.e., a definite sequence of nucleotides) the reading is done simply by starting with the first triplet on one side of the sequence. This

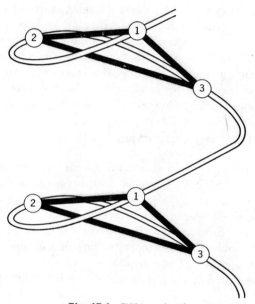

Fig. 17.6 RNA molecule.

would mean that, in the process of protein synthesis, various amino-acid molecules are not incorporated into the template simultaneously or at random at its different parts, but that the process progresses in good order through the amino acids, one by one, starting from one end of the gene.

17.8 STATISTICAL ASSIGNMENTS

Since in the protein synthesis we apparently deal with a nonoverlapping code, the assignment of 20 triplets to 20 amino acids cannot be done as originally attempted by looking for a single asignment that would not contradict the existing sequences. In fact, the impossibility of finding such a single assignment constitutes a proof that the protein code is not overlapping. Apparently the only possible method of finding that assign-ment is based on statistical considerations. Call a, b, c, and d the relative amounts of the four different nucleotides in the cell of a certain species $(a + b + c + d = 1)$. Then, using $(aw + bx + cy + dz)^3$ as generating function (see Chapter 3), we can expect that the relative amount of various triplets would be proportional to a^3, $3a^2b$, $6abc$, $3b^2c$, etc. Comparing these quantities with the directly measured percentages of various amino acids in the cell protein of the same species, we can hope to find which amino acid is determined by any given triplet. The easiest case to study would involve the relative amounts of the four bases in the nuclei of the glands that produce some simple protein, such as for example the silk glands. Silk, like all other "structural" proteins, shows a very simple amino-acid sequence consisting of monotonously repeating Ala and Gly. It could therefore be expected that the nucleotide composition of silk-gland RNA contains only those bases that determine the above two amino acids, or that at least this composition shows a marked difference from other cells in the organism. The unpublished experiment of Martynas Yčas, who analyzed nucleic acid extracted from the glands of caterpillars common in central Africa, led however to a negative result. This negative result is probably due to the fact that living cells contain a large number of different nucleic acids and enzymes and that silk-producing instructions occupy only a very short section of one of the nucleic-acid molecules. Thus the fact that a given cell *does* produce silk is not reflected in an essential way in the over-all RNA composition.

A much more promising method is to apply the foregoing principle to the viruses, which are much simpler living beings than cells. In fact, virus particles are believed to consist of single-type nucleic-acid molecules and single-type proteins. Unfortunately, the number of different viruses for which the compositions of both nucleic acid and proteins are known is very limited. Martynas Yčas and the author [7] tried to analyze the data existing for four viruses: tobacco mosaic, turnip yellow, tomato bushy stunt, and influenza. Although some guesses concerning the correlation between amino acids and nucleotide triplets could be made, the existing material does not represent a sufficiently large statistical sample to arrive at definite conclusions.

The most recent progress in this direction was made by M. W. Nirenberg and J. H. Matthaei [14], who have announced that by adding polyuridylic acid (that is, an RNA of which the bases are all uracils) to a cell-free system that can synthesize proteins, they succeeded in producing polyphenylalanine (that is, a protein formed entirely of phenylalanine molecules). This discovery implies that the phenylalanine amino acid is defined by a triplet of uracil nucleotides. It can be hoped that further studies in this direction will soon lead to a complete understanding of the relation between nucleotide triplets and amino acids.

17.9 MONTE CARLO METHODS

Another method of studying the distribution of amino acids in protein molecules was proposed some years ago by Martynas Yčas and the author [7]. This method does not require detailed knowledge of amino-acid

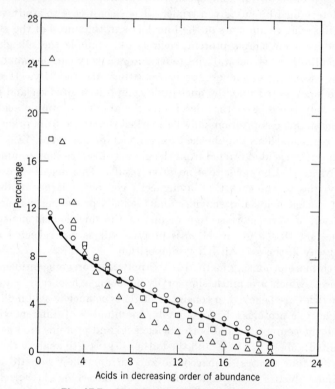

Fig. 17.7 Abundance curves of 20 amino acids.
Black circles: observed in 22 proteins.
Squares: von Neumann's formula.
Triangles: Monte Carlo method.
Open circles: triplets based on 7 RNA's.

sequences—which is available only in comparatively few cases—and can be used whenever the relative amounts of different amino acids in the protein molecules are known.

If we arrange the amino acids of each protein in the order of decreasing abundances and take the mean values of the most abundant, second most abundant, etc., amino acids, regardless of their identity, we obtain a curve shown by black circles in Fig. 17.7. In order to see whether this curve corresponds to a random distribution, we compare it with a mathematical model obtained in the following way. A segment of unit length is divided at random into 20 sections, and the lengths of the longest, second longest, etc., segments are averaged over a large number of such divisions. What will be the average lengths of the longest segment, of the second longest segment, ..., of the shortest segment? This problem was solved analytically, especially for this purpose, by J. von Neumann.[*] His argument runs as follows: Consider a unit length randomly divided into n parts so that the lengths of individual sections, as they follow from left to right, are $x_1, x_2, x_3, \ldots, x_n$. The values x_i are subjected to the conditions

$$\sum_{i=1}^{n} x_i = 1, \qquad 0 < x_i < 1.$$

Now let us define n numbers y_j as

$$y_1 = \text{smallest of all } x\text{'s},$$
$$y_2 = \text{2nd smallest of all } x\text{'s},$$
$$\cdot \quad \cdot \quad \cdot \quad \cdot \quad \cdot \quad \cdot \quad \cdot \quad \cdot$$
$$y_n = \text{largest of all } x\text{'s}.$$

The values y_j are subject to the conditions

$$\sum_{j=1}^{n} y_j = 1, \qquad 0 < y_1 \leq y_2 \leq \cdots \leq y_n < 1.$$

Considering the problem in n-dimensional space, we have a statistical weight proportional to

$$\begin{aligned}
d\pi &= dy_1\, dy_2 \cdots dy_{n-2}\, dy_{n-1} \\
&= dy_1\, dy_2 \cdots dy_{n-2}\, dy_n \\
&= \cdots \cdots \cdots \\
&= dy_1\, dy_3 \cdots dy_{n-1}\, dy_n \\
&= dy_2\, dy_3 \cdots dy_{n-1}\, dy_n.
\end{aligned}$$

The problem is to find the mean values of $y_1, y_2, \ldots, y_n$ for all possible divisions of the unit length. Put

$$z_j = y_j - y_{j-1}, \qquad j = 1, 2, \ldots, n,$$

where $y_0 = 0$.

Then, clearly,

$$y_j = \sum_{k=1}^{k=j} z_k,$$

and the restricting conditions on the z_k become

$$nz_1 + (n-1)z_2 + \cdots + z_n = 1, \qquad z_k > 0.$$

* Private communication.

The statistical weight is now proportional to

$$
\begin{aligned}
d\lambda &= dz_1 \, dz_2 \cdots dz_{n-2} \, dz_{n-1} \\
&= dz_1 \, dz_2 \cdots dz_{n-2} \, dz_n \\
&= \;.\quad .\quad .\quad .\quad .\quad .\quad . \\
&= dz_1 \, dz_3 \cdots dz_{n-1} \, dz_n \\
&= dz_2 \, dz_3 \cdots dz_{n-1} \, dz_n .
\end{aligned}
$$

Put

$$
w_k = (n + 1 - k)z_k, \qquad k = 1, 2, \ldots, n.
$$

Then the restrictions become

$$
\sum_{k=1}^{n} w_k = 1, \qquad w_k > 0,
$$

and the statistical weight is proportional to

$$
\begin{aligned}
d\mu &= \frac{1}{n!} \, dw_1 \, dw_2 \cdots dw_{n-2} \, dw_{n-1} \\[1mm]
&= \frac{2}{n!} \, dw_1 \, dw_2 \cdots dw_{n-2} \, dw_n \\[1mm]
&= \;.\quad .\quad .\quad .\quad .\quad .\quad .\quad . \\[1mm]
&= \frac{n-1}{n!} \, dw_1 \, dw_3 \cdots dw_{n-1} \, dw_n \\[1mm]
&= \frac{n}{n!} \, dw_2 \, dw_3 \cdots dw_{n-1} \, dw_n .
\end{aligned}
$$

Because of the symmetry of the restricting conditions in w-space, we have

$$
\bar{w}_1 = \bar{w}_2 = \cdots = \bar{w}_n,
$$

and, since

$$
\sum_{k=1}^{n} w_k = 1,
$$

it follows that

$$
\bar{w}_1 = \bar{w}_2 = \cdots = \bar{w}_n = \frac{1}{n}.
$$

Thus we obtain

$$
\bar{z}_k = \frac{\bar{w}_k}{n + 1 - k} = \frac{1}{n} \cdot \frac{1}{n + 1 - k},
$$

and therefore

$$
\begin{aligned}
\bar{y}_j &= \sum_{k=1}^{j} \bar{z}_k = \sum_{k=1}^{j} \frac{1}{n} \cdot \frac{1}{n + 1 - k} \\[1mm]
&= \frac{1}{n}\left(\frac{1}{n} + \frac{1}{n-1} + \cdots + \frac{1}{n - j + 1} \right),
\end{aligned}
$$

which is the desired result.

The results obtained from this formula for $n = 20$ are indicated by squares in Fig. 17.7, and it can be seen that they deviate quite considerably from the empirical data (black circles).

A possible explanation of the discrepancy may lie in the nonrandomness introduced by the translation procedure from the intrinsically random distribution of nucleotides of RNA molecules into the amino-acid sequences of proteins. In order to test this possibility, it became necessary to study, by the Monte Carlo method, the protein sequences formed from a random distribution of nucleotides by use of a definite triplet code. Let us denote the relative amounts of the four different nucleotides by A, B, C, D, with the normalization condition

$$A + B + C + D = 1.$$

and the relative amounts of the various amino acids by

$$\alpha_1, \alpha_2, \alpha_3, \ldots, \alpha_{20},$$

with the normalization condition

$$\sum_{i=1}^{20} \alpha_i = 1.$$

Furthermore, let us introduce the translation table

Amino acid	Nucleotide triplet
α_1	AAA
α_2	AAB
. . .	. . . .
. . .	. . . .
α_{20}	DDD

Now we take a long random sequence of nucleotides, as, for example,

$$CDA \quad DCB \quad BAB \quad AAB \ldots,$$

and, using the translation table, we turn it into a sequence of amino acids. We now apply to that sequence the already-described procedure for calculating the most abundant amino acid, the second most abundant, etc., and average them over a large number of runs in order to get a curve comparable to the empirical one. The result of 3000 runs of that kind, carried out by G. Fermi and N. Metropolis [5] on the Los Alamos electronic computer, is also plotted (triangles) in Fig. 17.7; it deviates from the observed distribution (black circles) even more than the results given by von Neumann's formula.

As a last resort, we took the actual percentages of nucleotides in seven known RNA molecules and calculated the frequencies of different triplets on that basis. The result, shown by open circles in Fig. 17.7, is in very good agreement with the observed curve, indicating that the deviation from randomness in amino-acid sequences of proteins is the result of the non-randomness in the distribution of four nucleotides in RNA molecules. The degree of that nonrandomness and its relation to biological evolution, caused by mutations and the survival of the fittest, represent an interesting subject for further studies.

17.10 EXPERIMENTAL RESULTS*

While the problem of protein coding by nucleic acids was under vigorous theoretical attack, which was rather difficult because of the scarcity of empirical data, experimental biochemists were shaking their test tubes to get the answer in a purely empirical way. The author of this chapter believes that, at least to a certain extent, the increased interest of biochemists in the problem of the relation between nucleic acid and protein sequences was prompted by his theoretical studies. It is easier for a biochemist, armed with all kinds of chemical equipment, to study biological structures, than it is for an interested theoretician having at his disposal only a pencil, paper, and perhaps a computing machine.

Thus in 1961, eight years after the appearance of the author's first article suggesting that the sequence of amino acids in protein molecules is determined by the sequence of base triplets in the molecules of nucleic acid, there appeared two papers, one by Severo Ochoa et al. from the New York University School of Medicine [12], and another by Marshall W. Nirenberg et al. from the National Institute of Metabolic Diseases [14], who by a long series of experiments (see the list of references at the end of this chapter) established the correlation between amino acids and protein sequences. Their findings completely sustained the original triplet hypothesis. The relations found experimentally by them are shown in Table 17.2.

Inspecting this table, we notice that although in general it follows the original triplet hypothesis, it shows a number of irregularities. For example, Ala and Arg are both determined by the same triplet UCG. On the other hand, some amino acids (Asp and Thr) can be determined by two different triplets, and Leu may have three different origins. Also,

* Added in proof, June 1964.

the triplets AAA, CCC, and GGG do not determine any amino acids. The possibility of these so-called substitutions was suspected in the previous theoretical work, but these irregularities so confused the issue that no purely theoretical relation could be found. Nature is tricky, but still the number of existing amino acids equals the number of triplets that can

TABLE 17.2
Correlation between Amino Acids and Base Triplets

Amino acid	Base triplet*	Amino acid	Base triplet*
Phe	UUU	Ileu	UUA
Ala	UCG	Leu	UUC; UUG; UUA
Arg	UCG	Lys	UAA
Asp	UAC	Met	UAG
Aspn	UAA; UAC	Pro	UCC
Cys	UUG	Ser	UUC
Glu	UAC	Thr	UAC; UCC
Glun	(probably) UCG	Try	UGG
Gly	UGG	Tyr	UUA
His	UAC	Val	UUG

* For convenience, the symbols Ad, Gu, Cy, and Ur are now replaced by A, G, C, and U, respectively.

be formed from four different units. Maybe sometime in the future we will understand why this is so.

MULTIPLE-CHOICE REVIEW PROBLEMS

1. Hereditary information is carried and replicated by

  (a) RNA, (b) DNA, (c) ATP, (d) RDX.

2. A protein molecule is composed of

  (a) nucleic acids, (b) amino acids,
  (c) benzene rings, (d) antibodies.

3. The number of various triplets formed from four different units, disregarding the order of units in each triplet, is

  (a) 12, (b) 20, (c) 36, (d) infinity.

4. In Problem 3, if the order within each triplet is important, the number of different possibilities is

  (a) 24, (b) 55, (c) 64, (d) 137.

5. The Poisson formula, as used in this chapter, gives

(a) the probability of getting singlets, doublets, triplets, etc., in a random distribution of n elements into m subsets,
(b) the relation between electric charge and potential,
(c) the number of primes less than N,
(d) a French recipe for cooking fish.

REFERENCES

1. Brenner, S., L. Barnett, F. H. C. Crick, and A. Orgel, "The Theory of Mutagenesis," *J. Mol. Biol.* **3** (1961), 121–124.

2. Crick, F. H. C., "The Recent Excitement in the Coding Problem," *Progress in Nucleic Acid Research*, Academic Press, New York, 1963.

3. Gamow, G., "Possible Relations between Deoxyribonucleic Acid and Protein Structures," *Nature* **173** (1954), 318.

4. Gamow, G., "Possible Mathematical Relation between Deoxyribonucleic Acid and Proteins," *Kgl. Danske Videnskab. Selskab. Biol. Medd.* **22**, 3 (1954), 1–13.

5. Gamow, G., and N. Metropolis, "Numerology of Polypeptide Chains," *Science* **120** (1954), 779–780.

6. Gamow, G., A. Rich, and M. Yčas, "The Problem of Information Transfer from the Nucleic Acids to Proteins," *Advances in Biological and Medical Physics* **4** (1956), Academic Press, 1956.

7. Gamow, G., and M. Yčas, "Statistical Correlation of Protein and Ribonucleic Acid Composition," *Proc. Nat. Acad. Sci. U.S.A.* **41** (1955), 1011–1019.

8. Jones, Oliver W., Jr., and Marshall W. Nirenberg, "Qualitative Survey of RNA Codewords," *Proc. Nat. Acad. Sci. U.S.A.* **48** (1962), 2115–2123.

9. Kaziro, Yoshito, Albert Grossman, and Severo Ochoa, "Identification of Peptides Synthesized by the Cell-Free *E. Coli* System with Polynucleotide Messengers," *Proc. Nat. Acad. Sci. U.S.A.* **50** (1963), 54–61.

10. Krakow, Joseph S., and Severo Ochoa, "Ribonucleic Acid Polymerase of Azotobacter Vinelandii, I. Priming by Polyribonucleotides," *Proc. Nat. Acad. Sci. U.S.A.* **49** (1963), 88–94.

11. Leder, Philip, Brian F. C. Clark, William S. Sly, Sidney Pestka, and Marshall W. Nirenberg, "Cell-Free Synthesis Dependent upon Synthetic Oligodeoxynucleotides," *Proc. Nat. Acad. Sci. U.S.A.* **50** (1963), 1135–1143.

12. Lengyel, Peter, Joseph F. Speyer, and Severo Ochoa, "Synthetic Polynucleotides and the Amino Acid Code, I–V," *Proc. Nat. Acad. Sci. U.S.A.* **47** (1961), 1936–1942; **48** (1962), 63–68, 282–284, 441–448, 613–616.

13. Matthaei, J. Heinrich, Oliver W. Jones, Robert G. Martin, and Marshall W. Nirenberg, "Characteristics and Composition of RNA Coding Units," *Proc. Nat. Acad. Sci. U.S.A.* **48** (1962), 666–677.

14. Nirenberg, M. W., and J. H. Matthaei, "The Dependence of Cell-free Protein Synthesis in *E. Coli* upon Naturally Occurring or Synthetic Polyribonucleotides," *Proc. Nat. Acad. Sci. U.S.A.* **47** (1961), 1588–1602.

15. Nirenberg, Marshall W., J. Heinrich Matthaei, and Oliver W. Jones, "An Intermediate in the Biosynthesis of Polyphenylalanine Directed by Synthetic Template RNA," *Proc. Nat. Acad. Sci. U.S.A.* **48** (1962), 104–109.

16. Sanger, F., "The Arrangement of Amino Acids in Proteins," *Advances in Protein Chemistry* **7** (1952), 1–67; Academic Press, New York, 1952.

17. Singer, Maxine F., Oliver W. Jones, and Marshall W. Nirenberg, "The Effect of Secondary Structure on the Template Activity of Polyribonucleotides," *Proc. Nat. Acad. Sci. U.S.A.* **49** (1963), 392–399.

18. Tsugita, A., H. Fraenkel-Conrat, Marshall W. Nirenberg, and J. Heinrich Matthaei, "Demonstration of the Messenger Role of Viral RNA," *Proc. Nat. Acad. Sci. U.S.A.* **48** (1962), 846–853.

19. Tsugita, A., D. T. Gish, J. Young, H. Fraenkel-Conrat, C. A. Knight, and W. M. Stanley, "The Complete Amino Acid Sequence of the Protein of Tobacco Mosaic Virus," *Proc. Nat. Acad. Sci. U.S.A.* **46** (1960), 1463–1469.

20. du Vigneaud, V., "Hormones of the Posterior Pituitary Gland: Oxytocin and Vasopressin," *Harvey Lectures* **50** (1954–55), 1–26.

21. Weissmann, Charles, Lionel Simon, and Severo Ochoa, "Induction by an RNA Phage of an Enzyme Catalyzing Incorporation of Ribonucleotides into Ribonucleic Acid," *Proc. Nat. Acad. Sci. U.S.A.* **49** (1963), 407–414.

22. White, Abraham, Philip Handler, and Emil L. Smith *Principles of Biochemistry* 3rd edition, McGraw-Hill Book Company, New York, 1964.

CHAPTER 18

Appendices*

HERMANN WEYL

MARSCHALLIN (*looking into her hand-mirror*):
Wie kann das wirklich sein,
dass ich die kleine Resi war
und dass ich auch einmal die alte Frau sein werd'?
. . . .
Wie macht denn das der liebe Gott?
Wo ich doch immer die gleiche bin.
Und wenn er's schon so machen muss,
warum lässt er mich denn zuschaun dabei
mit gar so klarem Sinn! Warum versteckt er's nicht vor mir?
Das alles ist geheim, so viel geheim . . .
H. VON HOFMANNSTHAL, *Der Rosenkavalier*, Act I.

18.1 ARS COMBINATORIA

1. Perhaps the philosophically most relevant feature of modern science is the emergence of abstract symbolic structures as the hard core of objectivity behind—as Eddington puts it—the colorful tale of the subjective storyteller mind. In Appendix A [of *PMNS*] we have discussed the structure of mathematics as such. The present appendix [Sec. 18.1] deals with some of the simplest structures imaginable, the combinatorics of aggregates and complexes. It is gratifying that this primitive piece of symbolic mathematics is so closely related to the philosophically important problems of individuation and probability, and that it accounts for some of the most fundamental phenomena in inorganic and organic nature.

* Reprinted from *Philosophy of Mathematics and Natural Science* (hereafter abbreviated *PMNS*) by the late Hermann Weyl, by permission of Princeton University Press. Copyright, 1949, by Princeton University Press; all rights reserved. The four sections 18.1 to 18.4 of this chapter consist, respectively, of Appendices B to E of *PMNS*.

The same structural viewpoint will govern our account of the foundations of quantum mechanics in Appendix C [Sec. 18.2]. In a widely different field J. von Neumann's and O. Morgenstern's recent attempt to found economics on a theory of games is characteristic of the same trend. The network of nerves joining the brain with the sense organs is a subject that by its very nature invites combinatorial investigation. Modern computing machines translate our insight into the combinatorial structure of mathematics into practice by mechanical and electronic devices.

It is in view of this general situation that we are now going to insert a few auxiliary combinatorial considerations of an elementary nature concerning aggregates of individuals. The reader should be warned beforehand that in their application to genetics the lines are drawn somewhat more sharply than the circumstances warrant. In the progress of science such elementary structures as roughly correspond to obvious facts are often later recognized as founded on structures of a deeper level, and in this reduction the limits of their validity are revealed. This hierarchy of structures will be illustrated in Appendix D [Sec. 18.3] by the theory of chemical valence.

An aggregate of white, red, and green balls may contain several white balls. Generally speaking, in a given aggregate there may occur several individuals, or *elements*, of the same *kind* (e.g. several white balls) or, as we shall also say, the same *entity* (e.g. the entity white ball) may occur in several *copies*. One has to distinguish between *quale* and *quid*, between equal (= of the same kind) and identical. To the question of individuation thus arising, Leibniz gave an *a priori* answer by his *principium identitatis indiscernibilium*. Physics has recently arrived at a precise and compelling empirical solution as far as the ultimate elementary particles, especially the photons and electrons, are concerned. Closely related is the question of the conservation of identity in time; the identical 'I' of my inner experience is the philosophically most significant instance.[1] Our decision as to what is to be considered as equal or different influences the counting of 'different' cases on which the determination of probabilities is based, and thus the problem of individuation touches the roots of the calculus of probability. It is through the combinatorial theory of aggregates that these things find their exact mathematical interpretation, and there is hardly another branch of knowledge where the relationship of idea and mathematics presents itself in a more transparent form.

The simplest combinatorial process is the *partition* of a set S of n elements into two complementary subsets $S_1 + S_2$. For the sake of identification

[1] The riddle of the identical ego, that is an onlooker at what is done to him and by him, is movingly expressed by the above lines from the *Rosenkavalier*.

and recording, we attach arbitrarily chosen distinct marks p to the elements. Only such relations and statements have objective significance as are not affected by any change in the choice of the labels p; this is the principle of relativity. Its abstract formulation reveals its triviality. An 'individual' subset S_1 is characterized by stating for each element, marked p, whether it is a member of S_1, $p \in S_1$, or of the complementary subset S_2. As the construction of S_1 thus depends on the decision of n alternatives ($p \in S_1$ or $p \in S_2$ for each of the n elements p) there are 2^n individually distinct possible subsets (including the vacuous null set as well as the total set S). However, this number is reduced to $n + 1$ if the n elements are considered as indiscernible. For then a subset S_1 is completely characterized by the number n_1 of its elements, a number that is capable of the $n + 1$ values $0, 1, \cdots, n$; and the partition $S = S_1 + S_2$ is characterized by the decomposition $n = n_1 + n_2$ of n into a sum of two terms (n_1, n_2 being the numbers of elements in S_1, S_2 respectively)[2]. One will ask how many individually distinct partitions $S = S_1 + S_2$ lead to the same 'visible' partition as characterized by the decomposition $n = n_1 + n_2$. The answer is $\dfrac{n!}{n_1! \, n_2!}$. Consequently the total number 2^n of all individual partitions must equal the sum

$$\sum \frac{n!}{n_1! n_2!}$$

extending over the $n + 1$ different decompositions $n = n_1 + n_2$:

$$(1) \qquad 2^n = \frac{n!}{0!n!} + \frac{n!}{1!(n-1)!} + \frac{n!}{2!(n-2)!} + \cdots + \frac{n!}{n!0!}.$$

This simplest case affords but little interest. Moving closer to reality, let us now assume that there is a certain respect in which elements may be equal ($\sim$) or different.[3] Balls may be white, red, or green; electrons may be in this or that position; animals in a zoo may be mammals or fish or birds or reptiles; atoms in a molecule may be H, He, Li, ... atoms. The universal expression for such 'equality in kind' is by means of a binary relation $\alpha \sim b$ satisfying the axioms of equivalence: $a \sim a$; if $a \sim b$ then $b \sim a$; if $a \sim b$, $b \sim c$ then $a \sim c$. Various words are in use to indicate equivalence, $a \sim b$, of two arbitrary elements a, b under a given equivalence relation $\sim$: a and b are said to be the same *kind* or *nature*, they are said to belong to the same *class*, or to be in the same *state*. An *aggregate* S is a set of elements each of which is in a definite state;

---

[2] n_1 and n_2 are understood to be natural numbers ranging over $0, 1, 2, \ldots$.

[3] A symbol for negation is no longer needed. It is therefore hoped that no confusion will result when from now on we use the sign $\sim$ for equivalence.

hence the term aggregate is used in the sense of 'set of elements with equivalence relation.' Let us assume that an element is capable of k distinct states $C_1, \ldots, C_k$. A definite *individual state* of the aggregate S is then given if it is known, for each of the n marks p, to which of the k classes the element marked p belongs. Thus there are k^n possible *individual states* of S. If, however, no artificial differences between elements are introduced by their labels p and merely the intrinsic differences of state are made use of, then the aggregate is completely characterized by assigning to each class C_i ($i = 1, \cdots, k$) the number n_i of elements of S that belong to C_i. These numbers, the sum of which equals n, describe what may conveniently be called the *visible or effective state* of the system S. Each individual state of the system is connected with an effective state, and any two individual states are connected with the same effective state if and only if one may be carried into the other by a permutation of the labels; here the principle of relativity finds expression in the postulate of invariance with respect to the group of all permutations. The number of different effective states equals that of the 'ordered' decompositions $n = n_1 + n_2 + \cdots + n_k$ of n into k summands n_i, a number for which one readily finds the value

$$(2) \qquad \frac{(n + 1) \cdots (n + k - 1)}{1 \cdots (k - 1)} = \frac{(n + k - 1)!}{n!(k - 1)!} .$$

Nor is it difficult to ascertain how many distinct individual states are connected with the same visible state and thus to explain the discrepancy between the two numbers k^n and (2), just as the equation (1) explains the discrepancy between the values 2^n on the left and the number $n + 1$ of terms in the sum on the right.

The number of individually distinct possible partitions of S into two complementary subsets S_1, S_2, or the number of individually distinct sub-aggregates S_1, has been found to be 2^n; but since the elements are now discernible according to their 'kind,' an effective sub-aggregate S_1 is fixed by assigning to each class C_i the number $n_i^{(1)}$ of elements with which that class is represented in S_1. Since $n_i^{(1)}$ is capable of the $n_i + 1$ values $0, 1, \cdots, n_i$ there are

$$(3) \qquad (n_1 + 1) \cdots (n_k + 1)$$

different possible effective partitions $S = S_1 + S_2$. The number (3) is therefore of necessity smaller than or equal to

$$2^n (n = n_1 + \cdots + n_k).$$

The maximum 2^n is attained if all n_i have the value 0 or 1, i.e. if no two elements of S are ever found in the same class. Indeed, that being the case, the elements of S may be completely characterized by the classes to which they belong (by their state or their 'nature') and there is no need then for an artificial differentiation by labels. In this case we speak of a *monomial* aggregate.

The process inverse to the partition of an aggregate S into two complementary sub-aggregates S_1, S_2 is the *union* of two given (disjoint) aggregates S_1, S_2 into a whole $S = S_1 + S_2$. The combinatorial theory of aggregates and of the mutually inverse operations of partition and union finds a particularly important application in *genetics*. The development of two organisms may run a different course, owing to 'external circumstances,' even if they are of the same genetic constitution (have the same germ plasm or are of the same genotype, in Weismann's and Johannsen's terminologies). This duality of constitution and environment, 'nature and nurture,' is basic for our interpretation of the facts of inheritance. It may be called an *a priori* conception like the somewhat similar duality of inertia and force in mechanics. Roughly, the environmental factors are characterized as being external to the organism, (relatively) variable and controllable, in contrast to the internal, given and (relatively) stable constitution. Constitution is often inferred and thus of hypothetical rather than manifest nature, as for instance the atoms that constitute a chemical compound. What belongs to the social environment of an individual may be a constitutive characteristic of the society in which he lives. As in the case of other fundamental conceptions, the precise meaning for each field gradually unfolds with a theory of the relevant phenomena: on the basis of a somewhat vague but natural interpretation one discovers certain laws that surprise by their exact form and are welded together into a theory; by holding on to these simple laws and interpreting the ever increasing array of detailed facts in the light of the theory one succeeds in making the original conception more and more precise. In this sense there is an overwhelming amount of empirical evidence in biology for the distinction of nature and nurture, although it never becomes a perfectly sharp one.

By breeding experiments one has succeeded in dissolving the genetic constitution into an aggregate of individual *genes* or 'points,' much as chemistry dissolves a molecule into an aggregate of atoms. And as an individual atom may be in one of the various states (may be one of the various 'entities atom') indicated by the chemical symbols H, He, Li, $\ldots$, so are the genetic points capable of different discrete states called alleles. In the act of fertilization (syngamy) two aggregates S_1, S_2, the 'gametes' (sperm and egg), are united into a 'zygote' or germ cell $S = S_1 + S_2$. A gamete is produced by an organism, S_1 by Ω_1, S_2 by Ω_2 (Ω_2 is not necessarily distinct from Ω_1, both may be the same self-fertilizing plant). All body cells of the organism Ω that develops from the germ cell S are, notwithstanding their functional differentiation, as far as their genetic constitution is concerned, replicas of its zygote S. Part of the body cells at a certain stage of the life-cycle undergo the inverse process of partition into two complementary subaggregates (maturation division or meiosis);[4] the organism Ω is therefore capable of producing as many constitutionally distinct gametes as there exist effective different sub-aggregates of S. This interplay of syngamy and meiosis, union and partition, explains the essential features of heredity: *constancy prevails in so far as the sum of two aggregates is uniquely determined by both parts, variability prevails inasmuch as partition of an aggregate into two complementary parts may be performed in various ways.*

An organism produces the gametes S' contained as sub-aggregates in its zygote S with certain relative frequencies (probabilities) $\gamma = \gamma(S')$. The probabilities will be influenced by external circumstances, in particular by temperature, and are thus, in contrast to the discrete aggregates, capable of continuous variation. But it is evident *a priori* that the complementary gamete S'' must occur with the same frequency as S', $\gamma(S') = \gamma(S'')$. (Even if, as is the case for a $\female$ organism, one of the two complementary parts S', S'' is eliminated after maturation division by degenerating into the polar body, one will hold on to the assumption $\gamma(S') = \gamma(S'')$ for the *a priori* probabilities.) It is plausible that the probability of syngamy between a gamete S' produced by Ω and a gamete S'_* produced by Ω_* is the same for the various kinds of gametes S' and S'_* that are produced by the two organisms Ω, Ω_*. The probability that the pair of parents Ω, Ω_* beget a child with the zygote $S' + S'_*$ is therefore presumed to be $\gamma(S') \cdot \gamma_*(S'_*)$.

Returning to the abstract theory, let us pass to a discussion of *temporal changes of state* of a given aggregate S. As long as elements are capable of discrete states only, we are forced to dissolve time also into a succession of discrete moments, $t = \ldots, -2, -1, 0, 1, 2, \ldots$. Transition of the system from its state at the time t into its state at time $t + 1$ will then be a jump-like mutation. With the n elements individualized by their labels p, the changing state of affairs will be described by giving the state $C(p; t)$ of the element p at the time t as a function of p and t. This 'individual' description, by means of the function $C(p; t)$, is to be supplemented by the principle of relativity according to which the association between the individuals and their identification marks p is a matter of arbitrary choice; but it is an association for all time, and once established it is not to be

---

[4] The actual process (a two-step process accompanied by the longitudinal splitting of each chromosome into two chromatids) is slightly more complicated than this its combinatorial result.

tampered with. If, on the other hand, at each moment attention is given to the visible state only, then the numbers $n_1(t), \ldots, n_k(t)$ in their dependence on t contain the complete picture—however incomplete this information is from the 'individualistic' standpoint. For now we are told only how many elements, namely $n_i(t)$, are found in the state C_i at any time t, but no clues are available whereby to follow up the identity of the n individuals through time; we do not know, nor is it proper to ask, whether an element that is now in the state, say C_5, was a moment before in the state C_2 or C_6. The world is created, as it were, anew at every moment, no bond of identity joins the beings present at this moment with those encountered in the next. This is a philosophical attitude towards the changing world taken by the early Islamic philosophers, the Mutakallimûn. This nonindividualizing description is applicable even if the total number $n_1(t) + \cdots + n_k(t) = n(t)$ of elements does not remain constant in time.

Wherever in reality identification of the same being at different times is carried out, it is of necessity based on the observable state. For a continuous flow of time and a continuous manifold of states, the underlying principle is by and large to be formulated as follows: suppose there exists at a time t but one individual in a certain state C appreciably different from the states of all other individuals; if afterwards, especially if shortly afterwards, at a time t', one and only one individual is encountered in a state C' deviating but little from C, or 'typically similar' to C, then the presumption is justified that one is dealing with the *same* individual at both moments t and t'. Instead of t and t' one may have a whole sequence of moments $t, t', t'' \ldots$. Think of following a wave moving over the surface of the water! Even in recognizing people, we are dependent on such means (the famous scenes of recognition in world literature, from the Odyssey on, come to mind) as long as the inner certitude of the identity of one's own ego and communications based thereon ("I am the same man who once met you then and there") are left out of play.

2. We saw that in speaking of different kinds of states or classes, reference is made to an underlying notion of equivalence. It is a frequent occurrence that classes of elements break up into sub-classes; we prefer to speak of genus and class, rather than of class and subclass.[5] Every class C belongs to a definite genus $G = [C]$, and an individual by being a member of the class C belongs also to the genus $[C]$. Thus the animals of a zoo are divided into mammals, fish, birds, etc., and the mammals again into monkeys, lions, tigers, etc. States may coincide in a certain character;

---

[5] This is the terminology used in number theory; a genus of quadratic forms is wider than a class. Biological taxonomy with its graded hierarchy of kingdom, class, order, family, genus, species, variety, favors the opposite usage.

this character then corresponds to a genus, and the state to a class, of elements. The division into genera and classes is based on a coarser and on a finer notion of equivalence: $a \sim b$ and $a \approx b$, where $a \sim b$ implies $a \approx b$. In different fields of knowledge this graded division appears under different terminological disguises. The aggregates of genetics are an example in point. The genes correspond to the genera, the alleles to the classes; a gene may have two or several alleles. The fact that an element p of the aggregate belongs to the class C and thereby to the genus $[C]$ is here expressed by the words 'the point p is occupied by the allele C of the gene $[C]$.'

I mention here a few of the names given in genetics to the basic notions of the combinatorial theory of aggregates, and describe the special circumstances "normally" prevailing in procreation. An individual aggregate S is known if for every one of its points p the class C_p is known to which p belongs; p then belongs also to the corresponding genus $[C_p]$. Two individual aggregates, S and S^*, are of the same *constitution*, $S \approx S^*$, if the labels p employed for the points of the first aggregate can be mapped in a one-to-one way upon the labels p^* employed for the points of S^*, $p \rightleftarrows p^*$, such that homologous points p and p^* in the two aggregates always belong to the same class (isomorphic mapping). According to the principle of relativity, aggregates of the same constitution are to be considered as indiscernible. Under given external circumstances the zygote S completely determines the phenotype, the visible development of an organism; the phenotype is necessarily the same for zygotes of the same constitution. An effective aggregate S is described by assigning to each class C the number n_C of the points of S in C; the number n_G of points in a genus G then equals the sum Σn_C extending over those classes C for which $[C] = G$. Individual aggregates are connected with the same effective aggregate if and only if they are isomorphic, i.e., of the same constitution. Two individual aggregates S and S^* are said to be of the same *species* σ if, with regard to a suitable one-to-one mapping $p \rightleftarrows p^*$, homologous points p and p^* in S and S^* always belong to the same *genus*.[6] Coincidence of the numbers n_G and n_G^* for all possible genera G is the necessary and sufficient condition for this to be the case; the numbers n_G therefore contain a complete description of the species σ of an aggregate. An aggregate S was

[6] This natural but purely combinatorial concept is related to but not identical with the meaning of the word species current in biology. There is no doubt that the latter, in spite of the difficulty of giving a precise definition, corresponds to a fundamental fact. As an example indicative of the wide gap between the two notions I quote Dobzhansky's "dynamic" definition (*Philosophy of Science*, 2, 1935, pp. 344–355): "Species is that stage of evolutionary process at which the once actually or potentially interbreeding array of forms becomes segregated in two or more separate arrays which are physiologically incapable of interbreeding."

called monomial if different points of S never belong to the same class; it is called *haploid* if different points of S never belong to the same genus, i.e., if for each genus G the number n_G equals 0 or 1. The corresponding species then deserves the name haploid. If S contains two points of different classes but of the same genus, then S is said to be heterozygous (*hybrid*); if it contains two points of different classes but of the same given genus G, it is heterozygous with respect to G. Union of two aggregates S_1, S_2 into a whole $S = S_1 + S_2$ and the inverse process of partition of S into S_1, S_2 may be called *balanced* in case the parts are of the same species. This is what normally happens in syngamy and meiosis. *Under balanced syngamy and meiosis, species remain constant throughout the sequence of generations.* Indeed, let S, S^* be two gametes of the same species σ that have united to form the zygote $S + S^*$; if the latter splits by balanced meiosis into S_1, S_2, then S_1, S_2 are necessarily of the same species σ as S, S^*. This remains true even if mutations are admitted by which a point may change its class but not its genus (point mutations). In particular, if the game of 'balanced' reproduction starts with two gametes of the same haploid species σ, then only haploid gametes of that species (and diploid zygotes) will turn up in the successive generations; this is the most common case that Gregor Mendel was dealing with. Assuming the zygote of a self-fertilizing organism Ω to be non-hybrid, all direct and indirect descendants of Ω will have the same genotype as Ω. Differences of phenotype in such a 'pure line,' if they occur, must be due to different external circumstances, and thus the invariable genetic constitution is most clearly separated from the variable environmental factors (W. Johannsen's experiments with beans, 1903).

3. In physics one aims at making division into classes so fine that no further refinement is possible; in other words, one aims at a *complete* description of state. Two individuals in the same 'complete state' are indiscernible by any intrinsic characters—although they may not be the same thing. Classical mechanics takes the state of a point of given mass (and charge) to be completely described by position and velocity, because by taking this view it reaches agreement with the principle of causality, which asserts that the (complete) state of a mass point at one moment determines its state at all times. The simplest example of a mass point is the linear oscillator; it oscillates on a definite line (thus requiring a space of one dimension only) and has a definite frequency ν (= number of oscillations per 2π seconds).[7] The possible states of an oscillator as specified by position and velocity form a two-dimensional continuous

---

[7] It is unfortunate that in English the word "frequency" is used in two entirely different senses—for the number of occurrences in a statistical ensemble (German "Häufigkeit") and for the number of oscillations.

manifold. According to quantum mechanics, however, it is capable of a discrete variety of different states only, specified by a number n assuming the values $0, 1, 2, \ldots$. In the state n the oscillator has the energy $n \cdot h\nu$ where $h = 1.042 \times 10^{-27}$ erg $\times$ sec is Planck's action quantum (the number now usually designated by a crossed $\hbar$). Radiation in a room, the walls of which are perfect mirrors ('Hohlraum'), is equivalent to a superposition of harmonic oscillations, each marked by an index α and having a definite frequency ν_α. Hence the Hohlraum radiation may be considered as an aggregate of linear oscillators endowed with certain proper frequencies ν_α. (The static part of the electromagnetic field is here disregarded.) According to the quantum mechanics of the individual oscillator, the complete state of our field of radiation therefore assigns an integer ν_α to each oscillator α; in this state the oscillator α has the energy $n_\alpha h\nu_\alpha$, and the sum $\Sigma_\alpha n_\alpha \cdot h\nu_\alpha$ extending over all oscillators is the total energy. In the language of photons one expresses this by saying that n_α photons in the state α and of energy $h\nu_\alpha$ are present. In the language of oscillators the index α specifies the individual oscillator and the integer n_α its state; whilst, in the language of photons, α designates the state of a photon and n_α the number of photons in that state. After translation into the photon language, radiation appears as a gas of photons.

Since the possible complete states of an individual form a discrete manifold in quantum mechanics, application of statistics consists here in a mere counting of states. Once the question of complete description of states is solved, all probabilities are evaluated by simple enumerations, and the problem of comparing probabilities in a continuous 'phase space' by measurement does not arise. Since photons come into being and disappear, are emitted and absorbed, they are individuals without identity. No specification beyond what was previously termed the effective state of an aggregate is therefore possible. Hence the state of a photon gas is known when for each possible state α of a photon the number n_α of photons in that state is given (Bose-Einstein statistics of radiation).

While one need not penetrate deep into the nature of light before it reveals, by such phenomena as diffraction, interference, etc., its undulatory character, its corpuscular features are more concealed. For the electrons the opposite is true; they openly show their corpuscular nature by hitting here or there, whereas their undulatory features were discovered by the experimentalists only simultaneously with the development of quantum theory. Yet it is clear that matter, like radiation, is to be represented by a wave field, the laws of which will form a counterpart to Maxwell's equations of the electromagnetic field (de Broglie, Schrödinger, Dirac). Once this has been accomplished for the individual electron, the same considerations take place as applied above to the Hohlraum radiation;

a gas of electrons is described by giving for each state α the number of electrons n_α that exist in this state and possess the corresponding energy $h\nu_\alpha$. With a gas of innumerable free electrons we contrast the shell of the few electrons that are tied to a positively charged atomic nucleus, and together with that nucleus constitute an *atom*. The ideas of discrete energy levels and of the photon have scored their most brilliant success in their application to the latter situation; for they lead straight to Bohr's frequency rule, according to which the energy $h\nu$ gained by an electron jumping from a higher to a lower energy level in the atom is emitted as a photon of frequency ν. This rule gives the clue for the explanation of the vast array of very accurate observations accumulated by the spectroscopists concerning the emission of spectral lines by radiating atoms and molecules. But full agreement is reached only after adding the assumption that no two electrons are ever found in the same complete state (Pauli's exclusion principle). This is the decisive fact for an understanding of the so-called periodic system of chemical elements. The quantum theory of chemical bonds rests on the same principle (cf. Appendix D [Sec. 18.3]). Once deduced from the spectroscopical facts, the principle could be applied to such free electrons as take care of electric conduction in metals or knock about in the interior of stars; and here too the results were found to be in accordance with experience. The upshot of it all is that the electrons satisfy Leibniz's *principium identitatis indiscernibilium*, or that the electronic gas is a "monomial aggregate" (Fermi-Dirac statistics). In a profound and precise sense physics corroborates the Mutakallimûn; neither to the photon nor to the (positive and negative) electron can one ascribe individuality. As to the Leibniz-Pauli exclusion principle, it is found to hold for electrons but not for photons.

4. The aggregates considered so far have been without *structure*. But the aggregate of atoms in a molecule possesses a structure characterized in a schematic way by Kekulé's valence strokes. It is to be presumed that the aggregate of gene points that constitutes a gamete or zygote likewise is not without a structure. Experience has taught that this structure is based on a simple binary relation of 'neighborhood' between points. We say that two neighboring points are 'joined.' With a name borrowed from topology, an aggregate endowed with this kind of structure may be called a *complex*.[8] Two complexes K, K^* are of the same $\begin{Bmatrix} \text{constitution} \\ \text{species} \end{Bmatrix}$ if a one-to-one correspondence $p \rightleftarrows p^*$ can be established between the points p and p^* of K and K^* respectively such that (i) two homologous points

---

[8] The complexes of topology consist of elements without quality, whereas our elements possess different qualities in so far as they belong to different classes.

always belong to one and the same $\begin{Bmatrix} \text{class} \\ \text{genus} \end{Bmatrix}$ and (ii) p^*, q^* are neighbors in K^* if and only if p and q are neighbors in K. Call the correspondence c-isomorphic in the first, g-isomorphic in the second case. A complex K consists of the two *separate* parts $K_1 + K_2$ provided no point of K_1 is neighbor to a point of K_2; K is *connected* if no decomposition into two separate parts is possible (except the trivial one in which one part is vacuous and the other the whole K). In a unique manner any complex may be decomposed into separate connected components. According to the combined experiences of genetics and cytology these components are to be identified with the chromosomes in the nucleus of a cell, and we shall therefore call them by this name. A connected complex that decomposes into two separate parts after removal of any one of its joins is said to be a *tree*. (Incidentally the trees used in Appendix A [of *PMNS*] to depict formulas and demonstrations are of this kind.) Under given external circumstances the complex of points that constitutes the zygote of an organism Ω determines the phenotype of Ω; or more precisely, the phenotype is the same for two c-isomorphic zygotes. This implies that generally speaking the phenotype not only depends on the *aggregate K* but also on the structure of the *complex K;* the structural influence is known under the name of position-effect.

In carrying out union and partition of complexes no joins must be severed nor new joins be established. If this were the whole story chromosomes would behave as indivisible wholes and there would be no way of distinguishing between different genes in the same chromosome. Under these circumstances Mendel's rule of independent assortment would hold, asserting that the probabilities γ for the various constitutionally different gametes produced by a definite organism are all alike. Whereas Mendel is right in that two points in two different chromosomes are independent, it has been found that points in the same chromosome are not absolutely but only more or less tightly linked together. This phenomenon of linkage has been studied with paramount success by T. H. Morgan and his school for the fruit fly *Drosophila melanogaster*, and has resulted in detailed gene maps from which quantitative information can be drawn about the probabilities γ. Morgan has explained linkage by the process of *crossing-over*. Suppose a zygote $K + K^*$ has been formed by balanced syngamy from two gametes K, K^* of the same species σ that are related to each other by the g-isomorphism $p \rightleftarrows p^*$. Let a, b be a pair of neighboring points in K and a^*, b^* the homologous pair in K^*. Then a^*, b^* will be neighbors in K^*; the points a, b will lie in one chromosome K_0 of K and a^*, b^* in the homologous chromosome K_0^* of K^*. Crossing-over consists in breaking the joins ab and a^*b^* and joining instead a with b^* and b with

a^*. If K_0 is a tree then this process carries the disconnected pair (K_0, K_0^*) into a pair $(\bar{K}_0, \bar{K}_0^*)$ of chromosomes that are g-isomorphic to K_0. Points that before crossing-over were in the same chromosome K_0 may now be separated, one belonging to $\bar{K}_0$, the other to $\bar{K}_0^*$. Pairs of homologous chromosomes in the nucleus of a cell are seen to put themselves in a position for such an operation of crossing-over immediately before meiosis takes place; they extend side by side, each point in one chromosome opposite the homologous point in the other (synapsis). If afterwards balanced meiosis occurs, the new gametes $\bar{K}$, $\bar{K}^*$ will be of the same species σ as K, K^*. *Linkage between two points a, b of a chromosome will be the looser the more ways there exist to separate them by crossing-over.*

Complexes may undergo two sorts of *mutations*. Besides the point mutations in which the joins are not tampered with while the points p change their classes C_p (without changing their genera), we have structure mutations that leave the state of the points undisturbed but alter the joins.[9] The operation described above as crossing-over may be performed with *any* four distinct points a, b, a^*, b^* (and may then be called 'switching-over'). A simple break and this process of switching-over seem to play the role of elementary operations for structure mutations. Mutations are rare events, in contrast to crossing-over for which an opportunity is provided by synapsis before every meiosis.

The simplest connected complexes are the rod a_1—a_2— $\cdots$ —a_h (in which consecutive points a_i are connected by the joins —) and the ring. With few exceptions the chromosomes seem to be rods (T. H. Morgan's *law of linear arrangement*). However, switching-over when occurring in one rod (not between two rods) may produce a rod plus a ring (or reproduce the rod with an inverted section). A complex consisting of separate rods and rings will preserve this character under any breaks and switchings-over.

A chromosome has a centromere. If a structural change produces chromosomes with no or two centromeres then they are usually left behind when the cell divides, and thus *deficiencies* result. There are also several ways in which the whole chromosome outfit of a cell nucleus or an individual chromosome, or part of a chromosome, may be *duplicated*.

Here we have attempted to develop the formal scheme of genetics in such general form as to comprise all more or less irregular occurrences. Nowhere in this scheme was it necessary so far to speak of *sex;* but of course the fact cannot be ignored forever that syngamy between two gametes takes place only if one is a sperm, the other an egg. This is a

---

[9] On the basis of the position-effect, R. Goldschmidt has recently challenged the entire conception of gene and the distinction between point and structure mutations.

polarity (gamete sexuality) that has nothing to do with genes.[10] On the other hand, whether an individual organism is a sperm-producing male or an egg-producing female (zygote sexuality) is determined like all its other 'visible characters' by the genotype of its zygote — in conjunction with the external circumstances influencing development. Experience shows that it is not a single gene in the zygote, but a balance between many genes, that determines the sex. The sex chromosome (where it is distinguishable from the autosomes) merely tips the scales. This explains the phenomenon of intersexes and modifies the common belief that sees in sex the outstanding example of a non-quantitative, an either-or, character.

[5. Our remarks about entropy and statistics in Section 23B of the main text [$PMNS$] and in the following Appendix C [Sec. 18.2] on quantum physics will be made clearer if we say at this place, in parenthesis as it were, a few words about the foundations of *statistical thermodynamics*. Here quantum theory has introduced a decisive simplification. Indeed, in quantum physics a system Σ is capable of no more than a discrete series of (complete) states with definite energy levels

$$U_i \qquad (i = 0, 1, 2, \ldots).$$

In view of the conservation law for energy, let us distribute a large number N of systems Σ at random over its possible states i, yet so that the total energy of the N systems has a preassigned value $N \cdot A$ (A = average energy of the individual system). One finds that in the overwhelming majority of all distributions the relative frequencies N_i/N with which the several states i are represented is, in the limit for $N \to \infty$, proportional to $e^{-\alpha U_i}$. Here α denotes a constant that is to be determined in terms of the given average energy A. We therefore define the *canonic distribution of parameter* α by assigning the relative probability $w_i = e^{-\alpha U_i}$ to the state i. (Relative probabilities need not satisfy the normalizing condition $\Sigma_i w_i = 1$.) Any quantity Z_i dependent on the state i will then have the mean value

$$\langle Z \rangle_\alpha = \sum_i Z_i \cdot e^{-\alpha U_i} \Big/ \sum_i e^{-\alpha U_i},$$

and it is this value that we ascribe to the quantity in 'thermic equilibrium.' The parameter α is connected with the given mean energy A by the

---

[10] Denote by $\Omega_{\alpha\beta}$ an organism arising from syngamy of a sperm of genotype α and an egg of genotype β. The fact that there are cases when, even under equal circumstances, the reciprocal cross $\Omega_{\beta\alpha}$ differs in appearance from $\Omega_{\alpha\beta}$ is evidence that interpretation of organic development in terms of genes alone will not always suffice. Besides the genes in the chromosomes of the cell nuclei, other hereditary agents influencing the development have to be assumed in the cytoplasm. This problem however is still far from a satisfactory solution.

equation $\langle U \rangle_\alpha = A$. The systems occurring in nature are capable of states with arbitrarily high energy values; consequently α must be positive, and the standard distribution which assigns to every state i the same probability $w_i^0 = 1$ may only approximately be realized as long as the energy A stays finite (namely for large values of A and correspondingly low values of α). The reciprocal number α^{-1} has the dimensions of energy and may for the moment be called statistical temperature.

Because of the conservation law of energy, the canonical distribution is stationary in time. The possible states of a system Σ consisting of two parts Σ, Σ' is characterized by the pairs (i, k) formed from any state i of the system Σ and any state k of the system Σ'. Let U_i, U_k' designate the energy of Σ in the state i and of Σ' in the state k respectively; then the energy of Σ in the state (i, k) equals $U_i + U_k'$, provided no interaction takes place between the two parts. For the probability of the state (i, k) in thermic equilibrium we obtain

$$\mathbf{w}_{ik} = e^{-\alpha(U_i + U_k')} = e^{-\alpha U_i} \cdot e^{-\alpha U_k'} \, (= w_i \cdot w_k').$$

This means three things: (1) thermic equilibrium of the whole implies thermic equilibrium for the parts; (2) the law of statistical independence prevails for the combination of the parts, $\mathbf{w}_{ik} = w_i \cdot w_k'$; (3) the parameter α has the same value for both parts, with which the value for the total system also agrees. On account of this third point statistical temperature shares with ordinary observable temperature (a more-or-less rather than a quantitative character of bodies) the decisive property that bodies in contact level their temperatures.

An ideal gas, i.e., an aggregate of n particles the states of which are completely described by position and velocity and the interaction of which is negligible, will occupy a definite volume V under a definite pressure p. Application of classical physics and the canonical distribution to such a gas yields the value pV/n for its statistical temperature. Hence if T designates the (absolute) temperature read from a gas thermometer in contact with the system Σ and filled with an ideal gas, then the statistical temperature turns out to be kT where k is a universal factor of proportionality (Boltzmann's constant) that must be added in order to reduce the scale of temperature to the customary Celsius degrees ($100°C =$ difference of the boiling and freezing points of water under pressure of one atmosphere).[11] The entire theory of thermic equilibrium thus boils down to

[11] The mass $M = n \cdot \mu$ of the gas is proportional to the number n of particles. $v = V/M$ denoting the specific volume, the Gay-Lussac laws are obtained in the usual form $pv = RT$ with a constant $R = k/\mu$ characteristic for the gas. Consequently k is of atomistic smallness, and for different gases the product of R and the "molecular weight" has the same value.

this one principle holding in quantum as well as classical physics: the canonical distribution w arises from the standard distribution w^0 by means of the equation

$$w = w^0 \cdot e^{-U/kT}$$

where U denotes the energy (variable from state to state) and T the fixed temperature of the system (or of the heat bath in which the system is immersed).]

REFERENCES

C. H. Waddington, *An Introduction to Modern Genetics*, London, 1939.

E. Schrödinger, *Statistical Thermodynamics*, Cambridge, 1946.

R. Goldschmidt, Position Effect and the Theory of the Corpuscular Gene, *Experientia*, **2** (1946), pp. 197–203, 250–256.

J. H. Woodger, *The Axiomatic Method in Biology*, Cambridge, 1937.

von Neumann, J. and O. Morgenstern, *Theory of Games and Economic Behavior*, second ed., Princeton, 1947.

18.2   QUANTUM PHYSICS AND CAUSALITY

1. Modern quantum theory has done away with strict causal determination for the elementary atomic processes. It does not deny strict laws altogether, but the quantities with which they deal regulate the observable phenomena only statistically. Quantum theory is incompatible with the idea that a strictly causal theory of unknown content stands behind it—in the manner in which it may be true that strictly causal motion of individual particles stands behind the statistical-thermodynamical regularities of a gas consisting of many particles. What is thinkable for the laws of a collective is demonstrably impossible for the elementary quantum laws. The uncertainty of the outcome of an atomic experiment is not such as to be gradually reducible to zero by increasing knowledge of the determining factors. The reasons for the passage from classical to quantum physics are no less compelling than those for the relinquishment of absolute space and time by relativity theory; the success, if measured by the empirical facts made intelligible, is incomparably greater. True, a final stage has not yet been attained; certain serious difficulties remain unsolved. But whatever the future may bring, the road will not lead back to the old classical scheme.

One of the most poignant revelations of light's corpuscular nature is the photoelectric effect: a metal plate when irradicated with ultra-violet or X-ray releases electrons. Observation shows that strangely enough their energy is determined by the *color* of the incident radiation, namely equal to or less than h times its frequency. One thus arrives at the conception that light of frequency ν gets absorbed in discontinuous quanta (photons) of energy $h\nu$ (Einstein 1905). This energy is used for the emission of an electron (whose kinetic energy may be short of $h\nu$ on account of the work used in the liberation of the electron). The intensity of the radiation does not determine the energy of the individual electron, but the number of electrons released per time unit. The process sets in at once even if the radiation is so weak that it would take hours before the accumulated influx of field energy into the region of an atom reaches the amount $h\nu$ necessary for the ejection of an electron. What a continuous field theory would describe as *presence of a fraction* of the radiating energy $h\nu$ must in fact be interpreted as *small probability* for the presence of a whole photon of that energy. The process inverse to the photoelectric effect is the transformation of primary electrons into secondary X-rays in a tube whose anode stops the electrons. Since the stopping may take place in several steps, a continuous spectrum for the X-rays is to be expected with a sharp edge at the frequency $\nu = eVh$ (where $-e$ is the charge of the electron and V the voltage of the tube). Experience has confirmed these relations first predicted by Einstein, including the numerical value for h that has to agree with Planck's constant derived from the thermodynamical laws of Hohlraum radiation.

The problem of reconciling the conceptions of light wave and photon is perhaps best illustrated by polarization. Let a plane monochromatic light wave that propagates in a definite direction $\vec{a}$ be linearly polarized; the direction of polarization, represented by a vector $\vec{s}$ of length 1, is perpendicular to $\vec{a}$. Choose an arbitrary point O as origin and draw a 'cross' G consisting of two mutually perpendicular axes 1 and 2 through O that are perpendicular to $\vec{a}$. The cross as well as the vector $\vec{s}$ lies in the plane perpendicular to $\vec{a}$. Assume that the light ray passes through a Nicol prism of orientation G; it splits into two parts 1 and 2, the part 1 being linearly polarized in the direction 1, the part 2 in the direction 2. The relative intensities of the two partial rays with respect to the total ray are given by the squares (of the lengths of the projection of $\vec{s}$ upon the axes 1 and 2, or) of the components s_1, s_2 of the vector $\vec{s}$ with respect to the cross G. On assuming the light ray to consist of photons we are forced to conclude that photons of two distinct 'characters' 1 and 2 (white balls and black balls as it were) occur with the respective probabilities s_1^2 and s_2^2. (By Pythagoras's theorem, $s_1^2 + s_2^2 = 1$, the probability for a photon to be 'white or black' equals unity, as it should.) The characters 1 and 2 are relative to G. The photons of both kinds are separated by the Nicol prism that operates like a sieve catching 1 and letting 2 pass. One should therefore expect the ray that has passed the

Nicol prism and is polarized in the direction 2 to be more homogeneous than the original ray. Such however is not the case, as the polarized plane monochromatic light wave represents the highest degree of homogeneity obtainable for light. The partial ray polarized in the direction 2, when sent through a second Nicol prism of a different orientation G', will again split up according to the rule of intensity just described. Something similar to polarization happens when a ray of silver atoms is acted upon by a nonhomogeneous magnetic field. Using Cartesian coordinates x, y, z, let us assume that the field strength depends on x only. A silver atom is a small magnetic dipole with a vectorial magnetic momentum $\vec{m}$. The field should split the ray into various parts according to the various values of the x-component m_x of $\vec{m}$. Since only two partial rays (of opposite curvature) are observed, one must conclude that this component is capable of two values only, $+\mu$ and $-\mu$ ($\mu =$ 'magneton'). This must hold whatever direction the x-axis has. A vector, however, whose components in every possible direction are either $+\mu$ or $-\mu$ is geometric nonsense! The impossibility of simultaneously ascribing to the photon or silver atom the several characters that correspond to the various orientations G of the "sieve" lies clearly in the nature of things and is not due to human limitations. The sorting by a sieve of orientation G is destroyed by a subsequent sifting of orientation G'. But for the particles passed by the sieve G one may ask what probability they have to pass the sieve G'; this probability can be computed *a priori* in terms of the orientation of G' with respect to G.

By a prism as Newton used it, or by a grating, light is decomposed according to its various monochromatic constituents. A 'sieve' permitting separation not only of two but of several 'sizes of grain' may, therefore, be called a 'grating.' The photoelectric effect sifts photons according to the place where they hit. For good reasons Eddington says in his beautiful book *New Pathways in Science* (p. 267): "In Einstein's theory of relativity the observer is a man who sets out in quest of truth armed with a measuring-rod. In quantum theory he sets out armed with a sieve." By a certain operation M the mammals are sifted from the other animals in a zoo. Let F be the corresponding operation for fish. The iterated operation MM yields no other result than the simple M, whereas the catch of MF, that is of the operation M followed by F, is zero. In view of the equations

$$MM = M, \qquad FF = F, \qquad MF = FM = 0,$$

M and F are called mutually orthogonal idempotent operators. A grating in classical physics is nothing but a classification of the states that a given physical system is capable of. Here the states are considered as the elements of an aggregate. Assuming the number of states to be finite, let $E_1, \ldots, E_r$ denote the several classes. The operation sifting the states of class E_i from the rest may also be designated by E_i. The grating $G = \{E_1, \ldots, E_r\}$ may be refined by dividing each class into subclasses. There is a finest division in which each class contains only one member. Two gratings, the division into the classes E_i and the division into the classes E'_k, may be superposed. The operation $E_i E'_k$ sorts out the members common to the two classes E_i and E'_k; commutativity prevails, $E_i E'_k = E'_k E_i$, and the total aggregate is divided up into the classes $E_i E'_k$ (some of which may be vacuous). After associating distinct numbers α_i with the several classes E one may speak of an *observable* (or state quantity) A that assumes the value α_i when the state of the system belongs to the class E_i.

This classical scheme is now to be confronted with that of quantum physics as adumbrated by the typical example of polarization. The aggregate of n states has to be replaced by an n-dimensional Euclidean vector space **I**. Given a linear subspace **E** of **I**, any vector $\vec{x}$ may be orthogonally projected upon **E**; the projection is a vector $E\vec{x}$, and the operation E of projection is idempotent, $EE = E$. Two linear subspaces **E**$_1$, **E**$_2$ are said to be orthogonal if each vector in the one is orthogonal to each vector in the other. We may then form their sum **E** = **E**$_1$ + **E**$_2$ consisting of all sums of a vector $\vec{x}_1$ in **E**$_1$ and a vector $\vec{x}_2$ in **E**$_2$. The decomposition of a vector $\vec{x}$ of **E** into these two summands $\vec{x}_1$ and $\vec{x}_2$ is unique and effected by the orthogonal projections E_1, E_2. In this way a three-dimensional vector space, for instance, splits into a horizontal plane and a vertical line. The situation is hardly more complicated when we deal with more than two mutually orthogonal subspaces **E**$_1$, **E**$_2$, $\ldots$, **E**$_r$. If their sum is the total space, then every vector $\vec{x}$ splits into r component vectors lying in the several subspaces **E**$_i$, according to the formula

$$\vec{x} = E_1 \vec{x} + \cdots + E_r \vec{x},$$

and the total space is said to be split into **E**$_1$ + $\cdots$ + **E**$_r$. The projections E_i $(i = 1, \cdots, r)$ are idempotent and mutually orthogonal. Let $|\vec{x}|$ designate the length of a vector $\vec{x}$.

By a direction in **I**, or by a vector $\vec{x}$ laid off in that direction, quantum physics represents the *wave state* of the physical system under investigation (be it a simple particle, or an aggregate of many, or even of an indeterminate number of particles). A *grating* $G = \{E_1, \cdots, E_r\}$ is a splitting of the total vector space into mutually orthogonal subspaces **E**$_1$ + $\cdots$ + **E**$_r$. We speak of a *character* i, more precisely $(G; i)$, corresponding to any one of these subspaces **E**$_i$. If the system is in the wave state $\vec{x}$, then its probability of having the character i equals

(1) $$p_i = |E_i \vec{x}|^2 / |\vec{x}|^2.$$

Pythagoras's equation,

$$|E_1 \vec{x}|^2 + \cdots + |E_r \vec{x}|^2 = |\vec{x}|^2,$$

states that the sum of these probabilities p_i equals unity. No grating can furnish more than n distinct characters; thus our model, like the classical model it is compared with, corresponds to a situation where the number of character values is limited. That is not so in nature, but no essential features of quantum mechanics are lost by using the finite-dimensional model. It is clear wherein the *refinement* of a grating G consists: indeed every $\mathbf{E}_i$ may again be split into mutually orthogonal subspaces. A *finest* grating consists of n mutually orthogonal axes and is thus identical with a 'cross' (= Cartesian coordinate system). Let us from now on refer to the characters $(G; 1), \ldots, (G; r)$ as the r *quantum states* defined by the grating G and call them *complete* quantum states if G is a finest grating.

I cannot refrain from pointing out that, without thinking of quantum physics, I used this very model at the end of Section 17 [of *PMNS*] to illustrate the relationship between object, observer, and observed phenomenon. Its decisive difference in comparison to the classical model is the fact that gratings in vector space defy superposition. Suppose **I** has been split into orthogonal subspaces in two manners,

$$\mathbf{I} = \mathbf{E}_1 + \cdots + \mathbf{E}_r \quad \text{and} \quad \mathbf{I} = \mathbf{E}_1' + \cdots + \mathbf{E}_s'.$$

It is of course possible to split any vector $\vec{x}$ of $\mathbf{E}_i$ into its components $E_k' \vec{x}$ $(k = 1, \cdots, s)$; but they will in general no longer lie in $\mathbf{E}_i$. Only if the operator E_i commutes with the operators $E_1', E_2', \ldots$, will this be so. Thus combination of two gratings presupposes commutability of the r operators E_i with the s operators E_k', and, that condition satisfied, the order of combination, whether G is followed by G' or G' by G, is irrelevant. The strange thing about the quantum-physical sieves is exactly this feature, that two such sieves may, and usually will, be non-commutative because of their 'discordant orientations.' Characters i and k referring to two non-commutative gratings are incompatible, and in a situation where i is determined, k is not. In this sense, position and momentum of a particle are incompatible. If Δx, Δp are the uncertainties of the x-coordinate of a particle and of the x-component of its momentum respectively, then the product $\Delta x \cdot \Delta p$ necessarily exceeds h. This *principle of indeterminacy* due to Heisenberg embodies the idea of complementarity in a precise form. If color and shape of a body were such incompatible characters, it would make sense to ask whether a body is green, and it would also make sense to ask whether it is round; but the question "Is it green *and* round?" would make no sense. Here as in the classical model distinct numbers α_i may be assigned to the several quantum states $(G; i)$ and thereby an observable A be defined that is capable of the r values $\alpha_1, \ldots, \alpha_r$ and assumes the value α_i with the probability p_i, (1), provided the system is in the wave state $\vec{x}$.[1]

Decomposition by a grating G of the vector space $\mathbf{I}$ into the several subspaces $\mathbf{E}_i$ is in itself a purely ideal process. The actual application of the grating to a physical system, however, throws the system from the wave state $\vec{x}$ into one of the wave states $E_1\vec{x}, \ldots, E_r\vec{x}$. Which wave state cannot be foretold; only the relative probabilities $|E_i\vec{x}|^2$ of these r events are predetermined. In this sense every measurement or observation implies an encroachment on the phenomenon, with results of no more than statistical predictability. At no time have experimental physicists closed their eyes to the fact that every measurement is coupled with a reaction of the measuring instrument on the object under investigation. As long as the hypothesis seemed admissible that the instrument could be made infinitely more sensitive than the object, this involved no difficulty of principle. But what if the object itself is of atomic refinement, which cannot be surpassed by any instrument? Then the very idea of facts prevailing independently of observation becomes dubious.

2. Let us for a moment return to the classical model, and use it to depict the temporal succession of events. The simplifying hypothesis of a finite number of states forces us to operate within a discontinuous time. The dynamical law will then assert that from one moment t to the next $t + 1$ the n states $1, \cdots, n$ undergo a certain permutation s, the same at every moment $t = \cdots -2, -1, 0, 1, 2, \cdots$. If this permutation s is of 'order m,' i.e. if one reaches identity after having performed the permutation m times, then the system returns to its initial stage after each period of length m (eternal recurrence). Quantum physics does not force a discontinuous time upon us even if the number of quantum states separable by a grating is universally limited. During the infinitesimal time interval dt the vector space experiences a certain infinitesimal rotation imparting the increment $d\vec{x} = L\vec{x} \cdot dt$ to the arbitrary vector $\vec{x}$. This dynamical law

$$(2) \qquad\qquad d\vec{x}/dt = L\vec{x}$$

(in which the operation L is independent of t and $\vec{x}$) is expressed in terms

---

[1] According to their definition such quantities can be added and multiplied provided they belong to commuting gratings.

of Cartesian coordinates x_i by equations of the form

$$dx_i/dt = \sum_j l_{ij}x_j(t) \qquad (i, j = 1, \cdots, n),$$

with given constant antisymmetric coefficients l_{ij} ($l_{ji} = -l_{ij}$). The salient point is that the wave state $\vec{x}$ varies according to a strict causal law; its mathematical simplicity is gratifying. A grating $G = \{E_1, \ldots, E_r\}$ and the corresponding quantum states $(G; 1), \ldots, (G; r)$ are *stationary* if the subspaces $\mathbf{E}_i$ are invariant in time, i.e. if the linear operators E_i commute with the linear operator L.

What in Appendix B [Sec. 18.1] has been called state of a particle or of an aggregate is now to be more precisely interpreted as *quantum state*. A circumstance that may have caused some misgivings there, the relativity of the notion of photon with respect to the Hohlraum and its proper frequencies, now appears as a special instance of a general phenomenon: the distinction of quantum states is relative to a grating.

Measurement means application of a sieve or grating. One must not imagine the wave state as something given independently of such measurements. In fact the monochromatic polarized light ray that is sent through the Nicol prism had itself been sorted out by a grating from natural light of unknown quality. This is in accordance with the fundamental fact that only the *relative* position of one Cartesian coordinate system with respect to another may be characterized in objective terms. Given the gratings $G = \{E_1, \cdots, E_r\}$ and $G' = \{E'_1, \cdots, E'_s\}$, we are, however, entitled to ask questions of this type: 'If the first grating shows our particle to be in the quantum state $(G; i)$, between what limits does the probability lie that a test by the second grating G' finds it in the quantum state $(G'; k)$?' In geometric terms this amounts to the following question: (I) 'Between which limits does the quotient $|E'_k\vec{x}|^2/|\vec{x}|^2$ lie if $\vec{x}$ varies freely over the space $\mathbf{E}_i$?' Should time elapse between application of the first and the second grating, then the change of the wave state between the two moments as determined by the dynamical law (2) has to be taken into account.

A system is never completely isolated from its surroundings, and its wave state is therefore subject to perpetual disturbances. This is the reason why the secondary statistics of thermodynamics is to be superimposed upon the primary statistics dealing with a given wave state and its reaction to a grating. In Euclidean space there is an *a priori* probability for the random distribution of vectors of length 1 according to which regions of equal area on the unit sphere are of equal probability. This 'standard distribution' assigns to the r quantum states $i = 1, \cdots, r$ defined by a grating $G = \{E_1, \cdots, E_r\}$ the probabilities n_i/n where n_i is the dimensionality of $\mathbf{E}_i$, and in particular equal probabilities to the n

quantum states defined by a *complete* grating. The actual probability distribution (Gibbs ensemble) need by no means coincide with this standard distribution. At the end of the previous section it has been described how the particular canonical distribution for a system embedded in a heat bath of known temperature proceeds from the standard distribution. A general question somewhat different from the one considered above now arises, namely: (II) 'Suppose a grating G and the statistical distribution of wave states is given, what probabilities result therefrom for the several quantum states $(G; i)$?' A grating and a statistical ensemble, rather than two gratings, are here compared with each other. To find the answer one has to average the probability p_i, (1), which depends on $\vec{x}$, according to the given statistical distribution of vectors $\vec{x}$ over the unit sphere. In the same manner one may ascertain the average probability that a particle in the quantum state $(G; i)$ is encountered in the quantum state $(G'; k)$ if tested by another grating G'. Whether question (I) or (II) is posed, the interest will always be focused, especially when we are concerned with systems consisting of numerous particles, on such events as can be foretold with overwhelming probability. In splitting a light ray by a Nicol prism the fate of the individual photon is unpredictable. Predictable however are the relative intensities of the two partial rays with an accuracy that increases with the number of photons.

The description here given must be corrected throughout in one point: the coordinates x_j in the underlying n-dimensional vector space are not real but arbitrary *complex* numbers and as such have an absolute value $|\vec{x}|$ and a phase. The square of the length of the vector is expressed in terms of a Cartesian coordinate system by the sum of the squares of the absolute values of the coordinates. The simplest of all dynamical laws (2) in such a complex space is of the form

$$(3) \qquad d\vec{x}/dt = i\nu\vec{x} \qquad (i = \sqrt{-1}).$$

Here ν is a real constant. The wave state $\vec{x}$ then carries out a simple oscillation of frequency ν,

$$\vec{x} = \vec{x}_0\{\cos(\nu t) + i\sin(\nu t)\} \qquad (\vec{x}_0 = \text{const.}),$$

and hence the energy has the definite constant value $h\nu$ (Planck's law). But whatever the dynamical law (2), the space can always be broken up into a number of mutually orthogonal subspaces $\mathbf{E}_j$ $(j = 1, \cdots, r)$ such that an equation (3) with a definite frequency $\nu = \nu_j$ holds in $\mathbf{E}_j$. The grating $G = \{\mathbf{E}_1, \cdots, \mathbf{E}_r\}$ thus obtained is stationary and effects a sifting with respect to different frequencies ν_j and corresponding energy levels $U_j = h\nu_j$. Thermodynamics is based on this G.

Any vector $\vec{x}$ in E_j satisfies the equation $L\vec{x} = i\nu \cdot \vec{x}$ ($\nu = \nu_j$), and this fact is expressed in mathematical language by saying that $\vec{x}$ is an eigenvector of the operation L with the eigenvalue $i\nu$. The operator $H = \dfrac{h}{i} L$, called energy, has the same eigenvectors, but the corresponding eigenvalues are the energy levels $h\nu$. The general equation (2) now reads

$$\frac{h}{i}\frac{d\vec{x}}{dt} = H\vec{x} \quad \text{(Schrödinger's equation)}.$$

The 'physical process' undisturbed by observation is represented by a mathematical formalism without intuitive (anschauliche) interpretation; only the concrete experiment, the measurement by means of a grating, can be described in intuitive terms. This contrast of physical process and measurement has its analogue in the contrast of formalism and meaningful thinking in Hilbert's system of mathematics. As it is possible to formalize an intuitive mathematical argument, so it is true that measurement by a grating G may be interpreted as a physical process. In doing so one has to extend the original system Σ to a system Σ^* by inclusion of the grating G. But as soon as we want to learn something about Σ^* that can be told in concrete terms, then the undisturbed course of events as ruled by the dynamical law (2) must again be disrupted by subjecting Σ^* to the test of a grating outside Σ^*.

3. Given two systems Σ, Σ', their union $\Sigma = \Sigma + \Sigma'$ is capable of all states (i, k) consisting of a combination of an arbitrary state i of Σ and an arbitrary state k of Σ'. That is the prescription for combination given by classical physics. Quantum physics agrees provided state means quantum state; a (finest) grating for Σ together with a (finest) grating for Σ' yields a (finest) grating for $\boldsymbol{\Sigma}$. Assuming therefore that a wave state of the first system is represented by a vector $\vec{x} = (x_1, \cdots, x_m)$ in an m-dimensional Euclidean space S referred to a Cartesian coordinate system, and that the generic vector $\vec{y} = (y_1, \cdots, y_n)$ of an n-dimensional space S' has the same significance for the second system, we conclude that the wave state of the united system is represented by a vector

$$\vec{z} = (z_{11}, \cdots, z_{m1}, z_{12}, \cdots, z_{m2}, \cdots, z_{1n}, \cdots, z_{mn})$$

in an mn-dimensional 'product space' $\mathbf{S} = S \times S'$. A vector $\vec{x}$ in S and a vector $\vec{y}$ in S' determine the vector $\vec{z} = \vec{x} \times \vec{y}$ with the components

$$(4) \qquad z_{ik} = x_i y_k \qquad (i = 1, \cdots, m; k = 1, \cdots, n)$$

in $\mathbf{S}$. This fixes what rotation in $\mathbf{S}$ is induced by two arbitrary rotations of the coordinate systems in S and S'. Since (4) implies $|z_{ik}|^2 = |x_i|^2 |y_k|^2$ one finds the probabilities of the quantum states of the two parts Σ and Σ'

to be independent of each other in a wave state of Σ of the special kind $\vec{z} = \vec{x} \times \vec{y}$. But the manifold of the possible wave states of the joint system Σ is much larger than those representable by the combinations $\vec{x} \times \vec{y}$ of arbitrary wave states $\vec{x}$ and $\vec{y}$ of the two parts. In fact *every* vector $\vec{z}$ in the product space represents a possible wave state. In this very radical sense quantum physics supports the doctrine that *the whole is more than the combination of its parts*. In general the probabilities of the quantum states of the whole system cannot be determined from the probabilities of the quantum states of the parts by the product rule of statistical independence. And this is so even when both parts are not in dynamical interaction.

This consideration is of special importance for a pair of two *equal* systems Σ, Σ', e.g. for a pair of electrons. Then the wave states of both parts are represented by vectors in one and the same Euclidean space S, and among the vectors $\vec{z} = (z_{ik})$ of the product space $S \times S$ ('tensors') one may distinguish the antisymmetric ones satisfying the condition $z_{ki} = -z_{ik}$ and the symmetric ones with the property $z_{ki} = z_{ik}$. Once the pair is in an antisymmetric wave state, its wave state will remain antisymmetric; no external influences can alter that because equal particles enter into the law of action in a symmetric fashion. It is therefore to be expected that the wave state of a pair of electrons has a definite symmetry character, that it is either antisymmetric or symmetric. Experience proves the first alternative to be correct. For an antisymmetric vector z_{ik} the equation $|z_{ij}|^2 = 0$ holds, i.e. the probability that both electrons are found in the same complete quantum state i is zero; *the permanent anti-symmetry of the wave state thus explains Pauli's exclusion principle*. The statistical independence of the quantum states of two electrons could scarcely be denied in a more radical fashion than by this principle! The hydrogen molecule may be treated, at least in first approximation, as a system of two electrons circling around two fixed nuclei, and it is obvious that the restriction by antisymmetry of the wave state of the electronic pair must be of decisive influence upon the result of the computation of their motion. It leads indeed, as London and Heitler have shown, to a full explanation of the chemical binding of neutral atoms in a molecule. Under the reign of classical physics this had remained an inscrutable conundrum.

The fact of antisymmetry carries over from two to more electrons. Since statistical independence of several particles is at variance with this law, it is not the same whether again and again an electron of given wave state or a simultaneous shower of many electrons is sent through a grating. A similar remark applies to a shower of photons. Its wave state is to be restricted by the condition of symmetry rather than of antisymmetry. (We know indeed that the exclusion principle does not hold for photons!)

In its final form the theory does not require the number of particles to be constant. Not only may photons appear and disappear, but owing to a bold interpretation of Dirac's it also accounts for the process of mutual annihilation of a positive and negative electron under emission of a photon of corresponding energy ("Zerstrahlung") and the inverse process.

4. I summarize those features of quantum physics which seem to me of paramount philosophical significance.

(1) Observation is impossible without an encroachment the effect of which can be predicted only in a statistical sense. Thus new light is thrown on the relationship of subject and object; they are more closely tied together than classical physics had realized. It has been said in Section 20 [of *PMNS*] that quantitative results derived from the observation of reactions of a body with other bodies are ascribed as inherent characters to the body itself, whether or not the reactions are actually carried out. We now see that this 'Euler principle' has very serious limitations. There are obvious analogies to this situation in the domain of psychic self-observation.

(2) Characters referring to two different gratings cannot meaningfully be combined by 'and' or 'or.' Classical logic does not fit in with quantum physics and is to be replaced by a kind of 'quantum logic.'

(3) The principle of causality holds for the temporal change of the wave state, but must be dropped as far as the relation between wave and quantum states is concerned.

(4) The whole is always more, is capable of a much greater variety of wave states, than the combination of the parts. Disjoint parts in an isolated system of fixed wave state are in general not statistically ind··- pendent even if they do not interact.

(5) The Leibniz-Pauli exclusion principle, according to which no two electrons may be in the same quantum state, is made comprehensible by quantum physics as a consequence of the law of antisymmetry.

(6) There exists a primary probability, as a basic trait of nature itself, that has nothing to do with the observer's knowledge or ignorance. The probabilities $|x_i|^2$ of the individual complete quantum states i are derived from the components x_i of a vector quantity $\vec{x} = (x_1, \cdots, x_n)$ describing the 'wave state.' This seems to me to confirm the opinion expressed in the main text [of *PMNS*], namely that probability is connected with certain basic physical quantities and can in general be determined only on the ground of empirical laws governing these quantities.

It must be admitted that the meaning of quantum physics, in spite of all its achievements, is not yet clarified as thoroughly as, for instance, the ideas underlying relativity theory. The relation of reality and observation is the central problem. We seem to need a deeper epistemological analysis of what constitutes an experiment, a measurement, and what sort of language is used to communicate its result. Is it that of classical physics, as Niels Bohr seems to think, or is it the 'natural language,' in which everyone in the conduct of his daily life encounters the world, his fellow men, and himself? The analogy with Hilbert's mathematics, where the practical manipulation of concrete symbols rather than the data of some 'pure consciousness' serves as the essential extra-logical basis, seems to suggest the latter. Does this mean that the development of modern mathematics and physics points in the same direction as the movement we observe in current philosophy, away from an idealistic toward an 'existential' standpoint?

Aside from the riddles of epistemological interpretation, quantum physics is also beset by serious internal difficulties; we do not yet possess a really consistent and complete quantum theory of the interaction between electromagnetic radiation and (negative and positive) electrons, let alone the other elementary particles.

Returning to safer ground, let us add a word about the position of quantum physics towards the problem of *past and future* as discussed in Section 23C [of *PMNS*]. What one wishes to understand is why light is emitted only 'towards the future.' We saw that physics can account for this distinction of the future from the past half of the light cone by keeping merely the retarded part of the potential in the Liénard-Wiechert formula. Quantum theory describes the interaction between the electrons of an atom and the field of Hohlraum radiation as a sequence of individual acts in which a light-quantum is emitted or absorbed under a corresponding energy jump of the atom. The formula for the frequencies of these acts can be interpreted as indicating that the individual act is either spontaneous or enforced. The frequency of the enforced acts is proportional to the density of the radiation, while the spontaneous acts are independent of it. The enforced part is symmetric with respect to past and future. Not so the spontaneous part; there is only spontaneous emission, but no spontaneous absorption. This asymmetry is accounted for by probability arguments of the same sort as led to the law of increasing entropy. Hence the distinction of the future half of the light cone has its roots here in the statistical principles of thermodynamics rather than in any elementary laws.

REFERENCES

P. A. M. Dirac, *The Principles of Quantum Mechanics*, third ed., Oxford, 1947.
J. von Neumann, *Mathematische Grundlagen der Quantenmechanik*, Berlin, 1932.
H. Weyl, *Gruppentheorie und Quantenmechanik*, second ed., Leipzig, 1931.

G. Wentzel, *Einführung in die Quantentheorie der Wellenfelder*, Vienna, 1943.

Niels Bohr, *Atomic Theory and the Description of Nature*, Cambridge, 1934.

——Kausalität und Komplementarität, *Erkenntnis*, **14** (1937), p. 293.

M. Born, *Atomic Physics*, transl. by J. Dougall, second ed., London & Glasgow, 1937.

—— *Experiment and Theory in Physics*, Cambridge Univ. Press, 1943.

H. Reichenbach, *Philosophic Foundations of Quantum Mechanics*, Univ. of Calif. Press, 1944.

L. Rosenfeld, L'evolution de l'idée de causalité, *Mém. Soc. Roy. Sci. Lièges*, 4e sér., VI (1942).

P. A. Schilpp (ed.), *Albert Einstein: Philosopher-Scientist*, Library of Living Philosophers VII, Evanston, Ill., 1949, articles by Niels Bohr, M. Born, W. Heitler, H. Margenau, and W. Pauli.

18.3 CHEMICAL VALENCE AND THE HIERARCHY OF STRUCTURES

The symbolic structure in terms of which quantum theory explains the atomic phenomena may well be of a primitive and irreducible nature. In contrast, the aggregate of atomic points joined by valence strokes, at which Kekulé depicts a chemical molecule, is only of an intermediary character. Indeed the valence bonds are an abbreviated symbol for the actual quantum-physical forces acting between the atoms, which in themselves are complex dynamical systems. The Kekulé diagram is thus seen to be founded on a mere primary structure, that of quantum mechanics. This is one instance of what Hilbert generally describes as "Tieferlegung der Fundamente."

The theory of chemical bondage affords such a striking illustration of the hierarchy of structures that I cannot refrain from describing it in a little more detail. The electronic spin and the exclusion principle are those features made responsible by quantum mechanics for chemical valence. *Position* is certainly a character of the electron. If separation according to position were a finest grating then the wave state of an electron would be given by a (complex-valued) function $\psi(P)$ of an argument P ranging over all points in space [the square of the length of this vector being the integral of $|\psi(P)|^2$]; and the wave state of an aggregate of f electrons $1, 2, \ldots, f$ would be an antisymmetric function $\psi(P_1, \ldots, P_f)$ of their positions $P_1, \ldots, P_f$. The exclusion principle is a consequence of antisymmetry. Because of the inner likeness of all electrons no dynamic action is imaginable that would ever carry an antisymmetric ψ over into one that is not. A function of several arguments i, $\psi(i_1, \ldots, i_f)$, is symmetric if it stays unaltered under all $f!$ permutations of its f arguments; it is antisymmetric if all even permutations leave it unchanged, all odd permutations carry it into $-\psi$. The nature of the argument i does not matter. It may range over a finite number of values $i = 1, 2, \ldots, n$, as we assumed for simplicity's sake in our exposition of quantum mechanics, or range over a whole continuum, as P does. We have previously seen that the distinction between even and odd permutations is the combinatorial basis for the polarity of left and right; we now find that it lies at the root of the periodic system of chemical elements and of a number of decisive traits of the physical world that defy explanation by the notions of classical physics.

Spectroscopic experience has shown that, over and in addition to separation of electrons by position, a splitting into two beams takes place e.g. under the influence of a magnetic field. We have to conclude that the wave function of a single electron $\psi(P\rho)$ depends on two variables, the continuous variable of position P, and a second variable ρ, called spin, that is capable of two values $+1$ and -1 only. The two components $\psi(P, +1) = \psi_+(P)$ and $\psi(P, -1) = \psi_-(P)$ are relative to a Cartesian frame and, as W. Pauli first recognized, transform according to the spinor representation mentioned in Section 15 [of $PMNS$], when one passes by rotation to another such frame. The wave function of an aggregate of f electrons is an antisymmetric function $\psi(P_1\rho_1, P_2\rho_2, \ldots, P_f\rho_f)$ of f pairs $(P\rho)$.

A third circumstance besides spin and antisymmetry is relevant: with considerable approximation the dynamic influence of the spin may be disregarded. Let us assume that it is strictly nil, i.e. that the dynamical operator H of energy operates only on the positional variables P, not on the spin variables. At first one may think that then one could ignore the spin altogether. That this is not so is due to the condition of antisymmetry with respect to the pairs $(P\rho)$. Let $\eta(P_1, \ldots, P_f)$ be an eigenfunction of the operator H with the eigenvalue $h\nu$, $H\eta = h\nu \cdot \eta$, η thus representing a stationary wave state of energy $h\nu$. Assume η to be antisymmetric in its f arguments P. On taking the existence of the spin into account one obtains a whole linear manifold of wave functions ψ of energy $h\nu$,

$$\psi(P_{1\rho 1}, \cdots, P_{f\rho f}) = \eta(P_1, \cdots, P_f) \cdot \varphi(\rho_1, \cdots, \rho_f).$$

Here the second factor $\varphi(\rho_1 \ldots \rho_f)$ could be any function of the f spin variables ρ; but antisymmetry of ψ requires φ to be symmetric. A symmetric function φ of $\rho_1, \ldots, \rho_f$ assumes a definite value φ_g if a given number g of the arguments ρ are $+1$ and $f - g$ of them are -1; and the function is completely characterized by its $f + 1$ values $\varphi_f, \varphi_{f-1}, \ldots, \varphi_0$. Hence the linear manifold of the symmetric functions φ is $(f + 1)$-dimensional, and the existence of the spin has the effect that the energy level $h\nu$ or 'term' ν acquires the multiplicity $f + 1$. Only when the actually existing weak interaction of the spins is taken into account, this term of multiplicity $f + 1$ splits up into a *multiplet* of $f + 1$ different terms. Consider for a moment what would happen, on the other hand, if η were symmetric. Then φ must be antisymmetric. But an antisymmetric function vanishes whenever two of its arguments have equal values. Hence if the individual argument is capable of two values only the function will vanish identically provided $f > 2$, and the term ν corresponding to a symmetric η is wiped out, its multiplicity becomes 0. Owing to the low dimensionality 2 of the spin space the possible permutational symmetry characters of functions $\varphi(\rho_1 \ldots \rho_f)$ can be described by one number, the valence v, which is capable of all values $0 \leq v \leq f$ that differ from f by an even number. States in which η is antisymmetric (and hence φ symmetric) are of valence f. A term of valence v has, merely on account of the permutability of electrons, the multiplicity $v + 1$.

We consider a neutral atom as an aggregate of f electrons of charge $-e$ that move in the field of a nucleus of charge $f e$ fixed at a center O. Non-relativistic mechanics is applied to this model in taking into account only the electrostatic forces between these charges plus the kinetic energy of the electrons. Let $\eta(P_1 \ldots P_f)$ be an antisymmetric eigenfunction of the energy operator H corresponding to the term ν. The atom in this state η has the energy $h\nu$ and the highest possible valence f. (Any permutation of the f points $P_1, \ldots, P_f$ would change η into an eigenfunction for the same term ν; but because of antisymmetry this causes no permutational multiplicity.) It is also true that the effect of any common rotation about O of the points $P_1, \ldots, P_f$ transforms $\eta(P_1 \ldots P_f)$ into an eigenfunction for the same term ν. Hence if we wish to avoid 'rotational' multiplicity of ν we must assume that the function $\eta(P_1 \ldots P_f)$ of the constellation $P_1 \ldots P_f$ of the electrons is invariant with respect to all rotations (central symmetry). Such a stationary wave state is called an S-state in spectroscopy. Thus we assume the atom to have its highest valence f and to be in an S-state. The probability $\mathscr{P}(r)$ to find one of the electrons at a distance greater than r from the center O is determined by η, and it turns out that $\mathscr{P}(r)$ falls off exponentially with increasing r.

After introducing two 'indeterminates' x_+, x_- corresponding to the two values $\rho = +1$ and -1 of the spin, a symmetric function $\varphi(\rho_1 \ldots \rho_f)$ is conveniently represented by the algebraic form of x_+, x_- of degree f,

$$\sum_\rho \varphi(\rho_1 \cdots \rho_f) x_{\rho_1} \cdots x_{\rho_f} = \sum_g \frac{f!}{g!(f-g)!} \varphi_g x_+^g x_-^{f-g}$$

with the coefficients φ. The sum at the left consists of 2^f terms as each ρ takes on its two values $+$ and $-$ while the range of g at the right side is the sequence $f, f - 1, \ldots, 0$. Considering the indeterminates x_+, x_- as components of a vector x in a plane we submit them to an arbitrary linear transformation

(1) $$x_+ = \alpha x'_+ + \beta x'_-, \qquad x_- = \gamma x'_+ + \delta x'_-$$

of modulus $\alpha\delta - \beta\gamma = 1$. A form $F(x, y, \ldots)$ of several indeterminate vectors $x, y, \ldots$ which is of degree f_a in x, f_b in y, $\ldots$, is said to be an *invariant* if $F(x, y, \cdots) = F(x', y', \cdots)$ whenever x and x', y and y', $\ldots$, are connected by the same transformation (1) of modulus 1.

Envisage now a number of neutral atoms $a, b, \ldots$ of $f_a, f_b, \ldots$ electrons with their nuclei fixed at definite points in space $O_a, O_b, \ldots$, and suppose that each is in a stationary S-state of highest valence, their respective energy levels being $h\nu_a, h\nu_b, \ldots$. This implies that the combined system of these atoms has the energy $h\nu_0$, $\nu_0 = \nu_a + \nu_b + \cdots$, and that its state belongs to a linear manifold Π of $(f_a + 1)(f_b + 1) \cdots$ dimensions. Speaking in this way we have disregarded the mutual interaction of the atoms and thus violated the essential likeness of all $f = f_a + f_b + \cdots$ electrons by assigning f_a of them to the entourage of O_a and letting these f_a electrons interact only among themselves and with the nucleus O_a. We assume the mutual distances r of the nuclei at $O_a, O_b, \ldots$ to be large in comparison to the Bohr radius h^2/me^2. Taking now the interaction between the several atoms into account as a small perturbation one finds that the term ν_0 breaks up by 'permutational resonance' into a number of term systems of the molecule according to the various possible valences $v = f, f - 2, \ldots$. The states of valence v form a linear submanifold Π_v of n_v dimensions that, in the approximation of perturbation theory, as a whole is invariant with respect to the energy operator H and hence stationary. The corresponding n_v terms $\nu = \nu_0 + \Delta\nu$ and individual stationary states of the molecule are to be determined as the eigenvalues and eigenfunctions of H operating in Π_v. Since each of the n_v terms ν of the molecule in a state of valence v has the multiplicity $v + 1$, comparison of dimension leads to the equation

$$(f_\alpha + 1)(f_b + 1) \cdots = \sum_v n_v(v + 1).$$

The shifts $\Delta\nu = V(O_a, O_b \cdots)$ are functions of the constellation $O_a\, O_b \ldots$ of the nuclei which are found to be of the same type as the probability $\mathscr{P}(r)$ mentioned above, namely falling off exponentially with increasing distances r. This accounts for the fact that the homopolar bond between neutral atoms is a short range force. (The attraction of two ions of opposite charges at a distance r, the heteropolar bond, is no mystery at all. Its energy follows the Coulomb law $1/r$ and is thus of the long range type.)

Let an indeterminate binary vector $x = (x_+, x_-)$, $y, \ldots$ be associated with each of the atoms $a, b, \ldots$ and add one more 'free' vector l. Then

Π_v is best described as the linear manifold of all invariants $J(x, y, \ldots, l)$ depending on the indeterminate vectors $x, y, \ldots, l$ with the given degrees $f_a, f_b, \ldots, v$. The details do not matter here. But so much should be clear that the two-dimensional vectors and invariance with respect to linear transformations play a role because, owing to the spin, the state (φ_+, φ_-) of an electron is such a vector. The dimensionality n_v of Π_v is the number of linearly independent invariants (of degrees $f_a, f_b, \ldots$ in the vectors $x, y, \ldots$ and) of degree v in the free vector l.

The simplest invariant, linearly depending on two indeterminate vectors x, y, is the 'bracket factor' $[xy] = x_+ y_- - x_- y_+$. Any product of such bracket factors is called a monomial invariant. A monomial invariant is completely described by a diagram in which each of the argument vectors $x, y, \ldots, l$ is represented by a point and each bracket factor like $[xy]$ by a line joining the points x and y. (A bracket factor $[xl]$ involving the free vector l may instead be represented by a stroke issuing from x the other end of which remains free.) The degrees $f_a, f_b, \ldots, v$ of the monomial invariant are the numbers of strokes ending at the respective points $x, y, \ldots, l$. Hence the monomial invariants correspond completely to the Kekulé valence diagrams. We shall therefore call a state described by such an invariant a pure valence state. The first main theorem of the theory of invariants states that every invariant of given degrees is a linear combination of monomial invariants of those degrees.

For a molecule consisting of two atoms x, y of valences a and b, $a \geq b$, we find only one invariant

$$[xy]^d [xl]^{a-d} [yl]^{b-d}$$

for each of the possible molecular valences $v = a + b - 2d, d = 0, 1, \ldots, b$. This corresponds exactly to what the valence diagrams would have us expect; d is the number of valence strokes joining the two atoms, and $a - d, b - d$ are the numbers of free valence strokes issuing from x and y respectively. For the corresponding term $v_0 + \Delta v$ of the molecule one finds $\Delta v = \lambda \cdot V(r)$ where $V(r)$ is a function of the distance r of the two atoms that does not depend on d, while the form factor

$$\lambda = (a - d)(b - d) - d$$

depends on d but not on r. The function $V(r)$ is difficult to compute, but in the simplest cases turns out to be positive for large r. Assuming this to be generally true one obtains a force of attraction or repulsion according to whether the form factor λ is negative or positive. λ is negative for $d = b$; but since λ varies from $-b$ to ab while d assumes the values $b, b - 1, \ldots, 0$, the form factor will be negative only for the strongest binding $d = b$, or possibly for a few of the stronger bindings $d = b, b - 1, \cdots$.

The picture changes somewhat when more than two atoms come into play. Then the number n_v of linearly independent invariants is less than that of the possible diagrams with v free valence strokes, owing to the existence of linear relations among the monomial invariants. Moreover the individual stationary states with definite energy levels $v_0 + \Delta v$ do no longer coincide with any of the pure valence states. There are clear indications for this in chemistry. For instance, Kekulé's famous formula for the benzene ring, a regular arrangement of six CH groups, foresees two possibilities whereas the study of ortho-derivatives proves conclusively that there is only one in nature. The skeleton shown in the valence

(S)

diagram S has the full hexagonal symmetry which one expects for the benzene ring; but it leaves one valence electron in each C-atom un-attached. This conception of a fixed skeleton upon which the variable state of bondage between the remaining valence electrons is superimposed may be an unwarranted simplification, but is useful for a first orientation and reduces our problem (which in fact involves forty-two electrons) to that of six equal one-electron atoms arranged in a regular hexagon of side r. Here the states of valence 0 form the linear manifold of all invariants depending linearly on six argument vectors 1, 2, 3, 4, 5, 6. The five monomial invariants A, A'; B_1, B_2, B_3 shown on p. 569 represent a basis for that manifold. (Their diagrams should be superimposed upon the skeleton S; the corresponding five pure valence states are 'in reso-nance.') The term shifts $\Delta v = \lambda \cdot V(r)$ of the various stationary states η differ by the form factor λ; the potential function $V(r)$ or r, however, is a common factor of the short range exponential type. Here is a list of the stationary states with their form factors λ:

$$
\begin{array}{l|l}
A + A' & \lambda = 0 \\
\beta_1 B_1 + \beta_2 B_2 + \beta_3 B_3 \quad (\beta_1 + \beta_2 + \beta_3 = 0) & \lambda = 2 \\
6(A - A') - (1 + \sqrt{13})(B_1 + B_2 + B_3) & \lambda = 1 + \sqrt{13} \\
\eta = 6(A - A') - (1 - \sqrt{13})(B_1 + B_2 + B_3) & \lambda = 1 - \sqrt{13} < 0.
\end{array}
$$

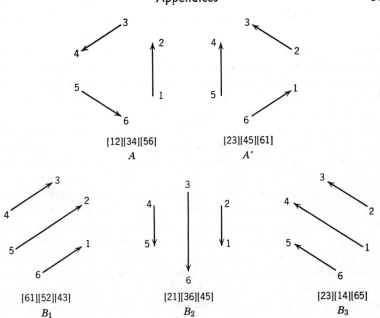

[12][34][56]
A

[23][45][61]
A'

[61][52][43]
B₁

[21][36][45]
B₂

[23][14][65]
B₃

As we know, a negative value of λ is suggestive of the existence of a stable molecule in the corresponding quantum state η. Only the η of the last line satisfies this condition, and thus it is this η that indicates the direction in which the quantum mechanical correction of Kekulé's benzene formula will lie: To the difference of the monomial invariants A, A' depicted by Kekulé's two diagrams is added a multiple of the sum of the three terms represented by the Dewar diagrams B_1, B_2, B_3. (Both $A - A'$ and $B_1 + B_2 + B_3$ change sign under the influence of a rotation of the hexagon by 60° and remain unaltered by reflection in one of the three diagonals.)[1]

The notion of quantum-mechanical resonance between states of (nearly) equal energy levels plays an important role in modern structural chemistry. At the same time one tries to hold on to well-tested and plausible valence schemes, keeping the modifications required by resonance to a minimum; and one is content in most cases to determine the perturbation energies by empirical observation rather than computation. This conservative procedure, illustrated here by the classical example of the benzene ring, has met with surprising success—surprising to the scrupulous mathematician who finds it hard to justify some of the 'plausible' assumptions of

[1] If one would treat the benzene molecule as a ring of six CH groups of valence 3, the number of resonant independent possibilities would increase from 5 to 34; for the linear manifold of binary invariants of six argument vectors and of degree 3 in each of them has the dimensionality 34.

approximative character on which it is based. (A more exact analysis may soon become feasible with the help of the high-speed computing machines now under construction.)

Finally I come to the lesson which I want to draw from this long excursus into quantum-mechanical chemistry. It concerns the hierarchy of structures. On the deepest level α we have the structure of quantum mechanics itself in terms of which we seem to be able to interpret all spectroscopic and chemical facts, all physical facts in short for which the inner constitution of the atomic nuclei is irrelevant. On the second level β the structure representing a molecule in its various possible states is the linear manifold of binary invariants. This picture has limited truth only. Above all it does not refer to the ready-made molecule but to the aggregate of its atoms with their nuclei fixed at distances large in comparison to the extension of the atoms. Moreover, as to the individual atoms, conditions as simple as possible are assumed with respect to the permutations of its electrons and rotation of their configuration in space. The structures which are used on the third level γ for the interpretation of chemical facts are the valence diagrams. In the light of β the picture γ is correct in one essential respect: all possible states of the molecule (all invariants) are indeed linear combinations of the pure valence states (monomial invariants). But it errs on three other counts: (1) There are not only a few discrete states, such as the pure valence states, but rather a whole linear manifold of wave states; this, of course, is the decisive contrast between classical and quantum mechanics. (2) The linear relations between the monomial invariants are ignored, and therefore too high a value is obtained for the number n_v of independent possibilities. (3) The n_v stationary quantum states coincide in general with none of the pure valence states but are certain linear combinations of them.

Contrary to our exposition, the historical order is that of descent to an ever deeper level, $\gamma \to \beta \to \alpha$. A. Kekulé developed his graphical representation of chemical structure in 1859. The intermediate level β was first reached by J. J. Sylvester in 1878[1] (he was later followed by the German

---

[1] Sylvester's paper published in the first volume of the *American Journal of Mathematics*, which he founded at Johns Hopkins, bears the title *On an application of the new atomic theory to the graphical representation of the invariants and covariants of binary quantics*. Of the opening sentences the first is such a characteristic statement of 19th century natural philosophy and the second such a charming piece of Sylvestrian prose that they may be quoted here. "By the *new* Atomic Theory I mean that sublime invention of Kekulé which stands to the *old* in a somewhat similar relation as the Astronomy of Kepler to Ptolemy's, or the System of Nature of Darwin to that of Linnaeus;—like the latter it lies outside of the immediate sphere of energetics, basing its laws on pure relations of form, and like the former as perfected by Newton, these laws admit of exact arithmetical definitions.—Casting about, as I lay awake in

invariant theorist P. Gordan and the Russian chemist W. Alexejeff). However in the absence of a physical interpretation for the addition of invariants and of dynamical laws by which to determine the binding forces and the actual stationary states the chemists stuck to their familiar valence diagrams. We can see today that only such radical departure as that of quantum mechanics could reveal the significance of the picture that Sylvester had stumbled upon as a purely formal, though very appealing, mathematical analogy.

The moral of this story is evident: do not take too literally such preliminary schemes as the valence diagrams, useful as they are as a first guide in a seemingly incoherent mass of facts. A picture of reality drawn in a few sharp lines can not be expected to be adequate to the variety of all its shades. Yet even so the draftsman must have the courage to draw the lines firm. There is no doubt that the gene aggregates of genetics with their linkages are structures of no less preliminary character than the valence diagrams of chemistry. The cytological study of cells reveals complicated motions of chromosomes and multiform physical processes whose details are capable of continuous variation and of whose result the discrete genetic diagrams are no more than abbreviated abstracts of limited validity. I should therefore not vouch too much for the adequacy of the primitive combinatorial scheme as depicted in Appendix B [Sec. 18.1], and yet is seemed best to make the picture itself, however limited its value, as definite as possible. (This is a principle which Nicolaus Cusanus stressed in *De docta ignorantia*: if the transcendental is accessible to us only through the medium of images and symbols, let the symbols at least be as distinct and unambiguous as mathematics will permit.)

The facts related in the next Appendix [Sec. 18.4] leave little doubt that the laws of inheritance are ultimately based on the same structure as the laws of chemistry: on the structure of quantum mechanics. A structure that could serve to mediate between the genetic diagrams and quantum physics should be one that takes into account the chemical complexity of the carriers of life. Perhaps the simplest combinatorial entity is the group

---

bed one night to discover some means of conveying an intelligible conception of the objects of modern algebra to a mixed society mainly composed of physicists, chemists and biologists, interspersed only with a few mathematicians, to which I stood engaged to give some account of my recent researches in this subject of my predilection, and impressed as I had long been with the feeling of affinity, if not identity of object, between the inquiry into compound radicals and the search for 'Grundformen' or irreducible invariants, I was agreeably surprised to find of a sudden distinctly pictured on my mental retina a chemico-graphical image serving to embody and illustrate the relations of these derived algebraic forms to their primitives and to each other, which would perfectly accomplish the object I had in view, as I will now proceed to explain."

of the $n!$ permutation of n things. This group has a different constitution for each individual number n. The question is whether there are nevertheless some asymptotic uniformities prevailing for large n or for some distinctive class of large n. Mathematics has still little to tell about such problems. One wonders whether a quantum theory of organic processes is tied up with their solution.

Whereas the quantum structure described in Appendix C [Sec. 18.2] has for the present been accepted by physics as the ultimate layer, the skeptic philosopher may wonder whether this reduction is more than one step, the last at the moment, in a *regressus ad infinitum*. But, so warns the scientist, nothing is cheaper and on the whole more barren than to play with such possibilities in one's thoughts before new discoveries place one before a concrete situation that enforces a further Tieferlegung of the foundations.

Physical phenomena are spread out in the continuous extensive medium of space and time; it was this aspect which dominated to a considerable degree the epistemological thoughts about natural science that the main part of this book [*PMNS*] tried to collect in 1926. This was historically justified and the accomplishments of general relativity, still very fresh at that time, lent additional emphasis to this point of view. In the last two decades, however, discontinuous and combinatorial structures underlying the natural phenomena have become of increasing significance. Here a deeper layer seems to come to light, for the description of which our ordinary language is woefully inadequate. The preceding Appendices bear witness to this changed outlook. However, we could not do much more than assemble relevant material; the philosophical penetration remains largely a task for the future.

REFERENCES

M. Born, *Chemische Bindung und Quantenmechanik*, Ergebnisse der exakten Naturwissenschaften, vol. 10, Berlin, 1931.

Linus Pauling, *The Nature of the Chemical Bond*, second ed., Cornell University Press, 1945.

W. G. Palmer, *Valency, Classical and Modern*, Cambridge, 1944.

18.4 PHYSICS AND BIOLOGY

1. One of the profoundest enigmas of nature is the contrast of dead and living matter. However one may characterize life phenomenologically: animate matter is obviously separated from inanimate by a deep chasm. Life dwells only in material systems that from a physicochemical standpoint are to be considered as highly complex. In a descriptive way and

without claiming completeness we enumerate some of the typical features of the living organism: its composition of cells, living units that are uniform in their more basic characteristics; wholeness as form (morphé, Gestalt) and as functional complex, with mutual adjustment of all cell differentiations to each other ("geprägte Form, die lebend sich entwickelt," Goethe); endowed through metabolism with the capacity of using alien matter as food and incorporating it into its own organization; development by assimilation of food, by growth and differentiation from relatively simple to more complicated states; in spite of inner lability, far-reaching though not unlimited capacity to maintain itself as this differentiated whole under changing external influences, in particular in the turmoil of the molecular heat motion, and to restore itself after disturbing encroachments; limitation of individual existence in time (birth and death); the capacity of propagation and of transmitting its specific constitution to its progeny. While dead matter is inert, the organism is a source of activity that bears the stamp of spontaneity ever more manifestly (with volitional action as its climax) the higher one climbs in the world of organisms. It is at the same time susceptible to stimuli (perceptions on the highest level) and endowed with the capacity of storing stimulative experiences (mneme). Life has unfolded into a vast multitude of species of typically diverse constitution, and the organisms are woven into a dense net of adaptations and relations to each other and their surroundings.

The last unit and its one basic property to which the essential characteristics of living matter seem to have been reduced by scientific analysis is the *gene* and its power of *self-duplication*. By this process a copy of the model gene is synthesized from the material available in the living cell. Incidentally, the gap between organic and inorganic matter has been bridged to a certain extent by the discovery of viruses. Viruses are submicroscopic entities that behave like dead inert matter unless placed in certain living cells. As parasites in these cells, however, they show the fundamental characteristics of life—self-duplication and mutation. On the other hand many viruses have the structure typical of inorganic matter; they are crystals. In size they range from the more complex protein molecules to the smaller bacteria. Chemically they consist of nucleo-protein, as the genes do. A virus is clearly something like a naked gene. The best studied virus, that of the tobacco mosaic disease, is a nucleo-protein of high molecular weight consisting of 95 per cent protein and 5 per cent nucleic acid; it crystallizes in long thin needles.

The elementary laws of matter that physics reveals and chemistry is ruled by are no doubt also binding on living matter. Hence such a profound change of physics as brought about by quantum theory must have its repercussions in biology. As long as progress from simple to more complicated configurations remains the methodologically sound way of science, biology will rest on physics, and not the other way around. The specific properties of living matter will have to be studied within the general laws valid for all matter; the viewpoint of holism that the theory of life comes first and that one descends from there by a sort of deprivation to inorganic matter must be rejected. It is therefore significant that certain simple and clearcut traits of wholeness, organization, acausality, are ascribed by quantum mechanics to the elementary constituents of all matter. A *rapprochement* between physics and biology has undoubtedly taken place in this regard. Structure and organization are not peculiar to living beings; physics is thoroughly familiar with this aspect and represents it by the symbolic apparatus of the theory that precedes all dynamical laws. The quantum physics of atomic processes will become relevant for biology wherever in the life cycle of an organism a moderate number of atoms exercises a steering effect upon the large scale happenings. (The radio tube is today the most familiar inorganic example of such a steering mechanism.)

On a broad empirical foundation, *genetics* furnishes the most convincing proof that organisms are controlled by processes of atomic range, where the acausality of quantum mechanics may make itself felt. At the turn of the century, when Planck introduced the action quantum into physics, de Vries discovered the jumplike mutations of the genetic constitution of Oenothera (the larger part of which, to be sure, are today recognized as structure rather than point mutations). For a physical understanding of mutations, their artificial generation by exposing chromosomes to X-rays has proved of momentous importance. The mere fact of such X-ray induced mutations proves that the genes are physical structures. When X-rays fall upon matter this or that photon relinquishes all or a large part of its energy to a fast secondary electron, and this in turn loses its energy in a number of steps by ionization (or excitation) of atoms. The average energy of ionization amounts to about 30 electron volts. By ingenious methods H. J. Muller, N. W. Timoféeff-Ressowsky, and others have succeeded in establishing simple quantitative laws concerning the rate of induced mutations. These results indicate that the mutation is brought about by a single hit, not by the concerted action of several hits, and that this hit consists of an ionization, and is not, as one might have thought, a process directly released by the X-ray photon or absorbing the whole energy of the secondary electron.

These facts suggest the hypothesis that a gene is a (nucleo-protein) molecule of highly complicated structure, that a mutation consists in a chemical change of this molecule brought about by the effect of an ionization on the bonding electrons, and thus allele genes are essentially isomeric molecules. The most elementary chemical changes which quantum physics can devise are localized two-step quantum jumps—first the molecule is lifted from an energy level 1 to a higher level 2, and from there it drops to a new stable state of energy level 3. The difference 2 minus 1 is the necessary activation energy U. The rate at which a specific quantum jump requiring the activation energy U occurs spontaneously at a given temperature depends essentially on U alone, but varies extremely strongly with U (according to an exponential law). At the temperature at present prevailing on the earth's surface such quantum jumps as correspond to values of U between say 1.4 and 1.7 would be occasionally occurring yet rare events. (For lower values of U the corresponding quantum jumps are so frequent that the statistical law of large numbers comes into power; they give rise to such ordinary chemical reactions as take place in the development of an organism.) Thus one is tempted to complete the picture by interpreting mutation as a rare quantum jump with an activation energy within the range just mentioned (Delbrück's model).[1] The observed absolute rate of mutations would be explained if a specific mutation requires that a hit occurs within a critical volume ('target') in the gene, the magnitude of which amounts to about 5–10 A cube (5–10 atomic distances cube). The physicist finds it, if not plausible at least acceptable, that a quantum jump at a specific point requiring an activation energy of about 1.5 is released by a hit of 30 electron volts within a sensitive volume of 5–10 A cube. The observed thermic variation of the spontaneous mutation rate (van't Hoff's factor) is in good quantitative agreement with the picture.

There are several methods for estimating size and molecular weight of a gene. Most of the radiation experiments are concerned with mutations called recessive lethals. A certain high percentage of these is due to an ionization depriving the gene of its reproductive power (while others are brought about by gross structure mutations). It is plausible to put the first kind of lethals in analogy to the inactivation of enzymes and viruses. For these latter processes, which are also due to single ionizations, one can determine the target size, either by means of the absolute dose of X-rays or by the relative efficiency of the various radiations. One finds target radii that are between one and five times as small as the radius of the enzyme molecule or the virus. Hence the target size of a gene for the totality of recessive lethal mutations 'of the first kind' ought not to be much smaller than the size of the gene. Another method for ascertaining the size of a gene or at least an upper bound for it is the following. The

[1] Cf. N. W. Timoféeff-Ressowsky, K. G. Zimmer, M. Delbrück, *Über die Natur der Genmutation und der Genstruktur* (Nachr. Gött. Ges. Wissensch., Math.-physik. Kl., Fachg. VI, 1), 1935, pp. 189–245.

greatly enlarged chromosomes of the salivary glands of Drosophila show a cross striation by bands of characteristically different width and design, and the parallelism of the genetic and cytological findings vindicates the hypothesis that these bands correspond to genes or small groups of genes in the many parallel threads of which the giant chromosome consists. The several methods agree in making it likely that the molecular weight of genes is of the order of magnitude of one million (times the atomic weight of hydrogen). This is exactly what one would have expected, considering that the weights of the threadlike molecules of nucleic acids range from fifty thousand to several hundred thousand while the weight of the individual tobacco mosaic virus molecule reaches the figure of forty millions.

The investigations of the last ten years have not been favorable to the special hypothesis that a mutation is due to a quantum jump localized in and restricted to a few atoms. Several complications have come to light. To give one extreme example, W. M. Stanley found that a certain spontaneous mutation of the tobacco mosaic virus changes its chemical composition by adding about one-thousand molecules of histidine. One is thus forced to think of some mechanism by which the individual ionization releases a chain of (enzymatic?) reactions with the complex mutation as its end result. Be this as it may, the direction in which our model points is hardly deceptive; the gene is to be considered as a complex molecule and mutations are closely connected with quantum jumps. The latter can be brought about by single ionizations, and thus one may conclude with P. Jordan, that "the steering centers of life are not subject to macrophysical causality but lie in the zone of microphysical freedom." Incidentally viruses that can be isolated and observed by means of the electron microscope are in many respects better objects for the investigation of the physical foundations of the mutation process than the invisible genes in the chromosomes of cells.

The nucleus of a fertilized egg is supposed to furnish by its genetic constitution the complete determinants for the development of the organism. In earlier times one often found great difficulty in harmonizing this view—so closely related to the issue of 'preformation' versus 'epigenesis'—with the vast manifold of animals and plants, all their various courses of development and all their minute differentiations. However, the fantastically high number of possible combinations of atoms in a gene molecule (cf. the characteristic numbers for combinations of symbols in Appendix A [of $PMNS$]) exceeds by far all that is needed for this purpose. It is thus not inconceivable that the miniature code contained in the gene molecules of the cell nucleus should precisely correspond with a highly complicated and specified plan of development and should somehow contain the means to put it into operation. In a famous experiment Driesch observed that the cut-off upper third of a Clavellina, its gill basket, reverts to an amorphous conglomerate of cells from which there develops a new complete Clavellina of reduced scale. He saw in this experiment a proof for the existence of an entelechy not expressible in terms of physical structure. Today we have a rather definite picture of the physical structure that can serve as such an entelechy. The question of selection among the combinatorial possibilities is something else; it points (i) toward the physicochemical problem of the stability of complex molecular compounds, (ii) to the mechanism by which the 'code' is translated into the development of an organism, and (iii) to the process of evolution.

The stability of a molecule stems from the chemical bonds between its atoms. As mentioned before, it was quantum physics that threw the first light upon the previously rather obscure nature of chemical bonds. A crystal (like diamond) is a regular pattern of atoms (C-atoms in the case of diamond) periodic in three independent spatial directions. Here the bonds extend between all atoms and thus the entire crystal is as it were a single molecule. The stability of solid bodies is that of crystals, and if the Delbrück model is basically correct, then the stability of genes rests on the same quantumtheoretic foundation. Yet while in a crystal the same building bricks are repeated periodically, each atom in the gene has its specific noninterchangeable place and role. Schrödinger therefore speaks of the gene as an aperiodic crystal and ascribes to it a higher degree of order and organization than to the periodic crystal. Whereas the macroscopic order and regularity of nature is based by statistical thermodynamics upon microscopic disorder, we encounter here in the crystals and the chromosomes of cell nuclei an order that is not overwhelmed by thermic disorder. In contrast to ordinary chemical reactions, the laws of which are obtained by averaging over an enormous number of molecular processes, mutations attack single genes. Inasmuch as (a) the order of the zygote is transmitted by self-duplication and mitosis to all somatic cells, and (b) the relatively minute speck of well-ordered atoms in the chromosomes of the cells controls the development of the living being, the dislocation of a few atoms in the mutant gene results in a well defined change in the macroscopic hereditary character of an organism. Before we have gained insight into the mechanism underlying the processes (a) and (b) we cannot claim to understand ontogenetic development.

While formal genetics has advanced by leaps and bounds during the last forty to fifty years, our knowledge in these fields is still very sketchy. As to the central problem of self-duplication, M. Delbrück has recently (1941) ventured to give a detailed but admittedly hypothetical picture of how amino-acids might conceivably be strung together in a pattern emulating a preexisting gene model by quantum mechanical resonance at the site of the peptide links. Connection between gene and visible character, e.g. between the wing form of Drosophila called jaunty and its gene, is certainly the resultant of a chain of intermediary actions. It is therefore an important step ahead that recently attention has been concentrated on genetic control of enzymatic action; many experiences point to a close relation between genes and specific enzymes (cf. the recent work of G. W. Beadle and others on Neurospora). When a tiny speck of solid crystalline substance causes a saturated solution of the same substance to crystallize, we witness how a germ of order is capable of spreading order. Although we are as yet unable to pursue this physical process in theoretical detail there is no doubt that it lies within the scope of our known physical laws. Science will press on to analyze the manifold processes on which the order in living organisms depends in fundamentally the same way, i.e. on the ultimate basis of quantum physics with its primary statistics. But there may be a bifurcation in the following sense: as order is derived from disorder by means of the secondary statistics of thermodynamics, so may a parallel but different type of macro-law account for the production of large-scale order from small-scale order in an organism (Schrödinger).

2. With the mutations a clearly recognizable non-causal element penetrates into the behavior of organisms. Whereas my perceptions and actions are in general the resultants of innumerable individual atomic processes and thus fall under the rule of statistical regularity, it is a noticeable fact that, if favorable circumstances prevail, a few photons (not more than 5 to 8) suffice to set off a visual perception of light. From here, from the quantum mutations in the gene molecule and the translation of a stimulus of a few photons into visual perception, it is still a long, long way to the full psychophysical reality with which man finds himself confronted, and to an integrated theoretical picture of it that would account for the facts of free insight and free will. "Although the door of human freedom is opened," says Eddington in *New Pathways in Science* (p. 87), "it is not flung wide open; only a chink of daylight appears. But it is no longer actually barred and efforts to prise it further open are encouraged." How far, we may ask, is the example of mutations representative, how far may organic processes be ascribed to the trigger action of small groups of atoms of unpredetermined behavior? The physicist P. Jordan has argued the point that to a considerable extent this is indeed the case, but has met with much opposition among biologists. Also Schrödinger warns that, without the almost complete precision and reliability of the macroscopic thermodynamical laws ruling the nervous and cerebral processes of the human body and its interactions with the surrounding world, perception and thought would be impossible. Niels Bohr, however, is inclined to

widen the domain of uncertainty by adding a specific biological principle of indeterminacy (the precise content of which is still unknown) to Heisenberg's well-established quantum-mechanical principle of indeterminacy. He has pointed out in this connection that an observation of the state of the brain cells exact enough for a fairly definite prediction of the victim's behavior during the next few seconds may involve an encroachment of necessarily lethal effect—and thereby make the organism predictable indeed. Bohr maintains that in this way analysis of vital phenomena by physical concepts has its natural limits; just as one had to put up with complementarity as expressed by Heisenberg's principle of indeterminacy in order to explain the stability of atoms, so are further renouncements demanded of him who tries to account for the self-stabilization of living organisms.

Such theoretical acts as the judgment that $2 + 2$ makes 4 have served in the main text to bring out the salient point of the problem of freedom. Thought as thought would be abrogated if it were denied that in my judging thus the mental fact that $2 + 2$ actually makes 4 gains power over an individual psychic act, and not only over the psychic act but also over the movements of my lips that form the corresponding words pregnant with meaning, or over the movements of the hand that, perhaps in the context of a mathematical proof, writes down the marks '$2 + 2 = 4$' on paper. Punching a hole in the strict causality of nature does not therefore suffice; a representation within the theory must be found for vital, psychic, and spiritual factors that in some way direct and steer the atomic process. It is certainly important in that grasp of the totality of nature that precedes all theory not to lose sight of such traits. In summing up his life work H. Spemann comes to the conclusion that the "processes of development . . . are comparable, in the way they are connected, to nothing we know in such a degree as to those vital processes of which we have the most intimate knowledge, viz., the psychical ones." Such voices as this—and it is not an isolated one—should be heeded. And yet, I believe that in a theory of reality the ideal factors which are here in question must be represented in basically the same way as the physical elementary particles and their forces, namely by a structure expressed in terms of symbols. I put no undue confidence in the suggestion made in the main text, that this purpose could be served by correlations between such atomic events as are treated as statistically independent by thermodynamics. Indeed the example of quantum mechanics has once more demonstrated how the possibilities with which our imagination plays before a problem is ripe for solution are always far surpassed by reality. Even so, the explanation of the chemical bond by Pauli's exclusion principle is perhaps a hint that the radical break with the classical scheme of statistical independence is an opening of the door as significant as the quantum mechanical complementarity.

Scientists would be wrong to ignore the fact that theoretical construction is not the only approach to the phenomena of life; another way, that of understanding from within (interpretation), is open to us. Woltereck, in a broadly executed *Philosophie der lebendigen Wirklichkeit*, has recently ventured to describe in some detail the "within" of organic life. Of myself, of my own acts of perception, thought, volition, feeling, and doing, I have a direct knowledge entirely different from the theoretical knowledge that represents the 'parallel' cerebral processes in symbols. This inner awareness of myself is the basis for the understanding of my fellow-men whom I meet and acknowledge as being of my own kind, with whom I communicate, sometimes so intimately as to share joy and sorrow with them. Even if I do not know of their consciousness in the same manner as of my own, nevertheless my 'interpretative' understanding of it is apprehension of indisputable adequacy. Its illuminating light is directed not only on my fellow men; it also reaches, though with ever increasing dimness and incertitude, deeply into the animal kingdom. Albert Schweitzer is right when he ridicules Kant's narrow opinion that man is capable of compassion, but not of sharing joy with the living creature, by the question, "Did he never see an ox coming home from the fields drink?" It is idle to disparage this hold on nature 'from within' as anthropomorphic and elevate the objectivity of theoretical construction. Both roads run, as it were, in opposite directions: what is darkest for theory, man, is the most luminous for the understanding from within; and to the elementary inorganic processes, that are most easily approachable by theory, interpretation finds no access whatsoever. For objective theory the understanding from within can serve as a guide to important problems although it cannot provide their objective solution. A recent example is provided by investigations about the direction of the instinctive behavior of animals by 'appetences.'

It is tempting to stretch Bohr's idea of complementarity far enough to cover the relation of the two opposite modes of approach we are discussing here. But however one may weigh them against each other, one cannot get around the following significant and undeniable fact: the way of constructive theory, during the last three centuries, has proved to be a method that is capable of progressive development of seemingly unlimited width and depth; here each problem solved poses new ones for which the coordinated effort of thought and experiment can find precise and universally convincing solutions. In contrast the scope of the understanding from within appears practically fixed by human nature once for all, and may at most be widened a little by the refinement of language, especially of language in the mouth of the poets. Understanding, for the very reason that it is *concrete* and *full*, lacks the freedom of the 'hollow symbol.' A biology from within as advocated by Woltereck will, I am afraid, be without that never-ending impetus of problems that drives constructive biology on and on.

REFERENCES

W. M. Stanley, Chemical properties of viruses, in *The Study of Man*, Bicentennial Conference, Univ. of Pennsylvania Press, 1941.

G. W. Beadle, The Gene, *Proc. Am. Phil. Soc.*, **90** (1946), pp. 422–431.

H. J. Muller, The Gene, *Proc. Roy. Soc.* **B134** (1947), pp. 1–37.

H. Spemann, *Embryonic Development and Induction*, Yale Univ. Press, 1938 (esp. p. 372).

E. Schrödinger, *What is Life?*, Cambridge and New York, 1945.

D. E. Lea, *Actions of Radiations on Living Cells*, Cambridge and New York, 1946.

R. Goldschmidt, *Physiological Genetics*, New York, 1938.

M. Delbrück, *Cold Spring Harb. Symp. on Quant. Biology*, **9** (1941), pp. 122–126.

Niels Bohr, Light and Life, *Nature*, **131** (1933), pp. 421, 457.

—— Biology and Atomic Physics, *Rend. gener. celebr. Galvani*, Bologna, 1938.

P. Jordan, *Naturwisssenschaften*, **20** (1932), p. 815; **22** (1934), p. 485; **26** (1938), p. 537.

—— *Erkenntnis*, **4** (1934), p. 215.

—— Zur Quanten-Biologie, *Biol. Zbl.*, **59** (1939), pp. 1–39.

—— *Die Physik und das Geheimnis des organischen Lebens*, Braunschweig, 1941.

R. Woltereck, *Philosophie der lebendigen Wirklichkeit*. Bd. 1: *Grundzüge einer allgemeinen Biologie*, Stuttgart, 1931; Bd. 2: *Ontologie des Lebendigen*, Stuttgart, 1940.

Answers to Multiple-Choice Review Problems

Chapter	Problem				
	1	2	3	4	5
1	d	b	b	c	b
2	c	a	b	a	d
3	b	a	c	b	a
4	b	d	c	a	a
5	a	b	b	d	d
6	b	c	c	d	c
7	d	a	b	c	d
8	b	c	c	c	a
9	c	b	c	b	b
10	c	c	b	d	a
11	b	a	b	c	d
12	c	c	c	c	d
13	c	c	a	b	b
14	c	a	d	c	b
15	a	b	b	d	a
16	c	a	d	d	c
17	b	b	b	c	a

AUTHOR INDEX

SUBJECT INDEX ———————————